AF748865

Macroecology

25443

Annual Symposia of The British Ecological Society

42 Dispersal Ecology
Bullock, Kenward & Hails, 2002

41 Ecology: Achievement and Challenge
Press, Huntly & Levin, 2001

40 The Ecological Consequences of Environmental Heterogeneity
Hutchings, John & Stewart, 2000

39 Physiological Plant Ecology
Press, Scholes & Barker, 2000

37 Dynamics of Tropical Communities
Newbery, Prins & Brown, 1997

36 Multitrophic Interactions in Terrestrial Systems
Gange & Brown, 1997

33 Genes in Ecology
Berry, Crawford & Hewitt, 1992

30 Toward a More Exact Ecology
Grubb & Whittaker, 1989

Special Symposia of The British Ecological Society

15 Genes in the Environment
Hails, Beringer & Godfray, 2003

14 Integrating Ecology and Evolution in a Spatial Context
Silvertown & Antonovics, 2001

13 Ecology of Arctic Environments
Woodin & Marquiss, 1997

12 Primary Succession on Land
Miles & Walton, 1993

Macroecology: Concepts and Consequences

The 43rd Annual Symposium of the British Ecological Society held at the University of Birmingham 17–19 April 2002

EDITED BY

TIM M. BLACKBURN
School of Biosciences
University of Birmingham
Birmingham, UK

AND

KEVIN J. GASTON
Department of Animal and Plant Sciences
University of Sheffield
Sheffield, UK

Blackwell Publishing

and published for them by
Blackwell Science Ltd
a Blackwell Publishing company

350 Main Street, Malden, MA 02148-5020, USA
108 Cowley Road, Oxford OX4 1JF, UK
550 Swanston Street, Carlton, Victoria 3053, Australia

First published 2003

Library of Congress Cataloging-in-Publication Data

British Ecological Society. Symposium (43rd : 2002 : University of Birmingham)
Macroecology : concepts and consequences : the 43rd Annual Symposium of the British Ecological Society, held at the University of Birmingham, 17-10 April 2002 / edited by Tim M. Blackburn and Kevin J. Gaston.
p. cm.
Includes bibliographical references.
ISBN 1-4051-0643-3 (alk. paper).—ISBN 1-4051-0642-5 (pbk. : alk. paper)
1. Macroecology—Congresses. I. Blackburn, Tim M. II. Gaston, Kevin J. III. Title.
QH541.15.M23 B75 2002
577—dc21

2002038484

A catalogue record for this title is available from the British Library.

Set in 10/12.5 pt Minion
by SNP Best-set Typesetter Ltd., Hong Kong
Printed and bound in the United Kingdom
by MPG Books Ltd, Bodmin, Cornwall

For further information on
Blackwell Publishing, visit our website:
http://www.blackwellpublishing.com

Contents

Contributors

Tim M. Blackburn
School of Biosciences, University of Birmingham, Edgbaston, Birmingham B15 2TT, UK

Jon R. Bridle
Institute of Zoology, Zoological Society of London, Regent's Park, London NW1 4RY, UK

James H. Brown
Department of Biology, University of New Mexico, Albuquerque, New Mexico 87131, USA *and* Santa Fe Institute, 1399 Hyde Park Road, Santa Fe, New Mexico 87501, USA

Roger K. Butlin
School of Biology, The University of Leeds, Leeds LS2 9JT, UK

Andrew Clarke
Biological Sciences Division, British Antarctic Survey, High Cross, Madingley Road, Cambridge CB3 0ET, UK

Hernán Cofré
Center for Advanced Studies in Ecology and Biodiversity, and Departamento de Ecologia, Pontificia Universidad Católica de Chile, Casilla 114-D, Santiago, Chile

J. Alistair Crame
Geological Sciences Division, British Antarctic Survey, High Cross, Madingley Road, Cambridge CB3 0ET, UK

Konrad Dolphin
Department of Biological Sciences, Imperial College London, Silwood Park, Ascot SL5 7PY, UK

Steinar Engen
Department of Mathematical Sciences, Norwegian University of Science and Technology, N-7491 Trondheim, Norway

Brian J. Enquist
Department of Ecology and Evolutionary Biology, University of Arizona, Tucson, AZ 85721, USA

Richard Field
School of Geography, University of Nottingham, Nottingham NG7 2RD, UK

Robert P. Freckleton
Department of Zoology, University of Oxford, South Parks Rd, Oxford OX1 3PS, UK

Kevin J. Gaston
Biodiversity and Macroecology Group, Department of Animal and Plant Sciences, University of Sheffield, Sheffield S10 2TN, UK

Adam T. Gawelczyk
Institute of Environmental Sciences, Jagiellonian University, Gronostajowa 3, 30-387 Cracow, Poland

Jennifer A. Gill
Tyndall Centre for Climate Change Research, Schools of Biological and Environmental Sciences, University of East Anglia, Norwich NR4 7TJ, UK

James F. Gillooly
Department of Biology, University of New Mexico, Albuquerque, New Mexico 87131, USA

Paul H. Harvey
Department of Zoology, University of Oxford, South Parks Road, Oxford OX1 3PS, UK

Stephen P. Hubbell
Department of Plant Biology, University of Georgia, Athens, GA 30602, USA *and* Smithsonian Tropical Research Institute, Box 2072, Balboa, Republic of Panama

David Jablonski
Department of Geophysical Sciences, University of Chicago, 5734 South Ellis Avenue, Chicago, IL 60637, USA

Masakado Kawata
Division of Ecology and Evolutionary Biology, Graduate School of Life Sciences, Tohoku University, Sendai, Japan

C. K. Kelly
Division of Biodiversity and Ecology, University of Southampton, Bassett Crescent East, Southampton SO16 7PX, UK

Juan E. Keymer
Department of Ecology and Evolutionary Biology, Princeton University, Princeton, NJ 08544-1003, USA

Marek Konarzewski
Institute of Biology, University of Bialystok, Swierkowa 20B, 15-950 Bialystok, Poland

Jan Kozłowski
Institute of Environmental Sciences, Jagiellonian University, Gronostajowa 3, 30-387 Cracow, Poland

Jeffrey Lake
Department of Plant Biology, University of Georgia, Athens, GA 30602, USA

Pablo A. Marquet
Center for Advanced Studies in Ecology and Biodiversity, and Departamento de Ecologia, Pontificia Universidad Católica de Chile, Casilla 114-D, Santiago, Chile

Brian A. Maurer
Department of Fisheries and Wildlife, Department of Geography, Michigan State University, East Lansing, MI 48824, USA

Sean Nee
Institute of Cell, Animal and Population Biology, University of Edinburgh, West Mains Road, Edinburgh EH9 3JT, UK

C. David L. Orme
Department of Biological Sciences, Imperial College London, Silwood Park, Ascot SL5 7PY, UK

Mark Pagel
School of Animal and Microbial Sciences, University of Reading, Whiteknights, Reading RG6 6AJ, UK

Andy Purvis
Department of Biological Sciences, Imperial College London, Silwood Park, Ascot SL5 7PY, UK

John D. Reynolds
Centre for Ecology, Evolution and Conservation, School of Biological Sciences, University of East Anglia, Norwich NR4 7TJ, UK

Ignacio Ribera
Department of Entomology, The Natural History Museum, London, SW7 5BD, *and* Department of Biological Sciences, Imperial College London, Silwood Park, Ascot SL5 7PY, UK

Michael L. Rosenzweig
Department of Ecology and Evolutionary Biology, University of Arizona, Tucson, AZ 85721, USA

Kaustuv Roy
Section of Ecology, Behavior and Evolution, Division of Biology, University of California at San Diego, 9500 Gilman Drive, La Jolla, CA 92093-0116, USA

Bernt-Erik Sæther
Department of Biology, Norwegian University of Science and Technology, N-7491 Trondheim, Norway

Van M. Savage
Santa Fe Institute, 1399 Hyde Park Road, Santa Fe, New Mexico 87501, USA *and* Los Alamos National Laboratory, T-8 Mail Stop B285, Los Alamos, New Mexico 87545, USA

James W. Valentine
Department of Integrative Biology, University of California, Berkeley, CA 94720, USA

Alfried P. Vogler
Department of Entomology, The Natural History Museum, London SW7 5BD, *and* Department of Biological Sciences, Imperial College London, Silwood Park, Ascot SL5 7PY, UK

Andrew R. Watkinson
Tyndall Centre for Climate Change Research, Schools of Biological and Environmental Sciences, University of East Anglia, Norwich NR4 7TJ, UK

Geoffrey B. West
Santa Fe Institute, 1399 Hyde Park Road, Santa Fe, New Mexico 87501, USA *and* Los Alamos National Laboratory, T-8 Mail Stop B285, Los Alamos, New Mexico, 87545, USA

Robert J. Whittaker
Biodiversity Research Group, School of Geography and the Environment, University of Oxford, Mansfield Road, Oxford OX1 3TB, UK

Katherine J. Willis
Biodiversity Research Group, School of Geography and the Environment, University of Oxford, Mansfield Road, Oxford OX1 3TB, UK

F. I. Woodward
Department of Animal and Plant Sciences, University of Sheffield, Sheffield S10 2TN, UK

History of the British Ecological Society

The British Ecological Society is a learned society, a registered charity and a company limited by guarantee. Established in 1913 by academics to promote and foster the study of ecology in its widest sense, the Society currently has around 5000 members spread around the world. Members include research scientists, environmental consultants, teachers, local authority ecologists, conservationists and many others with an active interest in natural history and the environment. The core activities are the publication of the results of research in ecology, the development of scientific meetings and the promotion of ecological awareness through education. The Society's mission is:

To advance and support the science of ecology and publicize the outcome of research, in order to advance knowledge, education and its application.

The Society publishes four internationally renowned journals and organizes at least two major conferences each year plus a large number of smaller meetings. It also initiates a diverse range of activities to promote awareness of ecology at the public and policy maker level in addition to developing ecology in the education system, and it provides financial support for approved ecological projects. The Society is an independent organization that receives little outside funding.

British Ecological Society
26 Blades Court
Deodar Road, Putney
London SW15 2NU
United Kingdom
Tel.: +44 (0)20 8871 9797
Fax: +44 (0)20 8871 9779
E-mail: info@britishecologicalsociety.org
URL: http://www.britishecologicalsociety.org

The British Ecological Society is a limited company, registered in England No. 15228997 and a Registered Charity No. 281213.

Preface

The past decade has seen the flowering of a bold and distinctive research programme in ecology that is concerned with thinking big. It is the ecology of wide expanses of space, long periods of time and large numbers of taxa. In a word, coined by Jim Brown and Brian Maurer, it is the discipline of macroecology.

Briefly (more detail is given in Chapters 1 and 21), macroecology is a way of studying relationships between organisms and their environment that involves characterizing and explaining statistical patterns of abundance, distribution and diversity. It aims to identify general principles or natural laws underlying the structure and function of ecological systems, expressed in the patterns of distribution and abundance of the entities that make up those systems. Macroecology explores the domain where ecology, biogeography, palaeontology and macroevolution come together. Although many of its concerns have a long and venerable history, stretching back over almost 200 years, it is an appreciation of the pervasive importance of large-scale effects on ecological systems at all scales, and the attempt to describe the fundamental principles underlying those effects, that mark out the macroecological approach as distinctive and new.

Following some initial misplaced scepticism, the macroecological research programme has been embraced enthusiastically by a growing band of ecologists. It was inevitable that this growth eventually would lead to the desire for an opportunity for practitioners to meet and discuss their interests in a dedicated forum. That opportunity came in April 2002 at the 43rd symposium of the British Ecological Society, held in Birmingham, UK, the first international symposium dedicated to macroecology. This volume is born of that meeting.

It appeared to us, as admittedly partial observers, that the development of interest in macroecology was not being matched by any development of unanimity within the field. Quite divergent opinions existed, and still exist, in the literature about the likely answers to a number of key questions. We think there are three primary reasons for this. First, the largely non-experimental approach necessitated by the scales of concern to macroecologists means that it is difficult, although far from impossible, conclusively to dispatch macroecological hypotheses. Thus, disagreements persist. Second, different workers couch the solution to macroecological questions in terms of different theoretical frameworks, such as speciation/extinction dynamics, energetics, niches or population dynamics. Different frameworks will lead to different conclusions, and so to disagreements. Third, people just plain disagree over the evidence. This is not peculiar to macroecology, or even science.

A BES symposium seemed to us to present the perfect opportunity to stimulate a dialogue amongst macroecologists about those questions that appeared to have generated the most, or most severe, disagreement in the discipline. With that aim, we planned the format of the meeting as a series of sessions, each one with speakers invited to talk on different perspectives on major questions of interest in macroecology. These questions are retained as section headings in this book.

Authors have taken a variety of approaches to the remit presented to them. Some sections and chapters more directly address the questions as a review of a certain perspective. Others use their chapter as an opportunity to present new ideas or analyses. The chapters are thus a stimulating mix of genesis and synthesis. Our principal concern (albeit one with which we would happily have dealt!), that the questions around which we based the symposium might have been conclusively answered between its inception and its execution, was not realized. The explicitly stated aim of the symposium was a reconciliation of different perspectives on macroecological questions. At the meeting, most of the sessions were marked by a degree of coming together of views, albeit that in some sessions this coming together was more in collision than union. We leave it up to the reader to decide the extent to which any reconciliation is achieved by the chapters herein.

The success of a symposium, and of the volume to which it subsequently gives rise, depends on much hard work by a dedicated group of people. Particular thanks are due to Hazel Norman and Richard English of the British Ecological Society, for organizing the symposium with a maximum of efficiency and, sensibly, a minimum of effort required by us. The vast majority of the praise for the meeting we received from the participants should rightfully have gone to them, and so we formally pass it on now. Their job was greatly eased by the assistance of Donna Willmetts, Julie Zacaroli and the conference support staff at the University of Birmingham and Chamberlain Hall. We would also like to thank Andy Dolman, Harprit Kaur and Claire Tyler for volunteering their help during the meeting to make it run smoothly, further relieving the burden on us.

No amount of organizational prowess will elevate a symposium if the scientific contributions are dull. We were lucky again. From start to finish, the invited speakers both stimulated and entertained the audience. Most of the remaining praise we received for the meeting is theirs, and that of those who so ably chaired the sessions. Thank you all. The symposium also benefited from the enthusiasm and interest of the other participants, and the fact that a high proportion of them took the time and trouble to present their research in poster form. It is a shame that there is no space here even to summarize those fascinating contributions, and that demands on our time during the meeting meant that we could talk to only a small fraction of the people that we would have liked to (particularly those who had travelled long distances to be there). Special thanks are due to Professor John Lawton for his contribution to the symposium. We were delighted and flattered that he took time out of his busy schedule as Chief Executive of NERC to be guest of honour at the formal conference dinner, and to regale the assembled diners with a speech of characteristic erudition and humour. John has been an incomparable influence on both our careers, as he has been on those of many ecologists around the world, and the meeting would not have been complete without his presence.

We believe that the quality of the symposium presentations has translated into the quality of the associated chapters in this volume. We thank the authors for the hard work they put into ensuring that this was the case, and for keeping up with the exacting schedule for preparation and submission of their written contributions. This process was aided by the willingness of the referees for these chapters to respond quickly, and constructively, to our requests for reviews. T.M.B. would also like to thank Pat Johnson for her assistance in the days before final manuscript submission. As always, it has been a pleasure to work with Blackwell Publishing, and we particularly thank Katrina McCallum, Delia Sandford and Sarah Shannon for their assistance with the production of this volume.

Finally, we would like to thank those people who make life outside work worthwhile too. T.M.B. particularly thanks Joanna for helping to build his fat reserves with fine food and wine, and Margareta for everything. K.J.G. particularly thanks Sian for her apparently boundless support of a globe-trotting obsessive, and Megan for her love of wild places.

Tim M. Blackburn
Kevin J. Gaston

Chapter 1
Introduction: why macroecology?

Tim M. Blackburn[*] *and Kevin J. Gaston*

Introduction

For most of human history, most people's knowledge of the world was limited to the characteristics of, and events in, their immediate surroundings. This was the source of food, water and other resources required for their survival. It was also the home for those other people whose actions influenced their lives. For many, the events occurring amongst people in areas only a few tens of kilometres away from their home village or home range could be as irrelevant to their existence as events on Mars are to us today. A classic illustration is provided by the diversity of languages that have survived to modern times in the highlands of Papua New Guinea, the development of which is indicative of the lack of interaction between groups in some cases living in neighbouring valleys, separated only by a few kilometres (albeit of terrain that is very difficult to cross). Phenomena such as the weather, determined by processes acting over much larger scales, and hence unpredictable with only local knowledge, were often a source of awe and a spur for myth.

Slowly, however, people's awareness of the world beyond individual domains has expanded. Development of and improvements in agriculture raised the carrying capacity of the land, and allowed population growth. Expanding populations met and coalesced (not always voluntarily), with the result that seats of authority became increasingly removed from the majority of the populace. People's lives came to be influenced more and more by events happening further and further away.

The administration of any social or political system requires information to be communicated to its constituents. Larger systems require more efficient methods of transmission. In many societies, word of mouth was augmented or replaced by written communication, allowing greater volumes of information to be transmitted more reliably. However, the rate of dissemination was still limited by the necessity for the physical movement of literature. For example, Brown (1989) noted how, even before the invention of the telegraph, better communications infrastructure meant that news dissemination in North America had speeded up, so that by 1841 'news

* *Correspondence address: t.blackburn@bham.ac.uk*

from Philadelphia covered the entire eastern seaboard, from Raleigh, North Carolina, to Pittsburgh, Pennsylvania, to Albany, New York, and north to Portland, Maine, within five days. Information from New York City was now arriving in nearly every city east of the Mississippi River within two weeks.' By modern standards, that is painfully slow. People corresponding between Britain and her furthest colonies could expect to wait several months for replies to their letters.

The invention of electronic communication has revolutionized the availability of information, a development that continues apace. Initially, telegraph and radio largely obviated the requirement for information to be disseminated by physical transport of documents. The scope of this revolution was broadened by the telephone, which personalized rapid communication. The need for physical document transport was reduced by the fax machine. Now, email allows manuscripts, figures and photographs to be transmitted across the world in seconds, rather than months. More powerful computers and capacious connections allow ever greater quantities of data to be processed and transmitted, and at ever greater speeds. As awareness of the world at large has expanded, so too have the means for communicating news of events of global import. This may now be widely available within minutes of an event's occurrence, or in real time (at least in those regions where sufficient freedom of the press is permitted). Events in New York City can now be watched as they happen.

The development of the science of ecology has been that of human society in microcosm. Since the early work of Elton (1927), the traditional focus has been on events and interactions occurring within a local community or species assemblage, and hence have concerned how ecological systems are structured by local-scale processes. Even today, the spatial extent of most ecological studies is on the order of square metres, rather than kilometres (Kareiva & Andersen 1988; May 1994; Baskin 1997; Lawton 1999; Gaston & Blackburn 2000). Studies are equally restricted in time, most being concerned with one or just a few seasons or years (Weatherhead 1986; Tilman 1989; Elliott 1994; Malmer 1994; Baskin 1997).

These points are well illustrated by the spatial and temporal scales with which papers appearing in the British Ecological Society publication *Journal of Animal Ecology* have been concerned (Figs 1.1 and 1.2). The number of papers published by the journal has grown ninefold from a low of 11 in 1945 (when matters other than ecology were of primary importance) to a high of 99 in 1999 (see also Shorrocks 1993). Most of that growth involves papers considering topics specifically addressing, or primarily relevant to, local ecological scales, or papers in the 'other' category, which principally comprise laboratory experiments and mathematical models (Fig. 1.1a). Studies specifically concerned with spatial scales larger than the local community (regional, national or biogeographical) have together shown some increase in number over time, particularly since the mid-1970s (see also Brown 1999). This growth over the past three decades is seen most clearly if the proportion of all studies in a given year is compared (Fig. 1.1b). However, also clear from Fig. 1.1b is that the growth is largely simply recouping ground lost in the 1950s and 1960s, which in Britain at least seem to have been the heyday of the local field study. Larger scale

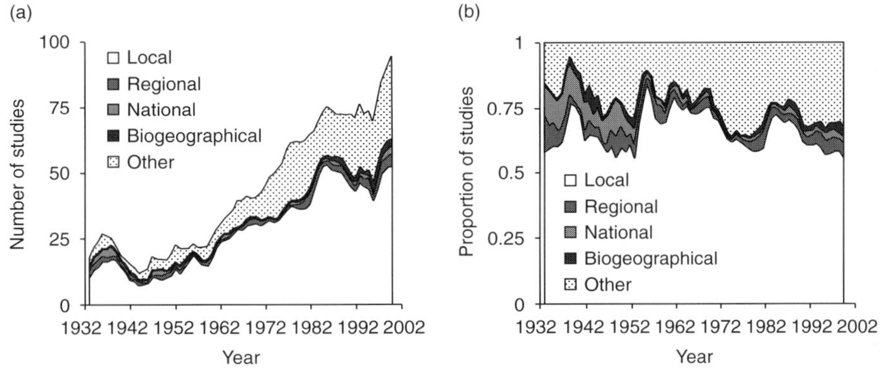

Figure 1.1 The number (a) and proportion (b) of papers published in *Journal of Animal Ecology* in each year between 1932 and 2001 (volumes 1–70, a total of 3042) that concern different spatial scales. The graphs have been smoothed using a 3-year running sum, and so the figures actually span the range 1933–2000. This classification does not include short communications (Comments or Forum papers). Studies were classified as 'local' if they were performed over restricted areas (e.g. at well-defined sites), or if they were performed at a few reasonably well separated sites but this separation was irrelevant to the aim of the study. 'Regional' studies were those concerning scales roughly equivalent to an English or American county, or a restricted part of a country. The 'national' scale refers to studies across regions roughly equivalent to whole countries, whereas 'biogeographical' studies consider multinational, continental or global scales. Studies that could not readily be assigned to any class in this scheme, or for which spatial scale was not relevant, were lumped into the 'other' category. These principally comprised papers reporting laboratory experiments or mathematical models, but also included reviews and some comparative analyses.

studies still represent a minority interest, if the composition of papers in *Journal of Animal Ecology* is a reasonable yardstick.

In terms of time, *Journal of Animal Ecology* has primarily published papers spanning relatively short temporal scales (Fig. 1.2). Most studies containing some field data tend to sample in only one to three different years. However, the number and proportion of longer term studies (4 years or more) published has increased over time, especially since the 1960s, such that these made up 37% of all papers published in the last three volumes. Studies of 7 years or more comprised one-quarter of all papers over these 3 years. Nevertheless, this general increase in temporal span should not be taken as indicative of a paradigm shift in ecological studies towards macroecological timescales. Rather, it represents the increasing availability of long-term data sets for local sites, as widespread recognition of the value of such data has resulted in ecologists extending their local studies as long as tenure, funding and longevity allow.

The increasing prevalence of long-term studies, albeit typically at small spatial scales, reflects an acknowledgement amongst ecologists of the importance of broader scale influences on local-scale communities, such as those mediated by the

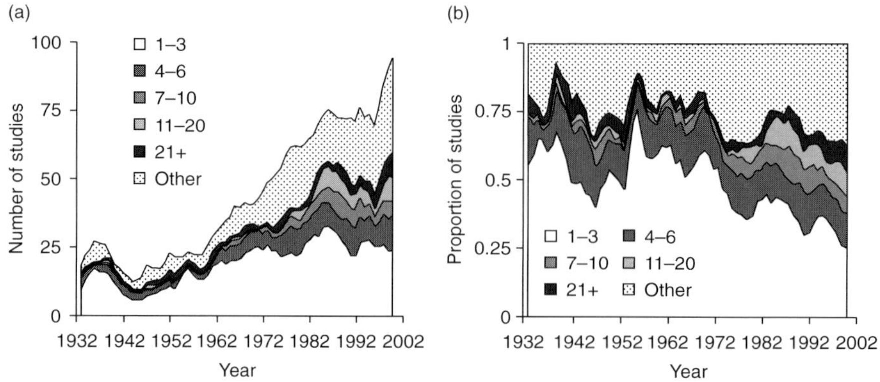

Figure 1.2 The number (a) and proportion (b) of papers published in *Journal of Animal Ecology* in each year between 1932 and 2001 (volumes 1–70, a total of 3042) that concern different temporal scales. The graphs have been smoothed using a 3-year running sum, and so the figures actually span the range 1933–2000. This classification does not include short communications (Comments or Forum papers). Studies were classified according to the number of calendar years in which field data used by the study were collected, or the range of years spanned by the data where this is relevant (e.g. data collected in 1982 and 1991 could be classified as a temporal scale of 2 years or 10 years, depending on whether or not the gap between the years was relevant for the aims of the study). Studies that could not readily be assigned to any class in this scheme, or for which temporal scale was not relevant, were lumped into the 'other' category. These principally comprised papers reporting laboratory experiments or mathematical models, but also included reviews and some comparative analyses.

weather and its temporal dynamics. However, significant effects of these outside influences often have been dismissed as 'unusual' events. For example, Weatherhead (1986) found that about 11% of studies in a sample of ecological journals reported unusual events, most of which referred to abiotic conditions. Unusual events were more frequent in shorter studies, suggesting that without the benefit of a longer perspective the importance of some events is overestimated. In other words, the effects of these outside events can be viewed as normal influences on local communities, which ecological studies need to incorporate, rather than events that disrupt the normal processes of ecological interactions.

At the same time as the attention of the general public was drawn towards the broad expanses of outer space, the early 1960s marked a rise in the attention paid by ecologists to the potential importance of larger scale influences on ecological systems. Publication in 1963 of the seminal paper on the equilibrium theory of insular zoogeography by MacArthur and Wilson (later generalized to the equilibrium theory of island biogeography: MacArthur & Wilson 1967) marked a milestone in this development. MacArthur & Wilson (1963, 1967) showed how the structure of ecological communities could be determined by immigration from a source pool of species, and local extinction within that community, with neither of these essential processes making reference to the identities of the species concerned. Although their model did incorporate biology, it showed that local-scale interactions were not required to generate realistic ecological communities. Similar processes were subsequently modelled by Levins (1969) to show how probabilistic immigration and extinction could determine the distribution of a species across an environment composed of discrete habitat patches.

Of course, large-scale patterns in ecological systems had been recognized long before the 1960s. For example, as early as 1807 Baron Alexander von Humboldt had noted that the variety of life increased from polar to tropical latitudes, describing what today is known as the latitudinal diversity gradient (Hawkins 2001). Arrhenius (1921) was apparently the first to point out that species richness generally increases with areal extent, and the observation that geographical range size may vary systematically with latitude was first made by Lutz (1921). These patterns ultimately must be the outcome of processes acting over large spatial and temporal scales—speciation, extinction and patterns of geographical range dynamics—albeit that local-scale interactions may influence and contribute. The importance of the insights of MacArthur and Wilson and others was to draw attention to the fact that the same large-scale processes also had the potential to drive local community structure. Indeed, they must, as local communities do not develop in isolation, but derive their constituents from the regional source pool. The composition of the pool constrains that of the local community.

Modern studies of large-scale effects in ecology arguably date to the beginning of the 1970s and publication of MacArthur's (1972) book *Geographical Ecology*, but growth since then has been slow. MacArthur's untimely death in that year may have been partly responsible for this, but the sheer difficulty of carrying out large-scale studies seems likely also to have contributed. Most such studies require large quantities of data, usually for multiple species occupying broad spatial extents. These data necessitate a massive effort and resources to collect, and are difficult to organize and analyse compared with those from studies of smaller spatial and temporal scale. All of these factors have counted against large-scale studies. In addition, the success of manipulative experiments in addressing ecological questions (e.g. Connell 1961; Paine 1966, 1974; Hairston 1989) seems to have developed a prejudice against the predominantly observational approaches that inevitably had to be adopted by ecologists interested in large-scale patterns and processes.

There is frequently a mismatch between what can be achieved with current levels of technological development, and what scientists would reasonably like to be able to achieve. This mismatch may be a spur for further such development, but equally it may be found that progress in one field of endeavour has serendipitous applications in others. So it has been with the information revolution and ecology. The rapid improvement in abilities to record, transmit, collate, store and process information provided by the development of digital computers, satellites and the internet has expanded the horizons of ecologists enormously. Many of the shackles that constrained the growth of large-scale ecology at the beginning of the 1970s have been broken. As a result, the past 15 years have witnessed a burgeoning of research into large-scale questions in ecology (for major reviews, see Hengeveld 1990; Ricklefs & Schluter 1993; Edwards *et al.* 1994; Gaston 1994; Brown 1995; Rosenzweig 1995; Maurer 1999; Gaston & Blackburn 2000), to the extent that a new term has been coined to help give this research coherence. This term is 'macroecology'.

What is macroecology?

Macroecology has been defined as concerned with understanding the division of food and space among species at large spatial (geographical) and temporal scales (Brown & Maurer 1989; Brown 1995). Perhaps more in keeping with definitions of ecology in general (e.g. Krebs 2001), it also can be regarded simply as being concerned with understanding the abundance and distribution of species at large spatial and temporal scales (Gaston & Blackburn 1999, 2000). As such, it can perhaps best be thought of as a field of study or a research programme. Brown *et al.* (this volume) emphasize the focus of macroecology on trying to describe and explain the statistical phenomenology of ecologically informative variables among large numbers of ecological 'particles', such as individuals within species or of species within communities. This approach to the study of ecology has a venerable history (e.g. Boycott 1919; Willis 1922; Hemmingsen 1934; Fisher *et al.* 1943; Preston 1948; Hutchinson & MacArthur 1959; MacArthur 1972). These definitions set out the principal aims of the macroecological research programme, which are to understand patterns in and determinants of the broad-scale distribution of life across the planet. Not explicit in these definitions are why such aims are sensible, nor what are their practical consequences.

The basis of the macroecological approach is to develop an understanding of ecological systems through the study of the properties of such systems in their entirety (MacArthur 1972; Brown 1995). This contrasts with the more traditional approach to ecology that seeks to develop such an understanding through reductionist study of the system's component parts (commonly referred to as a 'bottom-up' approach, as opposed to a 'top-down' one). Neither philosophy is inherently superior, albeit that the small-scale approach has to date predominated in ecology for the practical reasons described earlier. The macroecological view derives from the observation that complex systems, such as those of concern to ecologists, may exhibit properties or behaviours that arise from the interaction of their constituent parts, and so that are not evident in, or predictable from, knowledge of these parts alone (MacArthur 1972). Emergent properties make the behaviour of complex systems difficult to predict a priori. However, this difficulty can be circumvented easily by studying emergent properties directly. In ecology, this necessitates the large-scale approach that is the purlieu of macroecology.

Brown (1995) draws an analogy between the study of individual organisms or species and the study of gas molecules. It would be very hard, if not impossible, to understand the behaviour of gases from the study of individual molecules, although such studies would undoubtedly yield important insights. However, by studying the emergent statistical properties of large collections of molecules, scientists have discovered behaviours that are predictable enough to be enshrined as physical laws. This philosophy is not peculiar to macroecology. A complete understanding of most, if not all, scientific disciplines is likely to arise only by incorporating observations made from a range of viewpoints, or at a variety of scales (Dunbar 1995). The focus often changes as a field develops, and important gaps in knowledge are identified. In ecology, for example, the observation that species differ in their abundance led to attempts to describe the variation amongst large numbers of species, in studies that would today be termed macroecological (Fisher *et al.* 1943; Preston 1948, 1962). The descriptions initially were statistical, but mechanistic models of species–abundance distributions were developed subsequently (MacArthur 1960; Sugihara 1980; Tokeshi 1990, 1996; Hubbell 1997; Harte *et al.* 1999a). Models of species–abundance distributions have, in turn, resulted in studies aiming to validate assumptions or predictions of those models using small-scale data (e.g. Sugihara 1980; Harte *et al.* 1999b; Plotkin *et al.* 2000; Condit *et al.* 2002). Local ecological studies can be informed by macroecology, and vice versa. By following both small-scale 'bottom-up' and large-scale 'top-down' paths a better understanding has been reached than would have been derived from either approach alone. Both approaches are valid, and indeed necessary, tools for exploring the complexity of ecological systems.

The practical consequences of the macroecological approach are several-fold. Probably the most significant concern the methodology appropriate for answering questions in macroecology, statistical techniques, and the balance achieved between pattern and process in macroecological studies (Gaston & Blackburn 1999).

Methodology

Interest in large-scale questions largely precludes the use of manipulative experiments, as the scales involved are simply too great for most such experiments to be possible (practically or financially) or ethical. Manipulative experiments are not entirely precluded, however, as they can be used to address macroecological questions in systems where the manipulations are large scale for the organisms but not for the experimenter. Examples include the use of laboratory and field-based microcosms to investigate mechanisms underpinning interspecific relationships between abundance and distribution (Gaston & Warren 1997; Warren & Gaston 1997; Gonzalez *et al.* 1998). As manipulative experiments are probably the most powerful tool for differentiating between alternative hypotheses, such approaches are likely to become increasingly popular with macroecologists, albeit that the mechanisms structuring communities of organisms amenable to use in microcosms may be different to those for other groups, and issues of scaling between microcosms and geographical landscapes remain to be resolved (Peterson & Parker 1998).

Macroecological studies of larger bodied organisms typically rely on conclusions that can be drawn from natural experiments, the conceptually similar experiments in nature (Diamond 1986) or observational data (McArdle 1996). Natural experiments are changes in systems brought about by natural events, such as earthquakes and volcanic eruptions, whereas experiments in nature are changes in systems brought about as an intentional or accidental product of human activities, such as introductions of alien species. These are probably the only way in which experiments can be performed at the large temporal and spatial scales of concern to macroecologists. Their utility derives from their realism: unlike manipulative experiments, the study system is a natural entity. Thus, although natural experiments do not have the same discriminatory power as manipulative experiments, conclusions drawn from them may be more applicable to natural situations (Diamond 1986, 2001). Natural experiments have been a useful source of information on a variety of large-scale questions about ecological systems, such as those concerning extinction (e.g. Karr 1982; Pimm *et al.* 1988; Hinsley *et al.* 1995; Bellamy *et al.* 1996; Lomolino 1996; Manne *et al.* 1998; Spiller *et al.* 1998; Duncan & Young 2000; Terborgh *et al.* 2001) and invasion (Moulton & Pimm 1983, 1986; Moulton & Lockwood 1992; Thornton 1996; Veltman *et al.* 1996; Duncan 1997; Blackburn & Duncan 2001a, b).

In situations where neither manipulative nor natural experiments can be exploited, macroecologists generally must rely on observational data to generate and test hypotheses, and indeed this has been far and away the predominant approach to macroecological questions (see references in Gaston & Blackburn 2000). These data frequently derive from disparate unco-ordinated and unplanned sets of observations by multiple observers from different sites over a range of dates, so that their use must be tempered by caution if spurious and biased conclusions are to be avoided (Gotelli & Graves 1996; Blackburn & Gaston 1998, 2002). Observational data also generally lack the power to distinguish between similar hypotheses, which has lowered their respectability in recent years. Nevertheless, they form the foundation of most scientific disciplines, and with careful attention to their limitations can be valuable in hypothesis testing (McArdle 1996).

Statistical issues

The extent of the reliance of macroecology on unplanned natural experiments and observational data has two further consequences. First, because macroecological data usually lack controls, particular attention needs to be paid to the null hypothesis for any given data set or test. This issue has been discussed at length elsewhere in relation to macroecology (Gotelli & Graves 1996; Blackburn & Gaston 1998; Gaston & Blackburn 1999, 2000), although it is likely to remain a topic of fertile debate.

Second, a greater variety of statistical techniques generally need to be used to extract meaningful information from the raw data. When correctly formulated, controlled manipulative experiments allow the effects of all aspects of a system bar that under test to be held constant, so linking effects to causes (in practice, such an ideal situation can seldom be realized). The absence of controls in macroecological data can to some degree be compensated for by statistical approaches. For example, statistical null models can help identify significant features of macroecological data (Blackburn *et al.* 1990; Gotelli & Graves 1996; Gaston & Blackburn 1999; Colwell & Lees 2000; Gotelli 2000), and multivariate models can help factor out uncontrolled variation in natural systems (Mac Nally 1996; Thomson *et al.* 1996; Blackburn & Duncan 2001a). In many cases the data used by macroecologists, or the questions that they would wish to address, are difficult or impossible to analyse using traditional statistics (Blackburn & Gaston 1998). Macroecological studies frequently concern spatially or temporally explicit data, and so necessitate controls for associated autocorrelation (Lennon 2000; Lennon *et al.* 2001; Brewer & Gaston 2002). Similarly, because such studies often entail interspecific comparisons, as often required to elucidate the statistical phenomenology of ecologically informative variables among large numbers of ecological particles (Brown *et al.* this volume), autocorrelated responses of species owing to their shared phylogenetic inheritance need to be considered (Harvey & Pagel 1991; Harvey 1996; Blackburn & Gaston 1998; Gaston & Blackburn 2000). The techniques to allow these challenges to be met are only now being developed or discovered, their development motivated in part by the demands of the field (Purvis & Rambaut 1995; Carroll & Pearson 1998, 2000; Pagel 1999; Lennon 2000; Blackburn & Duncan 2001a).

Patterns and process

Finally, the constraints placed on macroecology by the methodological techniques available have had practical consequences for the theoretical development of the field. Whereas macroecological patterns are increasingly frequently reported, and hypotheses to explain them abound, unequivocal tests that link a given pattern to a specific process are relatively unusual. Although some hypotheses can be falsified for some systems (e.g. metapopulation dynamics cannot explain the positive abundance–occupancy relationship found for protist species across microcosms with no dispersal; Warren & Gaston 1997), it is much rarer for all bar one hypothesis to be falsified for any system (but see e.g. Gonzalez *et al.* 1998). It follows that understanding of process lags behind that of pattern in macroecology to a greater degree than in most other fields of ecology.

Gaston & Blackburn (1999) suggested two further reasons why an understanding of process in macroecology has been retarded, which are of particular relevance in the context of this symposium. First, different hypotheses of process may be rooted in different theoretical frameworks. For example, interspecific frequency distributions of body mass have been suggested to derive from speciation–extinction dynamics (Fowler & MacMahon 1982; Dial & Marzluff 1988; Maurer *et al.* 1992; Johst & Brandl 1997) or energetics (Brown *et al.* 1993; Maurer 1998), and abundance–occupancy relationships from resource usage (Brown 1984; Hanski *et al.* 1993; Gaston 1994) or population dynamics (Hanski *et al.* 1993; Holt *et al.* 1997). This means that hypothesized mechanisms may not be mutually exclusive, but instead represent different levels of explanation. Thus, the shape of a species–body-mass distribution must be explicable in terms of the differential speciation and extinction of species of different body size, but it may be patterns of energy usage that generate those differences in speciation and extinction rates. The multiplicity of frameworks means that macroecology currently lacks a central conceptual unification.

Second, even within a common theoretical framework, more than one mechanism may contribute to the observed form of a macroecological pattern. Ecology traditionally has been dominated by the search for single explanations for patterns, as typified by MacArthur & Connell's (1966) maxim that '(w)herever there is a widespread pattern, there is likely to be a general explanation which applies to the whole pattern'. However, although this view is both intrinsically appealing and satisfies Occam's razor, there is actually no necessary reason why it should be true. Indeed, given the complexity, variability and idiosyncrasy of ecological systems, it is perhaps more likely that a pattern that occurs across taxa, environments, or biogeographical regions may derive from the alignment of different contributory mechanisms, rather than the action of any given one (Wilson 1988; Warren & Gaston 1992; Blackburn & Gaston 1996; Lawton 1996; Gaston *et al.* 1997). The central question then becomes not which of any competing explanations is the correct one, but what is their relative importance, and when and where does this change. It also follows that identifying the mechanism underlying a given pattern in a given system does not mean that similar causal links apply to other systems (cf. Warren & Gaston 1997; Gonzalez *et al.* 1998). This considerably complicates the complete falsification of any macroecological hypothesis, unless that hypothesis is clearly context-specific.

Aims of this book

The combination of the need for a macroecological approach, the difficulty in implementing it, and the rapid development in the data and the tools available to do so, make this an exciting time for those involved in the field. Understanding is improving quickly. The present edited volume, the first dedicated to macroecology, brings together contributions from many of those at the forefront of these advances, with the primary goal of drawing out the divergent viewpoints on some of the major issues that macroecology has to address. The object here was not to expose these differences, as most have been well documented, but rather to seek to clarify the common ground they share, and thus to reveal some consensus about the shape of the natural world as viewed from a macroecological perspective.

The volume divides into eight main sections. In each of the first seven, two or more authors have been asked to address the same macroecological question from different viewpoints. None of these questions are novel, rather they for the most part reflect obvious characteristics of the natural world, such as the high diversity of species in the tropics, the predominance of small-bodied species, and the fact that most species are relatively rare. Nevertheless, answers to these questions are likely to prove central to a mature understanding of ecological systems, and so have captured the attention of macroecologists, and indeed ecologists more widely. They are also questions that have divided macroecological opinion, and so are ripe for the clarification and reconciliation that we hope that this volume will help advance.

The eighth and final section provides a series of statements about the relevance of the macroecological research programme. These statements show how macroecology is linked to, and can inform activities across, a broad swathe of environmental science. Given that we are part of the first generation in history that the information revolution allows to communicate and experience events over the entire planet, it is particularly appropriate that we should be helping to construct a large-scale view of the ecological systems that allow the planet to support us.

Acknowledgements

T. M. Blackburn thanks E. Cooke, R. Duncan, S. Duncan, K. Ferguson, C. MacLeod, M. Smalle and Lincoln University for kind hospitality during that part of the writing of this manuscript spent in New Zealand. This work was supported in part by the Darwin Initiative of the UK Department for Environment, Food and Rural Affairs.

References

Arrhenius, O. (1921) Species and area. *Journal of Ecology* **9**, 95–99.

Baskin, Y. (1997) Center seeks synthesis to make ecology more useful. *Science* **275**, 310–311.

Bellamy, P.E., Hinsley, S.A. & Newton, I. (1996) Local extinctions and recolonisations of passerine bird populations in small woods. *Oecologia* **108**, 64–71.

Blackburn, T.M. & Duncan, R.P. (2001a) Determinants of establishment success in introduced birds. *Nature* **414**, 195–197.

Blackburn, T.M. & Duncan, R.P. (2001b) Establishment patterns of exotic birds are constrained by non-random patterns in introduction. *Journal of Biogeography* **28**, 927–939.

Blackburn, T.M. & Gaston, K.J. (1996) A sideways look at patterns in species richness, or why there are so few species outside the tropics. *Biodiversity Letters* **3**, 44–53.

Blackburn, T.M. & Gaston, K.J. (1998) Some methodological issues in macroecology. *American Naturalist* **151**, 68–83.

Blackburn, T.M. & Gaston, K.J. (2002) Scale in macroecology. *Global Ecology and Biogeography* **11**, 185–189.

Blackburn, T.M., Harvey, P.H. & Pagel, M.D. (1990) Species number, population density and body size in natural communities. *Journal of Animal Ecology* **59**, 335–346.

Boycott, A.E. (1919) On the size of things, or the importance of being rather small. *Contributions to Medical and Biological Research* **1**, 226–234.

Brewer, A.M. & Gaston, K.J. (2002) The geographical range structure of the holly leaf-miner. I. Population density. *Journal of Animal Ecology* **71**, 99–111.

Brown, J.H. (1984) On the relationship between abundance and distribution of species. *American Naturalist* **124**, 255–279.

Brown, J.H. (1995) *Macroecology.* University of Chicago Press, Chicago.

Brown, J.H. (1999) Macroecology: progress and prospect. *Oikos* **87**, 3–14.

Brown, J.H., Marquet, P.A. & Taper, M.L. (1993) Evolution of body size: consequences of an energetic definition of fitness. *American Naturalist* **142**, 573–584.

Brown, J.H. & Maurer, B.A. (1989) Macroecology: the division of food and space among species on continents. *Science* **243**, 1145–1150.

Brown, R.D. (1989) *Knowledge is Power.* Oxford University Press, New York.

Carroll, S.S. & Pearson, D.L. (1998) The effects of scale and sample size on the accuracy of spatial predictions of tiger beetle (Cicindelidae) species richness. *Ecography* **21**, 401–414.

Carroll, S.S. & Pearson, D.L. (2000) Detecting and modelling spatial and temporal dependence in conservation biology. *Conservation Biology* **14**, 1893–1897.

Colwell, R.K. & Lees, D.C. (2000) The mid-domain effect: geometric constraints on the geography of species richness. *Trends in Ecology and Evolution* **15**, 70–76.

Condit, R., Pitman, N., Leigh Jnr, E.G., *et al.* (2002) Beta-diversity in tropical forest trees. *Science* **295**, 666–669.

Connell, J.H. (1961) The influence of interspecific competition and other factors on the distribution of the barnacle *Chthamalus stellatus. Ecology* **42**, 710–723.

Dial, K.P. & Marzluff, J.M. (1988) Are the smallest organisms the most diverse? *Ecology* **69**, 1620–1624.

Diamond, J. (1986) Overview: laboratory experiments, field experiments, and natural experiments. In: *Community Ecology* (eds J. Diamond & T.J. Case), pp. 3–22. Harper Row, New York.

Diamond, J. (2001) Dammed experiments! *Science* **294**, 1847–1848.

Dunbar, R.I.M. (1995) *The Trouble with Science.* Faber and Faber, London.

Duncan, R.P. (1997) The role of competition and introduction effort in the success of passeriform birds introduced to New Zealand. *American Naturalist* **149**, 903–915.

Duncan, R.P. & Young, J.R. (2000) Determinants of plant extinction and rarity 145 years after European settlement of Auckland, New Zealand. *Ecology* **81**, 3048–3061.

Edwards, P.J., May, R.M. & Webb, N.R. (1994) *Large-scale Ecology and Conservation Biology.* Blackwell Scientific Publications, Oxford.

Elliott, J.M. (1994) *Quantitative Ecology and the Brown Trout.* Oxford University Press, Oxford.

Elton, C. (1927) *Animal Ecology.* Sidgwick & Jackson, London.

Fisher, R.A., Corbet, A.S. & Williams, C.B. (1943) The relation between the number of species and the number of individuals in a random sample of an animal population. *Journal of Animal Ecology* **12**, 42–58.

Fowler, C.W. & MacMahon, J.A. (1982) Selective extinction and speciation: their influence on the structure and functioning of communities and ecosystems. *American Naturalist* **119**, 480–498.

Gaston, K.J. (1994) *Rarity.* Chapman & Hall, London.

Gaston, K.J. & Blackburn, T.M. (1999) A critique for macroecology. *Oikos* **84**, 353–368.

Gaston, K.J. & Blackburn, T.M. (2000) *Pattern and Process in Macroecology.* Blackwell Science, Oxford.

Gaston, K.J. & Warren, P.H. (1997) Interspecific abundance–occupancy relationships and the effects of disturbance: a test using microcosms. *Oecologia* **112**, 112–117.

Gaston, K.J., Blackburn, T.M. & Lawton, J.H. (1997) Interspecific abundance–range size relationships: an appraisal of mechanisms. *Journal of Animal Ecology* **66**, 579–601.

Gonzalez, A., Lawton, J.H., Gilbert, F.S., Blackburn, T.M. & Evans-Freke, I. (1998) Metapopulation dynamics, abundance, and distribution in a microecosystem. *Science* **281**, 2045–2047.

Gotelli, N.J. (2000) Null model analysis of species co-occurrence patterns. *Ecology* **81**, 2606–2621.

Gotelli, N.J. & Graves, G.R. (1996) *Null Models in Ecology.* Smithsonian Institution Press, Washington, DC.

Hairston, N.G. (1989) *Ecological Experiments: Purpose, Design, and Execution.* Cambridge University Press, Cambridge.

Hanski, I., Kouki, J. & Halkka, A. (1993) Three explanations of the positive relationship between distribution and abundance of species. In: *Species Diversity in Ecological Communities: Historical and Geographical Perspectives* (eds R.E. Ricklefs & D. Schluter), pp. 108–116. University of Chicago Press, Chicago.

Harte, J., Kinzig, A. & Green, J. (1999a) Self-similarity in the distribution and abundance of species. *Science* **284**, 334–336.

Harte, J., McCarthy, S., Taylor, K., Kinzig, A. & Fischer, M.L. (1999b) Estimating species–area relationships from plot to landscape scale using species spatial-turnover data. *Oikos* **86**, 45–54.

Harvey, P.H. (1996) Phylogenies for ecologists. *Journal of Animal Ecology* **65**, 255–263.

Harvey, P.H. & Pagel, M.D. (1991) *The Comparative Method in Evolutionary Biology.* Oxford University Press, Oxford.

Hawkins, B.A. (2001) Ecology's oldest pattern? *Trends in Ecology and Evolution* **16**, 470.

Hemmingsen, A.M. (1934) A statistical analysis of the differences in body size of related species. *Videnskabelige Meddelelser fra Dansk Naturhistorik Forening i Kobenhavn* **98**, 125–160.

Hengeveld, R. (1990) *Dynamic Biogeography.* Cambridge University Press, Cambridge.

Hinsley, S.A., Bellamy, P.E. & Newton, I. (1995) Bird species turnover and stochastic extinction in woodland fragments. *Ecography* **18**, 41–50.

Holt, R.D., Lawton, J.H., Gaston, K.J. & Blackburn, T.M. (1997) On the relationship between range size and local abundance: back to basics. *Oikos* **78**, 183–190.

Hubbell, S.P. (1997) A unified theory of biogeography and relative species abundance and its

application to tropical rain forests and coral reefs. *Coral Reefs* **16** (supplement), S9–S21.
Hutchinson, G.E. & MacArthur, R.H. (1959) A theoretical ecological model of size distributions among species of animals. *American Naturalist* **93**, 117–125.
Johst, K. & Brandl, R. (1997) Body size and extinction risk in a stochastic environment. *Oikos* **78**, 612–617.
Kareiva, P. & Andersen, M. (1988) Spatial aspects of species interactions: the wedding of models and experiments. In: *Community Ecology* (ed. A. Hastings), pp. 35–50. Springer-Verlag, Berlin.
Karr, J.R. (1982) Population variability and extinction in the avifauna of a tropical land bridge island. *Ecology* **63**, 1975–1978.
Krebs, C.J. (2001) *Ecology. The Experimental Analysis of Distribution and Abundance*, 5th edn. Benjamin Cummings, San Francisco.
Lawton, J.H. (1996) Patterns in ecology. *Oikos* **75**, 145–147.
Lawton, J.H. (1999) Are there general laws in ecology? *Oikos* **84**, 177–192.
Lennon, J.J. (2000) Red-shifts and red herrings in geographical ecology. *Ecography* **23**, 101–113.
Lennon, J.J., Koleff, P., Greenwood, J.J.D. & Gaston, K.J. (2001) The geographical structure of British bird distributions: diversity, spatial turnover and scale. *Journal of Animal Ecology* **70**, 966–979.
Levins, R. (1969) Some demographic and genetic consequences of environmental heterogeneity for biological control. *Bulletin of the Entomological Society of America* **15**, 237–240.
Lomolino, M.V. (1996) Investigating causality of nestedness of insular communities: selective immigrations or extinctions? *Journal of Biogeography* **23**, 699–703.
Lutz, F.E. (1921) Geographic average, a suggested method for the study of distribution. *American Museum Novitates* **5**, 1–7.
Mac Nally, R. (1996) Hierarchical partitioning as an interpretative tool in multivariate inference. *Australian Journal of Ecology* **21**, 224–228.
MacArthur, R.H. (1960) On the relative abundance of species. *American Naturalist* **94**, 25–36.
MacArthur, R.H. (1972) *Geographical Ecology: Patterns in the Distribution of Species.* Harper & Row, New York.
MacArthur, R.H. & Connell, J.H. (1966) *The Biology of Populations.* Wiley, New York.
MacArthur, R.H. & Wilson, E.O. (1963) An equilibrium theory of insular zoogeography. *Evolution* **17**, 373–387.
MacArthur, R.H. & Wilson, E.O. (1967) *The Theory of Island Biogeography.* Princeton University Press, Princeton, NJ.
Malmer, N. (1994) Ecological research at the beginning of the next century. *Oikos* **71**, 171–176.
Manne, L.L., Pimm, S.L., Diamond, J.D. & Reed, T.M. (1998) The form of the curves: a direct evaluation of MacArthur & Wilson's classic theory. *Journal of Animal Ecology* **67**, 784–794.
Maurer, B.A. (1998) The evolution of body size in birds. II. The role of reproductive power. *Evolutionary Ecology* **12**, 935–944.
Maurer, B.A. (1999) *Untangling Ecological Complexity.* Chicago University Press, Chicago.
Maurer, B.A., Brown, J.H. & Rusler, R.D. (1992) The micro and macro in body size evolution. *Evolution* **46**, 939–953.
May, R.M. (1994) The effects of spatial scale on ecological questions and answers. In: *Large-scale Ecology and Conservation Biology* (eds P.J. Edwards, R.M. May & N.R. Webb), pp. 1–17. Blackwell Scientific Publications, Oxford.
McArdle, B.H. (1996) Levels of evidence in studies of competition, predation, and disease. *New Zealand Journal of Ecology* **20**, 7–15.
Moulton, M.P. & Lockwood, J.L. (1992) Morphological dispersion of introduced Hawaiian finches: evidence for competition and a Narcissus effect. *Evolutionary Ecology* **6**, 45–55.
Moulton, M.P. & Pimm, S.L. (1983) The introduced Hawaiian avifauna: biogeographic evidence for competition. *American Naturalist* **121**, 669–690.
Moulton, M.P. & Pimm, S.L. (1986) The extent of competition in shaping an introduced avifauna. In: *Community Ecology* (eds J. Diamond & T.J. Case), pp. 80–97. Harper & Row, New York.
Pagel, M. (1999) Inferring the historical patterns of biological evolution. *Nature* **401**, 877–884.
Paine, R.T. (1966) Food web complexity and species diversity. *American Naturalist* **100**, 65–75.
Paine, R.T. (1974) Intertidal community structure. Experimental studies on the relationship between a dominant predator and its principal predator. *Oecologia* **15**, 93–120.
Peterson, D.L. & Parker, V.T. (eds.) (1998) *Ecological Scale: Theory and Applications.* Columbia University Press, New York.

Pimm, S.L., Jones, H.L. & Diamond, J. (1988) On the risk of extinction. *American Naturalist* **132**, 757–785.

Plotkin, J., Potts, M.D., Yu, D.W., *et al.* (2000) Predicting species diversity in tropical forests. *Proceedings of the National Academy of Sciences, USA* **97**, 10850–10854.

Preston, F.W. (1948) The commonness, and rarity, of species. *Ecology* **29**, 254–283.

Preston, F.W. (1962) The canonical distribution of commonness and rarity. *Ecology* **43**, 185–215, 410–432.

Purvis, A. & Rambaut, A. (1995) Comparative analysis by independent contrasts (CAIC): an Apple Macintosh application for analysing comparative data. *Computer Applications in the Biosciences* **11**, 247–251.

Ricklefs, R.E. & Schluter, D. (1993) *Species Diversity in Ecological Communities.* University of Chicago Press, Chicago.

Rosenzweig, M.L. (1995) *Species Diversity in Space and Time.* Cambridge University Press, Cambridge.

Shorrocks, B. (1993) Trends in the *Journal of Animal Ecology*: 1932–92. *Journal of Animal Ecology* **62**, 599–605.

Spiller, D.A., Losos, J.B. & Schoener, T.W. (1998) Impact of a catastrophic hurricane on island populations. *Science* **281**, 695–697.

Sugihara, G. (1980) Minimal community structure: an explanation of species abundance patterns. *American Naturalist* **116**, 770–787.

Terborgh, J., Lopez, L., Nuñez, V., *et al.* (2001) Ecological meltdown in predator-free forest fragments. *Science* **294**, 1923–1926.

Thomson, J.D., Weiblen, G., Thomson, B.A., Alfaro, S. & Legendre, P. (1996) Untangling multiple factors in spatial distributions: lilies, gophers, and rocks. *Ecology* **77**, 1698–1715.

Thornton, I.W.B. (1996) *Krakatau—the Destruction and Reassembly of an Island Ecosystem.* Harvard University Press, Cambridge, MA.

Tilman, D. (1989) Ecological experimentation: strengths and conceptual problems. In: *Long-term Studies in Ecology* (ed. G.E. Likens), pp. 136–157. Springer-Verlag, New York.

Tokeshi, M. (1990) Niche apportionment or random assortment: species abundance patterns revisited. *Journal of Animal Ecology* **59**, 1129–1146.

Tokeshi, M. (1996) Power fraction: a new explanation of relative abundance patterns in species-rich assemblages. *Oikos* **75**, 543–550.

Veltman, C.J., Nee, S. & Crawley, M.J. (1996) Correlates of introduction success in exotic New Zealand birds. *American Naturalist* **147**, 542–557.

Warren, P.H. & Gaston, K.J. (1992) Predator–prey ratios: a special case of a general pattern? *Philosophical Transactions of the Royal Society, London, Series B* **338**, 113–130.

Warren, P.H. & Gaston, K.J. (1997) Interspecific abundance–occupancy relationships: a test of mechanisms using microcosms. *Journal of Animal Ecology* **66**, 730–742.

Weatherhead, P.J. (1986) How unusual are unusual events? *American Naturalist* **128**, 150–154.

Willis, J.C. (1922) *Age and Area.* Cambridge University Press, Cambridge.

Wilson, D.S. (1988) Holism and reductionism in evolutionary ecology. *Oikos* **53**, 269–273.

Part 1

Why are some taxa more diverse than others?

Chapter 2

Evolutionary analysis of species richness patterns in aquatic beetles: why macroecology needs a historical perspective

Alfried P. Vogler and Ignacio Ribera*

Introduction

Species richness differs dramatically between groups of organisms, but explanations for these differences are elusive. For example, it is widely recognized that body size correlates with species richness, and richness is shifted towards small—but not the smallest—species (May 1973; Lawton 1995; Purvis *et al.*, this volume). This striking body-size distribution of species diversity is observed almost universally across many taxonomic groups. Even more pronounced are differences in species richness between major phyletic lineages, but these are less regular and predictable. Examples have been presented from vertebrates (Stiassny & Pinna 1994), plants (Sanderson & Donoghue 1994), insects (Mayhew 2002), tiger beetles (Vogler & Barraclough 1998) and many others. These differences in clade size are the result of variation in speciation and extinction rates in the history of a clade.

Similarly, geographical patterns of species richness vary greatly. General trends in species distributions have been described that apply widely to many groups and geographical regions, such as the species–area relationship or latitudinal gradients (Arrhenius 1921; Preston 1962; Williams 1964; MacArthur & Wilson 1967; Lawton 1999). Numerous studies attempted to establish the underlying causal mechanisms from counts of species numbers and their statistical correlates at regional and global scales (see Brown (1995) and Rosenzweig (1995) for reviews). Usually the existence of these patterns is explained as the result of physical parameters differing between areas, such as available energy and potential evapotransporation, or geographical parameters themselves, such as the area size (Brown 1995; Rosenzweig 1995).

Understanding the evolution and the ecological correlates of species richness ('macroevolution' and 'macroecology', respectively) has been the objective of many recent studies. However, both fields have not easily found a common ground, as the methodological and conceptual approaches are quite different. Yet it is clear that macroevolutionary processes of speciation and extinction in a clade depend greatly on the ecological arena in which these processes take place. Equally, macroecological

* *Correspondence address: a.vogler@nhm.ac.uk*

patterns are affected by the taxonomic richness of clades, resulting in diversity 'anomalies' where species richness is not following the expected geographical trends but is mostly determined by the local radiation of lineages. Hence, macroevolutionary and macroecological patterns cannot be considered in isolation.

This paper will mostly focus on the macroevolutionary approach to analyse and explain differences in species richness, and how ecological mechanisms have been invoked to explain macroevolutionary patterns (species diversification in lineages). This approach presents various problems, as we will demonstrate. We will also show the significance of evolutionary traits for patterns of geographical (macroecological) species richness, and how the geographical processes, in turn, might affect macroevolutionary rates of species diversification in a clade. It will become clear that neither macroevolutionary nor macroecological approaches in separation are satisfactory to explain richness patterns, and in particular, there is little hope to understand macroecological patterns without an evolutionary perspective. We will use our recent work in aquatic beetles to illustrate some of these issues.

Macroevolutionary approaches to understand species richness

Traditionally, the analysis of macroevolutionary patterns and processes has been addressed with two principal approaches. Each of these attempt to understand the processes of taxonomic diversification from an ecological perspective. They can be subdivided conveniently by the hierarchical levels on which they are proposed to operate. First, an explanation of species richness is derived from the correlation of species numbers and particular ecological traits. This idea gained momentum with the classic study of Mitter *et al.* (1988), who demonstrated an increased species richness in phytophagous insects relative to their non-phytophagous sister groups. With the wide use of (molecular) phylogenetics, this comparative approach has received wide application recently (Sanderson & Donoghue 1996; Givnish & Systma 1997; Barraclough *et al.* 1998; Farrell 1998) but it goes back to Simpson (1953), who suggested that the gain of a functional trait permits the exploitation of new resources, which directly or indirectly leads to species diversification. Such a 'key' trait enables its bearer to gain access to a new adaptive zone and the opportunity to diversify. Although in Simpson's writing the link to taxonomic richness was less explicit, the term 'key innovation' later gained widespread acceptance to refer to traits that confer an increase in the apparent speciation rate of a lineage resulting from the utilization of a new range of resources (Heard & Hauser 1995; Hunter 1998). The term has become synonymous with an approach using phylogenetic trees to study relative species diversity in a lineage in relation to ecological transitions.

The second approach to investigating lineage-specific diversity similarly invokes ecological mechanisms to increase species numbers, but it is concerned with the process of species origin and speciation itself. Diversification is considered at the level of the species, where ecological, behavioural and geographical factors generate differences between populations that lead to speciation, in a set of phenomena usually referred to as adaptive radiation (Lack 1947; Givnish & Systma 1997;

Schluter 2000). The hypothesis initially formulated by Lack (1947) on the basis of his work on Galapagos finches is that speciation results from changes in resource use in populations, in combination with allopatrically derived neutral traits and/or sexually selected and other behavioural differences. Under certain conditions, as frequently encountered on island archipelagos, lake systems and mountain ranges, the result is an ecologically driven increase of species diversity, coincident with greater variation in ecological traits.

In the following we will investigate macroevolutionary patterns in aquatic beetles based on these two principal approaches (clade level and species level). In particular, the transition from land to freshwater habitats in the Coleoptera (beetles) provides an example of a major ecological shift and the acquisition of a novel adaptive zone. We will establish general trends in species richness associated with this shift but it will be apparent that the clade-level approach is not suited to establishing the underlying mechanisms of how diversification rates are modulated. We will also present species-level analyses, in particular with regard to the differences observed in standing water (lentic) and running water (lotic) species. The latter provides an example where the transition between habitats also has profound implications for macroecological patterns, and by investigating these patterns in phylogenetically independent lineages, an assessment is possible of what caused the formation of these patterns.

The transition to aquatic habitats in Coleoptera

Aquatic beetles are taxonomically and ecologically highly diverse. They include many species with a hydrodynamically shaped body and modified legs with swimming hairs for improved locomotion. Beetles also differ in their swimming performance. Whereas some species essentially walk under water, crawl on submerged rocks or plants, or burrow at the bottom of the water bodies, others exhibit coordinated movements of the middle and hind legs, permitting great swimming speed and excellent manoeuvrability. In the extreme case of specialized aquatic modifications, the Gyrinidae (whirligig beetles) swim in the water surface film at high speed, being able to see simultaneously in the air and in the water column with two separate pairs of eyes.

The great morphological and behavioural diversity of aquatic beetles has sparked a debate about their origin (Beutel 1997; Hansen 1997; Shull *et al.* 2001). The two main coleopteran suborders, Adephaga and Polyphaga, include clades of aquatic beetles, and the suborder Myxophaga, a small group of only 56 known species of minute beetles (Crowson 1955), is entirely aquatic. Hence, within the primitively terrestrial (Crowson 1960) beetles at least three independent transitions from land to water can be inferred. In addition, phylogenetic analyses within the two main suborders revealed further land-to-water transitions. Whereas in Adephaga the aquatic lineages ('Hydradephaga') were found to be monophyletic (Crowson 1955; Shull *et al.* 2001; Ribera *et al.* 2002b) indicating a single transition, the situation in Polyphaga is more complex (Hansen 1997). Hydrophiloidea and Hydraenidae, two

Table 2.1 Sister comparisons of species richness in major aquatic lineages of Coleoptera. The number of species used in sister comparison of Polyphaga + Adephaga is shown in parentheses, as it also includes several aquatic groups.

Aquatic clade	Sister group	Species aquatic	Species terrestrial	Significance*
Myxophaga	Polyphaga + Adephaga	56	(350 000)	0.0003
Hydradephaga	Geadephaga	5500	24 000	0.37
Dryopoidea?	Byrrhoidea + Buprestoidea	5000	40 000	0.22
Hydrophiloidea	Histeroidea?	2800	4500	0.77
Hydraenidae	Ptilidiidae	1163	5000	0.38

* Modified Slowinski and Guyer test, $p < 0.001$.

aquatic groups in the Series Staphyliniformia, apparently are not closely related (Hansen 1997), and are also phylogenetically separated from Dryopoidea, a possibly monophyletic (Ribera *et al.*, unpubl.) group of a dozen or so families in the Series Elateriformia. This indicates three main independent origins of aquatic lifestyle in Polyphaga, and five in total (Table 2.1) (excluding Scirtidae and smaller aquatic groups in mainly terrestrial families, such as Chrysomelidae and Curculionidae).

With fairly complete knowledge of phylogenetic relationships in Coleoptera, sister groups can be compared for species richness. In all five sister comparisons, the aquatic lineage is less species rich than the terrestrial counterpart. This result is not significant in most individual cases when measured against a model of stochastic lineage separation (Slowinski & Guyer 1989) (Table 2.1). However, the modified Slowinski and Guyer test, which assesses species richness across several independent sister comparisons, is highly significant ($p < 0.001$) and lends support to the conclusion that species diversification is generally depressed in aquatic lineages.

Although there is a general trend towards reduced species richness in the aquatic lineages, it is not universal, as species numbers vary greatly between subclades. For example, in Hydradephaga there are three monogeneric families, Amphizoidae, Aspidytidae and Hygrobiidae, with a maximum of six species. Based on the most recent phylogenetic analysis of Hydradephaga (Ribera *et al.* 2002a), the six species of Hygrobiidae are sister to several thousand species in Dytiscidae (Fig. 2.1). Similarly, there are great discrepancies in clade size within these family-level taxa. For example, the single known species of Spanglerogyrininae is sister to over 1000 species of the remaining Gyrinidae; and the only species of the genus *Phreatodytes* is sister to the *c.*300 remaining members of Noteridae

Conceptual and methodological problems of clade size comparisons

The findings from the terrestrial–aquatic transition are contrary to the strict predictions of the key innovation hypothesis, in that the ecological transition resulted

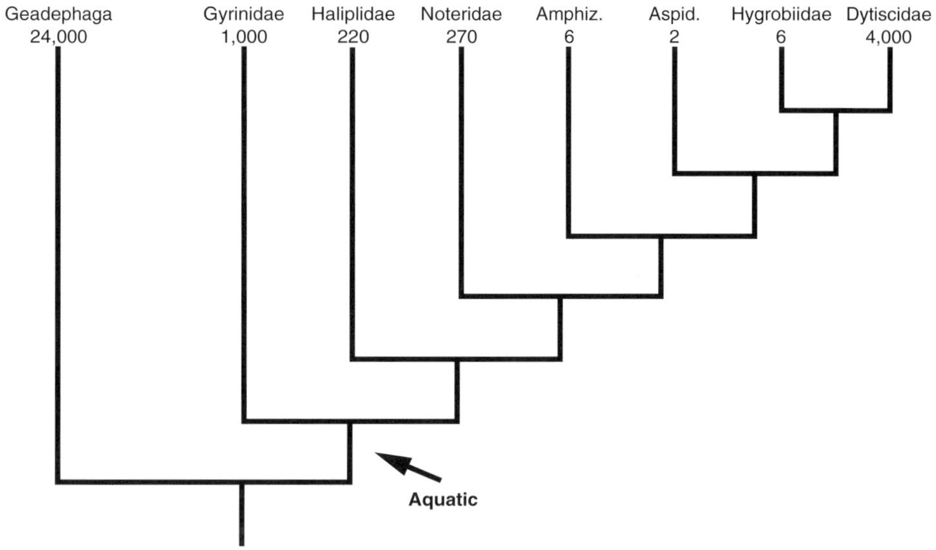

Figure 2.1 Phylogeny and species numbers in aquatic families of Hydradephaga. The tree is derived from 18S, 16S and COI gene sequences plus morphology, and is based on the analysis of several exemplars for all larger families (adapted from Ribera *et al.* 2002a). Amphi., Amphizoidae; Aspid., Aspidytidae.

in generally lower, rather than higher, species numbers. This is in contrast to the classic case of herbivory in insects (Mitter *et al.* 1988) where this trend is positive, but consistent with other 'innovations' such as leaf-mining in insects (Wiegmann *et al.* 1993), or the transition to endogaean (subsurface) habitat in beetles (Gibert & Deharveng 2002; Ribera, unpubl.). Whereas the strict correlation of evolutionary innovation and increased species numbers is perhaps only an addition of the more recent literature to the notion of key innovation, it is perhaps more revealing to consider the problems of drawing conclusions about ecological shifts and diversification rates.

First, the frequent practice of simply conducting comparisons of species numbers in sister taxa will produce ambiguous results, unless they can be seen in the context of the wider phylogenetic tree, as sister comparisons alone would not reveal the direction of change. Hence it will remain ambiguous whether the aquatic clades are depressed in net diversification rate, or the terrestrial groups are increased above a modal rate in the lineage, possibly correlated with an ecological difference that is not the focus of the investigation. Second, as differences in species richness within the aquatic lineages are much larger than the discrepancies in clade size with the terrestrial sister (Fig. 2.1), the common practice of calculating sister comparisons as the average over all nodes in the aquatic–terrestrial sister groups will fail to expose the most striking examples of rate differences. Because these differences in species richness are independent of aquatic versus terrestrial habitat associations, their explanation cannot be sought in these two alternative ecological categories.

In addition, there are practical problems with these analyses. Sister comparisons are only as good as the phylogenetic trees that they are based on, and frequently phylogenetic data are insufficient. For example, the ongoing debate about the monophyly of Hydradephaga (Beutel 1997) leaves a question mark in particular about the position of Gyrinidae. In the alternative hypothesis of Beutel and co-workers this family is the sister to all other Adephaga. Whereas this phylogenetic position would bias the aquatic–terrestrial sister comparison even more clearly against the aquatic lineage on one occasion, the exact position of the 4000+ remaining species of Hydradephaga relative to a particular terrestrial sister clade will affect the conclusions of clade size comparisons in another instance. Finally, the great discrepancy of species numbers in subclades highlights the importance of sampling the tree properly, in particular because small lineages may be more likely to remain unknown if taxonomic knowledge of a group is poor. For example, in the Adephaga the family Aspidytidae (Fig. 2.1) was discovered only recently in a specialized habitat rarely surveyed for aquatic beetles. To date we know only two species, from South Africa and China, but morphological and molecular data place them as sister to the large Dytiscidae plus Hygrobiidae (Ribera *et al.* 2002a). The general practice of using a single exemplar in a phylogenetic study to represent a large number of species is justified only with good knowledge of all major clades, and with solid taxonomic information about the total species number in each group. This is rarely available in studies of diversification rates, and fully sampled trees for larger groups are hardly ever to hand. This constitutes a great drawback to firm conclusions about where in a tree the change in net diversification rates occurs.

In summary, sister comparisons of ecologically defined groups, such as the aquatic–terrestrial taxa, in most cases can provide only a heuristic approach for detecting richness patterns in different habitats. Clearly, the transition from land to water is a major step, a syndrome of complex changes in physiology, behaviour and morphology, which were achieved by only a handful of insect groups. The Coleoptera are generally more plastic in life strategies and ecomorphology than most comparable groups and hence it is perhaps not surprising that they constitute one of the few groups that have made this transition multiple times independently. Even if diversification rates are low compared with terrestrial groups, these ecological shifts result in novel radiations and hence increased total species richness of Coleoptera. This effect of evolutionary novelty to produce greater species richness is an important contribution to overall species richness, but it has to be separated from the question about the speed-up of net diversification rates. More generally, the relationship between species richness and the evolution of novelty is complex, and both phenomena should be disentangled as much as possible.

The mechanisms of species diversification: investigations at the species level

As shown above, the simplistic approach of correlating key traits and clade size is unlikely to reveal the mechanisms of how species diversification rate is modulated. Cracraft (1990) remarked about key innovations that 'a more rigorous methodological approach to the study of innovation is required, one that begins at the hierarchical level of speciation analysis and extends downward to molecular developmental genetics.' Specifically, any trait invoked to cause differences in clade size will have to affect the dynamics of speciation and extinction rates. Hence, the study of these effects has to be conducted not with clades but at the hierarchical level of species, where many of the processes regulating macroevolutionary diversity are likely to operate.

In aquatic beetles it is generally assumed that speciation is largely the result of geographical factors, as beetles exist in a subdivided spatial matrix of freshwater habitats. Because these beetles are mostly confined to small water bodies such as ponds and creeks, some of which are temporary, many populations persist only by moving between patches of habitat, forming a network of populations with higher or lower levels of interaction. This spatial set-up of populations produces a potential for permanent (geographical and/or reproductive) separation between patches, and hence speciation. Conversely, the short-lived nature of some water bodies also increases the risk of population extinction unless species maintain some level of dispersal between isolated patches. The dynamics of speciation/extinction and the resulting rate of clade diversification may thus be seen in the context of spatial processes at the level of subdivided populations.

Water beetles, as well as many other aquatic invertebrates, exhibit strong preferences for either standing (lentic) or running (lotic) water bodies, with only a minority of species found in both types (Ribera & Vogler 2000). This habitat preference appears to be phylogenetically constrained, as closely related species of diving beetles usually do not differ in their preference. Nonetheless inferred transitions between either type are fairly frequent within a clade (Ribera & Vogler 2000). Hence, these changes between habitat types can be viewed in the same framework as other ecological transitions such as the terrestrial–aquatic transition discussed above, although in this case they appear much more labile.

We found a remarkable difference in the average sizes of geographical ranges of aquatic beetles depending on their habitat type. Lentic species generally occupy much larger ranges, sometimes corresponding to continent-size areas, whereas their lotic counterparts are frequently confined to a smaller area (Fig. 2.2). This observation is valid for all four clades of polyphagan and adephagan aquatic beetles (Table 2.1), and also in comparisons of smaller clades differing in habitat preference within these major lineages (Ribera & Vogler 2000). Hence, evolutionarily this difference has arisen independently every time a clade experienced habitat transitions. The proposed, but still unproven, explanation for this observation is that either habitat type selects for different life strategies: standing water bodies such as small ponds tend to fill in and disappear within decades or centuries, and populations are forced

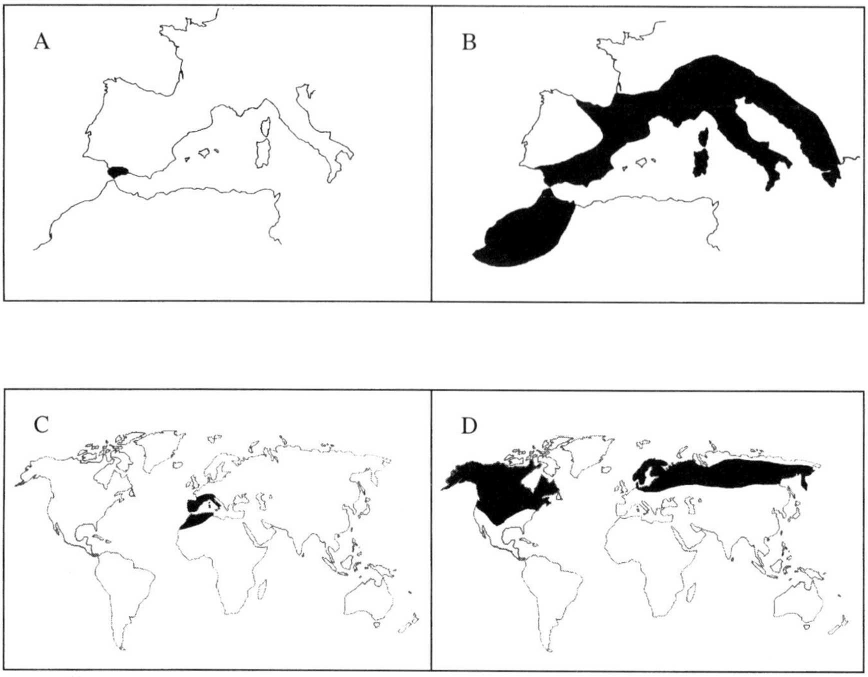

Figure 2.2 Examples of distributional ranges for lentic and lotic species. The top two maps represent the ranges of the lotic *Deronectes algibensis* (A) and *D. moestus* (B), whereas the bottom figures show the distribution of the lentic *Ilybius meridionalis* (C) and *I. angustior* (D). The ranges shown represent species with the smallest and the largest range sizes in their respective lineages, and they illustrate the vast differences in range for species of each habitat group, although exceptions are found in either category.

to disperse, whereas running water bodies provide greater temporal stability, even if the water course changes, as beetles can track the habitat continuously. Hence both types would result in different dispersal strategies, resulting in dispersive lentic and static lotic species (Ribera & Vogler 2000)*.

If it is correct that the dynamics of speciation and extinction in a lineage is affected by the spatial and temporal patterns of aquatic habitat 'islands', lineages occurring in either of the two main habitat types may differ with regard to these parameters and hence differ in diversification rates. As the ranges in lentic species are larger, presumably owing to their greater propensity for dispersal, they should also have a greater connectedness of populations. Hence speciation resulting from pop-

* *For the same reasons of differences in the propensity for dispersal, the local diversity of communities of running water species will be low, but the regional (or 'metacommunity') richness will be high, and vice versa for standing water species (Hubbell 2001, p. 218; Ribera et al., 2003, and unpubl. data).*

ulation subdivision is expected to be infrequent†. In contrast, speciation in lotic species is expected to be at a higher frequency, largely in allopatry and by subdivision of (small) geographical ranges. The difference in population structure also has an expected effect on extinction: the greater propensity for dispersal reduces the risk of any population going extinct, and hence lentic species are long-lived. In lotic species, populations and newly formed species are more prone to extinction, resulting in short-lived species with small ranges (Hubbell 2001, p. 256; Ribera *et al.* 2001, 2003).

How does this difference in temporal stability and the correlated difference in dispersal ability affect diversification rates? We performed a test of diversification rates on the basis of species-level phylogenetic analyses of a lotic and a lentic lineage of dytiscid water beetles (Fig. 2.3). Surprisingly, lineages-through-time plots of both groups showed a remarkable similarity in net diversification rates. We also tested for differences in the degree of sympatry, as it could be expected that widespread and dispersive lentic species would gain a high level of sympatry, whereas cladogenesis in lotic species should give rise to multiple allopatric isolates. However, plots of sympatry against node level show substantial overlap in geographical ranges even between close relatives in both groups, indicating a high proportion of range movement also in lotic species (Ribera *et al.* 2001). These results need to be confirmed based on larger samples, but with the data at hand the observed differences between lotic and lentic species are mostly related to geographical scale, rather than the precise spatial arrangement‡.

Effects on macroecological patterns

Although we did not detect clear differences in diversification rates (macroevolution) between habitat types, the macroecological consequences of the habitat shift were substantial. In a comparison of species numbers in aquatic beetles in western Europe based on distributional ranges of 813 species (Ribera *et al.* 2003), lotic and lentic species differed with regard to species turnover and latitudinal gradients. Lotic species followed the expected latitudinal gradient, exhibiting much greater species richness in southern compared with northern Europe. Their species richness in a given locality could be predicted with high accuracy from latitude only, whereas in lentic species no latitudinal gradient was apparent. Species richness in lentic

† *There is an alternative prediction from the application of a Wrightian model of genetic drift which would suggest that large ranges result in higher, rather than lower, rates of speciation owing to the greater chance of allele fixations across the entire range. In addition, the larger ranges may increase the probability that geographical barriers to dispersal form within the area, increasing the probability of allopatric speciation (Rosenzweig 1995). However, these propositions ignore the greater effect of dispersal that is needed to build up and maintain large ranges, and would counteract the chances of drift or allopatric speciation. If barriers form frequently, large ranges could not be established in the first place.*

‡ *There is an inherent problem to empirically demonstrating the predicted patterns: if poor dispersers have smaller ranges, lower abundances and shorter lifespans, they are less likely to be sampled, introducing a strong bias in any possible comparison (Hubbell 2001, p. 258; Ribera et al. 2001).*

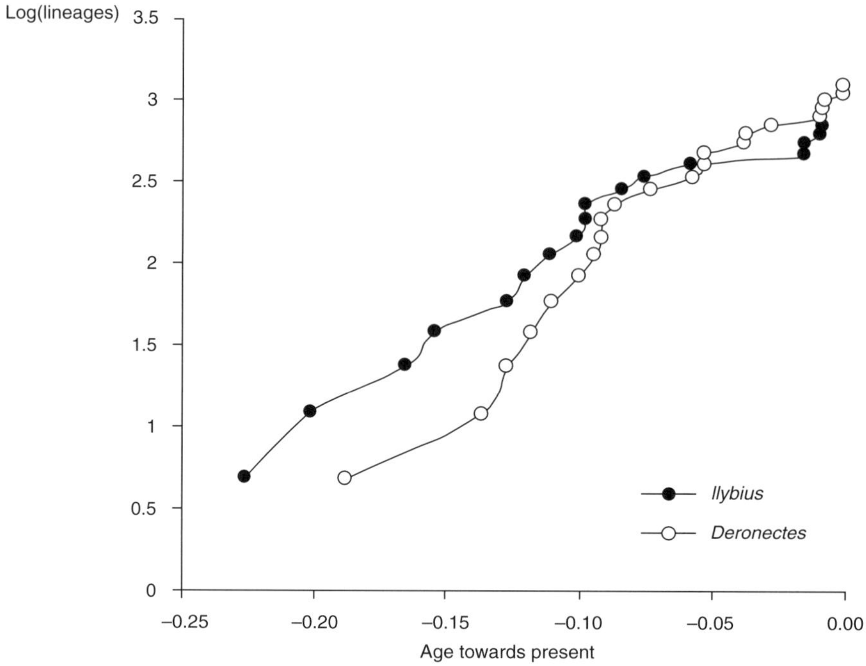

Figure 2.3 Lineages-through-time plots for *Deronectes* and *Ilybius*. The *y* axis is the log number of lineages. The *x* axis represents the relative age towards present, based on maximum likelihood (ML) scaled branch lengths from a mtDNA-based tree. Both plots are presented together for comparison, although this implies equivalent rates of molecular variation in both genera, an assumption necessary for a direct comparison of relative rates of species turnover. From Ribera *et al.* (2001).

species was almost entirely predicted by 'connectivity', a geographical measure of potential dispersal between areas. The nested species–area accumulation curve for all species had a *z* value near the canonical 0.26, but when separated according to their habitat type, species turnover for lotic species ($z = 0.38$) was much greater than for lentic species ($z = 0.21$) (Ribera *et al.* 2003). These findings are consistent with the observation of extended range sizes of lentic species, resulting from greater dispersal ability, which reduce spatial turnover and increase distributional ranges into northern latitudes to sites more distant from the glacial refugia. Their lotic counterparts, with poorer dispersal abilities, are confined mostly to the southern refuge areas (Ribera *et al.* 2003).

In the case of aquatic beetles, the macroecological and macroevolutionary patterns presented above are consequences of a habitat transition, but mediated through a third factor—the change in dispersal propensity. This means an additional degree of freedom, with the possibility that the consequences could change in an unpredictable way. In other freshwater organisms the same type of transition could have different effects. For example, in mayflies (Ephemeroptera), which have terrestrial adults, species occurring in running water tend to fly upstream, to compensate for the general downstream movement of the larvae carried by the current, and hence rely heavily on dispersal (albeit over short distances in some cases), in contrast to lotic species of beetles.

Implications for the macroecological approach

Macroecology is concerned mostly with the distributional patterns and general trends in species richness, testing plausible hypotheses by statistical analysis but without recourse to evolutionary history. However, the units of this analysis, species and populations, are the result of evolutionary processes, a fact to which macroecological research has paid little attention to date. It is likely that in many cases the discovery of a macroecological pattern is possible only by boldly counting species or abundances in a given area, community or habitat. There is no doubt, however, that the entities of the analysis are not equivalent to each other, as they may not share a common evolutionary history, and hence certain traits (Freckleton *et al.*, this volume). These traits are important determinants of ecological parameters such as dispersal, range size, habitat, climatic amplitude and, ultimately, the propensity for species origination and extinction.

Aquatic beetles provide examples of the problems associated with a non-historical approach to explaining the causes of macroecological diversity patterns. First, species richness and the propensity for species proliferation varies greatly between major lineages. Hence species numbers in a given area are strongly influenced by the presence or absence of certain lineages. The presence of such fast-diversifying clades may be the result of evolutionary history, including biogeographical history, antiquity of an area or habitat, or historical accident, rather than parameters of the environment such as available energy or potential evapotransporation; geographical parameters such as area size; or other parameters commonly used to explain macroecological patterns. Although geographical or environmental parameters may have an influence on rates of diversification, their effect on the different lineages is far from universal and hence they may have contradictory effects on the diversification of local radiations.

Second, closely allied to the above, the underlying mechanisms generating macroecological patterns vary greatly between lineages, which acquire particular traits at some stage in their history. In the case of aquatic beetles in Europe, the highly dispersive lentic species will tend to reflect a local equilibrium determined by current geographical or environmental variables; whereas distributional patterns in the lotic species, with their small ranges and short evolutionary species persistence, mostly reflect local species origination and extinction and hence their diversity will be highly dependent on historical factors. Ignoring the lineage history in a purely descriptive approach is unlikely to reveal the causes of species richness patterns, as they are not necessarily correlated with current conditions. The areas in which the species originated, with their associated geographical and environmental conditions affecting speciation and extinction rates, are not necessarily the same as the areas in which species are encountered at the present (Janzen 1985). Descriptive correlations with present conditions therefore may not be related to the processes that led to greater or lower diversity in a lineage. The fact that lentic species have relatively greater species richness in northern latitudes provides little information with regard to the areas of higher or lower speciation rates. Similarly, the fact that species diversity is generally greater in the tropics may provide little information as to whether lineages currently with tropical distribution exhibit comparatively greater diversification rates.

Third, not taking into account the phylogenetic affinities of taxa in macroecological studies greatly reduces the power of the analysis of species richness patterns (see also Blackburn & Gaston, this volume; Purvis *et al.*, this volume; Freckleton *et al.*, this volume). The lotic–lentic differences have arisen independently multiple times in all four major aquatic lineages of the Coleoptera. The multiple and independent occurrence of the associated patterns allows tests of hypotheses on causal mechanisms, but this would not have been possible in a non-evolutionary approach by pursuing explanations based on descriptive macroecological correlations.

In conclusion, non-historical approaches to studying macroecological patterns have shortcomings. Species richness is the result of past diversification in evolutionary lineages, historical range movements, as well as equilibrium responses to environmental and geographical parameters. The challenge for the study of species richness in the future will be to integrate existing statistical approaches with phylogenetic information.

Summary

Using aquatic beetles as a case example, we show how macroecological patterns are necessarily affected by the evolutionary history of species and clades. To start with, lineages often differ greatly in their species richness, resulting in highly unbalanced phylogenetic trees. These differences may or may not be correlated with ecological shifts, as, for example, a depressed total species number in the aquatic lineages of beetles in comparison to their terrestrial counterparts, but in any case they will have general effects on species richness patterns. Differences in traits acquired along the evolutionary history of lineages may also affect macroecological patterns. In the case of aquatic beetles, one of these traits is the type of habitat. Aquatic beetle species usually are confined to either standing or running water bodies, and although multiple transitions exist, the trait is phylogenetically conserved. We found that total range size, latitudinal gradients in species richness and species turnover are very different between species living in the two types of habitat. Differences in the resulting macroecological patterns may thus become properties of entire clades, and the comparison of sister groups is of great help in revealing the underlying causes of these macroecological differences. Non-historical analyses of species richness patterns by statistical correlations of species numbers with geographical and physical factors of the environment ignore the constraints of evolutionary history, and hence one of the most powerful analytical tools for explaining the origin of macroecological patterns.

Acknowledgements

Our work on water beetles was funded by the Natural Environment Research Council (UK) and the Leverhulme Trust. We thank Tim Barraclough, Richard Davies and Garth Foster for discussions.

References

Arrhenius, O. (1921) Species and area. *Journal of Ecology* **9**, 95–99.

Barraclough, T.G., Vogler, A.P. & Harvey, P.H. (1998) Revealing the factors that promote speciation. *Philosophical Transactions of the Royal Society, London, Series B* **353**, 241–249.

Beutel, R.G. (1997) Über Phylogenese und Evolution der Coleoptera (Insecta), insbesondere der Adephaga. *Abhandlungen des Naturhistorischen Vereins Hamburg* **31**, 1–164.

Brown, J.H. (1995) *Macroecology.* University of Chicago Press, Chicago.

Cracraft, J. (1990). The origin of evolutionary novelties: pattern and process at different hierachical levels. In: *Evolutionary Innovations* (ed. M.H. Nitecki), pp. 21–44. University of Chicago Press, Chicago.

Crowson, R.A. (1955) *The Natural Classification of the Families of Coleoptera.* Nathaniel Lloyd, London.

Crowson, R.A. (1960) The phylogeny of Coleoptera. *Annual Review of Entomology* **5**, 111–134.

Farrell, B.D. (1998) 'Inordinate fondness' explained: why are there so many beetles? *Science* **281**, 555–559.

Gibert, J. & Deharveng, L. (2002) Subterranean ecosystems: a trunctated functional biodiversity. *BioScience* **52**, 473–481.

Givnish, T. & Systma, K. (1997) *Molecular Evolution and Adaptive Radiation.* Cambridge University Press, Cambridge.

Hansen, M. (1997) Evolutionary trends in 'staphyliniform' beetles (Coleoptera). *Steenstrupia* **23**, 43–86.

Heard, S.B., & Hauser, D.L. (1995) Key evolutionary innovations and their ecological mechanisms. *Historical Biology* **10**, 151–173.

Hubbell, S.P. (2001) *The Unified Neutral Theory of Biodiversity and Biogeography.* Princeton University Press, Princeton, NJ.

Hunter, J.P. (1998) Key innovations and the ecology of macroevolution. *Trends in Ecology and Evolution* **13**, 31–36.

Janzen, D. (1985) On ecological fitting. *Oikos* **45**, 308–310.

Lack, D. (1947) *Darwin's Finches.* Cambridge University Press, Cambridge.

Lawton, J. (1995). Population dynamic principles. In: *Extinction Rates* (eds J.H. Lawton & R.M. May), pp. 147–163. Oxford University Press, Oxford.

Lawton, J.H. (1999) Are there general laws in ecology? *Oikos* **84**, 177–192.

MacArthur, R.H. & Wilson, E.O. (1967) *The Theory of Island Biogeography.* Princeton University Press, Princeton, NJ.

May, R.M. (1973) *Stability and Complexity in Model Ecosystems.* Princeton University Press, Princeton, NJ.

Mayhew, P.J. (2002) Shifts in hexapod diversification and what Haldane could have said. *Proceedings of the Royal Society, London, Series B* **269**, 969–974.

Mitter, C., Farrell, B. & Wiegmann, B. (1988) The phylogenetic study of adaptive zones: has phytophagy promoted insect diversification? *American Naturalist* **132**, 107–128.

Preston, F.W. (1962) The canonical distribution of commonness and rarity. *Ecology* **43**, 185–215.

Ribera, I. & Vogler, A.P. (2000) Habitat type as a determinant of species range sizes: the example of lotic–lentic differences in aquatic Coleoptera. *Biological Journal of the Linnean Society* **71**, 33–52.

Ribera, I., Barraclough, T.G. & Vogler, A.P. (2001)

The effect of habitat type on speciation rates and range movements in aquatic beetles: inferences from species-level phylogenies. *Molecular Ecology* **10**, 737–750.

Ribera, I., Beutel, R.G., Balke, M. & Vogler, A.P. (2002a) Discovery of Aspidytidae, a new family of aquatic beetles. *Proceedings of the Royal Society, London, Series B* **269**, 2351–2356.

Ribera, I., Hogan, J.E. & Vogler, A.P. (2002b) Phylogeny of hydradephagan water beetles inferred from 18S rRNA sequences. *Molecular Phylogenetics and Evolution* **23**, 43–62.

Ribera, I., Foster, G.N. & Vogler, A.P. (2003) Habitat type and large scale diversity patterns in Western European water beetles. *Ecography*, in press.

Rosenzweig, M.L. (1995) *Species Diversity in Space and Time.* Cambridge University Press, Cambridge.

Sanderson, M.J. & Donoghue, M.J. (1994) Shifts in diversification rate with the origin of angiosperms. *Science* **264**, 1590–1593.

Sanderson, M.J. & Donoghue, M.J. (1996) Reconstructing shifts in diversification rates on phylogenetic trees. *Trends in Ecology and Evolution* **11**, 15–20.

Schluter, D. (2000) *The Ecology of Adaptive Radiation.* Oxford University Press, Oxford.

Shull, V.L., Vogler, A.P., Baker, M.D., Maddison, D.R. & Hammond, P.M. (2001) Sequence alignment of 18S ribosomal RNA and the basal relationships of adephagan beetles: evidence for monophyly of aquatic families and the placement of Trachypachidae. *Systematic Biology* **50**, 945–969.

Simpson, G.G. (1953) *The Major Features of Evolution.* Columbia University Press, New York.

Slowinski, J.B. & Guyer, C. (1989) Testing the stochasticity of patterns of organismal diversity: an improved null model. *American Naturalist* **134**, 907–921.

Stiassny, M.L.J. & Pinna, M.C.C.D. (1994). Basal taxa and the role of cladistic patterns in the evaluation of conservation priorities; a view from freshwater. In: *Systematics and Conservation Evaluation* (eds P.L. Forey, C.J. Humphries & R.I. Vane-Wright), pp. 235-249. Clarendon Press, Oxford.

Vogler, A.P. & Barraclough, T.G. (1998). Reconstructing shifts in diversification rate during the radiation of tiger beetles (Cicindelidae).In: *XXth International Congress on Entomology. Phylogeny and Classification of Caraboidea (Coleoptera: Adephaga)* (eds G.E. Ball, A. Casale & A. Vigna-Taglianti), pp. 251–260. Museo Regionale di Scienze Naturali, Atti, Torino, Italy.

Wiegmann, B.M., Mitter, C. & Farrell, B. (1993) Diversification of carnivorous parasitic insects: extraordinary radiation or specialized dead end? *American Naturalist* **142**, 737–754.

Williams, C.B. (1964) *Patterns in the Balance of Nature.* Academic Press, London.

Chapter 3

The unified phenomenological theory of biodiversity

*Sean Nee**

Introduction

The study of the abundances of organisms is fundamental to ecology: the study of patterns of species abundances is fundamental to macroecology and has a venerable literature (for review see Magurran 1988; Tokeshi 1999). Starting with Fisher's derivation of the log-series distribution (Fisher *et al.* 1943), numerous other distributions have been compared with observed patterns of species relative abundances, such as the log-normal distribution (Preston 1962) and the geometric series (e.g. May 1975). In addition, numerous models have been proposed as possible explanations of these different distributions. Many are 'stick breaking' models, in which a hypothetical stick is broken into pieces according to some procedure, and the lengths of the resulting pieces are taken to be proportional to the abundances of each species in a community (Motomura 1932; MacArthur 1957; May 1975; Tokeshi 1999). It is then suggested that the 'stick' is 'niche space' that is being divided up, but one could argue that 'niche space' is no more biologically informative a concept than 'stick'. Population biological models also have been proposed to explain these distributions (e.g. Kendall 1948; May 1975). In addition, numerous distinct methods of representing abundance data have been developed (e.g. Southwood & Henderson 2000).

Although May (1975) showed how the numerous ways of representing the data were all related to each other, the field has the characteristics of a hotch-potch, in which you mix and match a model to data from a bag of unrelated possibilities.

Recently, Hubbell (2001) has proposed an idea that, like all great original ideas, seems obvious in retrospect. He has suggested a unified approach to the study of patterns of relative abundance. Via his neutral theory, he arrives at a flexible, three-parameter distribution that can be fitted to a large variety of data sets. Once these parameters have been fitted, they can, if the investigator so chooses, be interpreted in terms of the neutral theory. In this theory, the abundances of species in a metacommunity fluctuate completely at random, subject to the constraint of a fixed number of individuals of all species, and individuals may mutate into the first individual of

* *Correspondence address: sean.nee@ed.ac.uk*

an entirely new species: the metacommunity is connected by migration to a local community in which species abundances also fluctuate entirely at random, subject to the constraint of a fixed number of individuals, in total. The total size of the metacommunity, multiplied by the mutation rate, provides one parameter: the migration rate and the total size of the local community provide the other two. Bell (2001) has also proposed a related neutral theory.

Here, I propose an alternative two-parameter distribution, which I call the Yule distribution, that also fits a wide range of data (see below for the provenance of this name). I derive the distribution with no real regard for any possible biological mechanism generating it—hence, it is phenomenological. The virtue of this is argued in the discussion. The parameters may be interpreted as quantitative descriptions of the pattern of variation in abundance and, so, may be thought of as a two-dimensional diversity index. I then compare these alternatives—the neutral and the phenomenological—as candidates for a single descriptive framework for abundance patterns.

The Yule distribution: derivation

A fundamental stochastic process modelling changes in population size is the birth–death process. This model assumes that all individuals in a population are equivalent with respect to their probabilities of giving birth to a new individual, or dying, per unit of time, although these probabilities may vary over time. Under this model, the family size of an individual after some period of time is a geometrically distributed random variable (Kendall 1949). I will take this distribution as my starting point.

For the geometric distribution, the probability of a family size of n individuals, $\Pr\{N = n\}$, assuming that there are any individuals at all, is given by:

$$\Pr\{N = n\} = q(1-q)^{n-1} \qquad (3.1)$$

with $0 < q < 1$. I will now suppose that q is itself a random variable, drawn from the beta distribution. This is a very flexible two-parameter continuous distribution taking values between 0 and 1 and its density, $p(q)$, is given by (e.g. Feller 1966):

$$p(q) = \frac{1}{B(a, b)} q^{a-1}(1-q)^{b-1} \qquad (3.2)$$

where $B(a, b)$ is the Beta function and $a, b > 0$. Randomizing q with this distribution gives us a new two-parameter distribution, whose parameters are the a and b in equation (3.2). Calling this the Yule distribution, Yule(a, b), it is given by:

$$\Pr\{N = n\} = \frac{b\Gamma(a+b)}{\Gamma(a)} \frac{\Gamma(a+n-1)}{\Gamma(a+b+n)} \qquad (3.3)$$

where Γ is the Gamma function. Although this may look inelegant, it is actually very easy to work with. So, for example, its cumulative distribution function is given by:

$$\Pr\{N \le n\} = 1 - \frac{\Gamma(a+b)}{\Gamma(a)} \frac{\Gamma(a+n)}{\Gamma(a+b+n)} \tag{3.4}$$

which easily allows one to draw random numbers from the Yule distribution.

I am deliberately presenting the simplest, non-mechanistic, derivation of this distribution. Just as for the log-series distribution (Boswell & Patil 1971), countless complicated mechanistic—population biological—mechanisms could be proposed, leading to the same end-point, equation (3.3). In fact, Simon (1955) presented a population biological derivation for the special case of Yule(1, *b*). He christened this the Yule distribution, as previously Yule had derived a probability model that had Yule(1, *b*) as its limiting form, and I am retaining the name for the more general distribution. I feel that the worth of such distributions lies in their descriptive abilities (next section) and not in their manner of derivation.

The Yule distribution: fitting to data

The parameters of the Yule distribution were fitted to a variety of data sets by maximum likelihood. The goodness of fit was then evaluated with a χ^2 test. Degrees of freedom are $n - 3$, where n is the number of classes being fitted, two degrees being subtracted for the two parameters being estimated from the data. I note in passing that degrees of freedom may have been overestimated in previous work in this area, as they have not taken into account the fact that there has been a preliminary process of choosing a model that appears to fit the data. Data sets were chosen arbitrarily.

Corbet trapped butterflies in Malaya in the 1930s and 1940s. Some of his species–abundance data are illustrated in Fig. 3.1 together with the fitted Yule distribution. For his analysis (Fisher *et al.* 1943), Corbet discarded his data for all species represented by more than 24 individuals in his collection. The analysis in Fig. 3.1 is of his complete data set, which was retrieved from Williams (1964). It is obvious that the fit is very good. Furthermore, the value of the χ^2 statistic is almost the same as that reported for the fit to the log-series distribution. This sort of data typically has a very long, sparse tail to the right which is not possible to fit on a page: this is why I collapse the tail to a single data point. In any case, the statistical analysis uses the collapsed data. A further point: in any particular study, the data would be transformed some way or other.

Figure 3.2 shows diatom species abundance data taken from Patrick (1968) and the fitted Yule distribution. Visually and statistically, the fit is good.

Figure 3.3 shows the diatom data in Preston's (1962) logarithmic abundance class form: it has the appearance of a 'veiled' log-normal distribution (see Nee *et al.* (1991) for futher discussion and another example of the log-normal distribution). It is known theoretically that log-series distributions do not exhibit such a 'humped' distribution when represented in this way (e.g. Pielou 1977), hence the Yule is capable of fitting qualitatively different patterns of relative abundance.

Another example of data to which a log-normal distribution previously had been

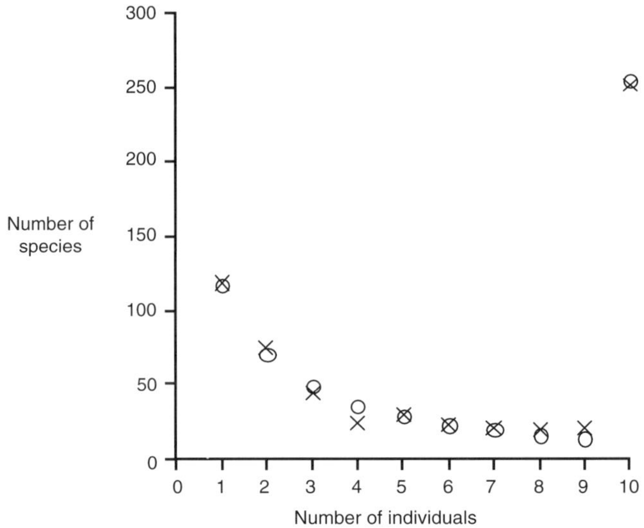

Figure 3.1 Corbet's butterfly data (crosses), showing the number of species represented by 1, 2, 3, etc., individuals. The final data point shows the number of species represented by 10 or more individuals. The circles show the fitted Yule distribution, Yule (2.42, 0.55). $\chi^2 = 8.6$ (n.s., d.f. = 7).

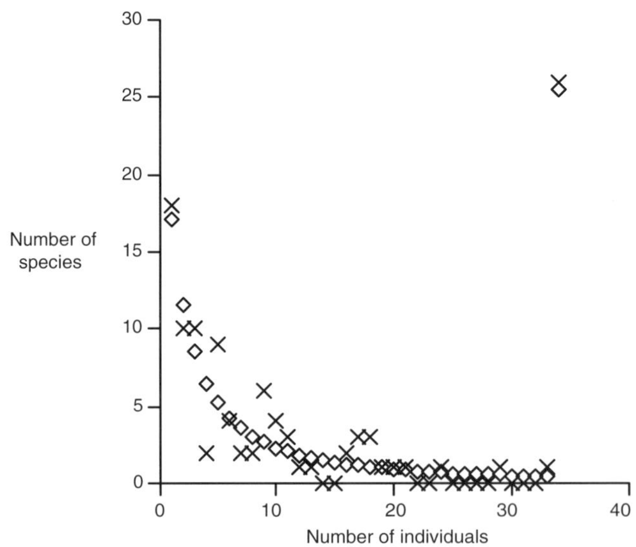

Figure 3.2 Diatom data from Patrick (1968) denoted by crosses. The diamonds show the fitted Yule distribution, Yule (3.35, 0.61). $\chi^2 = 13.2$ (n.s., d.f. = 9; classes were grouped to make the expected frequency in each class greater than 5).

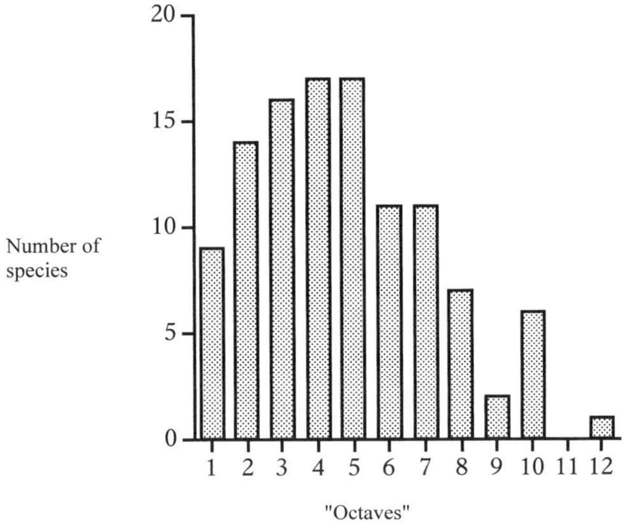

Figure 3.3 Diatom data from Fig. 3.2: this is a frequency histogram showing the numbers of species represented by the numbers of individuals in successive octaves. An octave is simply a logarithmic abundance class, where the logarithm is to the base 2: so, a bin with associated x-axis digit n corresponds to numbers of individuals between 2^{n-1} and 2^n. A species with exactly $2n$ individuals contributes 0.5 to the two adjacent octaves.

fitted (Southwood 1996) is given in Fig. 3.4: these are insect species–abundance data from oak trees, *Quercus robur*. The Yule distribution fits very well.

Figure 3.5 shows a data set—insect species–abundance data from *Quercus ilex*—that previously had been shown to fit a power series distribution very well, although being a poor fit for the log-normal and log-series distributions (Southwood 1996). The Yule distribution again performs very well.

Finally, there are data that the Yule distribution does not fit, at least in a statistical sense: word abundance data in Shakespeare's plays and poetry (data from Efron & Thisted 1976). The χ^2 test rejects the best-fit Yule distribution. However, Fig. 3.6 shows that, visually, the distribution fits very well indeed. We must bear in mind that with an enormous amount of data, theoretical distributions will almost always be rejected, as they are idealizations. Statisticians advise us, when we have a lot of data, to reject models only if we have specific alternatives in mind. (Efron & Thisted (1976) find that the negative binomial distribution provides an excellent fit to the data and consider subtleties and assumptions involved in analysing this sort of data that I have ignored here. The negative binomial distribution, like the Yule distribution, also can be derived from the geometric distribution—it is the sum of several random variables that are geometrically distributed.)

The negative binomial distribution, also a two-parameter distribution, is not a candidate distribution for the unified phenomenological theory. Simon (1955) was

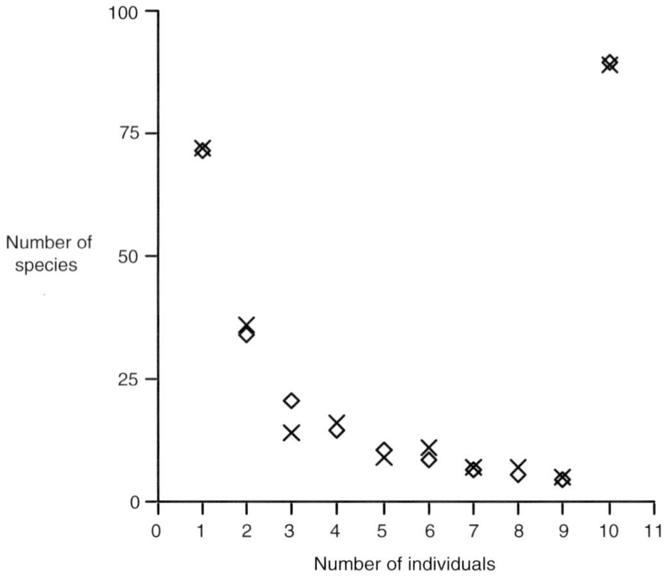

Figure 3.4 Phytophagous insects on *Quercus robur*. Conventions as in Figs 3.1 and 3.2. Yule (1.35, 0.5), $\chi^2 = 4$ (n.s., d.f. = 7).

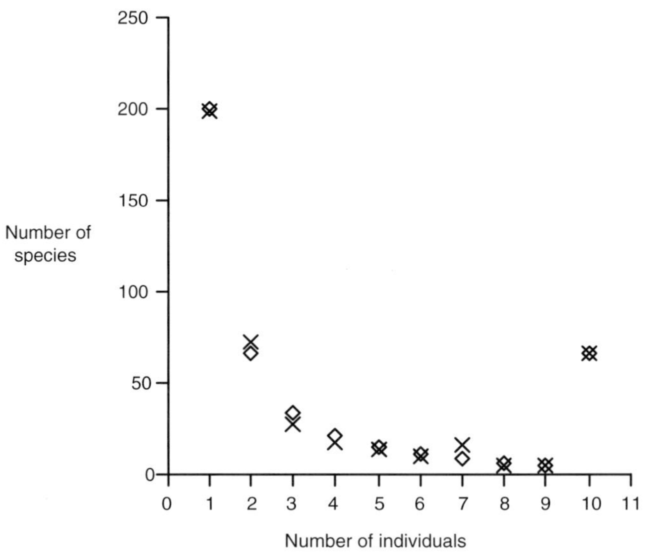

Figure 3.5 Insects on *Quercus ilex*. Conventions as in Figs 3.1, 3.2 and 3.4. Yule (0.84, 0.72), $\chi^2 = 10.3$ (n.s., d.f. = 7).

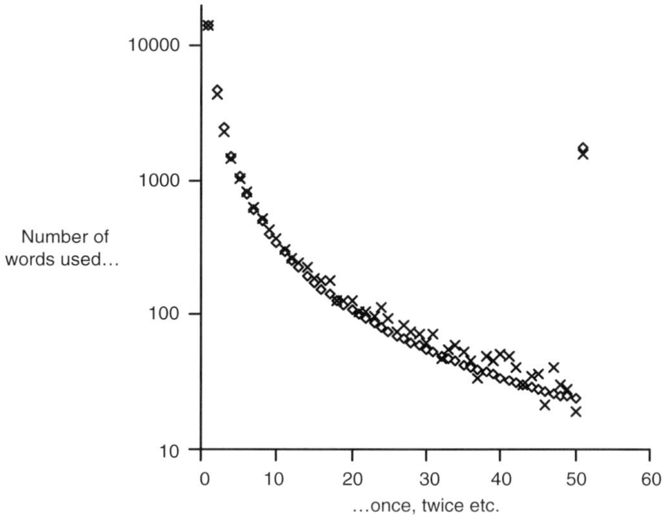

Figure 3.6 Distribution of the numbers of words used once, twice, etc., in Shakespeare's writings. Note the logarithmic vertical axis which is used for clarity: data display typical 'hollow curve' shape of the other data sets when viewed on arithmetic axes. Yule (0.82, 0.67), $\chi^2 = 188$ ($p < 0.005$, d.f. $= 47$). The data appear linear on double-logarithmic axes, but it is visually harder to distinguish observed (×) from predicted (◇).

motivated to discover Yule(1, b) by the following fact: the ratio—(number of species represented by two individuals)/(number of species represented by one individual)—is often close to 1/3. This is the case for the data in Fig. 3.5. The negative binomial cannot exhibit this behaviour if it is fitted to data with a long tail.

There is another, technical, objection to the negative binomial. Call the parameters of the negative binomial p and r: the distribution, modified because there is no zero term, is given by

$$\Pr\{N = n\} = \frac{\binom{n+r-1}{n} p^r (1-p)^n}{1 - p^r} \tag{3.5}$$

The probabilities generated by the singletons, $\Pr\{N = 1\}$, clearly play a large part in likelihood analysis of data of this sort. However, with numerous singletons, the estimate of p will be close to 1 and the maximum of the probability as a function of r is not found in the allowable space, $0 < r$. Hence, maximum likelihood procedures cannot be used for estimation.

The Yule distribution's failure: the geometric distribution

We know that the Yule distribution fails to fit data that accord with the geometric distribution, which is an important distribution. This can be seen from its derivation, where I randomized its parameter, q, with the beta distribution. In order for the Yule to mimic the geometric, it would be necessary for the beta density to be able to 'spike', i.e. to be concentrated on a narrow range of q, essentially picking out a particular value, and it does not behave this way. This is not entirely true: with a suitable choice of parameters, the beta density is able to 'spike' towards the extreme values of 0 and 1 and, so, the Yule can mimic the geometric at these extremes.

The nature of the Yule distribution's failure is readily seen for intermediate values of q, which generate geometrically distributed data with a small average size. Such data have the typical 'hollow curve' shape that we have repeatedly seen, but look 'stumpy'—they lack the typical long tail that I see in my data sets: in biological terms, if species abundances are drawn from a geometric distribution with a small average size, there are no species that are highly abundant. The Yule distribution is unable to cope with the lack of this tail in the distribution. Even my numerical procedures used for generating the maximum likelihood estimates of the Yule's parameters run into difficulties.

The geometric distribution has been used extensively in palaeontology and molecular phylogenetics. This is because it is the distribution of family sizes according to the birth–death stochastic process (Kendall 1949): just as we have been studying here the distribution of the numbers of individuals per species, one can go higher and study the distribution of the numbers of species per genus, the numbers of genera per family or the numbers of subtaxa per taxon—and the birth–death process is an obvious choice for a null model. It has been used successfully for years in numerous contexts (e.g. Raup *et al.* 1973; Nee *et al.* 1992, 1996; Sepkoski & Kendrick 1993; Nee 2001): in fact, it was in a macroevolutionary context that I first derived the Yule distribution (Nee *et al.* 1992), not suspecting that it might provide a unified framework for studies of species–abundance distributions.

Recently, the birth–death process model has been deemed to be unsuitable for phylogenetic studies (Hubbell 2001). This is because Hubbell (2001) observes an interior mode in his histogram of numbers of species per family in Sibley and Ahlquist's phylogeny of birds (Sibley & Ahlquist 1990) and believes this to be incompatible with the postulate that the numbers of species per family may be geometrically distributed. This is a misapprehension: Fig. 3.7 shows that the geometric distribution is fully capable of generating family sizes that exhibit an interior mode. Of course, the geometric distribution may not statistically be a good description of the bird data (although Nee *et al.* (1992) found that it was, at least where they were looking), and that is when it displays its considerable utility as a null hypothesis (Nee *et al.* 1996).

Discussion

The Yule distribution has several advantages over Hubbell's neutral distribution as a single framework for the description of species–abundance patterns.

1 It has two parameters, rather than three and, so, is a more parsimonious model.

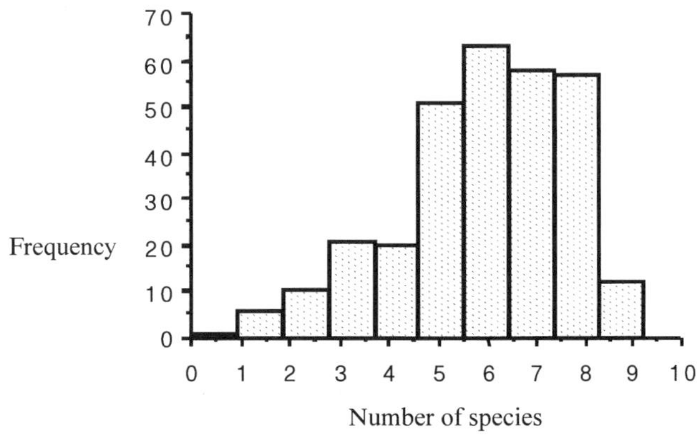

Figure 3.7 Simulated data: the numbers of species in 500 families were drawn from a geometric distribution with an average of 100. The data were $\log_2$ transformed prior to the construction of the default Statview histogram in order to be comparable to Hubbell's. The data display a clear internal mode. As is to be expected for a geometric distribution with a large mean, there are very few families with just one species, contrary to Hubbell's assertion that: 'The longer the time period sampled, the larger the number of possible descendent lineages and the [geometric] frequency distribution becomes flatter; but the highest frequency category remains the singleton category.' (Hubbell 2001). The highest *single* frequency category remains the singleton category—an important qualification. As the mean of the distribution increases, the probability of a singleton vanishes.

2 It is easy to fit to data using maximum likelihood, and maximum likelihood procedures—to evaluate parameter confidence intervals, for example—are readily conducted. In contrast, Hubbell's distribution is not easy to fit and requires computer simulations.

3 There are philosophical grounds for preferring a phenomenological model.

We know that fitting distributions to data is uninformative about the mechanism generating that data. Sokal (see below for reference) may have been the first to make this explicit for ecologists in the specific context of the log-series distribution (LSD). Commenting on a paper by Boswell & Patil, which presented numerous, distinct, mechanisms generating the LSD, Sokal wrote: '. . . I would like Drs Boswell and Patil to establish for the ecological audience the implications of their very interesting findings. . . . it would appear that a variety of probabilistic models can give rise to the LSD. Consequently, fitting LSDs to ecological or biogeographical data can only serve for purposes of calibration of sampling devices, empirical prediction and the like. If we are to use the LSD as an hypothesis-generating device, i.e. if a fit to an LSD is to be considered support for an . . . ecological hypothesis, it would seem that Boswell and Patil's work faces us with a multiplicity of models, choice among which would be impossible.' (Comments follow Boswell & Patil (1971).) These comments apply to any distribution, not just the LSD.

As Sokal makes clear, the fact that distribution is uninformative about mechanism does not undermine the utility of such distributions. A striking example of the predictive utility of phenomenological theory occurred in a rather different context. Efron & Thisted (1976) studied the distribution of the number of words Shakespeare used once, twice, thrice, etc., in his writings. This distribution is well fitted by a negative binomial distribution—the same distribution describing the distribution of parasitic worms among people, which obviously is generated by a different mechanism to that generating Shakespeare's works. They then developed the theory necessary to predict the following: if a new work by Shakespeare was discovered, how many words should appear in it that have never been used before by Shakespeare, used once before, twice before, etc. In the 1980s, this theory was used to authenticate a newly discovered sonnet in the Bodleian Library, Oxford (Kolata 1986). It would be interesting to make predictions about, for example, further collections of Malaysian butterflies. Departures from these predictions could be informative about the degradation of the fauna since Corbet's time.

Finally, however, there is an important circumstance in which Hubbell's neutral distribution is to be preferred. If you believe that the neutral theory is true, then fitting the distribution is desirable as the parameters have biological meaning and, so, you would want to estimate them.

References

Bell, G. (2001) Neutral macroecology. *Science* **293**, 2413–2417.

Boswell, M.T. & Patil, G.P. (1971) Chance mechanisms generating the log arithmetic series distribution used in the analysis of number of species and individuals. In: *Statistical Ecology* (eds G.P. Patil, E.C. Pielou & W.E. Wates), pp. 99–130. Pennsylvania State University Press, Philadelphia, PA.

Efron, B. & Thisted, R. (1976) Estimating the number of unseen species: how many words did Shakespeare know? *Biometrika* **63**, 435–447.

Feller, W. (1966) *An Introduction to Probability Theory and its Applications*, Vol. *2*. Wiley, New York.

Fisher, R.A., Corbet, S. & Williams, C.B. (1943) The relation between the number of species and the number of individuals in a random sample of an animal population. *Journal of Animal Ecology* **1943**, 42–58.

Hubbell, S.P. (2001) *The Unified Neutral Theory of Biodiversity and Biogeography*. Princeton University Press, Princeton, NJ.

Kendall, D.G. (1948) On some modes of population growth leading to R.A. Fisher's logarithmic series distribution. *Biometrika* **35**, 6–15.

Kendall, D.G. (1949) Stochastic processes and population growth. *Journal of the Royal Statistical Society B* **11**, 230–264.

Kolata, G. (1986) Shakespeare's new poem: an ode to statistics. *Science* **231**, 335–336.

MacArthur, R.H. (1957) On the relative abundance of bird species. *Proceedings of the National Academy of Sciences, USA* **43**, 293–295.

Magurran, A. (1988) *Ecological Diversity and its Measurement*. Chapman and Hall, London.

May, R.M. (1975) Patterns of species abundance and diversity. In: *Ecology and Evolution of Communities* (eds M.L. Cody & J.M. Diamond), pp. 81–120. Belknap Press of Harvard University Press, Cambridge, MA.

Motomura, M. (1932) On the statistical treatment of communities. *Zoological Magazine, Tokyo* **44**, 379–383.

Nee, S. (2001) Inferring speciation rates from phylogenies. *Evolution* **55**, 661–668.

Nee, S., Harvey, P.H. & May, R.M. (1991) Lifting the veil on abundance patterns. *Proceedings of the Royal Society of London, Series B* **243**, 161–163.

Nee, S., Mooers, A.Ø. & Harvey, P.H. (1992) Tempo and mode of evolution revealed from molecular

phylogenies. *Proceedings of the National Academy of Sciences, USA*, **89**, 8322–8326.
Nee, S., Barraclough, T. & Harvey, P.H. (1996) Temporal changes in biodiversity: detecting patterns and identifying causes. In: *Biodiversity* (ed. K.J. Gaston), pp. 230–252. Blackwell Science, Oxford.
Patrick, R. (1968) The structure of diatom communities in similar ecological conditions. *American Naturalist* **102**, 173–183.
Pielou, E.C. (1977) *Mathematical Ecology*. Wiley, New York.
Preston, F.W. (1962) The canonical distribution of commonness and rarity. Part 1. *Ecology* **43**, 185–215.
Raup, D.M., Gould, S.J., Schopf, T.J.M. & Simberloff, D.S. (1973) Stochastic models of phylogeny and the evolution of diversity. *Journal of Geology* **81**, 525–542.
Sepkoski, J.J. & Kendrick, D.C. (1993) Numerical experiments with model monophyletic and paraphyletic taxa. *Paleobiology* **19**, 168–184.
Sibley, C.G. & Ahlquist, J.E. (1990) *Phylogeny and Classification of Birds*. Yale University Press, New Haven, CT.
Simon, H.A. (1955) On a class of skew distribution functions. *Biometrika* **42**, 425–440.
Southwood, T.R.E. (1996) The Croonian lecture, 1995—natural communities: structure and dynamics. *Philosophical Transactions of the Royal Society of London, Series B* **351**, 1113–1129.
Southwood, T.R.E. & Henderson, P.A. (2000) *Ecological Methods*, 3rd edn. Chapman and Hall, London.
Tokeshi, M. (1999) *Species Coexistence*. Blackwell Science, Oxford.
Williams, C.B. (1964) *Patterns in the Balance of Nature*. Academic Press, London.

Part 2
Why are most species rare?

Chapter 4
The neutral theory of biodiversity and biogeography, and beyond

Stephen P. Hubbell[*] and Jeffrey K. Lake

Introduction

The neutral theory of biodiversity and biogeography extends the theory of island biogeography (MacArthur & Wilson 1967) to describe the expected distribution of relative species abundance that arises at steady state among speciation, dispersal and extinction (Hubbell 2001). As in the original theory, which is also neutral, species are treated as identical in their vital rates of birth, death and migration. Unlike the theory of island biogeography, however, the neutrality assumption is made at the individual level, not the species level, a crucial change in the theory that allows species to differentiate in relative abundance through ecological drift (demographic stochasticity). The lifespans of species under drift are then dictated by their abundances, so that the extinction rate is a true prediction of the theory, not a free parameter as it was in the original theory of island biogeography. In the neutral theory, the 'metacommunity' replaces the 'source area' concept of the theory of island biogeography. The metacommunity is defined as the evolutionary biogeographical unit within which most member species originate, live and become extinct, i.e. the region within which most species spend their entire evolutionary lifetimes. The theory also tells us how to measure the size of the metacommunity, in contrast to island biogeography theory, in which the size of the source area was undefined.

In the present neutral theory, ecological drift is not completely free and unfettered, but occurs under zero-sum dynamics: no species can increase in abundance without a matching decrease in the collective abundance or biomass of all other species. Therefore, it is not accurate to characterize the neutral theory as competition-free, as did Abrams (2001). On the contrary, competition is intense because the total size of the ecological community remains constant, despite the continually drifting relative abundances of member species. Because of the zero-sum rule, the theory applies (in its current form) to communities of trophically similar species that actually or potentially compete for the same or similar resources.

* *Correspondence address: shubbell@dogwood.botany.uga.edu*

For the zero-sum rule to apply, species must be facultative, flexible and overlap broadly in their use of limiting resources. Thus, neutral theory asserts that trophically similar species are less specialized than standard niche theory would suggest. For example, if one species should happen to become extinct in a particular community, the remaining species in the community are expected fully to exploit the resources that are freed up. The extent to which species are unspecialized and zero-sum dynamics apply in actual communities needs to be subjected to greater scrutiny to test the theory rigorously. The principal theorems of the current neutral theory were derived under this assumption, but how robust the theory is to violations of this assumption is not yet fully known.

Under neutrality, the species richness and relative species abundance in communities on local to global spatial and temporal scales can be predicted from knowledge of just a few things. Unfortunately, these things are not generally easy to measure directly, at least with current methods. The key parameters are: the speciation rate ν, the size of the metacommunity J_M and the mean dispersal rate m of individuals over the metacommunity landscape. Parameter J_M is defined as the sum of the population sizes of all species in the metacommunity. The theory produces a dimensionless biodiversity number, $\theta = 2J_M\nu$, which according to theory is fundamental because it, along with the mean dispersal rate, fully controls patterns of species richness, relative species abundance, species–area relationships, and even phylogeny and phylogeography under neutrality. Remarkably, it can be proven (Hubbell 2001) that θ is asymptotically identical to Fisher's α (Fisher *et al.* 1943), the oldest, most famous and most widely used measure of species diversity in ecology (Magurran 1988).

In this paper we first review a few of the main results and predictions of the neutral theory. Then we discuss several recent published efforts to evaluate elements of the theory, adding our own further analyses to the mix. We conclude with a brief prospect for the future of neutral theory in community ecology, and beyond.

Predictions of the neutral theory

Despite its simplicity, the neutral theory generates a surprisingly rich array of biologically interesting and testable hypotheses about many macroecological patterns of species diversity and biogeography. In many cases, the neutral theory fits macroecological patterns as well as or better than current niche assembly theory does. Here, we highlight a number of the theory's more salient, non-obvious, and often quite accurate predictions for relative species abundance and species–area relationships. Further predictions on these and other topics can be found in Hubbell (2001).

Relative species abundance

Neutral theory predicts that the distribution of relative species abundances in local communities will follow a new statistical distribution called a zero-sum multinomial (Hubbell 1997, 2001). This distribution is similar to the log-normal (Bell 2000), but differs in often being strongly negatively skewed, with a long tail of very rare species. The degree of skewing is predicted to be a function of the immigration rate. To our knowledge, the involvement of immigration in a model of the distribution of relative species abundance is completely novel. In fact, the moments of the zero-sum multinomial distribution, and therefore its shape, are all functions of the immigration rate (as well as functions of θ and the size of the community) (Hubbell 2001, unpubl.). As the immigration rate falls, the mean number of species decreases, the mean and variance of abundance among the remaining species increases and the rarest species become ever rarer relative to common species. Although other statistical distributions may be discovered that yield similar shapes (e.g. Nee, this volume; Plotkin & Muller-Landau 2002), to our knowledge no other hypothesis besides the neutral theory directly connects the moments of the relative abundance distribution to parameters that have an immediate population biological interpretation.

At the spatial scale of the metacommunity, the distribution of relative species abundance depends on the mode of speciation, according to neutral theory, again a previously unsuspected relationship. If new species arise like rare point mutations ('point mutation' speciation), the limiting metacommunity relative abundance distribution is Fisher's log-series (Fisher *et al.* 1943; Hubbell 2001). However, if new species arise by the random partition of an ancestral species ('random fission' speciation), then the expected metacommunity distribution is a zero-sum multinomial. As we will show below, these two modes of speciation are the extremes of a continuum, with intermediate distributions of relative species abundance between the log-series and the zero-sum multinomial, depending upon the mean size of species populations at origination.

The zero-sum multinomial can yield remarkably precise fits to observed relative abundance data. An example is the extremely good fit to the abundance data on 1175 tree species in a 50 ha plot in Lambir Hills National Park, Sarawak, here shown as a dominance–diversity curve (Fig. 4.1). Note that the observed curve for the local community (in the 50 ha plot) deviates from the metacommunity curve (estimated for a much larger regional area) at the rare-species end of the distribution (Fig. 4.1). Such local community departures from the metacommunity distribution are normal and expected according to the theory whenever there is dispersal limitation. Rare species are more extinction-prone, and once they become locally extinct, they are slower to re-immigrate from the metacommunity than common species. Because of an interaction between dispersal limitation and local extinction, rare species will be rarer, and common species commoner in local communities or islands than expected from a random sample of the metacommunity. Thus, more isolated local communities or islands will exhibit greater dominance by common species and lower species richness, given otherwise equal θ values. According to theory, the greater abundance of common species in this case does not result from the higher mean fitness of these species, but arises as a simple consequence of zero-sum ecological drift operating under restricted immigration.

The neutral theory of relative species abundance brings to light potential problems with certain approaches to testing neutral hypotheses that are in widespread use in community ecology. Because of the dependence of the expected

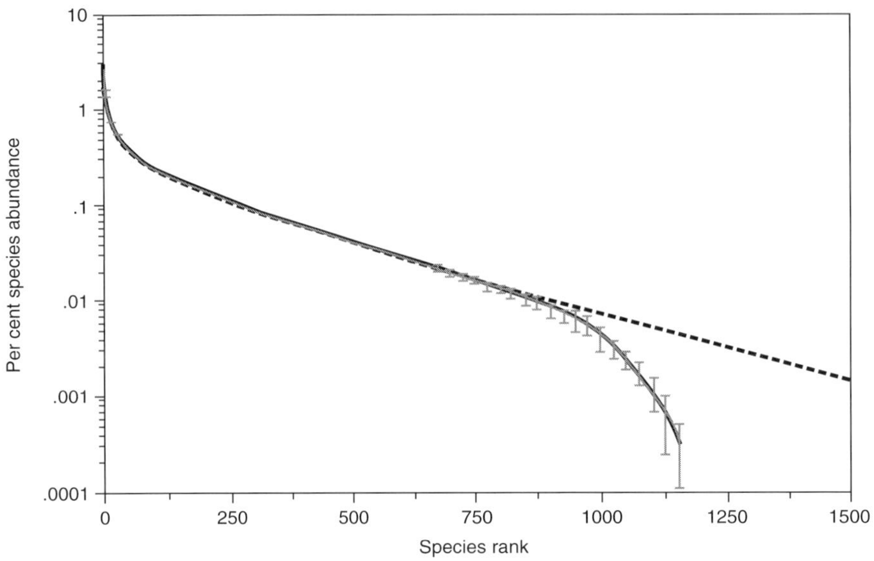

Figure 4.1 Dominance–diversity curve for a sample size of 324 592 trees and shrubs of 1175 species in a 50 ha forest plot in Lambir Hills National Park, Sarawak, Borneo. The dashed black line is the curve expected in a much larger area (the metacommunity), with an estimated value of $\theta = 310$. The solid black line is the observed dominance–diversity curve. The grey line with error bars (± 1 standard deviation of the mean) is fitted for an immigration rate $m = 0.18$ per birth. Rare species are rarer than predicted from the metacommunity dominance–diversity curve because they are more extinction-prone locally than common species, and once locally extinct, they take longer to re-immigrate.

relative species abundance distribution in a local community on the immigration rate, this distribution cannot be found simply by randomly sampling species abundances in the metacommunity. Even more problematic are attempts to test neutral models of community composition by drawing metacommunity species at random with equal probability, because this approach fails to take either relative species abundance or dispersal limitation into account. Common metacommunity species on average are less dispersal limited than rare species. Thus, sampling species with equal probability will tend to increase the rate of Type II errors (false rejections of neutrality) because rare species will be oversampled.

Species–area relationships

To our knowledge, with the exception of the model of Durrett & Levin (1995), the neutral theory is the only dynamic model of species–area relationships, and the only theory that connects demographic processes of birth, death, migration and speciation to landscape and regional patterns of species abundance and diversity (Bell 2001; Hubbell 2001). The most widely accepted relationship between species and area is the Arrhenius power law, $S = cA^z$, where S is the number of species tallied in area A, and c and z are parameters. Under the zero-sum assumption of a linear relationship between individuals J and area A, $J = \rho A$, where ρ is the mean density of individuals per unit area, then the number of species can be expressed as a simple power law of the number of individuals sampled, $S = J^z$, when we let $c = \rho^z$.

Many novel and qualitatively accurate predictions about species–area relationships are made by the neutral theory. One of the most significant is that it explains why the Arrhenius power law does not work over all spatial scales (Williamson 1988; Rosenzweig 1995), contrary to the arguments by Harte *et al.* (1999). On local to global spatial scales, the species–area relationship is expected to be triphasic because there are different scaling rules on different spatial scales. Moreover, the theory demonstrates that the effect of dispersal limitation switches from having a negative effect on the rate of species accumulation on small spatial scales, to having a positive effect on large spatial scales (Hubbell 2001).

On very small spatial scales, the species–area relationship is curvilinear on a double-log plot. This curvilinearity arises because the rate of species accumulation on small scales is dominated by the sampling of local relative species abundance (common species are sampled faster than rare species). On landscape to regional scales within a single metacommunity, however, the Arrhenius log–log species–area power law always holds, according to the theory. On regional scales, the rate of species addition is less sensitive to species abundance and more sensitive to the rate of encountering species ranges at steady state among speciation, dispersal and extinction. The theory also predicts that the Arrhenius power law parameters c and z will both be functions of the fundamental biodiversity number θ and the dispersal rate m. When θ is large (speciation rate high or large metacommunity) relative to the dispersal rate, then the log–log slope of the Arrhenius relationship will be steep. Conversely, when species disperse over the metacommunity landscape rapidly relative to the rate of speciation, then the slope of the relationship will be shallow. This is because when θ is small, mean species abundances in the metacommunity are large and species have plenty of time to disperse all across the metacommunity before the next species originates (Hubbell 2001).

One of the most profound results of neutral theory for species–area relationships is that it predicts the existence of a natural length scale in biogeography. This natural length scale defines the mean size of metacommunities that comprise the global biogeographical landscape. Thus, in contrast to the vagueness of the source–area concept in the theory of island biogeography, the neutral theory makes a prediction about the actual size of the evolutionary biogeographical units within which most species spend their entire evolutionary lifetimes. The theory predicts that on very large spatial scales the Arrhenius power law will break down, and the species–area relationship will exhibit an upward-bending inflection. This inflection point defines the correlation length of speciation–dispersal–extinction processes within the metacommunity—the natural length scale of the dynamic biogeographical process. On spatial scales larger than the area at the inflection point, these processes become spatially increasingly independent of one another. In theory when these processes are completely independent and decoupled, the limiting slope of the species–area curve will be unity (Durrett & Levin 1995; Hubbell 2001). The size of the geographical area at which this inflection point occurs is a function of the fundamental biodiversity number θ and the dispersal rate m. Metacommunities are small for large θ and small m, and large for small θ and large m. Knowledge about the natural length scales of diverse metacommunities will be important to conservation biology and reserve planning.

The neutral theory and beyond

Given the considerable success of the neutral theory in explaining many patterns in macroecology, the often remarkable precision of its fits to relative abundance data, and the derivation of Fisher's α and the log-series from the neutral theory, we hope that the theory will stimulate a deep reassessment and new hypotheses and theory about how ecological communities are assembled. As the neutral theory was published more than a year ago, nearly a dozen reviews have appeared, mostly favourable. However, there have been several critical papers, and we have chosen three for discussion: Chave *et al.* (2002), Condit *et al.* (2002) and Ricklefs (2003). Chave *et al.* argued that at least some of the patterns generated by the neutral theory also could be explained by niche assembly theory. Condit *et al.* argued that the observed similarity of neotropical forests at different separation distances was inconsistent with neutrality. Finally, Ricklefs argued that the two modes of speciation discussed in Hubbell (2001), 'point mutation' and 'random fission', lead to estimates of species lifespans that are either too short or too long respectively. Here we comment on these three papers in some detail, adding some of our own analyses to the discussion.

Density dependence and dispersal limitation

Chave *et al.* (2002) put density dependence and Gaussian dispersal (i.e. diffusion) into spatially explicit models that were similar to the toroidal metacommunity models studied by Hubbell (2001) (e.g. a process of speciation was included). Apart from ultimate population size limits imposed by zero-sum dynamics, there is no density dependence *per se* in the current neutral theory of Hubbell (2001). In the Chave models, each grid cell was occupied by a single individual, whereas in the Hubbell model, each cell contained a local community of J individuals. Density dependence in the Chave models affected the probability of recruitment in cells adjacent to cells occupied by a conspecific individual. In some of their simulations, Chave *et al.* also studied a trade-off between survival and competitive ability. We will not discuss the trade-off model here other than to note that one of the trade-offs they studied seemed biologically unreasonable (species with lower survival rates were better competitors).

Chave *et al.* (2002) described their models as being, in essence, niche assembly models, but except for the versions containing trade-offs, their models are neutral by the definition of Hubbell (2001). All of the species in their models experienced the same density dependence and dispersal, so that they were all governed by the same interaction rules – which is the definition of neutrality given by Hubbell. One of the authors of the paper, S. A. Levin (pers. comm.), prefers to call such models 'symmetric', but we note that the interaction rules under Hubbell's neutral theory are also symmetric. Where are the unique niches of species in a symmetric model? (There are none.) It seems to us that the crucial distinction should be between symmetric and non-symmetric models, and that all symmetric models are, in fact, neutral.

This distinction is more than a matter of semantics because Chave *et al.* argued on the basis of their simulations that non-neutral models can do just as well as the neutral theory. They found that adding recruitment density dependence and dispersal limitation allows more species to coexist than in their absence. The positive effect of dispersal limitation on species richness was previously analysed by Hubbell (2001), with similar results. However, Chave *et al.* went further and showed that adding in even fairly modest density dependence in recruitment can increase the steady state species richness in the metacommunity dramatically. In their model 'modest' density dependence meant spatially very local density dependence, affecting the probability of colonizing only the four grid cells immediately adjacent to (sharing a side with) a cell containing an individual of the same species. They argued that qualitatively similar dominance–diversity curves were obtained to those from the neutral model without density dependence. However, we have explored the Chave model further, and dispute the robustness of this claim.

To explore Chave *et al.*'s conclusions about density dependence more thoroughly, we developed a metacommunity model that allowed us to study the effects of varying the strength of dispersal limitation and density-dependent recruitment success, as well as potential interactions between these two factors. We modelled a metacommunity on a 40 401 cell (201×201) torus to avoid edge effects. We initialized all simulations with abundance distributions calculated for a metacommunity with no dispersal limitation or density dependence, following the methods given in Hubbell (2001). Deaths in the metacommunity occurred individually and at random, with immediate replacement by a new individual, maintaining ecological saturation at all times. Our simulations ran for 50 complete metacommunity turnovers ($> 2 \times 10^6$ individual replacements) at each level of density dependence and dispersal limitation tested. We previously determined that 50 turnovers guaranteed adequate time to reach a steady state. We tested three levels of community diversity ($\theta = 5.0$, 10.0 and 50.0), and we obtained qualitatively similar results at all three levels.

We modelled dispersal limitation using an exponentially decaying dispersal kernel, with probability density function, de^{-dx}, where d is the dispersal coefficient and x is dispersal distance. We tested four values of d: 0.1, 0.5, 1.0 and 2.0. Increasing coefficients cause increasingly local dispersal (Fig. 4.2). For density dependence, we used a modified version of the algorithm of Chave *et al.* (2002). We used the eight nearest neighbours, including the diagonal cells, plus the focal cell (site of prior occupant). The probability of successful establishment of an individual is given by: $w(q) = 1 - aq$, where q is the proportion of the nine cells occupied by con-

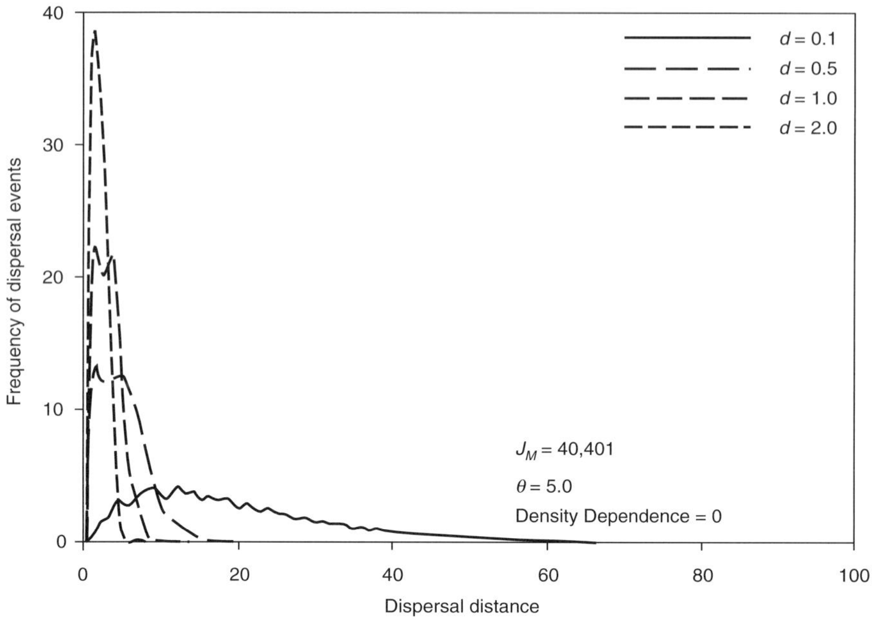

Figure 4.2 Frequency of dispersal distances as a function of level of dispersal limitation. Increasing dispersal limitation increases the percentage of recruits near the parent. Increased area in more distant annuli around the parent explains the increased frequencies at greater distances from the parent.

specifics of the potential replacement individual, and *a* is the per capita conspecific effect. We considered five values of *a*: 0, 0.1, 0.2, 0.5 and 1.0, representing increasingly strong density dependence. Although this new model incorporates two key biological interactions, they operate entirely symmetrically on a per capita basis, maintaining the fundamental assumption of neutrality that all individuals in the community be subject to the same interaction rules.

The results were as follows, and were sometimes surprising. First, if dispersal limitation became stronger (higher *d*), acting alone with no density dependence, then diversity increased with additional rare species in the metacommunity (Fig. 4.3). These results are qualitatively similar to the results of Hubbell (2001) and Chave *et al.* (2002), but differ quantitatively because different dispersal kernels were used in each study. Second, at low dispersal limitation, increasing the strength of conspecific density dependence in recruitment success reduced the most common species and allowed additional rare species to persist in the metacommunity (Fig. 4.4). Greater evenness in abundance resulted from increased density-dependent reduction in recruitment success.

The interaction of density dependence and dispersal limitation produced a number of unexpected, non-obvious results. At a fixed, strong level of density depen-

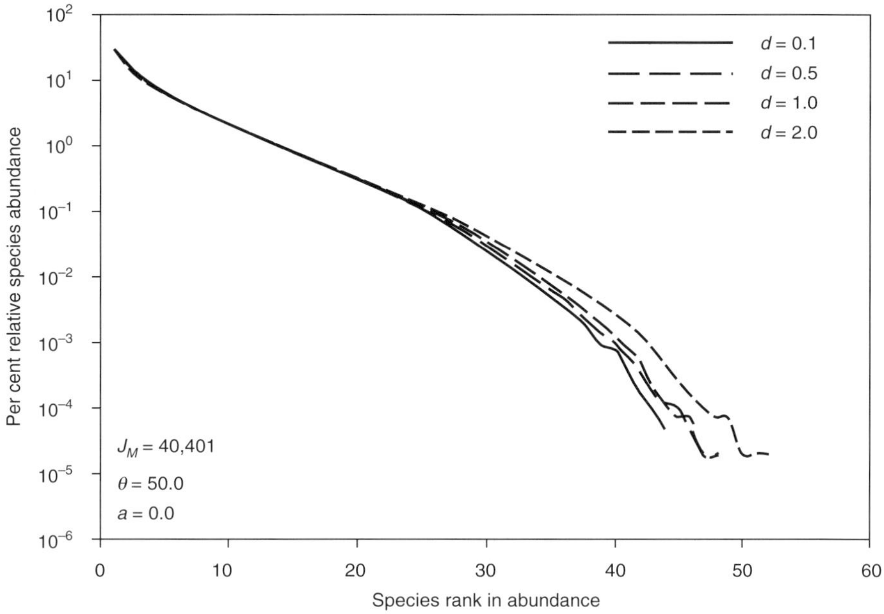

Figure 4.3 Dominance–diversity curves resulting from no density dependence in recruitment (*a*) at various levels of dispersal limitation (*d*). As expected, increasing dispersal limitation allows additional rare species to persist in the metacommunity and increases the evenness of the community somewhat.

dence ($a = 1.0$), increasing dispersal limitation once again allowed additional rare species. What was surprising, however, was that at high levels of density dependence, increasing dispersal limitation actually increased the expected abundances of the most common species in the metacommunity at the expense of species of intermediate abundances (Fig. 4.5). Contrary to our expectations, under the scenario of both strong, symmetric dispersal limitation and strong, symmetric negative density-dependence in recruitment, both the commonest and the rarest species benefit.

This effect, in hindsight, can be explained by differential mass effects. The commonest species are able to disperse to most empty cells owing to their high abundance, and the sheer mass of their propagule numbers overwhelms the negative effects of density dependence on their recruitment success. In contrast, species of intermediate abundance are sufficiently abundant to suffer strong density dependence but not to produce enough propagules to swamp out the negative density dependence in their recruitment success. Finally, the rarest species benefit because they are seldom subject to density dependence (they rarely encounter conspecifics), and they benefit from the depressed abundance of the intermediate species.

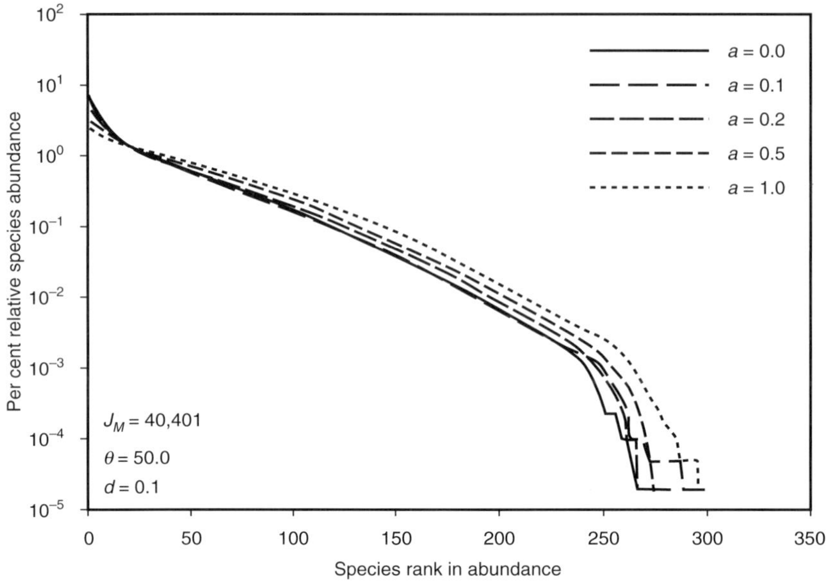

Figure 4.4 Dominance–diversity curves resulting from low dispersal limitation (d) at various levels of negative density dependence in recruitment (a). As expected, increasing density dependence reduces the abundance of the most common species, and allows additional rare species to persist in the metacommunity.

Perhaps the most significant conclusion from these simulations is that density dependence, separately and in combination with dispersal limitation, does not improve fits of the theory to empirical data. This was especially true when both density dependence and dispersal limitation were strong. For example, the shape of the curves in Fig. 4.5 bear little resemblance to the curves obtained in any of the 50 ha tropical forest plots that are in the network of the Center for Tropical Forest Science (e.g. Fig. 4.1). Indeed, only under quite weak density dependence did any of the curves approximate the S-shape found in every one of the tropical forest plots. When density dependence is strong, the dominance–diversity curves are too even, becoming flattened on the common-species end of the distribution. The problem may be with the symmetry argument—and therefore with neutrality. If not all species in actual ecological communities are subjected to the same degree of density dependence or dispersal limitation, then symmetry is broken. Elsewhere, we have published strong evidence that not all species experience the same level of density dependence, at least in the tropical forest on Barro Colorado Island (BCI), Panama (Harms *et al.* 2000; Hubbell *et al.* 2001), and that the strength of dispersal limitation varies significantly among BCI tree species (Hubbell *et al.* 1999; Harms *et al.* 2000).

The significance of density dependence in explaining higher species diversity in tropical forests remains controversial. Janzen (1970) and Connell (1971) indepen-

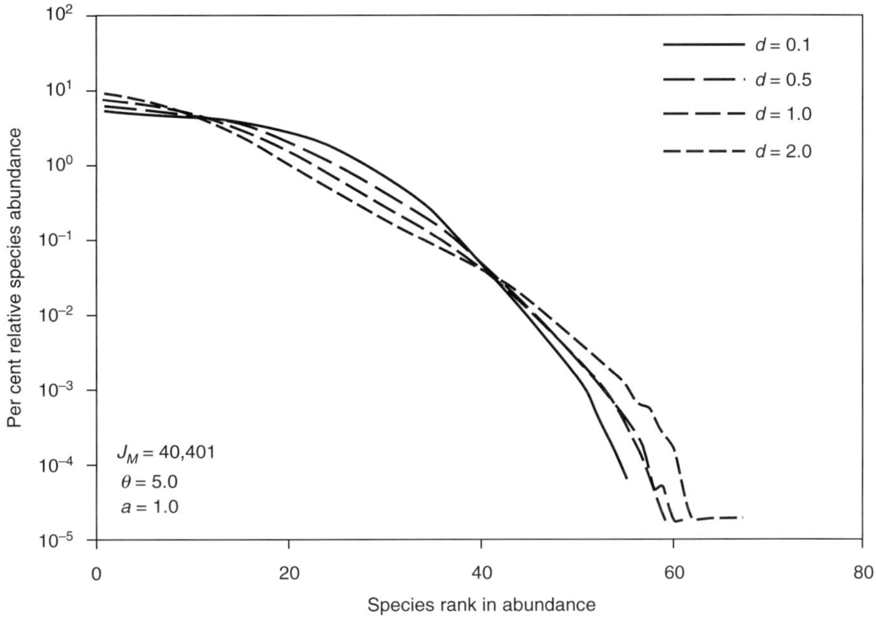

Figure 4.5 Dominance–diversity curves resulting from strong negative density dependence in recruitment (a) at various levels of dispersal limitation (d). As expected, increasing dispersal limitation allows additional rare species to persist in the metacommunity. However, the most common species also were more abundant under increasing dispersal limitation, at the expense of species of intermediate abundance.

dently hypothesized that density-dependent pest pressure, interacting with seed dispersal limitation, could explain the higher tree species richness in tropical forests. Density dependence may help explain higher levels of α diversity in local communities by allowing more rare species to coexist cheek by jowl with common species. However, if it is true that density dependence is not any stronger in tropical forests than in temperate forests (Packer & Clay 2000; Lambers *et al.* 2002), then the ultimate explanation for latitudinal gradients in tree diversity may lie elsewhere, probably at the larger biogeographical scales at which the metacommunity balance between speciation and extinction is established (Hubbell *et al.* 1999). Hubbell (2001) has argued that the gradient may be explained by latitudinal variation in the fundamental biodiversity number θ. Allen *et al.* (2002) have suggested that Rohde's evolutionary speed hypothesis might explain latitudinal variation in θ.

Long-distance dispersal events

Condit *et al.* (2002) have also challenged the neutral theory. Adapting a model of Nagylaki from population genetics, they estimated the probability, under neutrality, that two randomly chosen trees in two neotropical forests separated by distance r

would be of the same species, given a parent–offspring dispersal distance σ. They estimated dispersal distances on local scales within the 50 ha plot on BCI, among 1 ha plots across the Isthmus of Panama, spanning a distance of *c.*70 km, and among plots in Amazonian Ecuador and Peru, separated by 900–1000 km. Condit *et al.* (2002) found that the estimated parent–offspring dispersal distances needed to obtain the correct observed probabilities increased as the distance separating the forest stands increased. There was rapid decay of similarity between communities at very short distances (< 100 m), slower decay at mid-range distances, and very slow decay in similarity at long distances. Moreover, community similarity decayed faster with distance in Panama than in Ecuador–Peru. That the similarity of stands declined more rapidly locally and more slowly regionally cannot be explained by a single dispersal rate at all spatial scales.

Condit *et al.* argued that this observation is sufficient evidence to reject a purely neutral dispersal hypothesis, but this conclusion may be premature. It is quite possible that the neutral theory may fail because of higher habitat heterogeneity in Panama than in western Amazonia. On the other hand, there are at least a couple of possibilities that are consistent with neutrality. One possibility is that the slower rate of distance decay of similarity in Ecuador–Peru primarily results from larger metacommunity sizes in western Amazonia than in Panama. There is certainly a much larger area in western Amazonia than in Panama, which would be expected to support a larger continuous metacommunity. Chris Dick (Smithsonian Tropical Research Institute, pers. comm.) has suggested, based on phylogeographical evidence, that Panama is still in phylogeographical non-equilibrium since the formation of the isthmus *c.*3 Ma, resulting in a high degree of metacommunity fragmentation. In addition to this possibility, the Condit *et al.* estimates of the probability that two trees are conspecific were based mainly on the commonest species because only data from small (1 ha) plots were used. This tends to bias the finding of greater community similarity because common species are more widespread in the metacommunity (Hubbell 2001). Common species in a landscape mosaic composed of small metacommunities would be less widespread, and this might help explain the difference between Panama and western Amazonia.

Another explanation consistent with the neutral theory is that it is undoubtedly true that dispersal distances in a community are distributed, i.e. characterized by some statistical distribution, rather than by a single dispersal distance, which was a simplification in the original exposition of the theory (Hubbell 2001). Including a distributed dispersal kernel in the theory need not violate symmetry or neutrality. One interesting possibility is that dispersal kernels are non-Gaussian Lévy-stable distributions. Condit *et al.* did not comment on the fact that a power law relationship exists between their estimated parent–offspring dispersal distances and the distance between forest stands (Fig. 4.6). This suggests that a systematic underestimation of true dispersal distances occurs in local communities separated by short distances. The theoretical method of estimating dispersal used by Condit *et al.* assumes dispersal by Gaussian diffusion. If, on the other hand, true dispersal kernels are more accurately characterized as non-Gaussian Lévy-stable distribu-

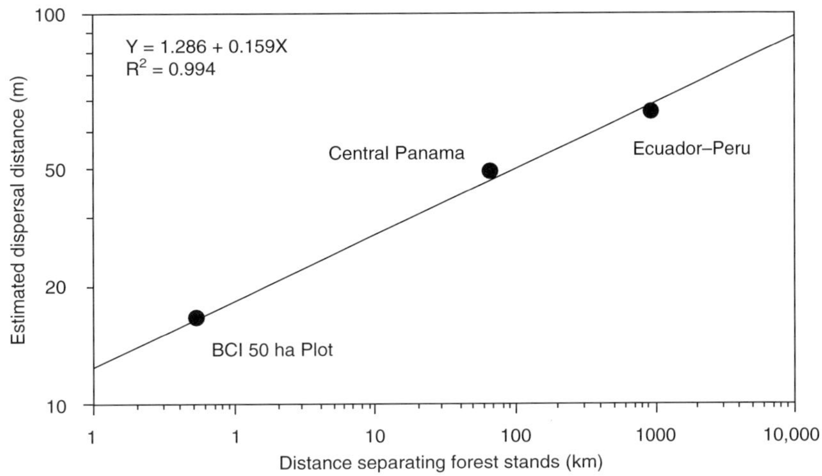

Figure 4.6 The log–log power law relationship between estimated mean parent–offspring dispersal distance (from Condit *et al.* 2002) and the distance separating the forests being compared. This observation suggests that true dispersal distances may have been underestimated by the assumption of dispersal by Gaussian diffusion. This power law relationship also suggests that the true dispersal kernels may be Levy–stable distributions.

tions, which have no defined moments (the tails of the distributions are so-called 'fat' tails that are described by power laws), then the Nagylaki method will systematically underestimate the frequency of long-distance dispersal events from data on communities separated by short distances. The power law relationship of the dispersal distance estimates in Fig. 4.6 strongly suggests that this is a real possibility.

Speciation and phylogeny

As originally proposed, the neutral theory considered two modes of speciation: point mutation and random fission. Point mutation speciation is based on the premise that speciation events ultimately can (in theory at least) be traced to individual-level variation and that new species arise in essence as singleton lineages. This mode of speciation is consistent with sympatric speciation. Under point mutation speciation, metacommunity relative abundances follow Fisher's log-series (Hubbell 2001). At the other extreme is random fission speciation, in which an ancestral species is randomly cleaved into two daughter species of arbitrary population size summing to the population size of the ancestral species. We consider random fission speciation to be the analogue of allopatric speciation. This mode of speciation results in a wide and highly variable range of initial population sizes at origination and leads to a metacommunity relative abundance distribution that is zero-sum multinomial.

In a critique of the neutral theory, Ricklefs (2003) points out that under point mutation speciation there is a plethora of short-lived species that become extinct very quickly owing to their initial rarity. He calculated mean lifespans of species under this mode of speciation to be just a few generations. In contrast, random fission speciation produces many very long-lived species because the initial population size usually is much larger than 1, so these new species take much longer to become extinct. In very large metacommunities, Ricklefs argued that new species would be so abundant that they would take much too long to become extinct by ecological drift alone.

We respond to Ricklefs as follows. First, it is not clear whether Ricklefs realized that species lifespans are computed in terms of the total number of deaths of all species in the metacommunity until extinction of the focal species. In a given timespan, absolutely more deaths will occur in a large metacommunity than in a small metacommunity. This makes the lifespans of species appear to be much longer than they really are on an absolute timescale as metacommunity size increases. Second, we have always regarded, and indeed originally described, point mutation speciation and random fission speciation as two extremes of a speciation continuum (Hubbell 2001). For example, one can reasonably hypothesize that new species often start as small populations >1 individual, but not as large as the mean size of a random fraction of the total parent species population. Here we examine such a hypothesis as a third and intermediate mode of speciation, which we have dubbed 'peripheral isolate speciation'. Under this mode of speciation, small subpopulations become isolated from other parts of the metapopulation, and undergo speciation. These peripheral isolate populations are less variable in size than those produced under random fission speciation, and provide a continuum of starting population sizes between point mutation and random fission speciation.

To simulate peripheral isolate speciation, we specified a mean and a variance of the initial population sizes of new species. The speciation procedure was as follows. An individual from the metacommunity was picked at random, and its species was determined. If the abundance of the selected species was greater than twice the calculated size of the peripheral isolate, then speciation proceeded; otherwise, no speciation event occurred. One could choose a variety of algorithms for speciation under peripheral isolate speciation that might give quantitatively slightly different answers. However, our main qualitative point is unlikely to change: under peripheral isolate speciation, a complete continuum in equilibrium metacommunity dominance–diversity curves arises between the low-diversity extreme under point mutation speciation (equivalent to a peripheral isolate size of unity with no variance), and the high diversity extreme under random fission speciation.

In Fig. 4.7 we illustrate two cases of peripheral isolate speciation, one in which the mean isolate size ± one standard deviation is 10 ± 2, and a second case in which the mean size is 100 ± 10. The dominance–diversity curves for these two cases are intermediate between the curves for point mutation and random fission speciation. In Fig. 4.8, we illustrate how the distribution of lifespans of new species is affected by the size of the species population at origination. For larger mean isolate sizes at origination, the lifespan distribution moves further to the right on the *x* axis. The mean lifespans for peripheral isolates of known initial sizes can be calculated analytically

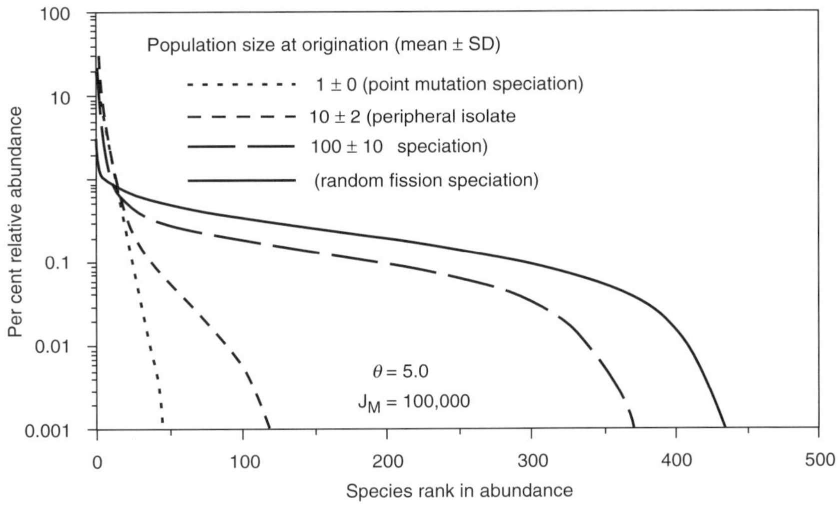

Figure 4.7 Dominance–diversity curves resulting from different initial population sizes and different modes of speciation. Point mutation speciation results in relatively few rare species persisting in a steady state metacommunity because most species (with starting population size of 1) tend to become extinct quickly. Random fission speciation produces many more rare species that persist for longer in the metacommunity owing to their larger starting population size. Peripheral isolate speciation, with intermediate initial population sizes (with some mean and variance), produces dominance–diversity curves with intermediate numbers of rare species persisting. Initial population size at origination leaves a signature in the shape of the metacommunity relative abundance distribution.

from formulae given in Hubbell (2001, chapter 8). As in the case of point mutation and random fission speciation, there should be a signature of peripheral isolate speciation in the metacommunity dominance–diversity curves, which should be intermediate between Fisher's log-series and the zero-sum multinomial.

Our third response to Ricklefs is to defend point mutation speciation as reasonable. Part of the problem lies in the semantics of naming a species. Under point mutation speciation, in principle, one should be able, at least retrospectively, to trace the ancestry of a species back to a single individual (the analogue of 'Lucy'). However, whether this is technically possible or not, our point is that if the origin of species truly lies in the variation among lineages founded by individuals, then what we are really describing under point mutation speciation is the survival of lineages, a few of which survive long enough and become sufficiently abundant and differentiated to be discovered and recognized by taxonomists as species. When this is realized, it is no longer at all unreasonable to postulate mean lineage lifespans of just a few generations. Under this mode of speciation, the geometry of clades is fractal, and all surviving lineages are represented in the clade. We are left with the realization that there is a complete continuum of variation from the individual level all the way up to the full

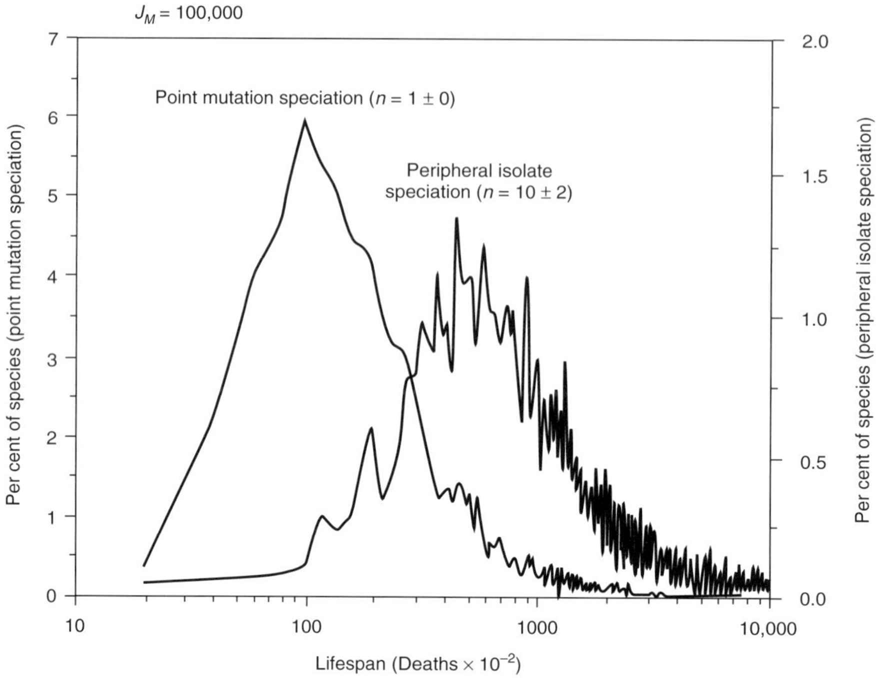

Figure 4.8 The expected lifespan distributions for species originating under point mutation and peripheral isolate speciation. Point mutation speciation results in most species having much shorter lifespans than those originating under a peripheral isolate mode of speciation. Cases illustrated are the distributions of lifespans for a starting population of unity and of 10 individuals. Larger initial population sizes would have distributions located to the right of those illustrated. Note the log transformation of the lifespan axis.

metacommunity or phylogenetic clade. Therefore the typological distinction of what we call 'species' breaks down completely.

Unfortunately, current empirical data on metacommunity relative abundances are not yet sufficient in most cases to evaluate rigorously these different modes of speciation. However, if Fisher's log-series remains the best fit to metacommunity relative abundances, then the mean size of species populations at origination must be quite small, although perhaps not at the point-mutation extreme. However, only very small incipient population sizes can produce such a log-series pattern of metacommunity relative abundances.

Conclusions and the next steps

In this paper, we have offered a synopsis of many of the main predictions and results of the neutral theory to illustrate not only the power of the theory, but also to indicate how the theory may be challenged. We do not expect neutral theory to fit all communities, and indeed we already know that it does not (Hubbell 2001). It is our hope that the specific ways in which the theory fails will provide useful insights into the mechanisms structuring actual ecological communities. Nevertheless the neutral theory has proven tougher to falsify than we expected, and we have found it to be a rich source of hypotheses and to generate remarkably good fits between its predictions and empirical data from many communities.

Neutral macroecological theory, according to Bell (2001), may be evaluated against either 'soft' or 'hard' standards. Against the soft standard, one can evaluate the theory by how well it characterizes empirical macroecological patterns of diversity, irrespective of whether it correctly captures the underlying mechanisms. Against the hard standard, one evaluates neutral theory not only by its ability to fit data, but also by whether it accurately describes the mechanisms. Although neutral theory may fail according to both standards in many instances, the theory nevertheless describes many macroecological patterns of diversity remarkably well (Hubbell 2001), so that by a number of measures, Bell's soft standard is often met. If the assumptions of neutral theory are not met, then the more challenging question is why the theory performs as well as it does. It is remarkable that a theory comprised of basically only three parameters can produce results that often fit empirical macroecological patterns qualitatively and quantitatively extremely well and, we still claim, better than most niche assembly theories. Why does it work so well given strong evidence for niche differentiation among trophically similar species in nature? Hubbell (2001) suggested that the dynamics of ecological communities may be governed by rules of far lower dimensionality than their species richness might suggest. He argued that life history trade-offs are fitness invariance rules that reduce the dynamic complexity of ecological communities by making species much more similar in fitness, and thereby opening a window for ecological drift to be dynamically important.

We remain to be convinced either that non-neutral mechanisms will prove to be universally necessary, or that neutral explanations will be fully sufficient, to understand ecological communities. Our best guess is that most ecological communities contain a few ecological dominants that are overly common relative to neutral predictions. However, we also predict that the dynamics of many species of middling abundance in communities may be well described by neutral hypotheses. One can easily construct model communities in which a few species have higher relative fitnesses and so achieve abundances higher than expected under neutrality. These same communities, however, can also have lots of species that do not enjoy any fitness advantage relative to each other, but whose collective abundance is depressed by the higher fitness common species. If these models are explicitly spatial with dispersal limitation, then the higher fitness common species cannot competitively exclude the remaining species (Tilman 1994; Hurtt & Pacala 1995). We believe that further exploration of symmetric as well as asymmetric biotic interactions, such as those of Chave *et al.* (2002) and ours, will prove fruitful in moving toward a more synthetic theory for community ecology. Through such work we can develop a new theoretical framework that incorporates both ecological drift and deterministic interspecific interactions, including trophic interactions. We believe that such a synthesis may resemble the synthesis in population genetics of genetic drift and selection theory. This synthesis could be a powerful tool for evaluating under what conditions ecological drift is important in structuring ecological communities.

References

Abrams, P.A. (2001) A world without competition. *Nature* **412**, 858–859.

Allen, A.P., Brown, J.H. & Gillooly, J.F. (2002) Global biodiversity, biochemical kinetics and the energetic equivalence rule. *Science* **297**, 1545–1548.

Bell, G. (2000) The distribution of abundance in neutral communities. *American Naturalist* **155**, 606–617.

Bell, G. (2001) Neutral macroecology. *Science* **293**, 2413–2416.

Chave, J., Muller-Landau, H.C. & Levin, S.A. (2002) Comparing classical community models: theoretical consequences for patterns of diversity. *American Naturalist* **159**, 1–23.

Condit, R., Pitman, N., Leigh Jr., E.G., *et al.* (2002) Beta-diversity in tropical forest trees. *Science* **295**, 666–669.

Connell, J.H. (1971) On the role of natural enemies in preventing competitive exclusion in some marine animals and in rainforest trees. In: *Dynamics of Populations* (eds P.J. den Boer & G.R. Gradwell), pp. 298–312. Centre for Agricultural Publishing and Documentation, Wageningen, The Netherlands.

Durrett, R. & Levin, S.A. (1996) Spatial models for species-area curves. *Journal of Theoretical Biology* **179**, 119–127.

Fisher, R.A., Corbet, A.S. & Williams, C.B. (1943) The relation between the number of species and the number of individuals in a random sample of an animal population. *Journal of Animal Ecology* **12**, 42–58.

Harms, K.E., Wright, S.J., Calderón, O., Hernández, A. & Herre, E.A. (2000) Pervasive density-dependent recruitment enhances seedling diversity in a tropical forest. *Nature* **404**, 493–495.

Harte, J., Kinzig, A. & Green, J. (1999) Self-similarity in the distribution and abundance of species. *Science* **284**, 334–336.

Hubbell, S.P. (1997) A unified theory of biogeography and relative species abundance and its application to tropical rain forests and coral reefs. *Coral Reefs* **16**(Supplement), S9–S21.

Hubbell, S.P. (2001) *The Unified Neutral Theory of Biodiversity and Biogeography*. Princeton University Press, Princeton, NJ.

Hubbell, S.P., Foster, R.B., O'Brien, S.T., *et al.* (1999) Light-gap disturbances, recruitment limitation, and tree diversity in a neotropical forest. *Science* **283**, 554–557.

Hubbell, S.P., Ahumada, J.A., Condit, R. & Foster, R.B. (2001) Local neighborhood effects on long-term survival of individual trees in a neotropical forest. *Ecological Research* **16**, S45–S61.

Hurtt, G.C. & Pacala, S.W. (1995) The consequences of recruitment limitation: Reconciling chance, history, and competitive differences between plants. *Journal of Theoretical Biology* **176**, 1–12.

Janzen, D.H. (1970) Herbivores and the number of tree species in tropical forests. *American Naturalist* **104**, 501–528.

Lambers, J.H.A., Clark, J.S. & Beckage, B. (2002) Density-dependent mortality and the latitudinal gradient in species diversity. *Nature* **417**, 732–735.

MacArthur, R.H. & Wilson, E.O. (1967) *The Theory of Island Biogeography*. Princeton University Press, Princeton, NJ.

Magurran, A.E. (1988) *Ecological Diversity and its Measurement*. Princeton University Press, Princeton, NJ.

Packer, A. & Clay, K. (2000) Soil pathogens and spatial patterns of seedling mortality in a temperate tree. *Nature* **404**, 278–281.

Plotkin, J.B. & Muller-Landau, H.C. (2002) Sampling the species composition of a landscape. *Ecology* **83**, 3344–3356.

Ricklefs, R.E. (2003) A comment on Hubbell's zero-sum ecological drift model. *Oikos* **100**, 186–193.
Rosenzweig, M.L. (1995) *Species Diversity in Space and Time.* Cambridge University Press, Cambridge.
Tilman, D. (1994) Competition and biodiversity in spatially structured habitats. *Ecology* **75**, 2–16.
Williamson, M. (1988) Relationship of species number to area, distance and other variables. In: *Analytical Biogeography* (eds A.A. Myers & P.S. Giller), pp. 91–115. Chapman and Hall, New York.

Chapter 5

Breaking the stick in space: of niche models, metacommunities and patterns in the relative abundance of species

Pablo A. Marquet, Juan E. Keymer and Hernán Cofré*

Introduction

'The available evidence apparently does suggest that a great many biological "universes" have the logarithmically-Gaussian form . . . If birds in a valley, moths in a trap, plants in a quadrat, insects in a sweep net, and micro-organisms in a suspension of hay do in fact all agree in this, it would seem that some very general law must lie behind it all.' (Preston 1948)

The search for general laws underlying the structure and functioning of ecological systems has been a long-term goal for ecology. This search, which started with the early detection of general patterns in the diversity, distribution and abundance of species, has fructified in recent years as a consequence of the establishment of the macroecological research programme within ecology (Brown 1995, 1999; Maurer 1999; Gaston & Blackburn 2000; Marquet 2002) and because of the emergence of the science of complex systems (e.g. Waldrop 1992; Cowan *et al.* 1994; Levin 1999). At the core of both macroecology and complex-system science is the search for generalities in the form of power-law behaviour and scale invariance in patterns, such as the distribution of city sizes, body sizes, words in natural languages, abundance within natural communities (e.g. Marquet *et al.* 1990; Schroeder 1991), the growth of firms, countries or natural populations (e.g. Keitt & Stanley 1998; Stanley *et al.* 2000). However, the discovery of patterns and their statistical description is the beginning of a process that ends with the proposition of hypotheses about mechanisms that potentially could give rise to the observed phenomena. The history of ecology tells us that the time it takes to traverse this path is usually very long and is marked by bursts of activity reflected in the generation and coexistence of several alternative models and hypotheses that can explain the same phenomenon. Although this is true for most macroecological patterns (e.g. Gaston & Blackburn 1999), it is paradigmatic in the case of those related to the commonness and rarity of species (see Fig. 5.1).

* *Correspondence address: pmarquet@genes.bio.puc.cl*

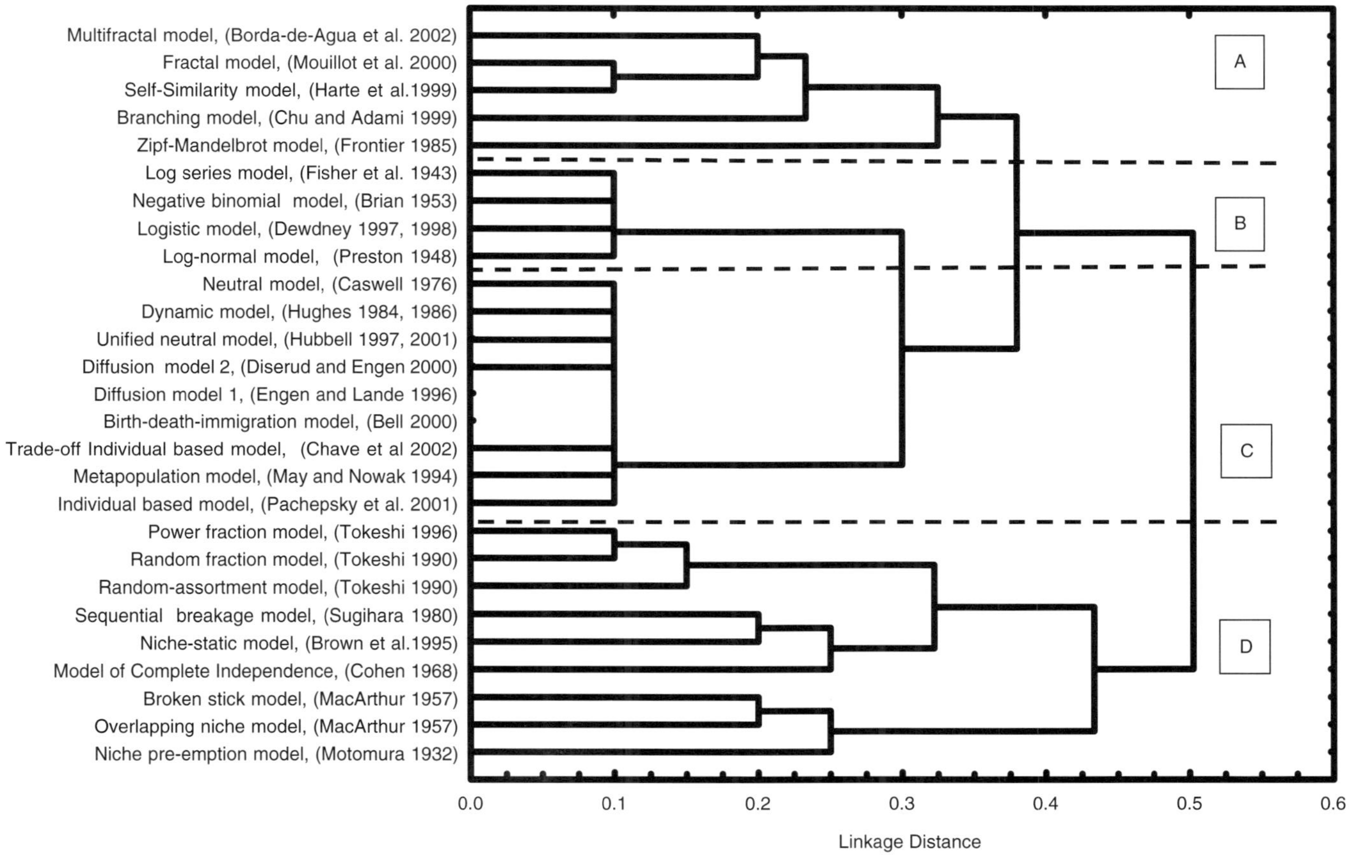

Figure 5.1 Models proposed to explain patterns in the relative abundance of species. For the purpose of graphical representation we carried out a cluster analysis using the UPGMA algorithm and the percentage disagreement distance metric. To construct the distance matrix, models were classified according to the inclusion (1) or not (0) of different features: metacommunity dynamics, birth–death processes, speciation process, niche apportionment processes, ecological equivalency, spatially implicit, spatially explicit, phenomenological, fractal theory and year of publication (scaled to 1). Cluster A groups models based on fractal processes or theory, B identifies statistical models, C groups dynamic models based on birth–death processes and D groups niche-based models (see text for discussion).

In this chapter, we attempt to show part of the scientific history underlying patterns in the distribution and abundance of species, and its place within the macroecological research programme, with emphasis on a particular class of explanations based on niche apportionment mechanisms, generally known as niche models. We will characterize these models, show that their key assumption that the apportionment of resources and individuals are equivalent processes holds, and finally take niche apportionment models one step further by making them spatial, stochastic and dynamic, and mechanistically linked to species life-history attributes. We do not expect to provide a final answer for the empirical reality we address; our aim is to contribute with a valid alternative explanation that potentially could serve this purpose. Foremost, however, our effort should be understood as an attempt to unify niche models with models based on birth–death processes (e.g. Hubbell 1997, 2001; Bell 2000). We do this by connecting resource acquisition with reproductive output in a spatially explicit metacommunity model. We conclude that niche models provide a valid framework to understand patterns in the distribution of abundance.

The pattern at issue

For more than 60 years it has been recognized that species are not equally abundant in samples from local communities. The existence of regularities in the distribution of abundance prompted the development of several models that supposedly captured the pattern. Among them, those that have received most attention probably are Fisher's log-series model (Fisher *et al.* 1943) and Preston's log-normal model (Preston 1948), especially in its canonical form (Preston 1962), but there are several others (see Fig. 5.1). Fisher's sampling model was based on the assumption that the true distribution of abundance was a gamma distribution (Fisher *et al.* 1943; Brian 1953; Kempton & Taylor 1974), whereas the sample distribution of individual abundances was a Poisson series, such that the compound distribution was a negative binomial, which in the infinite limit gave rise to a simple model to predict the expected number of species with *n* individuals, known as the log-series. Under this model rare species with one individual (singletons) are the most abundant (Fig. 5.2b). Preston (1948, 1958) on the other hand assumed that the true distribution of abundance or 'universe' was log-normal, and showed that by graphing species abundance in classes of doubling number of individuals (called octaves), the resulting distribution of abundance in samples was humped with few rare and common species and many species of intermediate abundance (Fig. 5.2a). Interestingly, although Preston's universe was log-normal, samples in local communities were not. Preston (1948) noticed that the sample will tend to look like the hypothetical universe with the left-hand end missing, or veiled, because of the underrepresentation of very rare species in small samples (which he suggested will resemble a log-series on an arithmetic scale). This implies that as the sample gets larger the distribution or universe will be progressively unveiled and eventually will become log-normal. Preston's (1958) analysis of the Audubon Christmas counts and Williams' (1953) analysis of British birds partially confirm this prediction (see also Whittaker 1965).

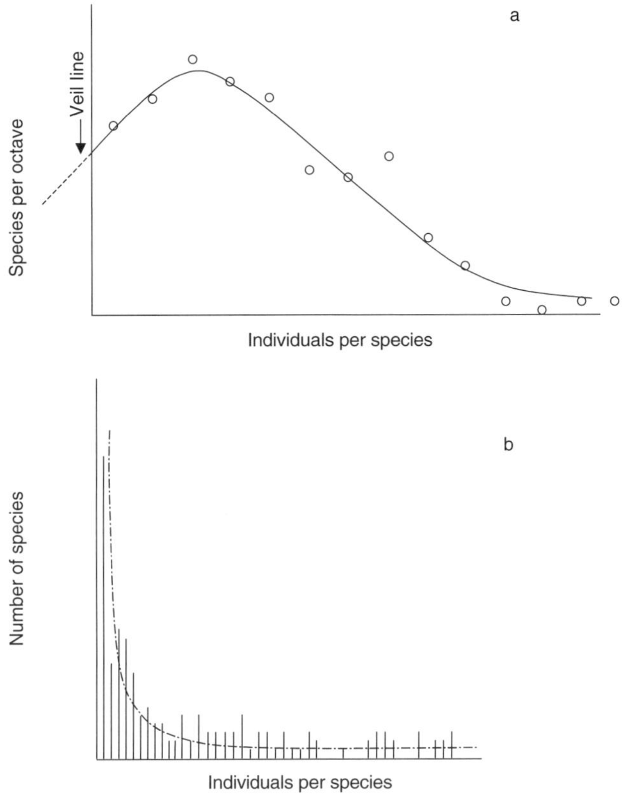

Figure 5.2 Illustration of (a) Preston's log-normal model fitted to data on moths caught in light traps (after Preston 1948) and (b) Fisher's log-series model fitted to data on Rothamsted's Lepidoptera surveys (after Fisher *et al.* 1943).

Fisher's and Preston's models can be dubbed as phenomenological or inductive (e.g. Hubbell 1997, 2001) as they lack any explicit reference to generating mechanisms, although some have been proposed (e.g. Kendall 1948; MacArthur 1960; May 1975; Caswell 1976). Further, the statistical assessment of the fit provided by the log-series and log-normal models to different data sets has produced mixed results (e.g. Hughes 1986; Magurran 1988). However, beyond these problems is the critical issue about the assumptions they make regarding their derivation and in particular on the nature of the parent distribution or universe from where samples were taken. Both Fisher and Preston assumed that the universe corresponded to a humped distribution with either a zero (log-normal) or positive skew (gamma). Only recently have high-quality data on the abundance of species at large spatial scales become available, mostly for birds in North America, Britain and other European countries. This has allowed the possibility to access the 'universes' that Preston and Fisher assumed in their models. One of the first analyses of this issue was carried out by Nee *et al.* (1991a). These authors, using data on the abundance of British birds, reported that Preston's unveiled distribution was not log-normal but distinctly and significantly left skewed. However, a subsequent analysis by Gregory (1994) of the same but updated data base confirmed the left skew of the distribution but showed that this was not different from zero, implying that it could arise if we assume that the abundance of birds represents independent random draws from a log-normal distribution. A recent analysis by Gregory (2000) confirms the pattern of negatively skewed distributions in most European countries (40 out of 48, and significantly so in 10) and for both resident and migrant species, although this was less frequent in the latter (see also Gaston *et al.* 1997a). In Fig. 5.3a, we show the same pattern but for North American birds at the continental level, and at the level of states within North America (Fig. 5.3b–f). In all cases we used data from the Breeding Bird Survey (Peterjohn 1994). The distributions shown in Fig. 5.3 are significantly left skewed (one-tailed *t*-test $P < 0.05$), except for the states of Alaska and British Columbia.

The evidence discussed above suggests that the universe might not be as originally assumed by Preston (1948) or by Fisher *et al.* (1943), implying that there are many more rare species than previously believed, substantiating the claim that most species are rare. But how can we explain this pattern? Several explanations have been proposed in the literature, starting with Whittaker (1965) and the McNaughton & Wolff (1970) dominance model, and continuing with niche models (Tokeshi 1996, 1999), neutral models (Hubbell 1997; Bell 2000), self-similar models (Harte *et al.* 1999) and spatially explicit models (Chave *et al.* 2002). In what follows we will focus mostly on how niche models can bear on this pattern to show later how these models can be framed so as to be consistent with neutral models focused on birth–death processes in a spatially explicit framework.

Niche models

Niche models have a long tradition in ecological sciences, and several ecologists have devoted a large research effort, attracted by the beauty of their simplicity, whereas others have been discouraged by their restrictive assumptions.

The first model known to western scientists was MacArthur's (1957) broken-stick model, the immediate ancestor of which is found in Motomura's (1932) geometric series model (Fig. 5.1; see Whittaker 1965). Niche models simplify reality by considering that the environment, or niche space available for a group of species, can be represented by a stick of unit length, which can be broken following a particular rule. Once the stick is broken, subsequent species invading the local community must choose among the pieces available (following a choosing rule) and break it again. It is assumed that the length of the segment allocated to each species is proportional to its abundance within the community. Depending on the rule followed for choosing and breaking and on its nature (deterministic or stochastic), several models can be identified (for a review see Tokeshi 1993, 1997, 1999).

Whereas Motomura's model entails the fixed and sequential division of niche

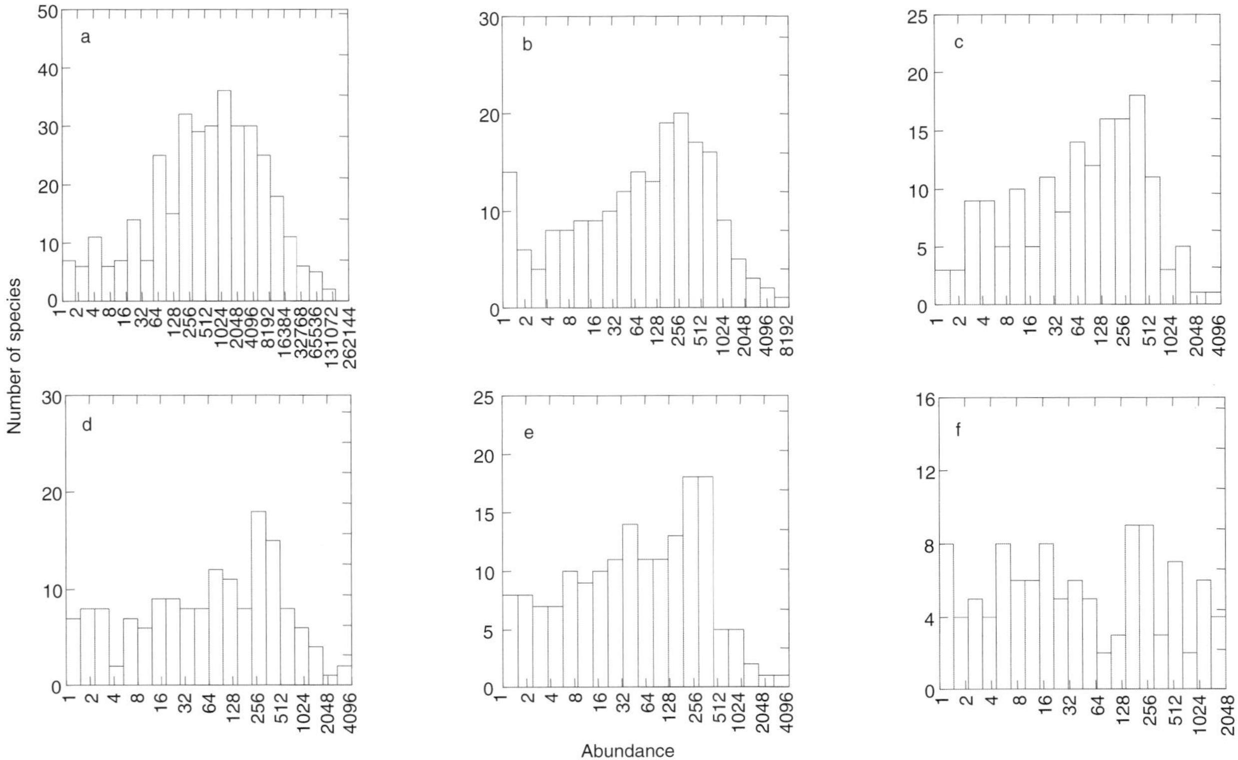

Figure 5.3 Distribution of abundance for 352 species of North American birds censused in June 2000 along 2366 routes of the Breeding Bird Survey (a) and State-level analysis of the distribution of abundance for 273 species of North American birds censused in June 2000 along 331 routes, in the States of California (b), Oregon (c), Washington (d), British Columbia (e) and Alaska (f).

space, such that each species invading the community takes a fraction k of the unused niche space, MacArthur's broken-stick model (MacArthur 1957) envisages a simultaneous process such that the stick of unit length is simultaneously partitioned at $n - 1$ points thrown at random. A similar model was later proposed by Sugihara (1980), however, this model was sequential and multidimensional, and purported to represent the breakage of several different niche axes. In testing his model, Sugihara (1980) showed that random selection of sticks and sequential breakages following a 0.75:0.25 proportionality instead of being random (Sugihara's fixed fraction model, SFF), fit the data better than MacArthur's model and also lead to a canonical log-normal distribution of abundances, which is not obtained if breakages are random in the interval (0,1). Sequential breakage, as originally envisaged by Motomura (1932), provides a more sensible analogy for how communities are built up, envisaging the successive carving out of niche space by species. A further addition to sequential niche models was introduced by Tokeshi (1990). This author recognizes six additional sequential models depending on how sticks are chosen and subsequently broken. Among them are the random fraction model (RF), where both processes are random, and the MacArthur fraction model (MF), which is the sequential version of the original broken-stick model. In this last model, the breakage is random but the stick selection probability is positively related to stick length. As shown by Tokeshi (1996), the RF and MF models are extremes of a more general model, the power fraction (PF) model, where the probability of stick selection is proportional to the sizes of existing niche fragments raised to the exponent k. A PF model with $k = 0$ corresponds to the RF model, whereas when $k = 1$ it corresponds to the MF model.

Niche models assume a zero-sum game, where the total number of individuals of all species together is essentially constant, such that increases in the population of some species result in corresponding decreases of others (MacArthur 1960). This means that most abundance distributions show a negative skew (Fig. 5.4), which is most apparent for the MF model, where rare species are more likely to accumulate because of the biased stick selection process it entails. In general, asymmetric segment division tends to exacerbate the negative skewness (Novotny & Drozd 2000).

Thus, negatively skewed distributions can be explained easily by niche models. As shown by Tokeshi (1996), the PF model with $k = 0.05$ gives rise to distributions that do not differ from that observed for British birds, and for k values in the range 0–0.2 the model fits data from a wide variety of species-rich assemblages. However, what is the precise biological meaning of the parameter k? Before attempting to answer this question we will examine the validity of one of the central assumptions of niche models. That is, that the apportionment of resources (stick fragments) is equivalent to the apportionment of individuals (abundance).

How do species divide resources?

Sugihara (1980) clearly established the main assumption of niche apportionment models (p. 773) '. . . because a niche translates ultimately into number of organisms (or biomass), observed abundance patterns can offer a useful standard measure of

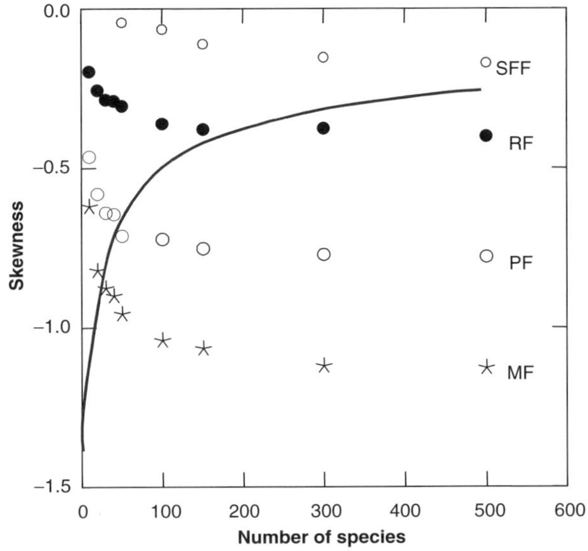

Figure 5.4 Changes in average skewness with number of species. Each value corresponds to the average calculated over 500 simulations for the random fraction (RF), MacArthur's fraction (MF), Sugihara's fixed fraction (SFF) and power fraction (PF, $k = 0.5$) models. The solid line shows the critical value ($\alpha = 0.05$) of skewness.

niches . . .' As is clear from this statement, niche models assume that the abundance of a species is a reflection of the amount of limiting resources it controls, such that knowledge of how resources are divided among species within communities would allow us to predict abundance patterns. The first test of this assumption was attempted by Harvey & Godfray (1987). These authors point out that Sugihara's model assumes that the way individuals and resources are apportioned to communities are equivalent processes, which would work if individuals of all species are equivalent. However, most species are likely to differ in body mass, which scales with both population density and per capita resource requirements such that both processes might not be equivalent. They notice that if population density scales with body mass (e.g. Damuth 1981, Peters 1983) as

$$N \propto M^{-x} \tag{5.1}$$

and that according to Kleiber's law per individual energy requirements (P) increase with body mass as

$$P \propto M^{y} \tag{5.2}$$

then substituting equation (5.1) into equation (5.2), and considering that according to (5.1) $M \propto N^{-\frac{1}{x}}$ per individual energy requirement can be expressed as

$$P \propto \left(N^{-\frac{1}{x}}\right)^{y} \propto N^{-\frac{y}{x}} \quad (5.3)$$

Then, the amount of energy used per unit area by a population ($E \propto P \times N$) would be equal to $E = c \times N^{\left(1-\frac{y}{x}\right)}$, where c is a normalization constant. Therefore

$$\log(E) = \log(c) + \left(1 - \frac{y}{x}\right)\log(N) \quad (5.4)$$

If $\log(N)$ is normally distributed across species, then $\log(E)$ also will be, and its variance will be

$$V(\log E) = \left(1 - \frac{y}{x}\right)^2 V(\log N) \quad (5.5)$$

From equation (5.5) it is apparent that if $x \approx y$ then $V(\log E)$ will tend to zero, and the expected value of $\log(E)$ will be constant and independent of population density (as has been reported under the Energetic Equivalence rule; Damuth 1987; Nee *et al.* 1991b; Marquet *et al.* 1995; Enquist *et al.* 1998). Further, as x is less than $2y$ in most studies, then $V(\log E)$ will be lower (more equitable) than $V(\log N)$. Hence, from this perspective the apportionment of individuals and resources might not be equivalent processes, which is the main assumption of niche models. However, as pointed out by Sugihara (1989) these authors assume that the allometric equations they used had no error variance (no scatter), implying that this has no impact upon the total variance estimate for biomass and E. Interestingly, once error variance is considered, error propagation also becomes a problem. As demonstrated by Taper & Marquet (1996) the estimates of $V(\log E)$ derived by Harvey & Godfray (1987), Sugihara (1989) and later by Pagel *et al.* (1991) were all derived under the implicit assumption that population density influences body size, which in turn influences metabolic rate and consequently population energy use. Taper & Marquet (1996) show that if we solve for M in equation (5.2) and substitute it to equation (5.1) we arrive at an alternative expression for $V(\log E)$, but this time assuming that P causes or determines M, which in turn determines or causes N. These authors identify three potential models, with different expressions for $V(\log E)$, and specify different pathways of error propagation, which they label N causal, P causal and M causal models.

Although there is no objective way of deciding which model best describes reality, traditionally body size has been assigned a causal role in the study of a variety of physiological, ecological and evolutionary phenomena (e.g. Peters 1983; Calder 1984; Brown & Maurer 1986; Brown *et al.* 1993), thus it seems biologically reasonable to use the M causal model to assess if the apportionment of individuals and resources are equivalent processes.

Under this model $V(\log E)$ is expressed as

$$V(\log E) = \left(1 + \frac{y}{x}\right)^2 [V(\log N) - V(d_n)] + V(d_n) - V(d_p) \quad (5.6)$$

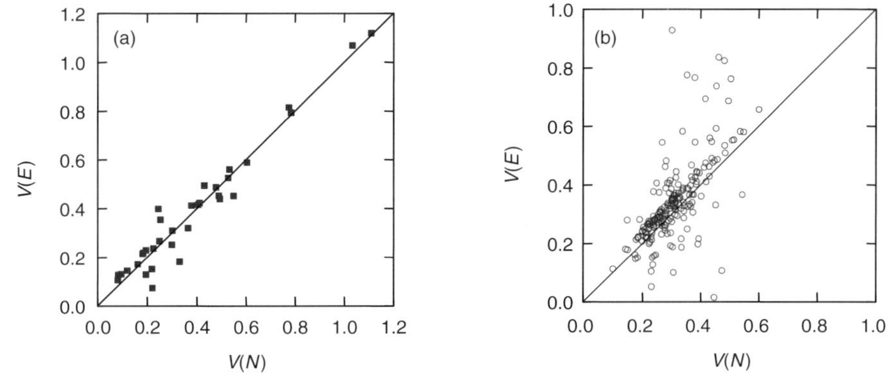

Figure 5.5 Relationship between the variance in population energy use $V(E)$ and variance in population density $V(N)$. (a) Relationship calculated for 41 bird communities reported by Brown & Maurer (1986) and recalculated from Taper & Marquet (1996) using $V(d_p) = 0.03$. (b) Relationship calculated for 204 Breeding Bird Survey routes that satisfied the criteria of having more than 20 species with a relationship between body size and abundance of $R^2 > 0.1$.

where $V(d_n)$ and $V(d_p)$ are the error variances associated with equations (5.1) and (5.2) respectively. The analysis of 41 local bird communities compiled by Brown & Maurer (1986) shows an almost perfect agreement between $V(\log E)$, estimated by equation (5.6), and $V(\log N)$ (see Fig. 5.5a). Further, an expanded analysis using the data compiled by the Breeding Bird Survey for those routes in the states of California, Oregon, Washington, Alaska and British Columbia (Fig. 5.5b) shows that there is a good correlation between both variables ($r = 0.66$, $P < 0.001$, $n = 204$), despite the fact that these data are considerably less accurate than those compiled by Brown & Maurer (1986). Thus, based on this evidence we can conclude that the apportionment of individuals and resources are equivalent processes, which validates the central assumption of niche models. However, their usefulness ultimately will depend on the biological plausibility of the sequential apportionment process they entail. This returns to the question of the meaning of k.

On the meaning of *k*

According to Tokeshi (1996), k has an ecological and evolutionary meaning. In ecological terms it implies that communities are built up by a sequential niche apportionment process, whereby the probability of niche division or invasion of existing niche space is slightly higher for species with larger niches or higher abundance. Evolutionarily, k is related to the probability of speciation, such that $k > 0$ means that species with higher abundance, which also tend to have larger geographical ranges (see review by Gaston *et al.* 1997b; Gaston & Blackburn 2000), are more likely to generate new species. Interestingly, the distribution of range sizes is also left-skewed after logarithmic transformation, which suggests a potential connection between both patterns (Gaston 1998; Gaston & Blackburn 2000; Gaston & He 2002). However, although this argument is plausible, because larger ranges are more likely to be intercepted by barriers, the evidence at best is equivocal and depends critically on assumptions about the distribution and type of barrier sizes (Gaston 1998).

In any case, is not obvious that the PF model can account for the positive correlation between distribution and abundance. To see this, assume that we have a large number of communities in a large geographical area, and that each species that invades a community will reach a local abundance as dictated by the PF model. In this scenario, for a positive relationship between range size and abundance to arise would require an almost perfect spatial synchronization in the sequence of invasion, such that each species invades all local communities at about the same time, and the existence of a lower threshold of abundance such that smaller sticks are associated with higher probability of local extinction. Clearly, such a process would entail the use of spatial models incorporating the dynamic processes of invasion and extinction. In this regard, metapopulation models (e.g. Levins 1969; Hanski 1999) represent a natural point of departure.

Breaking the stick in space

Niche models emerged as a statistical approach to understand patterns of abundance in samples of local communities. The original emphasis was on proposing mechanisms of resource division that might account for the observed patterns. However, since their inception in ecology, much emphasis has been put on testing how adequately the models fit reality (e.g. Tokeshi 1990; Naeem & Hawkins 1994; Bersier & Sugihara 1997) and less on trying to understand the biological meaning of their parameters and underlying assumptions. In particular, the PF model proposed by Tokeshi (1996) is flexible enough to be able to fit most of the observed patterns in abundance by varying the parameter k, the meaning of which, as already discussed, is not yet clear. However, in addition to this constraint on explanatory power and biological relevance, available evidence indicates that the basic assumption of equivalence between the processes of apportionment of individuals and resources holds. As we will show below, this fact permits a link between resource acquisition, resource conversion and space occupation processes. As for niche models, there is still much room for improvement. Among other things, niche models are static, the number of species is not a result of the model, but a free parameter specified by the investigator, and because they are not spatial it does not matter how species arrive and persist in local communities, hence they do not incorporate any mechanistic link with the basic processes of fecundity and mortality (Hubbell 2001). Interestingly, metapopulation models of the patch-occupancy type (Levins 1969) have some key ingredients that might serve the purpose of making niche models dynamic and linked to species life-history attributes. On the other hand, multispecies metapopulation models (i.e. metacommunity models) might take advantage of the simplicity of the niche apportionment process to generate within-patch dynamics in abundance.

Metapopulation models, of the patch-occupancy type, assume that the environment is composed of a large number of identical patches that either can be empty or occupied by a local population. In these models, the proportion of patches (p) that a species occupies in a given area is driven by the rates of colonization (f) and local extinction (e)

$$\frac{dp}{dt} = fp(1-p) - ep \tag{5.7}$$

At equilibrium, the proportion of occupied patches is given by $\hat{p} = 1 - \frac{e}{f}$, thus, a positive proportion of occupied patches at equilibrium will exist whenever $\frac{f}{e} > 1$. Taking into account that the average lifetime of an occupied patch is $\frac{1}{e}$, the above expression represents the total number of secondary colonizations produced by an occupied patch during its lifetime. This threshold parameter corresponds to the basic reproductive number used in epidemiology (usually denoted by the symbol R_0), where it measures the number of secondary infections that a single infectious individual produces when introduced in a completely susceptible population (Diekman *et al.* 1990; Hernández-Suárez *et al.* 1999). If this number is greater than 1, the disease spreads in the host population. The basic reproductive number is therefore an invasion criterion: it determines if a pathogen will be able to survive in a host population once it is introduced. In a metapopulation context, it determines if a landscape composed of a set of empty patches will be colonized successfully, and also determines its long-term persistence (Marquet & Velasco-Hernández 1997). To appreciate the importance of R_0 in affecting metapopulation persistence, we can rescale time in equation (5.7) by taking as a unit the average time to extinction $\frac{1}{e}$. With this rescaling, Levin's original model becomes

$$\frac{dp}{d\tau} = R_0 p(1-p) - p \tag{5.8}$$

where τ stands for the new rescaled time (Hernández-Suárez *et al.* 1999). It is clear in equation (5.8) that if $R_0 < 1, p \rightarrow 0$ and if $R_0 > 1, p \rightarrow 1$.

In this model, if the site can accommodate only one individual, p becomes the regional abundance of a species (Tilman 1994), the colonization rate f becomes the rate of propagule production by one individual or fecundity, and the extinction rate becomes an individual mortality rate (m). As before, the equilibrium abundance for a species with a given f and m becomes $\hat{p} = 1 - \frac{m}{f}$. Multispecies versions of this model incorporating competitive interactions among species, as well as their spatial analogues, have been widely analysed in the literature (e.g. Hastings 1980; Tilman 1994; Kinzig *et al.* 1999). As a point of departure let us analyse the simplest multispecies model, one where species compete for space, such that once an individual of species i arrives at a site it makes it unavailable for others until it dies.

This model can be written as

$$\frac{dp_i}{dt} = f_i p_i \left(1 - \sum_{j \neq i} p_j - p_i\right) - m_i p_i \tag{5.9}$$

The first term on the right-hand side gives the colonization of empty habitat patches (i.e. individuals) and the last term the extinction of habitat patches. This mean-field equation can be converted easily to a spatially explicit model, which we will call the basic contact process with no-sticks. First, however, it should be noticed that in this model the equilibrium abundance of species will depend on their $R_0 = \frac{f}{m}$, but weighted by the availability of resources (empty space); we will call this weighted R_0 the realized R_0. To see this we can rescale equation (5.9) (see Kinzig *et al.* 1999; Dushoff *et al.*, in press) by defining the quantity $\theta = 1 - \sum_{j \neq i} \hat{p}_j$ as the number of sites not occupied by species other than *i* after each species *j* has reached its equilibrium abundance ($\hat{p}_j$).

Using this quantity the equilibrium abundance of species *i* can be expressed as

$$\hat{p}_i = \left(\frac{\theta R_0 - 1}{R_0}\right)$$

Unlike equation (5.7), now the equilibrium abundance is a function of the species life-history and the abundance of resources in the environment, such that persistence will be possible only if the condition $\theta > \frac{1}{R_0}$ is satisfied. This relationship between life-history attributes, resource availability and persistence is essential to forge a link between the resource apportionment process and birth–death processes.

To construct a spatial version of this model let us assume a regular homogeneous lattice of 256 × 256 sites in a torus such that each lattice site has the same number of neighbours. Each site can accommodate one individual, and can be either occupied or empty, such that the number of occupied sites translates to regional abundance (*N*). The model dynamics are given by the species fecundity (f) and mortality (m) rates. Fecundity rate is the expected number of birth events by an individual during a unit time interval. Newborns are sent to nearest-neighbour sites with equal probability, but establish only if the landing site is empty. Similarly, each individual dies at a constant rate (i.e. the expected number of death events in a unit time interval). Thus species follow a contact process type of dynamics (Durret & Levin 1994a,b). The model is initialized by randomly seeding the landscape with *n* species, each in 0.1% of the sites; no further seeding is allowed.

As is well-known for these models (Neuhauser 1992), in the long-term, the species with the largest R_0 will reach the highest abundance, and eventually displace all others, achieving complete occupancy if its reproductive number is greater than 1. However, the transient dynamics towards this equilibrium can be very long and give rise to left-skewed distributions of abundance (Fig. 5.6). Thus, under non-equilibrium conditions this model predicts distributions of abundance qualita-

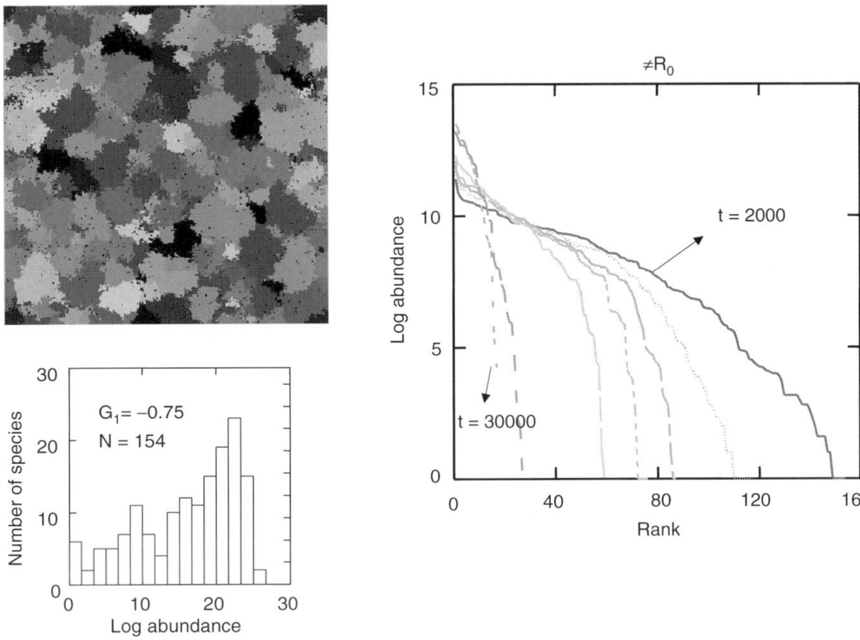

Figure 5.6 Transient dynamics in the spatially explicit pre-emptive competition model run in a lattice of 256 × 256 sites. The model was started with 10 000 species randomly seeded in 1% of sites. Each species was randomly assigned a fecundity and mortality value drawn from a uniform distribution between 0 and 1. The resulting transient spatial pattern is shown on the upper left panel for $t = 2000$, where different intensities of grey correspond to different species, and empty cells, owing to the death of an individual, are shown in black. The panel in the upper right shows the rank–abundance plots for different time-steps. The lower left panel shows the distribution of abundance for 154 species for the time-step = 2000. Notice that the distribution is strongly left-skewed, as was also observed for all other time-steps.

tively compatible with empirical data as a result of all but one species going randomly extinct in time. A similar pattern emerges when all species have equal R_0 (Fig. 5.7; $f = 0.7$, $m = 0.2$), with the important distinction that in this case any species can win and the dynamics is a random walk towards either extinction or occupation of all available space, which is proportional (but not exactly equivalent, because of spatial correlations) to $\hat{p} = 1 - \frac{m}{f}$. Under these conditions species follow a zero-sum ecological drift (Hubbell 2001).

This model is suitable for understanding the dynamics of species undergoing competition for space in a closed community (i.e. without immigration from outside the system). Our next step will be to build a metacommunity model of species competing for food resources, as assumed in traditional niche models. Our aim here

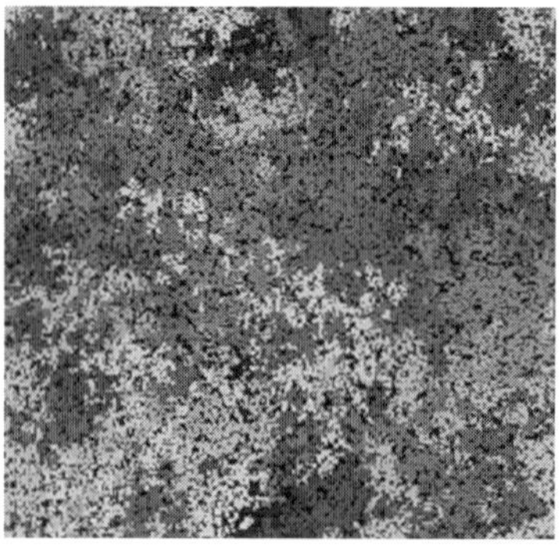

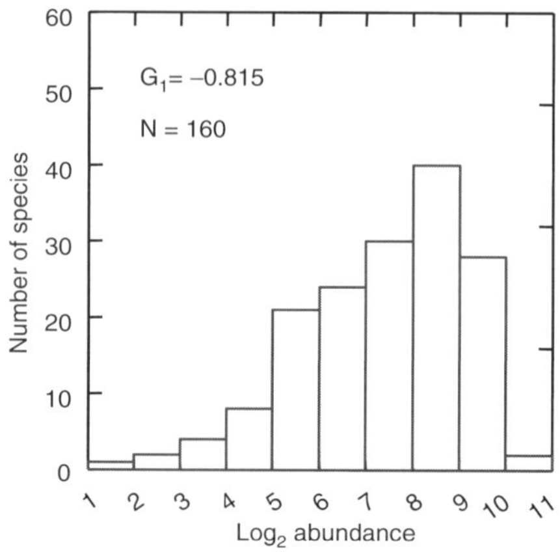

Figure 5.7 Same as in Fig. 5.6 but this time under conditions of ecological equivalence where each species had the same fecundity ($f = 0.7$) and mortality ($m = 0.2$). The results shown correspond to the time-step = 3500.

is to link niche models with those based on birth–death processes (e.g. Hubbell 2001).

Our model (Keymer & Marquet, unpubl.) links together the acquisition of energetic resources (E_1) with the appropriation of space through a simple transformation inspired in a model originally developed by Brown *et al.* (1993), which envisages organisms as operating a transformation on the environment

$$E_1 + I_0 \rightarrow I_1 \tag{5.10}$$

$$I_1 \rightarrow I_0 + b \tag{5.11}$$

$$b + S \rightarrow I'_0 \tag{5.12}$$

Under this model, an organism without resources for reproduction (I_0) acquires them from the environment, forming an unstable individual–resources complex (I_1) (equation 5.10) and decaying back to an individual without resources but using the captured energy to make a potential copy of itself (b) (equation 5.11) at a rate f which will later react with a position in space (S) to become a full self-reproducing copy (I'_0).

This framework allows us to couple resource acquisition processes (niche breakage) with space occupation. Assuming equal mortality and longevity and no resource limitation, the intrinsic reproductive number of species i becomes $R_{0,i} = f_i$. However, if resources are in fact limiting (i.e. the length of the stick is finite as assumed in niche models) the reproductive number of species i will be given by the amount of resources it can capture (stick length, x_i). In turn for each species i we can define a realized reproductive number $R_{0,i}$ that will be a function of the stick length

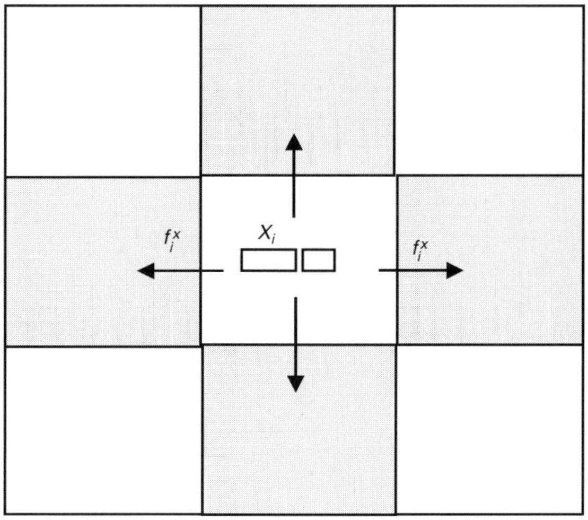

Figure 5.8 Schematic representation of the lattice model.

$$R_{0,i}(x_i) = R_i \times x_i \tag{5.13}$$

Keymer & Marquet (unpubl.) describe in detail the spatial model resulting from letting this process unfold in space. Here we will provide a summary description.

Consider a two-dimensional regular lattice, wherein each cell represents a local community holding a stick of unit length to be apportioned among invading species (Fig. 5.8). Once a species arrives it will break the stick (according to Tokeshi's (1996) PF model with parameter k), appropriating a fraction x_i of available resources and producing propagules according to equation (5.13). Each propagule will disperse randomly to one of its four local neighbours (contact process-like) and invade the community located therein. At each local community populations can become extinct at a constant rate μ, and the released stick is proportionally reassigned among the locally extant populations. Under the assumption of ecological equivalency (equal $R_{0,i}$) and $\mu > 0$ the system has a long transient and eventually one species reaches total dominance for any k. In the transient phase, distributions of abundance are left-skewed (Fig. 5.9) as would be expected if species randomly walk to extinction. However, when $\mu = 0$ and $k = 0$ (RF model), species do not go extinct but maintain a statistical steady state and give rise to symmetrical log-normal distributions of abundance (Fig. 5.10a). A similar situation occurs for $k = 1$ (MF model), but this time the distribution is left-skewed (Fig. 5.10b) as a consequence of a biased stick-breakage process.

Spatial niche apportionment processes can generate a wide spectrum of distributions of abundance, some of which are in agreement with the patterns observed for the non-spatial models. This especially is evident when there is no local extinction of populations and under the assumption of ecological equivalency. When $\mu > 0$, the

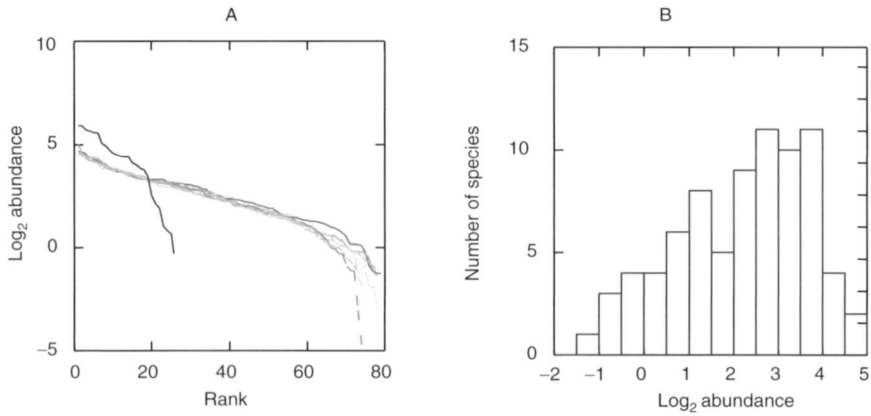

Figure 5.9 Results of the spatially explicit stochastic niche model for 100 species in a system of 1024 local communities (i.e. 32 × 32 lattice) with $k = 1$, which corresponds to the MacArthur fraction niche apportionment model. Local extinction probability was $\mu = 0.01$, and fecundity $f_i = 1$. The rank–abundance plot (A) is shown for the model at different time-steps (10, 20, 40, 60, 80, 100 and 160). The number of species present in the system decreases in time owing to extinction. The frequency distribution of abundance (B), as shown for $t = 60$, was left-skewed for all time-steps.

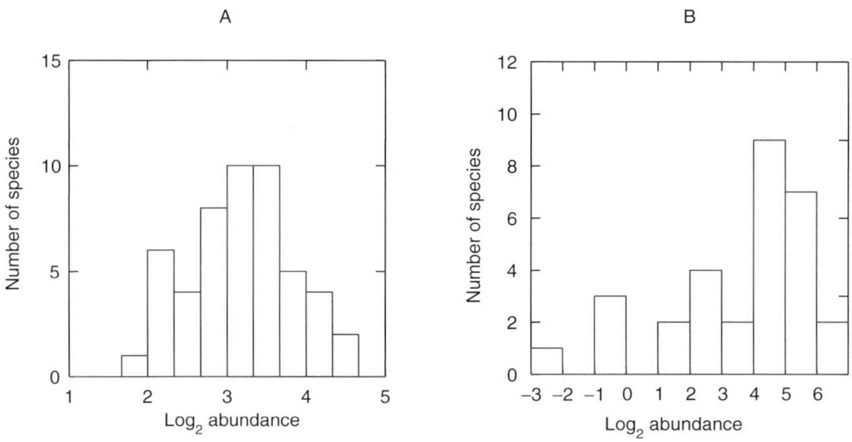

Figure 5.10 The same as in Fig. 5.9, but for 50 species with $\mu = 0$, and (A) $k = 0$ and (B) $k = 1$.

dynamics do not reach an attractor and one species eventually wins. Local extinction allows the liberation of resources and their capitalization by those species that already hold large sticks, a process that in the long term leads to a random walk to extinction for most species and the dominance of one. These simple spatially explicit stochastic models allow us to put traditional niche models in a dynamic context, linking resource acquisition with the fundamental processes of birth and death, thus showing that niche models can be fruitfully expanded to account for community patterns in the distribution of abundance. Although the patterns that emerge qualitatively resemble empirical data, our emphasis has been on providing a framework for reconciling these two divergent perspectives on the emergence of patterns. We showed that this is possible by linking resource and individual apportionment processes in a spatial context.

Macroecology was originally defined by Brown & Maurer (1989) as the search for patterns in the division of food and space among species at continental scales. Although in this definition it is debatable that macroecology is restricted only to the analysis of patterns and processes operating at the level of continents, the way in which organisms divide food and space resources seems to be fundamental to our understanding of the organization of ecological systems at any scale, and provides the basis for connecting local-scale phenomena, linked to niche apportionment processes, with patterns emerging at the level of metacommunities.

Acknowledgements

We gratefully acknowledge support from Grant FONDAP-FONDECYT 1501–0001. Comments by Kevin Gaston and an anonymous reviewer were greatly appreciated.

References

Bell, G. (2000) The distribution of abundance in neutral communities. *American Naturalist* **155**, 606–617.

Bersier, L.F. & Sugihara, G. (1997) Species abundance patterns: the problem of testing stochastic models. *Journal of Animal Ecology* **66**, 769–774.

Borda-de-Agua, L., Hubbell, S.P. & McAllister, M. (2002) Species–area curves, diversity indices, and species abundance distributions: a multifractal model. *American Naturalist* **159**, 138–155.

Brian, M.V. (1953) Species frequencies in random samples from animal populations. *Journal of Animal Ecology* **22**, 57–64.

Brown, J.H. (1995) *Macroecology*. University of Chicago Press, Chicago.

Brown, J.H. (1999) Macroecology: progress and prospect. *Oikos* **87**, 3–14.

Brown, J.H. & Maurer, B.A. (1986) Body size, ecological dominance and Cope's rule. *Nature* **324**, 248–250.

Brown, J.H. & Maurer, B.A. (1989) Macroecology: the division of food and space among species on continents. *Science* **243**, 1145–1150.

Brown, J.H., Marquet, P.A. & Taper, M.L. (1993) Evolution of body size: consequences of an energetic definition of fitness. *American Naturalist* **142**, 573–584.

Brown, J.H., Mehlman, D.W. & Stevens, G.C. (1995) Spatial variation in abundance. *Ecology* **76**, 2028–2043.

Calder, W.A. (1984) *Size, Function and Life History*. Harvard University Press, Cambridge, Mass.

Caswell, H. (1976) Community structure: a neutral model analysis. *Ecological Monographs* **46**, 327–354.

Chave, J., Muller-Landau, H.C. & Levin, S.A. (2002) Comparing classical community models: theoretical consequences for patterns of diversity. *American Naturalist* **159**, 1–23.

Chu, J. & Adami, C. (1999) A simple explanation for taxon abundance patterns. *Proceedings of the National Academy of Sciences, USA* **96**, 15017–15019.

Cohen, J.E. (1968) Alternate derivations of a species–abundance relation. *American Naturalist* **102**, 165–172.

Cowan, G.A., Pines, D. & Meltzer, D. (1994) *Complexity. Metaphors, Models, and Realities*. Perseus Books, Reading, MA.
Damuth, J. (1981) Population density and body size in mammals. *Nature* **290**, 699–700.
Damuth, J. (1987) Interspecific allometry of population density in mammals and other animals: the independence of body mass and population density. *Biological Journal of the Linnean Society* **37**, 193–246.
Dewdney, A.K. (1997) A dynamical model of abundances in natural communities. *Coenoces* **12**, 67–76.
Dewdney, A.K. (1998) A general theory of the sampling process with applications to the 'Veil Line'. *Theoretical Population Biology* **54**, 294–302.
Diekman, O., Hesterbeek, J.A.P. & Metz, J.A.J. (1990) On the definition and computation of the basic reproductive ratio R_0 in models for infectious diseases in heterogeneous populations. *Journal of Mathematical Biology* **28**, 365–382.
Diserud, O.H. & Engen, S. (2000) A general and dynamic species abundance model, embracing the lognormal and the gamma models. *American Naturalist* **155**, 497–511.
Durrett, R. & Levin, S.A. (1994a) The importance of being discrete (and spatial). *Theoretical Population Biology* **46**, 361–394.
Durrett, R. & Levin, S.A. (1994b) Stochastic spatial models: a user's guide to ecological applications. *Philosophical Transactions of the Royal Society, London, Series B* **343**, 329–350.
Dushoff, J., Worden, L., Keymer, J. & Levin, S.A. (in press) Scale invariance in aspect space and community assembly. *Theoretical Population Biology*.
Engen, S. & Lande, R. (1996) Population dynamic models generating the lognormal species abundance distribution. *Mathematical Biosciences* **132**,169–183.
Enquist, B.J., Brown, J.H. & West, G.B. (1998) Allometric scaling of plant energetics and population density. *Nature* **395**, 163–165.
Fisher, R.A., Corbet, A.S. & Williams, C.B. (1943) The relation between the number of species and the number of individuals in a random sample from an animal population. *Journal of Animal Ecology* **12**, 42–58.
Frontier, S. (1985) Diversity and structure of the benthos in aquatic ecosystems. *Oceanographic Marine Biology Annual Review* **23**, 253–312.
Gaston, K.J. (1998) Species–range size distributions: products of speciation, extinction and transformation. *Philosophical Transactions of the Royal Society, London, Series B* **353**, 219–230.
Gaston, K.J. & Blackburn, T.M. (1999) A critique for macroecology. *Oikos* **84**, 353–368.
Gaston, K.J. & Blackburn, T.M. (2000) *Pattern and Process in Macroecology*. Blackwell Science, Oxford.
Gaston, K.J., Blackburn, T.M. & Gregory, R.D. (1997a) Abundance–range size relationships of breeding and wintering birds in Britain: a comparative analysis. *Ecography* **20**, 569–579.
Gaston, K.J., Blackburn, T.M. & Lawton, J.H. (1997b) Interspecific abundance–range size relationships: an appraisal of mechanisms. *Journal of Animal Ecology* **66**, 579–601.
Gaston, K.J. & He, F. (2002) The distribution of species range size: a stochastic process. *Proceedings of the Royal Society, London, Series B* **269**, 1079–1086.
Gregory, R. (1994) Species abundance patterns of British birds. *Proceedings of the Royal Society, London, Series B* **257**, 299–301.
Gregory, R. (2000) Abundance patterns of European breeding birds. *Ecography* **23**, 201–208.
Hanski, I. (1999) *Metapopulation Ecology*. Oxford University Press, Oxford.
Harte, J., Kinzig, A. & Green, J. (1999) Self-similarity in the distribution and abundance of species. *Science* **284**, 334–336.
Harvey, P.H. & Godfray, H.C.J. (1987) How species divide resources. *American Naturalist* **129**, 318–320.
Hastings, A. (1980) Disturbance, coexistence, history, and competition for space. *Theoretical Population Biology* **18**, 363–373.
Hernández-Suárez, C.M., Marquet, P.A. & Velasco-Hernández, J.X. (1999) On threshold parameters and metapopulation persistence. *Bulletin of Mathematical Biology* **60**, 1–14.
Hubbell, S.P. (1997) A unified theory of biogeography and relative species abundance and its application to tropical rain forest and coral reefs. *Coral Reefs* **16**(Supplement), S9–S21.
Hubbell, S.P. (2001) *The Unified Neutral Theory of Biodiversity and Biogeography*. Princeton University Press, Princeton, NJ.
Hughes, R.G. (1984) A model of the structure and dynamics of benthic marine invertebrate com-

munities. *Marine Ecology Progress Series* **15**, 1–11.

Hughes, R.G. (1986) Theories and models of species abundance. *American Naturalist* **128**, 879–899.

Keitt, T.H. & Stanley, E.H. (1998) Dynamics of North American breeding bird populations. *Nature* **393**, 257–260.

Kempton, R.A. & Taylor, L.R. (1974) Log-series and log-normal parameters as diversity discriminants for the Lepidoptera. *Journal of Animal Ecology* **43**, 381–399.

Kendall, D.G. (1948) On some modes of population growth leading to R.A. Fisher's logarithmic series distribution. *Biometrika* **35**, 6–15.

Kinzig, A.P., Levin, S.A., Dushoff, J. & Pacala, S. (1999) Limiting similarity, species packing, and system stability for hierarchical competition–colonization models. *American Naturalist* **153**, 371–383.

Levin, S.A. (1999) *Fragile Dominion. Complexity and the Commons.* Perseus Books, Reading, MA.

Levins, R. (1969) Some demographic and genetic consequences of environmental heterogeneity for biological control. *Bulletin of Entomological Society of America* **15**, 237–240.

MacArthur, R.H. (1957) On the relative abundance of species. *Proceedings of the National Academy of Sciences, USA* **43**, 283–295.

MacArthur, R.H. (1960). On the relative abundance of species. *American Naturalist* **94**, 25–36.

Magurran, A.E. (1988) *Ecological Diversity and its Measurement.* Princeton University Press, Princeton, NJ.

Marquet, P.A. (2002) The search for general principles in ecology. *Nature* **418**, 723.

Marquet, P.A., Navarrete, S.A. & Castilla, J.C. (1990) Scaling population density to body size in rocky intertidal communities. *Science* **250**, 1125–1127.

Marquet, P.A., Navarrete, S.A. & Castilla, J.C. (1995) Body size, population density and the energetic equivalence rule. *Journal of Animal Ecology* **64**, 325–332.

Marquet, P.A. & Velasco-Hernández, J.X. (1997) A source–sink patch occupancy metapopulation model. *Revista Chilena de Historia Natural* **70**, 371–380.

Maurer, B.A. (1999) *Untangling Ecological Complexity: the Macroscopic Perspective.* University of Chicago Press, Chicago.

May, R.M. (1975) Patterns of species abundances and diversity. In: *Ecology and Evolution of Communities* (eds M.L. Cody & J.M. Diamond), pp. 81–120. Belknap Press of Harvard University Press, Cambridge, MA.

May, R.M. & Nowak, M.A. (1994) Superinfection, metapopulation dynamics, and the evolution of diversity. *Journal of Theoretical Biology* **170**, 95–114.

McNaughton, S.J. & Wolf, L.L. (1970) Dominance and the niche in ecological systems. *Science* **167**, 131–139.

Motomura, I. (1932) On statistical treatment of communities. *Japanese Journal of Zoology* **44**, 379–383. (In Japanese.)

Mouillot, D., Leprêtre, A., Andrei-Ruiz, M.C. & Viale, D. (2000) The fractal model: a new model to describe the species accumulation process and relative abundance distribution (RAD). *Oikos* **90**, 333–342.

Naeem, S. & Hawkins, B.A. (1994) Minimal community structure: how parasitoids divide resources. *Ecology* **75**, 79–85.

Nee, S., Harvey, P.H. & May, R.M. (1991a) Lifting the veil on abundance patterns. *Proceedings of the Royal Society, London, Series B* **243**, 161–163.

Nee, S., Read, A.F., Greenwood, J.J.D. & Harvey, P.H. (1991b) The relationship between abundance and body size in British birds. *Nature* **351**, 312–313.

Neuhauser, C. (1992) Ergodic theorems for the multitype contact process. *Probability Theory and Related Fields* **91**, 467–506.

Novotny, V. & Drozd, P. (2000) The size distribution of conspecific populations: the peoples of New Guinea. *Proceedings of the Royal Society, London, Series B* **267**, 947–952.

Pachepsky, E.J., Crawford, W.J.L., Brown, J.L. & Squire, G. (2001) Towards a general theory of biodiversity. *Nature* **410**, 923–926.

Pagel, M.D., Harvey, P.H. & Godfray, H.C.J. (1991) Species-abundance, biomass and resource-use distributions. *American Naturalist* **138**, 836–850.

Peterjohn, B.G. (1994) The North American Breeding Bird Survey. *Birding* **26**, 385–399.

Peters, R.H. (1983) *The Ecological Implications of Body Size.* Cambridge University Press, Cambridge.

Preston, F.W. (1948) The commonness, and rarity, of species. *Ecology* **29**, 254–283.

Preston, F.W. (1958) Analysis of the Audubon

Christmas counts in terms of the lognormal curve. *Ecology* **39**, 620–624.
Preston, F.W. (1962) The canonical distribution of commonness and rarity. I and II. *Ecology* **43**, 185–215, 410–432.
Schroeder, M. (1991) *Fractals, Chaos, Power Laws. Minutes from an Infinite Paradise.* W.H. Freeman and Company, New York.
Stanley, H.E., Amaral, L.A.N., Gopikrishnan, P., Ivanov, P.Ch., Keitt, T.H. & Plerou, V. (2000) Scale invariance and universality: organizing principles in complex systems. *Physica A* **281**, 60–68.
Sugihara, G. (1980) Minimal community structure: an explanation of species abundances patterns. *American Naturalist* **116**, 770–787.
Sugihara, G. (1989) How do species divide resources? *American Naturalist* **113**, 458–463.
Taper, M.L. & Marquet, P.A. (1996) How do species really divide resources? *American Naturalist* **147**, 1072–1086.
Tilman, D. (1994) Competition and biodiversity in spatially structured habitats. *Ecology* **75**, 2–16.
Tokeshi, M. (1990) Niche apportionment or random assortment: species abundance patterns revisited. *Journal of Animal Ecology* **59**, 1129–1146.
Tokeshi, M. (1993) Species abundance patterns and community structure. *Advances in Ecological Research* **24**, 111–186.
Tokeshi, M. (1996) Power fraction: a new explanation of relative abundance patterns in species-rich assemblages. *Oikos* **75**, 543–550.
Tokeshi, M. (1997) Species coexistence and abundance: patterns and processes. In: *Biodiversity. An Ecological Perspective* (eds T. Abe, S.A. Levin & M. Higashi), pp. 35–55. Springer-Verlag, New York.
Tokeshi, M. (1999) *Species Coexistence. Ecological and Evolutionary Perspectives.* Blackwell Science, Oxford.
Waldrop, M.M. (1992) *Complexity. The Emerging Science at the Edge of Order and Chaos.* Simon & Schuster, New York.
Williams, C.B. (1953) The relative abundance of different species in a wild animal population. *Journal of Animal Ecology* **22**, 14–31.
Whittaker, R.H. (1965) Dominance and diversity in land plant communities. *Science* **147**, 250–260.

Part 3

Why are there more species in the tropics?

Chapter 6
How to reject the area hypothesis of latitudinal gradients

Michael L. Rosenzweig

Introduction

More area means more species (von Humboldt 1807; Arrhenius 1921; Williams 1943, 1964; Dony 1963). Noticing the very large extent of tropical ecosystems compared with others, and recognizing the importance of bioprovincial areas to speciation and extinction rates, John Terborgh theorized that the latitudinal area-gradient produces the latitudinal species-diversity gradient (Terborgh 1973).

In place of the mathematics of Terborgh's theory, I substituted a theory that parameterizes both speciation and extinction rates in terms of species' geographical ranges (Rosenzweig 1975). I also claimed that data and theory combine to make a very strong case that area gradients do play a strong role in the latitudinal gradient (Rosenzweig 1995).

Yet, most ecologists have remained sceptical, probably because of the complexity of the issue. The latitudinal gradient has at least two first-order causes and area is but one. Looking for an explanation rooted in a single cause has confused more than it has helped. In addition, variations in species diversity can be observed at several different scales of space and these may be confounded.

Is there an *experimentum crucis*, such that its result would establish, for all reasonable scientists, the effect of area in causing the latitudinal gradient? Yes, but it would involve floating biotic provinces of various fixed (but substantial) sizes between tropics and high latitudes. We would do this slowly enough to allow diversity to track any changes in its steady state in each province. Observing diversity changes in provinces of each size would measure the effects of all causes associated with latitude but not area. Next we would compare continents of varying sizes arrayed along a latitudinal spectrum between tropics and high latitudes. The extent to which these continents deviated from the equal-area continents would measure the effects of unequal area. If the deviations turned out to be large, no one could deny area's ironclad right to a place at the table of explanations for latitudinal gradients.

Such experiments are too monumental to be possible. It may not even be possible

* *Correspondence address: scarab@u.arizona.edu*

to discern them in the fossil record as natural experiments. Are we thus doomed to an endless controversy? Not if we can reject the effect of area even with imperfect experiments. In this chapter, I will examine methods that might lead (or might have led) to a rejection of the area hypothesis. I will also discuss methods that superficially seem to be able to do so, but actually cannot.

In the end, the chapter will arrive at several conclusions. First, the evidence insists that area cannot be the sole cause of latitudinal gradients. Although stressed before, this bears reiteration. Second, area does play a strong role in setting the diversities of different tropical provinces. Third, we have not yet eliminated area as a principal cause of latitudinal gradients. In part at least, we have not done so because we lack a clear understanding of the other causes (which probably include productivity gradients). But I will describe some analytical methods that, if followed scrupulously, would lead to rejection of the area hypothesis if it is false, and do so even before we achieve a working knowledge of the other causes of the gradient. One can hope that if we try these methods and they fail to eliminate area, ecology will—after the manner of most science—stop arguing the point.

Theory

The area hypothesis is the prediction of a standard deductive theory. It is not simply the extrapolation up to the provincial scale of local species–area relationships. Such an extrapolation would be an argument by analogy, a style of argument not well suited to science. Designing a successful test of the area hypothesis demands that we be aware of the structure and probable weaknesses of its generating theory. So I begin with an outline of the theory itself.

Any dynamic theory of provincial species diversity must model diversity's production (speciation) and destruction (extinction). Although we cannot explicitly write down the equations for production and destruction, they must be differential equations. Happily, mathematics shows that we can achieve a measure of robust comprehension regarding differential equations despite not being able to write the explicit equations. We do it by deducing their qualitative dynamic behaviour from their known properties. The advent of useful numerical methods and careful computer simulation has supplemented this approach, but must not be allowed to eclipse it.

Unless diversity dynamics proves to be a run-away process—and we will see that much evidence now demonstrates the contrary—these differential equations will result in homeostasis. Thus we need to find a set of properties inherent in diversity that feedback negatively on its net rate of increase, so producing a self-regulating steady state. (That is not to say the dynamics of the steady state will be damped. I know of no theoretical reason to exclude many other possible trajectories including complex, chaotic ones.) Although many environmental and genetical variables are thought to influence speciation rates or extinction rates, most cannot lead to steady states. Only those that diversity influences can do that, because, by definition, only they are feedback variables.

In fact, I know only one variable likely to lead to homeostasis of species diversity. It is the 'average geographical range of the species in a bioprovince'. For a single province fixed in both area and latitude, the average species' geographical range will be inversely proportional to species diversity. More diversity will lead to more ecological vicars/competitors and more predators too. The theory assumes that such interactions restrict some species from some portions of their physiologically possible geographical ranges, thus reducing their realized geographical ranges.

Other things being equal, species with larger ranges ought to have both higher geographical speciation rates and lower extinction rates. So, 'average species geographical range' is a true parameter. Like the radius of a circle used to determine both the *x* and *y* coordinates of the points along the circle's circumference, average species geographical range can be used to predict the rates of both speciation and extinction in a bioprovince. (In the following, note that rates of increase may be negative.)

That completes a negative feedback loop:

1 more diversity ⇒ smaller ranges ⇒ lower speciation rates and higher extinction rates ⇒ smaller rate of increase in diversity;

2 less diversity ⇒ larger ranges ⇒ higher speciation rates and lower extinction rates ⇒ greater rate of increase in diversity.

The feedback relationships allow us to plot the total extinction-rate curve of a province against S (its total species diversity). The total extinction-rate curve will exhibit a positive second derivative. The feedback relationships also allow us to plot the total speciation-rate curve of the province against S. The total speciation-rate curve will exhibit a negative second derivative. But the opposition of these second derivatives to each other may not suffice to produce a steady-state diversity.

For a steady-state diversity, the rate curves of speciation and extinction must intersect over a positive value of S, say $\hat{S}$, the steady state. I have argued that such an intersection is supported empirically by the continued existence of life over billions of years, and by long periods of diversity without any obvious trend of increase or decrease (Rosenzweig 1995). But it is not a pure deduction.

Capitalizing on the empirical evidence and assuming the existence of a steady state, I extended the theory to allow for more than one trophic level (Rosenzweig 1975). I followed the principle of R. H. Whittaker (1972), i.e. that species are niches for other species, a principle explored in some depth by papers dealing with notions of enemy-free space and apparent competition. Imagine a two-dimensional space in which the axes are the diversities of two adjacent trophic levels. If a steady state exists in one trophic level, then Whittaker's principle means that the steady state will be larger in the presence of species from an adjacent trophic level. The more species in that adjacent level, the larger the steady state will be. The result is a pair of mutualistic isoclines with positive slope. They cross at a point where both levels have diversities enhanced by the mutualisms (as posited by Whittaker). But an equilibrium remains and its dynamics ought also to be stable (unless they are beset with the sort of parameter values that produce complex fluctuations).

This theory applies to the question of latitudinal gradients: at any given S, the species of smaller provinces will have smaller ranges than those of larger provinces. Hence a large province will have a larger steady state than a smaller province. A latitudinal gradient of bioprovincial size should yield a latitudinal gradient of species diversity.

This prediction is so straightforward that one must ask, 'How could Terborgh's hypothesis be rejected? How might area *not* be a factor in causing latitudinal gradients of diversity?' I will now describe the several answers to this question.

Assumptions of the theory could be wrong

The theory itself depends critically upon five assumptions:

1 geographical speciation must play a dominant role because other modes of speciation appear not to depend on geographical range size;

2 larger populations (compared with smaller populations) must have smaller extinction rates (other things being equal);

3 the evolution of larger diversities in a bioprovince must squeeze species into smaller average biogeographical ranges;

4 species diversities in different bioprovinces must be at or near their steady states;

5 species in smaller ranges must have lower speciation rates.

The first three assumptions are generally accepted, and evidence for the fourth mounts steadily (see below). Jablonski, largely responsible for palaeobiological evidence in support of the second assumption (Jablonski 1986), seems to have introduced evidence against the fifth assumption at this symposium (Jablonski *et al.*, this volume). But I shall argue (also below) that his data do not actually test it. So, I believe that Terborgh's hypothesis is unlikely to fall owing to inadequate theoretical assumptions.

Evidence for steady states in diversity (fourth assumption)

The text-books from which we teach introductory students often claim that marine-invertebrate or land-plant diversity has risen more or less steadily over the past half-billion years. However, at least with respect to species diversity, this conclusion errs for several reasons.

1 It suffers from sample-size bias.

2 It reports familial or generic diversities rather than species diversities.

3 It ignores the absence of a trend in marine-invertebrate familial diversity for some 200 Myr before the Cretaceous.

4 It does not consider the past 25 yr of work on the fossil record.

First consider the sample-size bias. Ecology has known for 60 yr that apparent diversity grows with sample size (Fisher *et al.* 1943). But the older the period for which diversity is being estimated, the less material palaeobiologists have to work with (Raup 1976). Moreover, unconsolidated deposits, which yield many more fossils of much better quality with far less effort than hard-rock deposits, are found mostly in Cenozoic strata (A. Miller, pers. comm.). So, for both reasons

—sample size and sample quality—the younger a period, the more we know about it.

The second problem, working with diversities at other than the specific level, cannot be wished away. Evidence about species diversity, whatever its weaknesses, is still evidence about species diversity. Evidence about familial diversity is not. We do not yet know how to transform one into the other. The problem may be more tractable than changing lead into gold, but a problem it is.

The third problem rather shocks me. The late Jack Sepkoski was one of the world's most successful, most respected palaeobiologists. His observation of apparent steady state in the number of marine-invertebrate fossil species for much of the Palaeozoic and Mesozoic is one of the best known of his contributions (Sepkoski 1978). Over the years, he reiterated and emphasized it often. How can it be ignored when teaching undergraduates (and ecologists) about the basics of the science?

Arthur Boucot was perhaps the first palaeobiologist to take the possibility of a steady state in species diversity seriously (Boucot 1975). For a long time other contributions in support of this idea were few (Bambach 1977; Rosenzweig & Duek 1979; Nichols & Pollock 1983), and most came from outsiders. The palaeobiological world itself was ruled by a paradigm established by what became known as the consensus paper (Sepkoski *et al.* 1981), a perfect example of how the very finest minds in a field may serve it best when working alone. If only one person had written that paper, the rest would have pulverized it and prevented it from cowing a whole generation of colleagues into ignoring the emperor's nakedness.

But that era is over. For the past decade or so, a new generation—most of them the children and grandchildren of the consensus authors—have been hard at work fixing things. I have summarized some of their work up to 1994 (Rosenzweig 1995), but a number of exciting papers have appeared since then. For example, it turns out that the increase in invertebrate diversity during the Ordovician occurred, not steadily, but in one or two fits of creativity (Miller & Foote 1996). Even more recently, Alroy has looked at the issue of steady state in both mammals and marine invertebrates (Alroy 1998; Alroy *et al.* 2001). Evidently, the dynamic elements of a steady state were present and active among mammals in North America during the Cenozoic. As their diversity fluctuated, speciation rate and extinction rate responded, keeping diversity within narrow bounds. The two rates intersected over a diversity that did not change significantly during the 65 Myr period (except in response to mass extinction). Alroy's work was not unprecedented (Van Valkenburgh & Janis 1993).

Speciation rate and species diversity (fifth assumption)

Alroy (1998) also showed that the per-taxon extinction rate did not vary significantly with diversity, but that the per-taxon speciation rate did. Per-taxon speciation rate declined with increases in *S*, thus implicating variations in speciation rate as the most responsive component of the system's dynamic stability. (But this conclusion may be scale dependent. Over a shorter period of only 0.5 Myr during the late Eocene, I showed that small mammal extinction rate was more sensitive to diversity than was speciation rate (Rosenzweig 1995). So, in the data I studied, extinction rate, not speciation rate, was primarily responsible for maintaining diversity at a steady state.)

Some genetic and chromosomal work also suggests that speciation rate will slow if range declines. This flows from the macroecological fact that species with small ranges contain fewer total member individuals (smaller-ranged species actually exist at smaller densities; Gaston & Blackburn 2000) and from the theory of centrifugal speciation (Brown 1957). Centrifugal speciation holds that small populations are evolutionarily frozen relics rather than engines of innovation, and consequently that speciation occurs to large-ranged species. Some evidence supports centrifugal speciation at least in some cases (Rosenzweig 1995). Meanwhile, other evidence, admittedly theoretical, suggests that small populations cannot evolve fast enough to avoid the evolutionary neutralization of most of their genetic opportunities for innovation (Walsh 1995). This work extends Fisher's belief that evolutionary rate is constrained by available genetic variety, a hypothesis Bradshaw (1984) called 'genostasis'.

So how is one to understand Jablonski's announcement in this symposium (Chapter 19) that speciation rate is inversely correlated to range size? Jablonski's data analyse the speciation rates of different species that have different size ranges. That does not tell us how the speciation rate of any single species will respond to a shrinking range. Different species have different life histories, and one of the most consequential life-history details for a marine invertebrate is where it will spend its larval stage. If it floats out into the open ocean, it tends to have a large range and a single panmictic population. Geographical speciation will have a hard time getting started. On the other hand, species that remain near the shore tend to have smaller ranges, and ought to be much more easily subdivided (Hansen 1980; Jablonski & Lutz 1980). This relationship must be a major portion of the signal in Jablonski's new analysis, but it does not tell us what happens as *S* fluctuates within a group of species all of whose larvae stay near adult habitats. (Nor, equivalently, what happens within a group of species all of whose larvae float out to the open ocean.) It does not tell us whether their ranges respond to *S* nor whether their speciation rates fluctuate accordingly. I would hope that Jablonski's data could be interrogated for answers to those questions.

Assumptions incomplete

Like any theory, the area theory of diversity ignores most variables. In particular, I restricted it to range size so as to allow for a homoeostatic feedback process. Nonetheless, perhaps one or more of these ignored variables can generate a process that dwarfs the influence of area on diversity across a latitudinal gradient. Fortunately, to reject Terborgh's hypothesis, we do not even need to know the identity of such a phantom variable let alone have an alternative theory based on it.

A perfect data set would contain a group of separate provinces arranged independently and orthogonally along the two axes of area and latitude (Fig. 6.1). Latitude

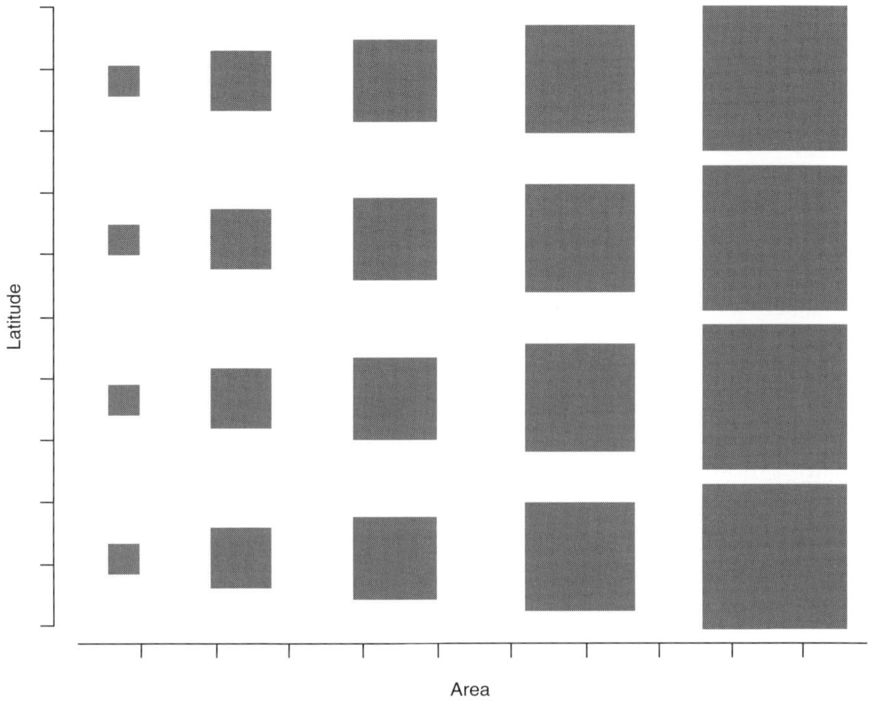

Figure 6.1 A perfect data set. The grey areas depict the locations and relative sizes of each of 20 separate bioprovinces.

would serve as a surrogate for the phantom variable. If no correlation of diversity with latitude existed, then there would be no phantom variable. If the effects of area and latitude were both significant, both Terborgh's hypothesis and the presence of at least one phantom variable would be confirmed. If no correlation of diversity with area existed, then Terborgh's hypothesis would be rejected. But we have no perfect data set.

An alternative way to reject Terborgh's hypothesis would be to find data that conflict with its predictions. As a matter of well known fact, we already do have data that use this method and show that area cannot be the sole causal variable in the latitudinal gradient. For example, mammal diversity declines with latitude in North America (Fig. 6.2) although, in America, northern boreal and tundra areas are, at the very least, as large as those of the subtropics and temperate zones. On the basis of area alone, they also should be equal in species diversity (Fig. 6.3).

Productivity (or something very closely allied to it) appears to be the principal 'phantom' variable behind latitudinal gradients. I wish we had a deductive theory for it because, for too long, we have taken its effect on diversity for granted. We do not really understand it even at the low range of productivities where no one doubts the existence of a positive correlation (Waide *et al.* 1999). Nonetheless, even total understanding of the relationship between productivity and diversity cannot elimi-

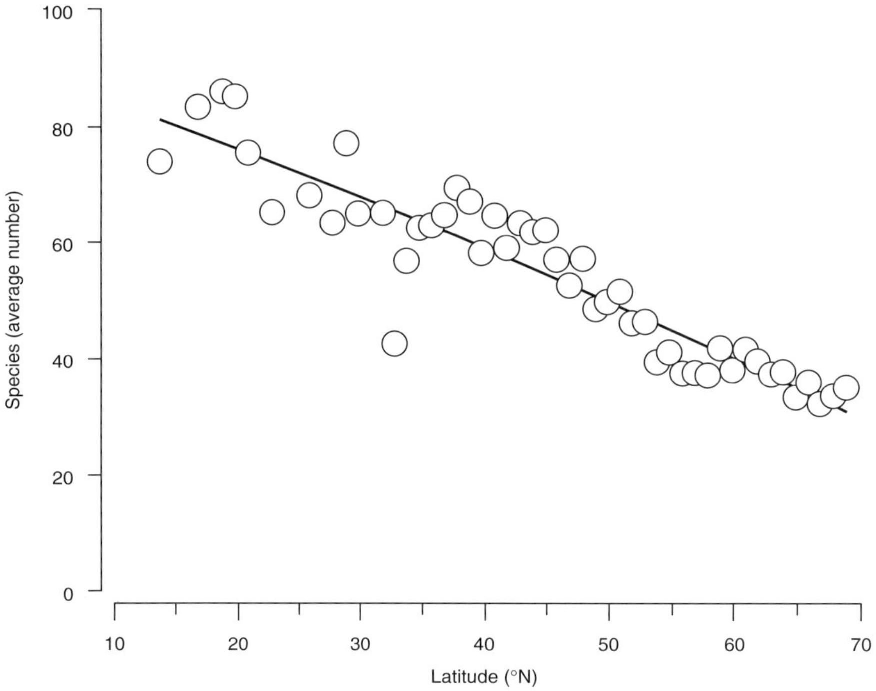

Figure 6.2 Quadrupedal mammal diversity declines with latitude in North America. Each point is the average species diversity of large, equal-area blocks with their centres at a single latitude. (Redrawn from Rosenzweig & Sandlin 1997.)

nate the importance of area or even relegate it to secondary status. In the first place, productivity may not correlate positively with diversity except where productivities are very low. As productivity rises beyond some threshold value, diversity may even decline for many or most taxa (Rosenzweig & Abramsky 1993). Second, even where productivity counts heavily, we have to understand how the two variables combine. To lose status, area will have to turn out to play a small role in the combination.

In combining area and productivity, we confront an almost pernicious, multi-scale, multivariable problem. Consider the following approach, which at first seems absolutely reasonable. We take a series of equal-area sample locations arranged along a gradient of productivity. We determine the number of (say) butterfly species in each location (e.g. Turner *et al.* 1987). Then we say that if there is a relationship, it must be attributable to something other than area. Perhaps we attribute the effect to climate, and through climate to productivity.

But area plays its role at different scales, and using equal-area sample locations eliminates only the two local scales (sample size and habitat heterogeneity). It does not eliminate the regional scale, the one that works through evolutionary time. This scale is evidenced by the relationship between local and regional diversities (Fig.

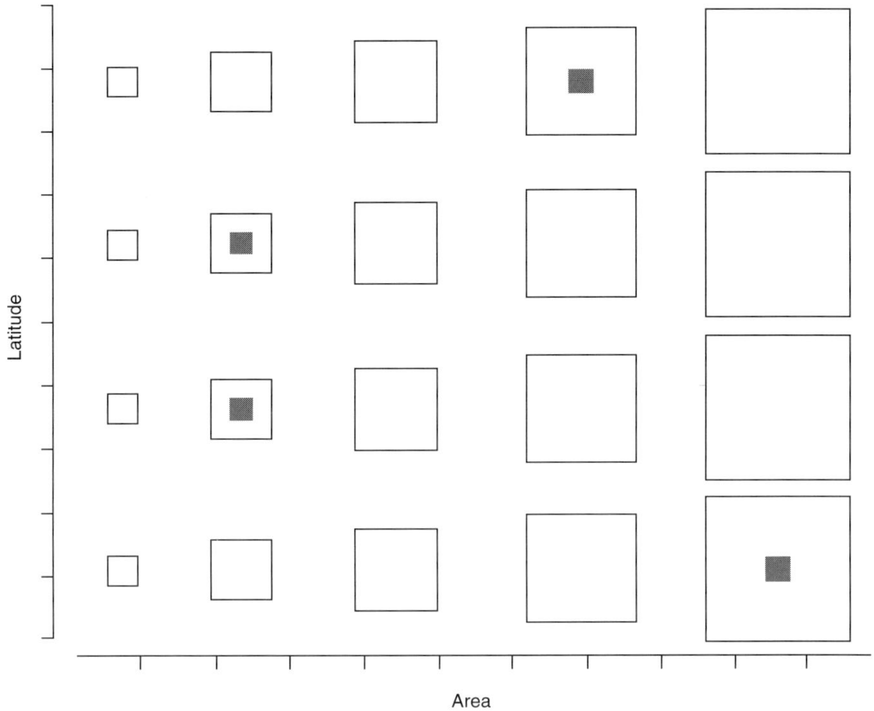

Figure 6.3 The underlying sampling design that led to the result of Fig. 6.2. Equal areas are sampled from assemblages of highly unequal area. If area alone determined species diversity, high-latitude diversities would be almost as high as tropical diversities.

6.4), a pattern first investigated by Terborgh & Faaborg (1980). Rosenzweig & Ziv (1999) have called such patterns 'echo patterns', because areas of the same size taken from regions of different diversity will echo the rank order of the diversities from which they are taken. Thus if there were to be a connected series of regions of different size (Fig. 6.5), and *because of the different sizes*, these regions evolved different sized pools of species, then a series of equal-area samples from the regions would show unequal diversities, and the cause would be the echo of the unequal areas of the regions.

So, ordinary methods, such as elaborate analyses of variance, will prove to be insufficient. We will need to use hybrid methods in which the quantitative results of the best empirical investigations are used to feed parameter values to computer models, or to evaluate the success of those models.

Note that Brown *et al.* (this volume) also caution us to doubt the power of mere correlative analyses in macroecological work. Brown *et al.*'s point is that we may need theory to suggest appropriate mathematical forms. My (complementary) point is that mere correlative analyses cannot automatically detect and compensate

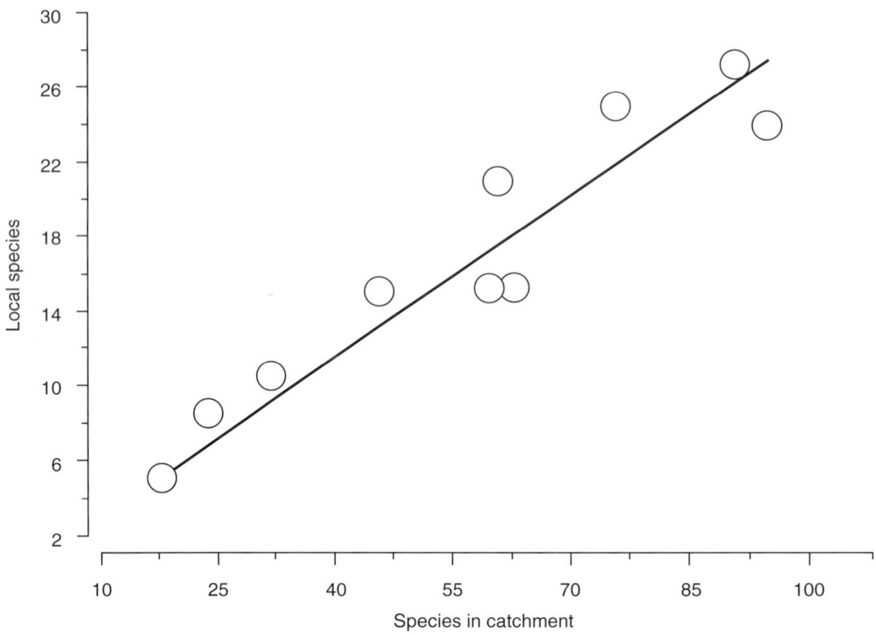

Figure 6.4 An echo pattern of species diversity. Fish diversity in equal-area samples echoes the diversity of the entire drainage from which the sample comes. (Redrawn from Rosenzweig & Ziv 1999.)

for more than one scale of a variable at a time. So, whenever we suspect multiscale relationships, we must be very, very wary about how we test for them.

Some useful data sets exist

Some data sets are not as good as others. They may confound the variables of area and productivity. They may confound space and time. They may intermingle island and provincial scales. They may be so taxonomically restricted that they contain too few species to reveal statistical tendencies. They may not deal with diversity at the species level. They may be fraught with uncorrected sampling biases. Uncritical meta-analyses are likely to confuse us more than they help.

Nevertheless, appropriate data sets do exist and currently tend to support a strong role for area. These data come from different-sized provinces with similar sets of abiotic conditions (Fig. 6.6). The swath of provinces cut across a single latitudinal zone virtually eliminates productivity's influence or, for that matter, the influence of any variable other than area. The diversities of such provinces reveal very steep species–area curves (Fig. 6.7). As a rule of thumb, an order of magnitude increase in provincial area is accompanied by a similar increase in species diversity.

One might object that the clearest cases so far examined deal only with tropical

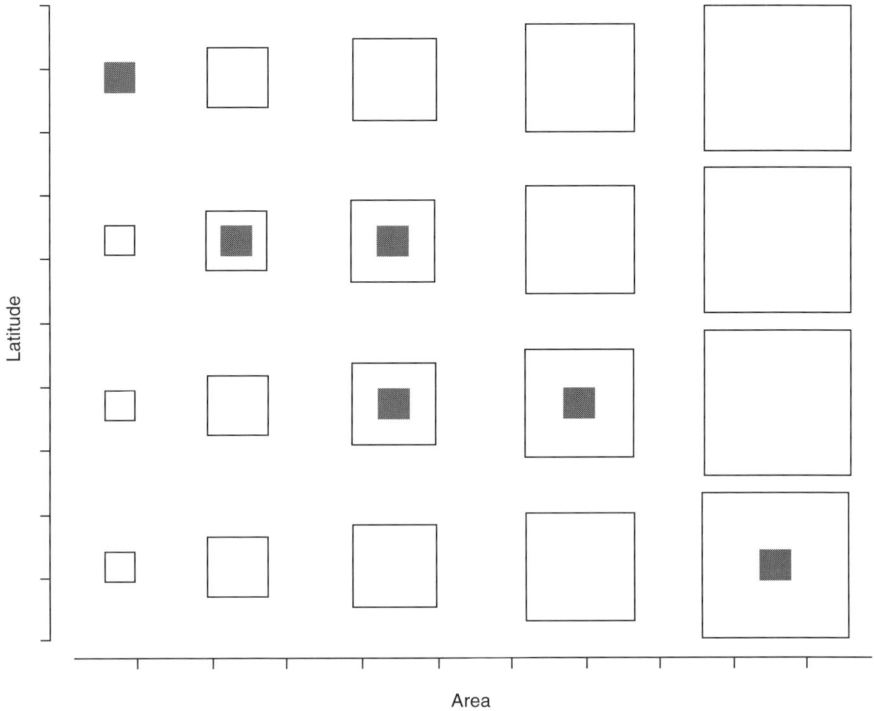

Figure 6.5 A series of equal-area samples from unequal-area regions. The diversity of each sample will echo the diversity of the region from which it comes. If those regions differ in area, local diversity will vary in proportion to the area of the region in which it is embedded despite the fact that the local samples are equal-area samples.

provinces and thus look at the area effect only at low latitudes. Perhaps, at higher latitudes, area loses its effect. Then it would not matter whether high latitude zones are small or large; their low diversities would emerge solely from their low productivities.

I cannot entirely discount such a possibility. However, there is one monumental data set that is *not* restricted to the tropics. It covers the entire Northern Hemisphere (Fig. 6.8). Land plants during the past 408 Myr appear to have had diversities that responded linearly to changes in land area (Tiffney & Niklas 1990; Rosenzweig 1998). Of course, tropical plant fossils might play such a dominant role in determining this relationship that one might not be able to detect the lack of area's effect in the poorer zones of the hemisphere. But if tropical plant diversities do dominate and area plays a dominant role in determining tropical diversities (which we already know), then what remains to prove?

The fossil plant pattern would seem to clinch the matter. As a rule of thumb, an order of magnitude increase in provincial area is accompanied by a similar increase in species diversity. Although this may not be able entirely to account for latitudinal gradients, it is far too strong to be ignored.

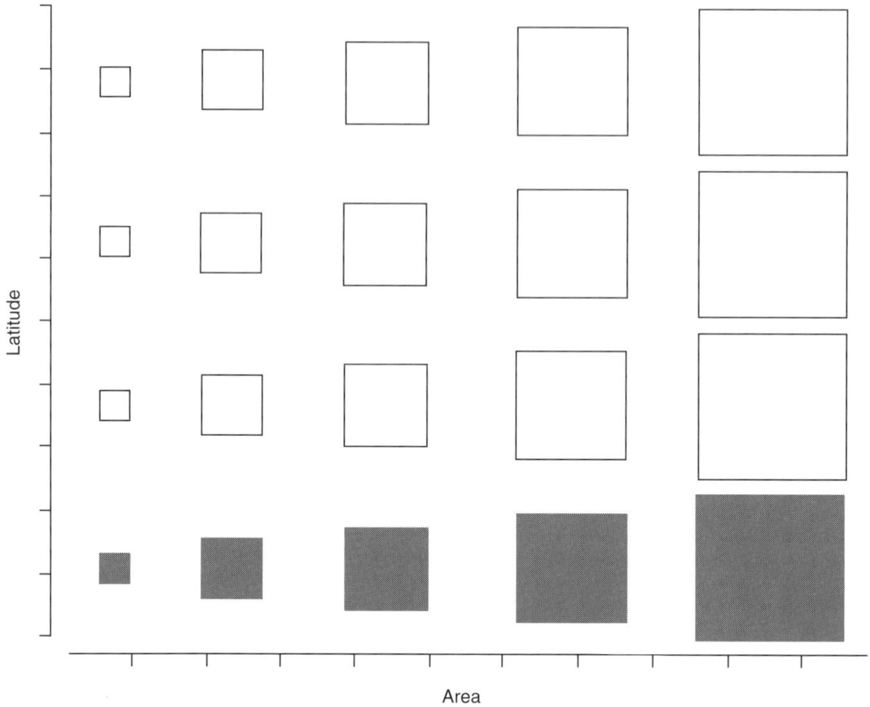

Figure 6.6 Diversities of a set of tropical provinces permit one to establish the role of area independent of latitude and the other variables associated with latitude.

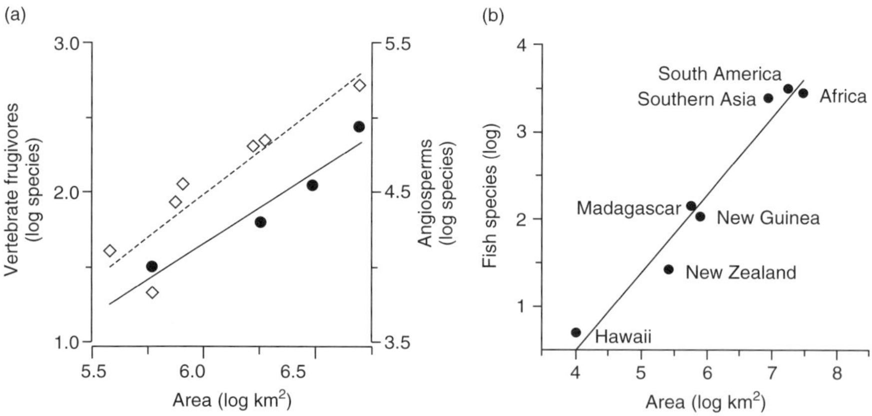

Figure 6.7 Three species–area relationships that follow the design of Fig. 6.6. (a) Vertebrate frugivores (diamonds, $y = 1.15x - 4.94$) and flowering plants (circles, $y = 0.97x - 1.67$). (b) Tropical freshwater fishes ($y = 0.89x - 2.91$). (Part (a) redrawn from Rosenzweig 1995; data for part (b) from Peter Reinthal.)

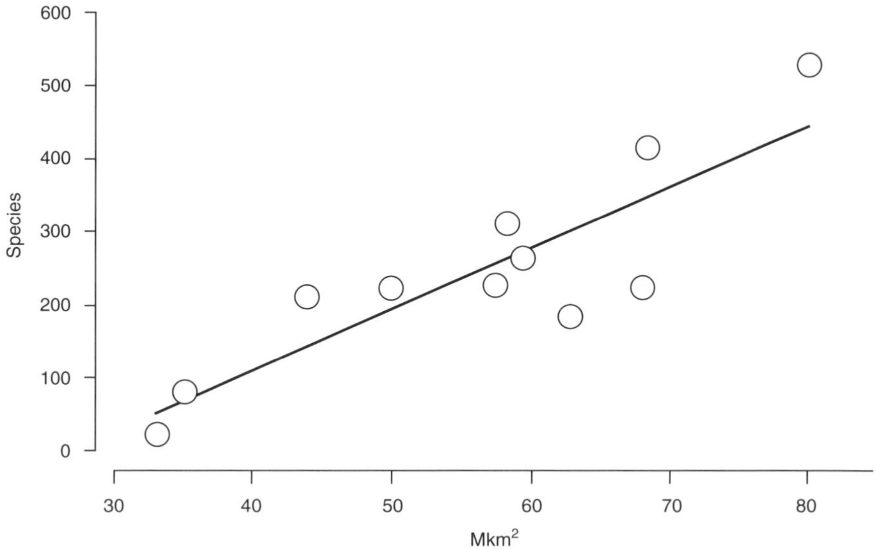

Figure 6.8 Plant diversity of the Northern Hemisphere in 11 periods with non-overlapping lists of fossil species. Diversity varied linearly with area of land, whether plotted by itself, as here ($y = 8.39x - 227.3$), or studied with a multiple regression to incorporate the effect of age ($y = 1.78 + 1.00x + 0.33z$; where y is the log number of species; z is the log median age of the period; and x is the log area of the period). (Data from Karl Niklas and Bruce Tiffney; graph redrawn from Rosenzweig 1998.)

Latitudes not in different provinces

Terborgh's original paper imagined the different latitudinal regions to be connected in a string from pole to pole. So too has every attempt to check the basic premise that actual area declines poleward. That includes the most recent attempt by Hawkins & Porter (2001), the attempts by Rohde (1992) and by Rahbek & Graves (2001), as well as the first attempt (Rosenzweig 1992). But the theory assumes that the areas being compared constitute separate biogeographical provinces.

The only hint that stringing biomes together might pose a problem for Terborgh's hypothesis comes in work to check the intensity of range bleeding, i.e. species spreading out from the latitudes of their origin to other latitudes in other biomes (Blackburn & Gaston 1997; Rosenzweig & Sandlin 1997; Fine 2001). Does range bleeding violate the basics of the theory enough to vitiate its prediction or interfere with its test? I suspect not.

The key is the initial climatic tolerance of a new species. Suppose there are two zones of very different area but the same diversity, and a new species arises in each zone. Further suppose that each of these new species is physiologically competent in, but limited to, the climatic zone in which it arises. Thus, the new species of the larger zone will have the larger physio-geographical range, the higher speciation rate and the lower extinction rate. In short, even if provinces are joined, all species whose ranges cannot bleed into the adjacent province will satisfy the theory.

But what if, as is likely, some new species are not physiologically limited to the zone in which they arise? That will reduce the disparity among zones, but not eliminate it entirely. Some of the species born in the small zone may escape its restrictions by being able to inhabit the large zone too, but some will not. But all the species born in the large zone will receive its areal benefits (low extinction and high speciation rates).

Imagine a world of *n* zones, each of a different area but all with the same initial species diversities as well as the same distributions of physiological tolerances. Small zones next to a large zone would have more speciation and less extinction than similarly small zones next to another small zone, introducing noise into our detection of the area signal. Small zones next to a large zone would also have more species arise as immigrants from their neighbouring zone, further increasing the noise. But the signal would remain: large zones should increase in net diversity faster than small ones, and generate larger steady states.

Notice that I began the previous paragraph by assuming equality of diversities among zones. That is because we must focus on rate curves, not rates. The differences between the curves will generate different steady states for each zone, and the extinction rate of the larger zone at its steady state may be considerably in excess of that in the smaller zone (Rosenzweig 1975). One cannot overemphasize the point: measuring rates at a single value of *S*, is different from measuring a rate curve over a range of *S* values.

Are the tropics larger?

Even if the area theory is correct, it might still not explain latitudinal gradients. The tropics might not be larger. In fact, most recently, Hawkins & Porter (2001) reach this very conclusion.

No one doubts the actual distribution of land as a function of latitude, but how should we subdivide it to study Terborgh's hypothesis? I relied solely on temperature because that is one of the three foundations of the hypothesis. Terborgh showed that tropical temperatures hardly change within tropical latitudes whereas those of other regions change linearly with latitude (Fig. 6.9). No temperature plateaux similar to that of the tropics show up in other latitudinal bands. How finely should we divide the sloped part of the relationship? I do not know, so I used a convention that seems realistic: subtropics; temperate; boreal; tundra (Fig. 6.10).

On the other hand, Hawkins & Porter (2001) divide the terrestrial surface of the Earth according to Whittaker's (1972) biomes. Now comparing biomes cannot reveal the only answer to latitudinal gradients because such gradients exist within biomes too (Fig. 6.11)(Davidowitz & Rosenzweig 1998). But Whittaker's biomes, in particular, lead to problems for the analysis.

Whittaker (1972) divides the tropics into four different biomes on the basis of aridity, but he does not do so for other latitudinal zones. Tundra, taiga and desert are

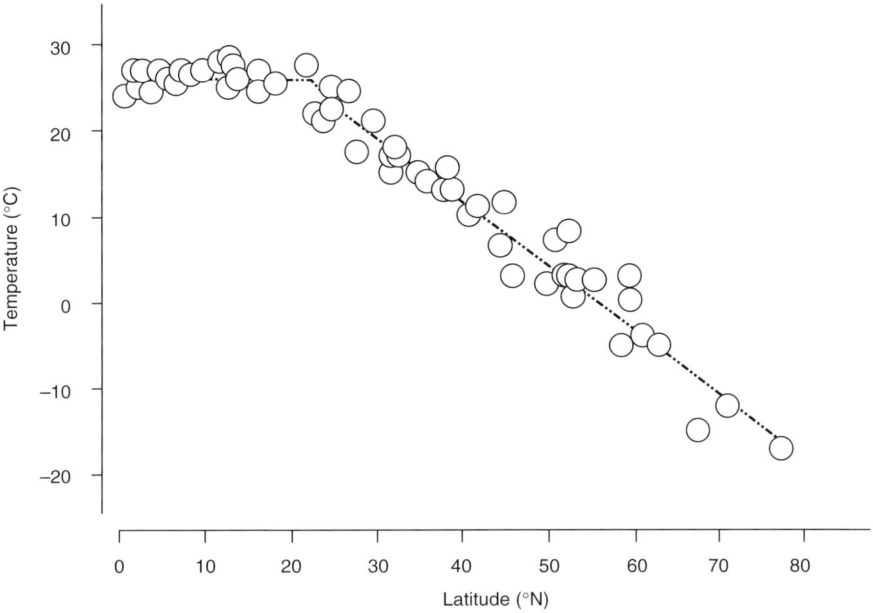

Figure 6.9 Mean annual temperature at various latitudes in the Northern Hemisphere. (Redrawn from Rosenzweig 1992.)

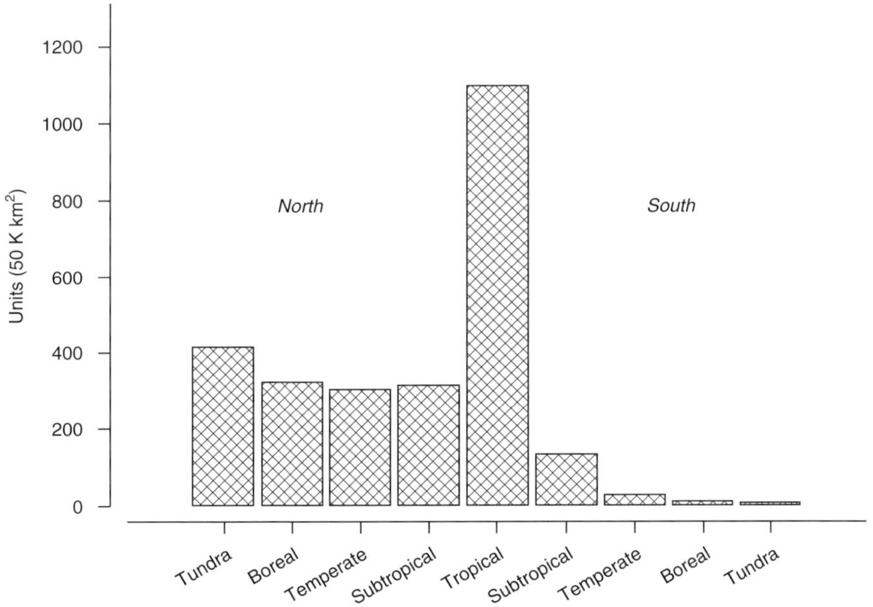

Figure 6.10 World-wide, land-area distribution in latitudinal zones. (Redrawn from Rosenzweig 1992.)

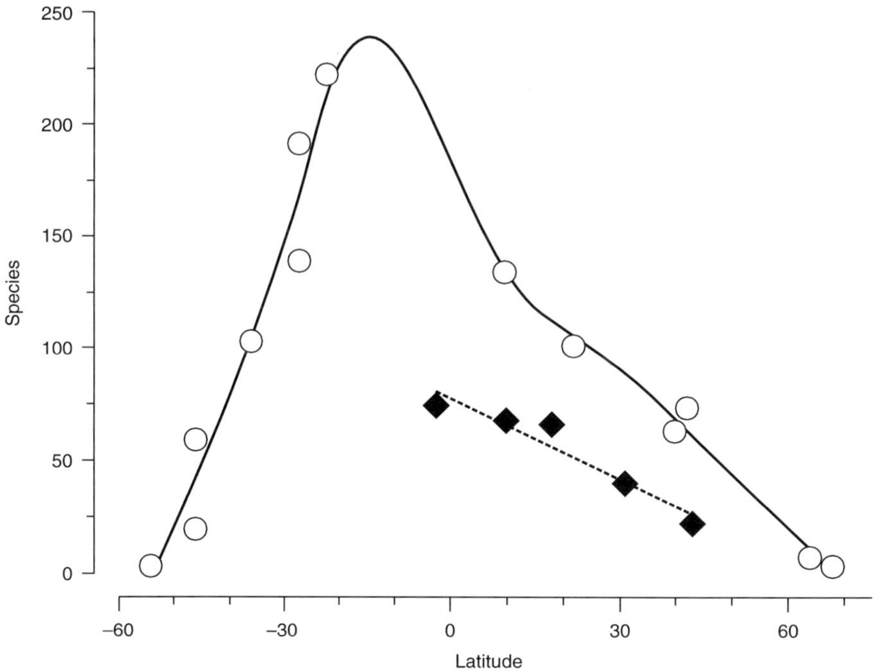

Figure 6.11 Species diversity within a single habitat type also shows a latitudinal gradient. This example is for ants. The upper curve shows the multihabitat gradient over a full range of latitudes. The shorter curve beneath it shows the gradient for one habitat. (Redrawn from Davidowitz & Rosenzweig 1998.)

not subdivided at all; the temperate zone is divided in two, and the subtropics contain basically just the very small biome, Mediterranean scrub. Whittaker's division therefore erects a bias that runs counter to latitude; the lower the latitude the more finely subdivided the region. Hawkins & Porter (2001) justify this treatment on the basis of lower diversity/broader niches in the higher latitudes, but that begs the question. Remember, we identify the kinds of biomes on the basis of their recognition by life, i.e. plant species. The more plant species, the more biomes we identify (see chapter 7 in Rosenzweig 1995).

Whittaker's (1972) scheme produces another consequential bias: low productivity is associated with large biome area. If Whittaker saw that a region is characterized by a serious limiting stress, i.e. low temperature or aridity, then he did not subdivide it according to the other factor. For instance, he combined all deserts in North America. But the desert in North America runs from about 50°N to about 30°N, and the higher latitude cold desert presents such a different set of biotic challenges compared with the lower latitude warm desert that they share almost no plant species. The deserts in Hawkins & Porter (2001) cover an even deeper range: they extend from the Equator to about 50°. Cold biomes raise a similar question. In addition to low temperatures, surely there is considerable ecological significance to their internal variation in water availability. Given the fact that Whittaker's (1972) scheme subdivides all the other regions by both temperature and water, it is clear why all of Hawkins & Porter's largest biomes are either cold or arid.

Terborgh presented evidence that a focus on temperature is the right focus. It is average temperature that changes very little over tropical latitudes. If one wishes to add aridity to the mix, one should first demonstrate the existence of broad regions of similar temperature–aridity. Until then, I shall continue to maintain that the world's largest regions are tropical: according to Hawkins & Porter's (2001) figures, the African tropics (*c.*16 million km^2) and the New World tropics (*c.*15 million km^2).

Despite the problems with Whittaker's (1972) scheme, most of Hawkins & Porter's (2001) data points (37/45 in their fig. 1B) show precisely the pattern predicted by Terborgh: equatorial biomes have the largest areas, and biome areas fall off regularly as biomes are located farther away from the Equator. (The exceptions are the northern cold biomes and the palaearctic desert.) Their linear regression through a highly non-linear set of points obscures that agreement.

Several other issues in Hawkins & Porter's (2001) analyses also must be addressed. In one analysis, they intermingle the zonal data of several biogeographical provinces, some very large and some very small—no problem if area does not matter. But if it does, then range bleeding fouls the area prediction for the zones within a province. Judging from the data of several tropical provinces (see above), intermingled provinces will have diversities characterized by the sum total size of the set of provinces.

In addition, based on results from within three of four provinces, they found that biome area has no effect on diversity within provinces. They did find a positive relationship within Africa. However, the provinces they used in their analysis did not include South America or the Oriental Province. Others have already extracted a strong species–area relationship (SPAR) from South America (Mares & Ojeda 1982), and I do believe one exists in the Oriental Province too. That would mean that, among regions with tropical systems, only Australasia lacks an intraprovincial SPAR, and that province is dominated by low productivity habitats.

Finally, in one analysis that neglects to separate productivity from area, Hawkins & Porter (2001) conclude that, both within and between provinces, area and diversity are unrelated. Everyone agrees that low productivity, or something closely correlated to it, greatly depresses diversity. This signal must be separated from the data before we can conclude that area or any other variable plays no role. It can be done. It should be done. I will be happy when it is done.

Summary

Species diversity ought to be self-regulating via the parameter 'average species geographical range'. Average range should diminish as diversity grows. Species with larger ranges ought to have both higher speciation rates and lower extinction rates.

That sets up a negative feedback, which could lead to steady-state diversities. Empirical evidence indicates that it has.

Because larger provinces have larger steady states, a latitudinal gradient of bioprovincial sizes should yield a latitudinal gradient of species diversity. Terborgh hypothesized that such size gradients are responsible for latitudinal diversity gradients. Concatenating provinces will add to the noise of this signal but not eliminate it.

Terborgh's hypothesis does not suffer from inadequate theoretical assumptions but more needs to be done to investigate several of them. Specifically, we need to make sure of several points: (i) that geographical speciation plays a dominant role in diversification; (ii) that evolution of larger diversities in a bioprovince squeezes species into smaller average biogeographical ranges; (iii) that species in smaller ranges have lower speciation rates.

Nonetheless, area cannot be the sole causal variable responsible for the latitudinal gradient. Something such as ecological productivity must be involved. But we have no theory for the effect of productivity at the provincial scale. And we do not know how to combine productivity with area in a unified model, not even a statistical one. Ordinary analysis of variance methods do not work because they ignore the multiscale influence of area. Doing the job will call for hybrid methods in which we perform simulations using dynamic computer models and empirically determined parameter values to try to generate known quantitative patterns of diversity.

Available data suggest that species diversity in provinces is proportional to their area, and nearly linearly so. Most data sets examine the effect of area on diversity in tropical provinces, but a single temporal study looks at the plant fossils of the entire Northern Hemisphere as a sequential set of independent provinces. It too shows a linear response of species diversity to area.

Some biogeographers suspect that the terrestrial tropics do not actually cover more area. If they do not, Terborgh's hypothesis could not apply. However, this claim depends on classifying the land into its biomes, a process with biases that have not been attended to. It also involves a research plan that does not flow from Terborgh's empirical discovery of a 50°-wide latitudinal plateau in average annual temperature, a plateau that centres on the Equator.

Howard Wettstein (2001, p. 342) offers some sage advice to those evaluating theories: 'Quite generally the problem is not whether a theoretical approach can with enough brilliance be shored up. Often, if not always, it can de done. Perhaps it can always be done—if one is willing to pay the price. The real question—the place one needs to train one's vision—is rather that of natural fit. One needs to focus not on "One could say . . . ," but rather on what is the most natural thing to say.'

In this case, the most natural thing to say is that both data and deductive reasoning currently tend to support a strong role for area. But there are many opportunities to improve our understanding of just how strong that role is compared with other variables that also play important roles.

References

Alroy, J. (1998) Equilibrial diversity dynamics in North American mammals. In: *Biodiversity Dynamics: Turnover of Populations, Taxa and Communities* (eds M.L. McKinney & J.A. Drake), pp. 232–287. Columbia University Press, New York.

Alroy, J., Marshall, C.R., Bambach, R.K. *et al.* (2001) Effects of sampling standardization on estimates of phanerozoic marine diversification. *Proceedings of the National Academy of Sciences, USA* **98**, 6261–6266.

Arrhenius, O. (1921) Species and area. *Journal of Ecology* **9**, 95–99.

Bambach, R.K. (1977) Species richness in marine benthic habitats through the Phanerozoic. *Paleobiology* **3**, 152–167.

Blackburn, T.M. & Gaston, K.J. (1997) The relationship between geographic area and the latitudinal gradient in species richness in New World birds. *Evolutionary Ecology* **11**, 195–204.

Boucot, A.J. (1975) *Evolution and Extinction Rate Controls.* Elsevier Scientific Publications, Amsterdam.

Bradshaw, A.D. (1984) The importance of evolutionary ideas in ecology—and vice-versa. In: *Evolutionary Ecology* (ed. B. Shorrocks), pp. 1–25. Blackwell Scientific, Oxford.

Brown, W.L., Jr. (1957) Centrifugal speciation. *Quarterly Reviews in Biology* **32**, 247–277.

Davidowitz, G. & Rosenzweig, M.L. (1998) The latitudinal gradient of species diversity among North American grasshoppers (Acrididae) within a single habitat: a test of the spatial heterogeneity hypothesis. *Journal of Biogeography* **25**, 553–560.

Dony, J.G. (1963) The expectation of plant records from prescribed areas. *Watsonia* **5**, 377–385.

Fine, P.V.A. (2001) An evaluation of the geographic area hypothesis using the latitudinal gradient in North American tree diversity. *Evolutionary Ecology Research* **3**, 413–428.

Fisher, R.A., Corbet, A.S. & Williams, C.B. (1943) The relation between the number of species and the number of individuals in a random sample of an animal population. *Journal of Animal Ecology* **12**, 42–58.

Gaston, K.J. & Blackburn, T.M. (2000) *Pattern and Process in Macroecology.* Blackwell Science, Oxford.

Hansen, T.A. (1980) Influence of larval dispersal and geographic distribution on species longevity in neogastropods. *Paleobiology* **6**, 193–207.

Hawkins, B. & Porter, E. (2001) Area and the latitudinal diversity gradient for terrestrial birds. *Ecology Letters* **4**, 595–601.

Jablonski, D. (1986) Background and mass extinctions: the alternation of macroevolutionary regimes. *Science* **231**, 129–133.

Jablonski, D. & Lutz, R.A. (1980) Larval shell morphology: ecology and paleoecological applications. In: *Skeletal Growth of Aquatic Organisms* (eds D.C. Rhoads & R.A. Lutz), pp. 323–377. Plenum, New York.

Mares, M.A. & Ojeda, R.A. (1982) Patterns of diversity and adaptation in South American hystricognath rodents. *Special Publication of the Pymatuning Laboratory of Ecology* **6**, 393–432.

Miller, A.I. & Foote, M. (1996) Calibrating the Ordovician radiation of marine life: implications for Phanerozoic diversity trends. *Paleobiology* **22**, 304–309.

Nichols, J.D. & Pollock, K.H. (1983) Estimating taxonomic diversity, extinction rates, and speciation rates from fossil data using capture-recapture models. *Paleobiology* **9**, 150–163.

Rahbek, C. & Graves, G.R. (2001) Multiscale assessment of patterns in avian species richness. *Proceedings of the National Academy of Sciences, USA* **98**, 4534–4539.

Raup, D.M. (1976) Species diversity in the Phanerozoic: an interpretation. *Paleobiology* **2**, 289–297.

Rohde, K. (1992) Latitudinal gradients in species diversity: the search for the primary cause. *Oikos* **65**, 514–527.

Rosenzweig, M.L. (1975) On continental steady states of species diversity. In: *Ecology and Evolution of Communities* (eds M.L. Cody & J.M. Diamond), pp. 121–140. Belknap Press of Harvard University Press, Cambridge, MA.

Rosenzweig, M.L. (1992) Species diversity gradients: we know more and less than we thought. *Journal of Mammalogy* **73**, 715–730.

Rosenzweig, M.L. (1995) *Species Diversity in Space and Time.* Cambridge University Press, Cambridge.

Rosenzweig, M.L. (1998) Preston's ergodic conjecture: the accumulation of species in space and time. In: *Biodiversity Dynamics: Turnover of Populations, Taxa, and Communities* (eds M.L. McKinney & J.A. Drake), pp. 311–348. Columbia University Press, New York.

Rosenzweig, M.L. & Abramsky, Z. (1993) How are diversity and productivity related? In: *Species Diversity in Ecological Communities* (eds R.E. Ricklefs & D. Schluter), pp. 52–65. University of Chicago Press, Chicago, IL.

Rosenzweig, M.L. & Duek, J.L. (1979) Species diversity and turnover in an Ordovician marine invertebrate assemblage. In: *Contemporary Quantitative Ecology and Related Ecometrics* (eds G.P. Patil & M.L. Rosenzweig), pp. 109–119. International Co-operative Publishing House, Fairland, MD.

Rosenzweig, M.L. & Sandlin, E.A. (1997) Species diversity and latitudes: listening to area's signal. *Oikos* **80**, 172–176.

Rosenzweig, M.L. & Ziv, Y. (1999) The echo pattern in species diversity: pattern and process. *Ecography* **22**, 614–628.

Sepkoski, J.J., Jr. (1978) A kinetic model of Phanerozoic taxonomic diversity I. Analysis of marine orders. *Paleobiology* **4**, 223–251.

Sepkoski, J.J., Jr, Bambach, R.K., Raup, D.M. & Valentine, J.W. (1981) Phanerozoic marine diversity and the fossil record. *Nature* **293**, 435–437.

Terborgh, J. (1973) On the notion of favorableness in plant ecology. *American Naturalist* **107**, 481–501.

Terborgh, J.W. & Faaborg, J. (1980) Saturation of bird communities in the West Indies. *American Naturalist* **116**, 178–195.

Tiffney, B.H. & Niklas, K.J. (1990) Continental area, dispersion, latitudinal distribution and topographic variety: a test of correlation with terrestrial plant diversity. In: *Causes of Evolution: a Paleontological Perspective* (eds R.M. Ross & W.P. Allmon), pp. 76–102. University of Chicago Press, Chicago, IL.

Turner, J.R.G., Gatehouse, C.M. & Corey, C.A. (1987) Does solar energy control organic diversity? Butterflies, moths and the British climate. *Oikos* **48**, 195–203.

Van Valkenburgh, B. & Janis, C.M. (1993) Historical diversity patterns in North American large herbivores and carnivores. In: *Species Diversity in Ecological Communities: Historical and Geographical Perspectives* (eds R. Ricklefs & D. Schluter), pp. 330–340. University of Chicago Press, Chicago.

Von Humboldt, F.H.A. (1807) *Essai sur la geographie des plantes.* von Humboldt, Paris.

Waide, R.B., Willig, M.R., Steiner, C.F., *et al.* (1999) The relationship between productivity and species richness. *Annual Reviews in Ecology and Systematics* **30**, 257–300.

Walsh, J.B. (1995) How often do duplicated genes evolve new functions? *Genetics* **139**, 421–428.

Wettstein, H. (2001) Against theodicy. *Judaism* **50**, 341–350.

Whittaker, R.H. (1972) Evolution and measurement of species diversity. *Taxon* **21**, 213–251.

Williams, C.B. (1943) Area and the number of species. *Nature* **152**, 264–267.

Williams, C.B. (1964) *Patterns in the Balance of Nature.* Academic Press, London.

Chapter 7

Climatic–energetic explanations of diversity: a macroscopic perspective

Robert J. Whittaker, Katherine J. Willis and Richard Field*

Introduction

A general theory of diversity must necessarily cover disparate phenomena, at various scales of analysis. For instance, spatial patterns in richness and in endemicity vary, indicating a need to develop separate models for their explanation. Those models, in turn, must be reconcilable within a general body of diversity theory, exhibiting the properties of a hierarchical theory (cf. Allen & Starr 1982), structured with respect to scale (Whittaker *et al.* 2001; Willis & Whittaker 2002; Table 7.1). Our principal concern herein is the existence and understanding of *predictable* macroscale patterns in species richness across the land surface of the globe. Our given brief was to consider 'energy' in relation to geographical variation in diversity, i.e. to focus on a dynamic, climatic–energetic explanatory framework (as e.g. Turner 1991; O'Brien 1993; Whittaker *et al.* 2001).

That climatic gradients underpin diversity gradients is an old idea within ecology, with varying contemporary expressions. Our theoretical starting point is water–energy dynamics (O'Brien 1993, 1998). This involves beginning with plants, and in particular with 'trees' (large woody plants). Trees are the dominant autotrophic life-forms of terrestrial ecosystems, and are comparatively long-lived organisms that might be best anticipated to reveal fundamental climatic controls on diversity. We make no claims to a neutral review and the limited space available precludes close attention to foundational literature (but see e.g. Brown & Lomolino 1998; Brown *et al.* 2003). Before advancing the case for 'energy', we set out some key general issues that hamper attempts to synthesize and build a general body of diversity theory.

* *Correspondence address: Robert.Whittaker@geog.ox.ac.uk*

Table 7.1 A hierarchical framework for diversity theory. Adapted from Willis & Whittaker (2002). The table aims to identify the main diversity phenomena of interest at particular spatial scales of analysis, some of the more prominent explanatory variables and the temporal scales over which their variation is particularly prominent.

Spatial scale	Diversity phenomena	Environmental variables predominant	Temporal scale at which processes occur
Local	Species richness within communities, within patch	Small-scale biotic and abiotic interactions, e.g. habitat structure, disturbance by fires, storms	*c.*1–100 years
Landscape	Species richness between communities; turnover of species within a landscape	Soils, topography, altitude, drainage	*c.*100–1000s years
Regional	Species-richness patterns across large geographical areas within continents	Water-energy dynamics, area effects (e.g. peninsula effect)	The past 10 000 years, i.e. since end of last glacial period
Continental	Differences in species lineages across continents	Aridification events, glacial/interglacial cycles of the Quaternary, mountain building episodes, e.g. Tertiary uplift of the Andes	The past 1–10 million years
Global scale	Differences reflected in the biogeographical realms, e.g. distribution of mammal families between continents	Continental plate movements, sea-level change	Tens of millions of years

Phenomenology and methodological issues

Area

As the number of individuals and the number of species increase with area, and as measurable environmental heterogeneity varies with area, it is necessary to control area in analyses of geographical patterns of species richness. The most satisfactory approach is to use sample units of fixed size: this avoids the danger that analyses become confounded by variables that happen to co-vary with the area of the underlying sampling units.

Scale, extent and focus

There are numerous metrics of diversity, and they have been applied on widely varying scales, often inconsistently. A crucial distinction is that between (i) the geographical extent of a study system, being the space over which observations are made, e.g. a hillside, a state, a continent or the tropics, and (b) the grain (focus) of the data, being the contiguous area over which a single observation is made, or at which data are aggregated for analysis, e.g. quadrat, plot, latitude–longitude grid cell (Palmer & White 1994; Whittaker *et al.* 2001). Many ecologists appear to use the term scale, or spatial scale, to refer to both of these quite distinct properties of their study systems, but they are not equivalent or substitutable (Palmer & White 1994; Nekola & White 1999). This is shown by analyses using nested sets of grid-cells that reveal different patterns by varying grain and holding extent constant (Rahbek & Graves 2000, 2001; Lennon *et al.* 2001; Koleff & Gaston 2002). Similarly, when scale (i.e. grain) is held constant, but geographical extent is varied, again different patterns can result (e.g. compare Clinebell *et al.* (1995) and Kay *et al.* (1997), as re-analysed by Mittelbach *et al.* (2001, appendices A and B)). A study that uses 0.1 ha plots (*alpha* scale *sensu* Whittaker 1977), be they distributed across different continents, remains a local scale (*sensu* Whittaker *et al.* 2001), or fine focus/grain study (*sensu* Nekola & White 1999), but may not be so described in the literature. The confounding of these two properties within the diversity literature has in our view caused confusion and hindered progress.

Inventory diversity and differentiation diversity at the macroscale

Areas rich in endemics are often also species rich, but deviations occur at regional scales, and patterns in richness and endemism are not necessarily positively related (see Huntley 1996; Brooks *et al.* 2001; Vetaas & Grytnes 2002). In general the explanation for biogeographical patterns of distinctiveness (endemicity) requires theories focusing on evolution and historical contingencies (e.g. Burgess *et al.* 1998), whereas species richness patterns can often be related to contemporary ecological processes and controls (e.g. Linder 2001). This is not to say that climatic correlates cannot be found for regional patterns in endemicity or that there is no historical signal to be detected in species-richness patterns. Rather, we hold that whereas there is some prospect of developing predictive climate-based models of species richness of global applicability, it is hard to conceive doing so for patterns of endemicity. This is because concentrations of relatively localized endemicity within regions are contingent upon historical geographical circumstances, many of which, such as those identified for Amazonia, are subject to conflicting historical hypotheses involving megatectonics and climate-change mechanisms (e.g. Bush 1994; Maslin & Burns 2000; Willis & Whittaker 2000; Colinvaux *et al.* 2001; Richardson *et al.* 2001a,b). At intercontinental scales, biogeographical differences reflect even longer term historical factors ranging across timescales of many millions of years (Table 7.1). Notwithstanding the importance of such processes to present-day patterns of endemicity, it appears that we can consider the species involved as essentially equivalent 'particles' when modelling species richness of ecological classes such as 'trees' (O'Brien 1998; see also Hubbell 2001) or guilds of mammals (Andrews & O'Brien 2000; Aava 2001).

Climate, energy, water–energy dynamics and tree species richness

That climate might provide the fundamental explanation for the 'latitudinal' diversity gradient is an idea that can be traced back at least to the early 19th century work of Baron Alexander von Humboldt (e.g. O'Brien 1998; Hawkins 2001), and the 20th century saw many important theoretical contributions focused on energetics (e.g. Hutchinson 1959; Connell & Orias 1964; MacArthur 1972; Brown 1981). Contemporary authors commonly refer to Wright's (1983) species–energy hypothesis, which postulates that the amount of available energy sets limits to the richness of the system. Recently, a theoretical model of plant species richness at the macroscale has been developed that is consistent with the ideas of species–energy theory, but which encapsulates a more fundamental relationship: the dynamic interaction between water and energy.

Trees are the largest autotrophic life forms, are generally long-lived, and dominate the most productive and diverse natural terrestrial ecosystems. They are lacking from the tops of the highest mountains, the highest latitudes, and from deserts. Where trees do occur azonally within desert regions it is alongside watercourses or where water from aquifers emerges at the surface. As we near the limits of tree growth, and especially the colder limits, few species can tolerate the conditions: the high-latitude tree lines contain typically only one or two species. Globally, these constraints underpin the distribution of biomes, vegetation types and the broad pattern of tree species richness. In essence, trees cannot balance their energy budgets where their requirements for light and heat are not met, where water is unavailable in liquid form for much of the year and where the effective growing season is too short. By extension, richness should be maximal where energy conditions (heat and light) and liquid water availability are optimal throughout the year, i.e. in the tropical rainforest biome. This is to argue that climate sets *the capacity for richness* (*sensu* Brown 1981), and that it should therefore be possible to develop climate-based models of richness to describe known patterns in richness and to predict patterns where richness is not adequately known. Clearly, other factors apart from climate are important and in particular circumstances can act to promote or reduce richness. These factors can best be analysed systematically once climatic variation has been accounted for.

More formally, for trees, as for other plants, two facets of energy matter: heat (ambient air temperatures) and light (the wavebands of photosynthetically active radiation). However, although heat and light are necessary, they are not sufficient, as plant functioning depends critically on water availability, i.e. the presence in the rooting layer of liquid water. It follows that photosynthesis and thus plant productivity must be related to both energy (heat/light) and water and, given the need for water to be available to plants in liquid form, the dynamic interaction between energy and water (O'Brien 1993, 1998). This dynamic is foundational to an understanding of spatial variation in photosynthesis and thus biological activity and, through this activity, spatial patterning in the capacity for plant species richness to be supported (Fig. 7.1). On coarse, cross-continental scales of analysis, O'Brien (1993, 1998) demonstrates that richness increases as a linear function of rainfall and a parabolic function of energy (heat/light). This has been shown for southern Africa at species (Fig. 7.2a), genus and family levels of analysis, for which a simple two-variable regression

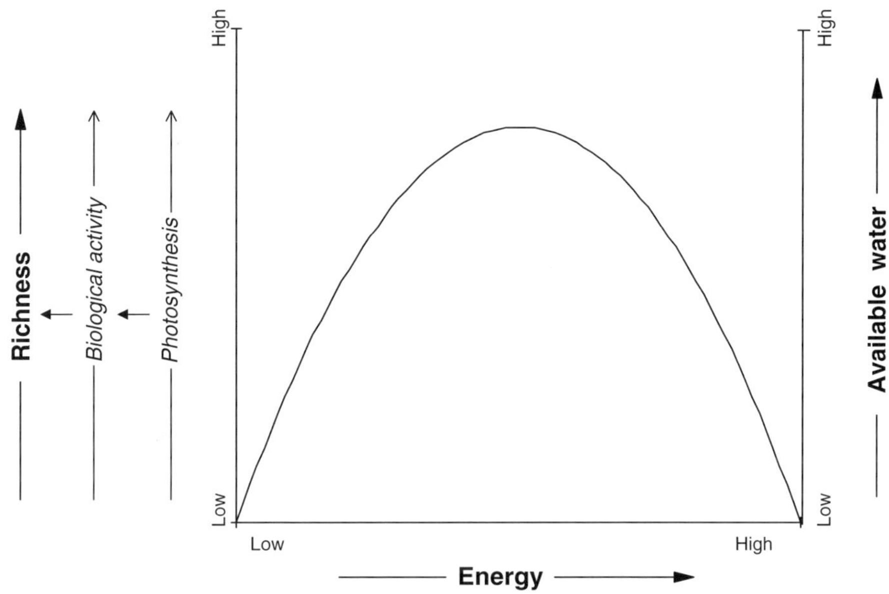

Figure 7.1 An idealized description of the relationship between terrestrial plant richness and climate-controlled water–energy dynamics (from O'Brien *et al.* 1998). Climate's contribution to richness is independent, and described by both the abscissa (energy; heat/light) and the right ordinate (liquid water). The left ordinate indicates the dependent biotic variables. Variability in species richness is depicted as being dependent on biological activity (over time), which depends on photosynthesis–productivity (over time), which depends on water–energy dynamics (over time). Empirically, richness and photosynthesis increase and decrease as a maximized, or linear, function of liquid water (right ordinate) and a parabolic function of energy (abscissa). A parabolic function means that the dependent variables (and liquid water) only operate/exist within a portion of the full range of energy values. This portion is defined by a minimum and maximum value, with an optimum in between (peak of curve). The physical state of water depends on ambient energy conditions: too cold, it is a solid; too hot, a gas. However, the distribution of liquid water on land can be independent of local energy conditions, being a function of a combination of climatic and physiographic factors (e.g. wind, terrain). The biotic variables are always dependent on water–energy dynamics. The curve describes their potential increase from zero to maximum values at the energy optimum and then decrease thereafter back to zero, as a function of energy alone. However, depending upon available liquid water, they can be greater or less than expected as a function of energy alone.

model, richness = PAN + (PEMIN – PEMIN^2), accounts for respectively 79%, 80% and 70% of the variation (O'Brien 1993; O'Brien *et al.* 1998; see Table 7.2 for definition of terms). The inclusion of a third, modifying variable, topographic relief (richness = PAN + (PEMIN – PEMIN^2) + *ln*TOPOG), raised the R^2 values to 86%, 87% and 82% respectively (O'Brien *et al.* 2000).

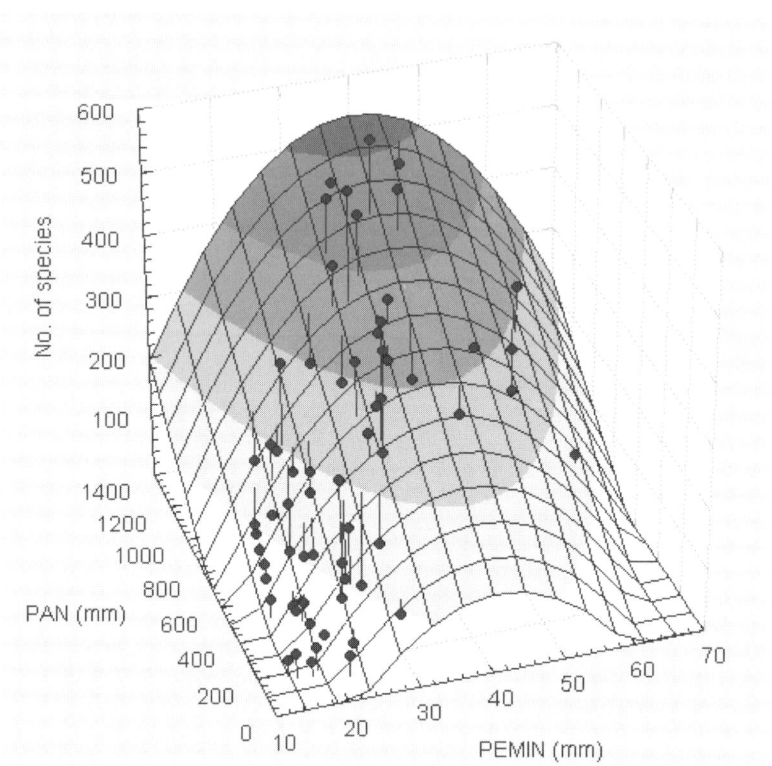

(a)

Figure 7.2 (a) Species richness of southern African trees as a function of PAN (annual rainfall) and PEMIN (potential evapotranspiration). The surface describes the regression model richness = PAN + (PEMIN – $PEMIN^2$), and the stalks connected to each data point indicate the residual variation. Data from O'Brien (1993). These analyses were based on systematic species-range maps, aggregated into cells ($n = 65$, cell size = 25 000 km^2). The climatic data took the form of climate station data assigned to cells: those cells lacking climate stations were omitted from the analysis. (b) The climatic potential for species richness globally, as depicted in O'Brien's (1998) interim general model (IGM). Note that the global energy optimum is higher than that for southern Africa. This model was based on the relationships established for southern Africa (Fig. 7.2a) redescribed with reference to climate data for the whole of Africa. It describes the climatic potential for richness at a scale of 25 000 km^2, across a range of PAN and PEMIN values that are globally representative. The heavy outlines represent the climate envelopes for particular regions: southern Africa study area (SAf), the whole of Africa (Af), South America (SAm), the USA and China (Ch). The values used for these envelopes represent the minimum observed and the 95% quantiles derived from climate station data from Thornthwaite & Mather (1965). The temperate subcontinental regions have a limited range of energy regimes and thus may not reveal the parabolic nature of energy's relationship to taxon richness. In practice, it should not be expected that richness should increase indefinitely as a function of PAN as, beyond a certain point, rainfall ceases to be limiting. Also, in cold regions of the world, heavy winter precipitation can run off without great influence on productivity.

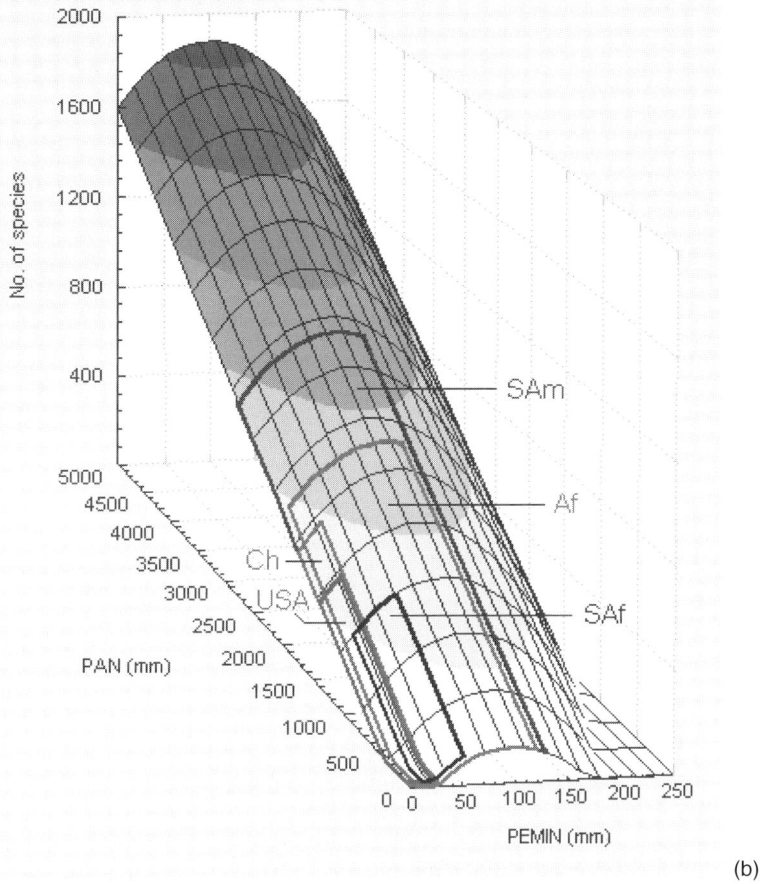

Figure 7.2 *continued*

Southern Africa provides an unusual opportunity in that systematic species-range data are available for a region containing sufficient variation in energy regime to reveal the parabolic form of the relationship. However, the climatic variation within southern Africa is insufficient to be globally representative. In developing an interim general model (IGM) of the climatic potential for richness, O'Brien (1998) therefore recalibrated her original two-variable species-richness model by reference to climate data for the whole of Africa. The resulting two-variable IGM (Fig. 7.2b) has been shown to reproduce known patterns of richness irrespective of whether they are latitudinal (elsewhere in Africa, China), longitudinal (USA) or a mixture of the two (South America) (O'Brien 1998). Two points should be recognized. First, this does not mean that better models cannot be developed to describe richness globally (or for particular regions), merely that the first-order pattern can be

Table 7.2 Glossary of key terms used in this paper (diversity and scale concepts after Whittaker *et al.* 2001, table 1).

Diversity concepts	
Species diversity	Varied meaning: e.g. number of species, or indices weighted by abundance distributions of species (equitability); implying of itself no standardization of sampling
Species richness	Number of species, implying of itself no standardization of sampling
Species density	Number of species in a standardized sample, e.g. per unit area; more precise than the above but less widely adopted
Species turnover, i.e. differentiation diversity	In the present context meaning compositional turnover in space between two inventory (typically local-scale) samples, expressed by a variety of indices or multivariate analyses, and thus qualitatively different from species richness or density
Endemism	An endemic is simply a species confined to a particular geographical area; a focus on areas of high numbers of endemics implies an interest in biogeographical distinctiveness (whether at species or other taxonomic level); often the focus of interest is on restricted-range endemics
Scale concepts	
Spatial scale	Should refer to the size of the base unit used in sampling and analysis, but in practice usage of this term varies, such that either or both of 'extent' and 'focus' may be meant; moreover, size of sample unit is very often not held a constant but is allowed to vary within a study
Extent (geographical)	The geographical space (distance) over which comparisons are made, whether they be using e.g. 1 m^2 or 10 000 km^2 sample units, i.e. of itself implying nothing about spatial scale in the strict sense
Focus	The spatial scale at which data are collected (also called the 'grain') or, in cases, at which data are aggregated for analysis (e.g. local or field scale to regional scale); this concept, unlike 'extent', can be synonymous with spatial scale
Other terms used	
PAN	Annual precipitation (mm)
PET and PEMIN	PET, potential evapotranspiration; PEMIN, minimum monthly PET (mm). Several formulations are in common usage that differ in their properties. Thornthwaite's PEMIN as used in our analyses of southern Africa is derived empirically from temperature and daylength, providing a simultaneous measure of the amount of incident solar energy and the potential loss of water into the atmosphere from surface evaporation and plant biological processes
AET	Actual evapotranspiration is empirically and theoretically distinct from PET, and describes the actual amount of water used to meet the environmental energy demand. AET can never exceed PET, but if rainfall > PET then AET is roughly equal to PET; if rainfall < PET, then AET is equal to precipitation; if precipitation greatly exceeds PET then AET can be less than both PET and rainfall, because of runoff
TOPOG	Maximum – minimum elevation (m), as used by O'Brien *et al.* (2000): from U.S. Geological Survey DEM, resampled to 0.1° resolution, $n > 200$ elevation points/grid cell

depicted (and predicted) by a simple, two-variable, climate-based model. Second, critical evaluation of this model is currently hindered by the lack of systematic species-range data for most of the globe.

Species richness and productivity: theoretical expectations and empirical findings

The relationship between productivity and richness is both crucial and contentious. Productivity can be variously defined, but what is relevant is the rate of energy capture by the target organisms. For autotrophs (including trees) a commonly used metric is thus net primary productivity (NPP). Productivity estimates made in the 1970s (e.g. Lieth & Whittaker 1975) established the broad picture of global variation in terrestrial NPP: it peaks in the tropical forests, with savanna and subtropical woodlands following, temperate and boreal forests lower, and tundra and deserts (hot and cold) declining to extremely low values. More recent and more sophisticated modelling efforts reaffirm this broad picture (e.g. Amthor *et al.* 1998; Zheng *et al.* 2001), and are consistent with a positive relationship between tree richness and productivity at the macroscale.

In contrast to the above, much of the literature concerning the relationship between productivity and plant species richness is empirically and conceptually framed at the local (alpha) scale of analysis (e.g. see Oksanen 1996, 1997; Grace 2001; Mouquet *et al.* 2002). It was at the local patch scale that unimodal productivity–diversity relationships were first noted, and from such studies has come the notion that the general form of the relationship may in fact be unimodal (Rosenzweig & Sandlin 1997; Huston 2001; see review by Abrams 1995). We predict that this generalization will turn out to be as poorly founded as the generalization of declining species richness with increasing elevation (Rahbek 1995, 1997; Lomolino 2001). The theoretical arguments for unimodal patterns focus largely on competition within patches (plus possible artefactual effects: Oksanen, 1997). Additionally, at the local scale, small patches may have atypically high productivity for a variety of reasons (e.g. (i) experimental application of fertilizers and (ii) successional immaturity), and in these instances, through different mechanisms, may have lower species richness than moderately productive patches; i.e. producing hump-shaped richness–productivity relationships. Mittelbach *et al.* (2001) provide an important (if flawed: see below) contribution in their meta-analysis of 257 richness–productivity data sets drawn from 171 published studies. This analysis demonstrates that there is variation in the form of the relationship, as a function of taxon choice and geographical extent of the analysis. Although unimodal patterns were common, positive, u-shaped and negative relationships were also reported, undermining the notion that unimodal patterns have generality.

At the macroscale, climatic explanations for species-richness patterns involve a more straightforward logic regarding productivity. Photosynthetic activity, i.e. productivity (NPP), is controlled fundamentally by climate: the greater the amount and duration of photosynthesis, the greater the biological activity, and in turn the greater is the capacity for richness to be generated and maintained within a region (O'Brien 1993, 1998; O'Brien *et al.* 1998). The expectation is thus for a positive, monotonic curve where diversity increases with productivity. Underlying theoretical arguments include: (i) increased productivity raises abundance of rare species, reducing their extinction rates; (ii) increased productivity increases abundance of rare resource combinations and conditions required by specialist species; (iii) increased productivity increases intraspecific density dependence (e.g. through pest-pressure mechanisms); (iv) over large geographical areas, cells of generally high productivity will contain scattered low productivity sites, and their species will contribute to the diversity measured across high productivity regions (Abrams 1995).

In short, we contend that a positive relationship should prevail for productivity–tree-species-richness relationships at the macroscale. Contrary to this expectation, the Mittelbach *et al.* (2001) meta-analysis includes seven tree data sets that they classified as (i) continental or regional in scale and (ii) demonstrating hump-shaped productivity–richness relationships, and they report one regional-scale study with a u-shaped relationship. The latter is in fact a misclassification based it seems on a transcription error (compare Williams *et al.* (1996) with the classification of these data by Mittelbach *et al.* (2001, appendix B)). Indeed, in all eight cases, we find reason to dismiss the classification of these studies as evidence for unimodal relationships at the macroscale. In brief, we give here just two general criticisms.

First, although Mittelbach *et al.* (2001) use the term 'geographical scale' for the classification of studies as regional (200–400 km) or continental to global (>4000 km), they have actually grouped the data sets not by the grain of the analysis (i.e. area of sampling units) but by the geographical extent over which the sample sites were distributed. Four of the data sets in question include plots of less than 1 ha in size, i.e. they are not macroscale in the crucial sense of focal scale: it is with the latter we are concerned in respect of geographical patterns of overlap in ranges, and thus in richness. Second, because NPP measurements are hard to attain, especially at a landscape scale, the meta-analysis relied on the use of surrogate variables, including biomass, annual rainfall (PAN) and AET (Table 7.2), assuming that each scales simply with productivity. This is a problematic assumption. For instance, there is evidence in the literature from which the meta-analysis has been compiled that systematic variations in energy regime in relation to rainfall can mean that unimodal relationships between rainfall and richness are consistent with—and even supportive of—an underlying positive, monotonic relationship between productivity and richness (e.g. Kay *et al.* 1997). Biomass is also an ambiguous indicator of productivity, as forest stands can have the same biomass with very different levels of stand turnover (cf. e.g. Phillips *et al.* 1994; Whittaker *et al.* 1999). Two of the eight studies cited used AET, a fairly straightforward surrogate for productivity, but only one of these satisfied our first criterion of being based on macroscale (grain) data. Whereas Mittelbach *et al.* (2001, appendix B) classified it as unimodal, the original analysis reported by Currie & Paquin (1987) shows a stronger fit for a logistic model in which peak richness corresponds to maximum AET. Thus, by using geographical extent instead of grain as their metric for geographical scale, and by using productivity surrogates in an oversimplified and in cases invalid fashion, the statistical models reported by Mittelbach *et al.* (2001) appear in practice to be unduly conservative of unimodal patterns. We are thus unconvinced that any published data support the existence of unimodal productivity–richness relationships for tree data sets at the macroscale (i.e. using units of analysis of hundreds or thousands of square kilometres).

As, however, very fine-scale data often demonstrate unimodal productivity–richness patterns, the question arises: can a unimodal local-scale pattern be reconciled with a positive pattern at the macroscale? In theory at least, the answer is 'yes'. By varying patterns of differentiation diversity (species turnover), different patterns can be obtained between inventory diversity and productivity at the local and macroscales (Fig. 7.3; Whittaker *et al.* 2001). For an empirical demonstration of this see Chase & Leibold (2002) and for evidence of the spatial dependency of the relationships between spatial turnover and richness at landscape–regional scales of analysis for birds see Lennon *et al.* (2001) and Koleff & Gaston (2002). Of the patterns depicted in Fig. 7.3, we consider panel (f) to be the most reasonable depiction for trees and panels (c) and (e) to be highly improbable.

Productivity is so difficult to measure directly across large areas that we can still only infer the macroscale patterns from existing data. Such clues as we have provide support in broad terms for the existence of a positive, monotonic relationship between tree richness and productivity, as indicated by the work of Lieth and others in the 1970s (Esser 1998; and see Williams *et al.* 1996). Advances in this area await the more refined application of satellite data, although the ground-truthing of satellite data remains problematic (Liu *et al.* 2002). Process-based models (e.g. Woodward *et al.* 1995) provide an alternative approach, although being based on climatic variables they do not provide an entirely independent means to evaluate climate–productivity relationships.

Linking mechanisms

> 'More dynamic forests are more species-rich because of their greater heterogeneity in space and time, permitting a greater mix of species to coexist. High rainfall and aseasonal lowland forests should be especially productive and therefore especially dynamic. And however complex the causal chain of events may be, a consistent result has emerged: the most species-rich forests in the world are ever wet and ever warm.' (Clinebell *et al.* 1995, p. 83)

We have argued above that climate-controlled water–energy dynamics provide the fundamental pattern in plant productivity, subsequent biological activity and thus the capacity for richness of trees. Next, we consider the mechanisms whereby these fundamental climate–biological drivers may be converted into richness patterns, which at the macroscale represent the patterns resulting from the differential overlap of species ranges. We envisage that there are numerous linking mechanisms that

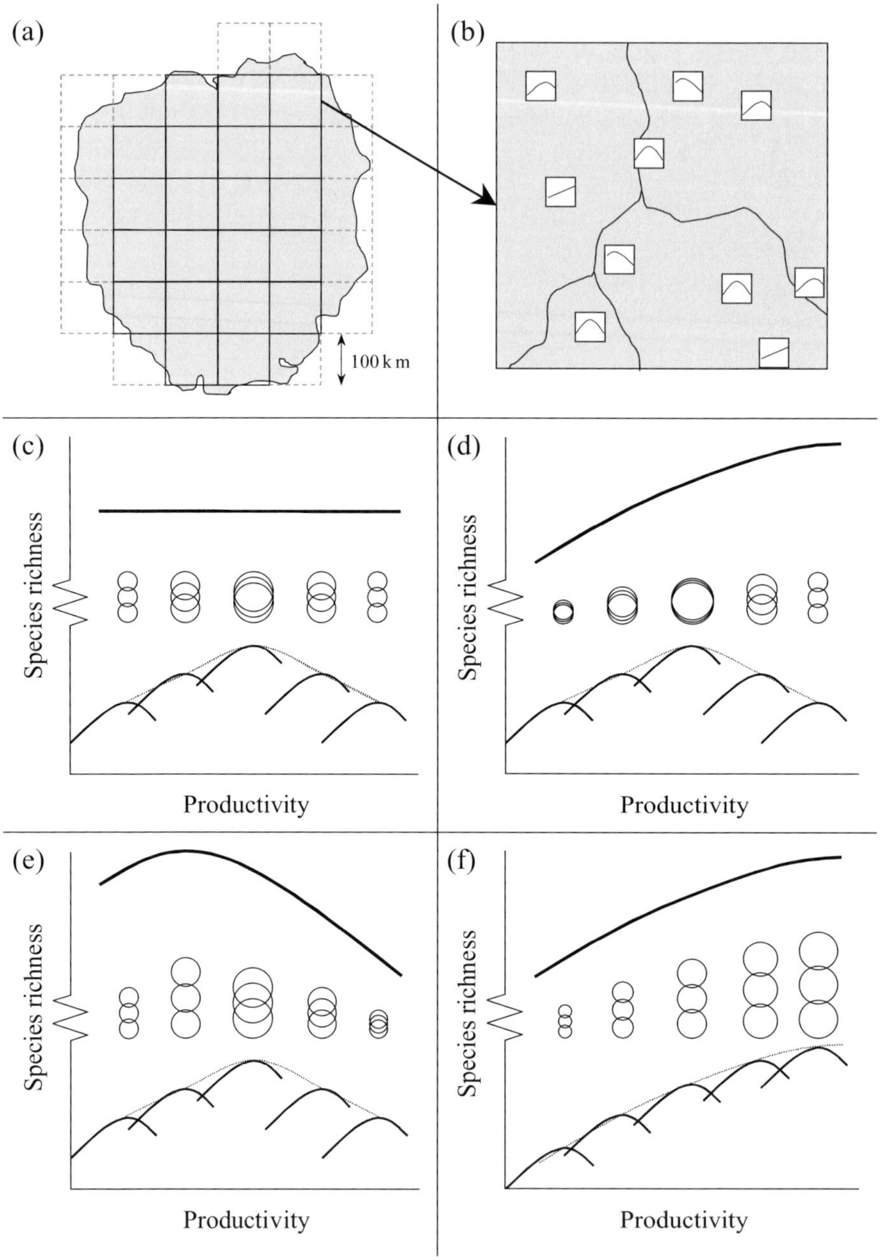

Figure 7.3 Hypothetical productivity–diversity relationships at different scales of analysis (from Whittaker *et al.* 2001). The figure indicates how different relationships theoretically could be obtained between inventory diversity and productivity at the local and macroscales, by varying patterns of differentiation diversity (species turnover) across landscapes or regions. Panel (a) depicts a hypothetical land mass, over which has been placed a grid-cell

continued

have operated over time to contribute to the accumulation of more species per unit area in the tropical forests than at higher latitudes. Our general case is that biological/ecological mechanisms are either local in scale of application, or if regional in impact, are dependent upon climate for their patternation, and thus constitute secondary or 'linking mechanisms' rather than in their own right constituting first-order independent controls setting the pattern (Whittaker *et al.* 2001). We briefly consider a few of these actual or hypothetical linking mechanisms.

Disturbance and turnover

Disturbance (phenomena leading to significantly enhanced mortality) is multifaceted, comprising many phenomena with varying spatial–temporal signals. Some forms are climatic (e.g. storm damage, droughts, aseasonal freezing events), others are to varying degrees independent of climate (e.g. megatectonics, slope failures, river channel migrations). The intermediate disturbance hypothesis (IDH) (Connell 1978) postulates that maximal diversity occurs at intermediate frequencies and intensities of disturbance. It has been operationalized largely at a local scale. As so many key disturbance agencies are climatically controlled and relate to the regional water and energy regimes, it is hard to see the IDH as a competing hypothesis for the grand clines of diversity, unless there can be shown to be measurable macroscale gradients in these disturbance properties across the globe that are

---

system of fixed shape and area cells. Species-richness counts are derived based on presence of native species within each cell, discounting cells that contain significant areas of ocean (as, e.g. O'Brien 1993). Panel (b) depicts a cell from this grid, in which a number of local-scale studies have been conducted, using small (e.g. 1 or 10 m^2) fixed-area quadrats. These hypothetical studies vary in their findings (shown as curves within small squares), but in general appear to show a unimodal response of richness to productivity variation. A river system is shown for the purposes of illustration. In panels (c) to (f): the lower (thinner) curves represent local-scale studies of inventory richness from small (e.g. 1 m^2 or 10 m^2) plots; the upper solid lines represent macroscale studies of inventory richness that might be derived using the macroscale (100 km^2) grid-cell data; and the circles in the middle of each panel indicate varying amounts of differentiation diversity, wherein circle size is scaled roughly to local inventory richness and circle overlap represents the proportion of species in common between local-scale plots. Panels (c) to (f) assume that a unimodal relationship for small plot inventory richness pertains, each solid curve below the break in the *y* axis showing the series of data points sampled within a small area: the series of these curves representing a set of comparative small-plot studies from across a large region. Panels (c) to (e) involve identical local-scale findings, but varying trends in differentiation diversity, and thus three radically different patterns at the macroscale. Panels (d) and (f) involve identical macroscale patterns, underlain by varying patterns from the overall series of small plot studies. For woody plants, according to O'Brien's (1993, 1998) water–energy dynamics theory, we would expect to find the macroscale pattern shown in panels (d) or (f) to pertain, with panel (f) the most likely general representation of the four panels given.

independent of climate. However, disturbance could indeed provide a linking mechanism (Molino & Sabatier 2001). Phillips *et al.* (1994) propose just such a linkage in their analysis of the dynamics of 25 forests from the major tropical regions. These analyses show a relationship between the population turnover and species richness of the forests. They postulate that forests of high productivity must have higher stem turnover as long as biomass is ultimately capped by constraining factors. Thus high productivity may help maintain high richness by promoting frequent, spatially unpredictable small-scale disturbance. This promotes the availability of a wide array of regeneration niches, frequent regeneration opportunities, and thus the sympatric coexistence of ecologically equivalent species.

Nutrients

Clinebell *et al.* (1995) analysed woody plant richness (trees and lianas) for 69 0.1 ha 'Gentry' plots from the neotropics. They generated a number of regression models, for different subsets, and found that annual rainfall and rainfall seasonality were the most important variables. The positive relationship of richness with rainfall appears to reach an asymptote at high values of PAN (*c.*4000 mm PAN). Soil variables were also correlated with precipitation, such that soils of drier forests were more nutrient-rich. Available soil nutrients were shown to have little explanatory power independent of climate-controlled variation in analyses of richness per plot. Interestingly, the number of individuals per plot was positively related to PAN and negatively related to soil fertility. Some authors argue that it is better to express richness as a function of number of individuals rather than of area sampled: when this was included in the analysis, PAN and rainfall seasonality retained their importance but the soil variables declined further in importance. Clinebell *et al.* (1995) interpret their findings as consistent with direct cycling of nutrients from organic matter being of greater importance with increased rainfall within tropical forests. Thus, they find that soil nutrient status is of relatively little independent significance to explaining variations in local richness of sites distributed across a large geographical extent.

Density dependence and the pest pressure hypothesis

Janzen's (1970) pest pressure hypothesis is an example of a density-dependent mechanism. It postulates that tree recruitment is depressed near conspecific adults because of host-specific predation, pest and pathogen attack. This suppresses the expression of competitive dominance, ensuring the sparse distribution of tree species. Condit *et al.* (2000) report analyses of spatial distributions in six tropical forest plots of 25–52 ha on two continents. Their analyses were undertaken as a test of density-dependent effects (including the pest pressure mechanism). They found that in general the tree species were distributed at lower densities than would be the case in temperate forests, but that the majority of species nonetheless demonstrated a significant degree of aggregation. Interestingly, the rarest species within each plot were the most clumped. There is evidence that herbivores and especially plant diseases (Wills *et al.* 1997) do play some role in reducing aggregation but most of this effect has been played out by the time individuals reach 1 cm stem diameter, and beyond 10 cm there is no further reduction in clumping. This study is a useful addition to an equivocal literature (e.g. Givnish 1999; Nichols *et al.* 1999; Harms *et al.* 2000) on the pest pressure hypothesis. It provides qualified support for density dependence being a general force within tropical forests, whereby forests of high NPP are influenced by high levels of biological interactions across trophic levels (fungal infections, herbivores, pollinators, dispersers). These interactions combine to reduce the tendency towards intraspecific clumping that naturally occurs owing to underdispersion of seeds (Webb & Peart 2001). These interactions therefore contribute to the low densities of tropical trees, and the corollary, high alpha diversity. However, none of the studies cited provide comparative evidence by which to assess the differential impact of Janzen's mechanism across different latitudes (Whittaker *et al.* 2001). This rests on variations in 'climate favourability', but Janzen's model provides no means of developing a predictive model as it lacks climatic parameterization.

Specialist mutualisms

Integral to the foregoing mechanism is the idea that ecological specialism is enhanced in the most speciose tropical forests. This is a long-standing concept, and includes not only pest and pathogens, but also seed-dispersal mutualists and pollinators (e.g. Terborgh 1986; Renner 1998; Webb & Peart 2001). The larger variety of seed transportation devices supported by productive tropical ecosystems provide in theory for more finely subdivided dispersal/regeneration niches. Wind pollination is reduced in significance in tropical forests compared with temperate ecosystems, with zoochory, and especially specialist insect pollinators of correspondingly enhanced importance (e.g. Bush & Rivera 2001). This mechanism constitutes a clearly evolutionary–ecological mechanism, in that it suggests a link between high productivity over time and high niche packing through successful speciation.

Niche differentiation

Again linked to the above density-dependent mechanisms, is the idea that the ecological niches of the plants themselves can be very finely subdivided in the high productivity tropics. This should be detectable most readily in the regeneration/recruitment phase of the life cycle. However, as with many of these ideas, it is often difficult to demonstrate the existence of fine subdivision of regeneration niches amongst sympatric congeners in the field (e.g. Brown & Jennings 1998; Svenning 2000). Nonetheless, it is possible to detect ecological differences amongst tree species in a variety of niche dimensions including regeneration requirements, seed dispersal, pollination, phenology, wood density, responses to light, water-logging, etc. These facets of evolutionary ecology, as with the other linking mechanisms, are essentially means by which the macroscale pattern is delivered rather than the driving force behind the pattern. They display where variation is permissible within a stand of co-occurring tree species.

Energy equivalence

Hubbell (2001; Hubbell & Lake, this volume) has developed a new body of theory, at the core of which is a redescription of the species abundance pattern, a new function that appears better to describe data, as exemplified by several tropical forest tree data sets. It is part of a dynamic model, which views the occupants of an area as being drawn from a regional pool, and subject to turnover through time. Hubbell's (2001) elegant analyses demonstrate that we can get a long way by considering trees as ecological particles, by ignoring what makes them different and ignoring phylogeny, and focusing instead on the idea that there is a capacity (tied to the productivity of the landscape) for a fairly well prescribed density of individuals and that these individuals will be distributed in a predictable fashion (the zero-sum multinomial) across a set of distinct species. As we know, these species do differ ecologically, but they are similar in one key respect, that is they must balance their budgets in terms of their capacity to fix energy into chemical form, or they will not persist in the stand or region. This insight, of energy equivalence among tree species, is demonstrated by Enquist *et al.* (1999) in an allometric analysis of 2283 trees of 45 species measured 20 years apart. These observations in turn are supported by advances in the understanding of the internal distribution networks of plants (e.g. West *et al.* 1999; Brown & West 2000; Niklas & Enquist 2001; Brown *et al.*, this volume; Enquist, this volume). The allometric analyses and Hubbell's neutral theory provide differing but in important respects congruent insights into the importance of energetics even at the level of the stand: a scale at which climatic differences are trivial and are masked by significant variation in other variables, such as drainage, slope, pedology, etc. However, the unified neutral theory as it stands does not provide predictive global models of richness variation: this requires climatic variables.

Climate, energy and other taxa

What form should the relationship take for 'other plants'?

We have concentrated on trees as they are the largest autotrophs, are generally long-lived and dominate where ecological conditions are optimal, i.e. in the systems of highest NPP, specifically lowland tropical forests. If Fig. 7.2b is valid, then it might be anticipated that the same form of relationship should emerge if the analysis is extended from just 'trees' to all plants. However, trees are on average very much larger, perhaps on average orders of magnitude larger, than herbs, grasses and even 'other woody plants': the physical space and resources taken up by these respective growth forms, must on average scale accordingly (Niklas & Enquist 2001). Thus, the number of individuals that can be assembled per unit area is much larger for these other synusia than for 'trees'. On the macroscale, this is likely to be accompanied by climatic variation that controls whether the region is tree dominated. The crossing of critical climatic thresholds for tree growth results in highly contrasting vegetational landscapes and associated local diversity patterns, despite the differences in climate being relatively slight. Therefore, where analyses of landscape and regional diversity patterns cross thresholds from tree-dominated to generally tree-less landscapes, the availability of space and resources to other, smaller, plant life forms (excepting epiphytes) is greatly enhanced (Specht & Specht 1989a,b). As these life forms subdivide space at a finer scale, the geographical pattern in total plant diversity across such a study area may be anticipated to differ in form from that encapsulated in Fig. 7.1. Thus, although Fig. 7.1 provides a theoretical statement of the climatic basis for variation in the capacity for richness, developing an IGM that extends from large woody plants ('trees') to incorporate other synusia is likely to require a more complex rule base to the model (see Specht & Specht 1989a,b; Ohlemüller & Wilson 2000).

Climate, energy and animal taxa

It also should be anticipated on ecological grounds that animal taxa will demonstrate differing patterns of richness from trees. For instance, accessible, useable energy supplies for terrestrial megafauna may be anticipated to be lower in closed tropical forest than in more open tropical habitats. Thus, although we contend that water–energy dynamics underpins species richness variation globally, it does not follow that the form of model derived for large woody plants should be mirrored in simple fashion in animal taxa (Andrews & O'Brien 2000). But, we can expect to find that climatic variables are directly, or indirectly, of central importance (e.g. Turner *et al.* 1988, 1996; Currie 1991; Fraser 1998; Boone & Krohn 2000; Aava 2001; Balmford *et al.* 2001; Rahbek & Graves 2001). Similarly, in the marine realm, macroscale patterns in richness may be anticipated to reflect the fundamental controls on productivity in the three-dimensional oceanic systems (e.g. Smith & Brown 2002).

Andrews & O'Brien (2000) analysed variation in mammal species richness for southern Africa using the same grid-based system and climate data as used by O'Brien (1993) for trees. They found that variability in tree richness accounted for 75% of the variability in mammal richness across cells whereas, of the climate variables, the best, thermal seasonality, accounted for 69%. Subdividing the mammal data by size, spatial and dietary guilds, they found that (i) strong correlations with annual temperature exist only for large mammals (60–67% variance), (ii) that small mammals are strongly correlated with plant richness and thermal seasonality, and (iii) that up to 77% of richness of arboreal, frugivorous and insectivorous species could be accounted for by tree richness, compared with just 38–48% for terrestrial herbivores. These striking ecologically structured differences point to the sensitivity of analyses to the initial selection of groups for inclusion and indeed to the range in climatic conditions within the study region. Andrews & O'Brien (2000) also point out that whereas grasslands may be less productive than forests, up to half their NPP may pass through the animal grazing food chain, such that mammal productivity may be high where overall NPP is not. Tropical forests have both greater biomass and higher productivity, but 90% of NPP passes through the detritus food chain, with comparatively little available for large mammals to eat. Thus, macroscale mammal-species-richness patterns across continents may not exhibit a simple positive correlation with NPP (see Aava 2001; Balmford *et al.* 2001).

Transcalar models of richness

If it is taken that the geographical gradients in richness are essentially climatic gradients, and that there is scale dependence in models and analyses of richness (for the latter see e.g. Cornell & Karlson 1996; Clarke & Lidgard 2000; Rahbek & Graves 2001; Whittaker *et al.* 2001; Chase & Leibold 2002; Condit *et al.* 2002; Koleff & Gaston 2002), it remains to develop more complete transcalar models of richness. As a general case, no matter how important we know an environmental factor to be to the ecology of a group of organisms, if that factor happens to exhibit no measurable variation across a series of fixed-area sites, it clearly cannot be causing any non-random pattern of richness that might be found across those sites: attention therefore should be given at the macroscale to those environmental variables of biological significance that vary at the macroscale. Some work has been undertaken on the fractal nature of landscapes but comparatively little seems to be known about how measurable heterogeneity varies with scale for many biologically important variables (see e.g. Palmer 1992). The importance of climate is seen in the success of even very simple one- or two-variable models (e.g. Currie & Paquin 1987; Adams & Woodward 1989). Recent research points to the importance of seasonality in terms both of water and energy regimes (e.g. O'Brien 1993, 1998; Barthlott *et al.* 1996; Andrews & O'Brien 2000), suggesting that more complete climatic models for a variety of taxa are within sight. In addition to purely climatic variables, topography and elevation are important modifying variables that display measurable heterogeneity at the macroscale and which have been found to contribute to model development (e.g. O'Brien *et al.* 2000; Rahbek & Graves 2000, 2001). Increasing topographic amplitude may have several effects: (i) modifying climate across a grid cell; (ii) increasing the diversity of meso- and microclimates and of habitat types; (iii) increasing effective area; (iv) provision of a series of isolates, encouraging local allopatry; (v) providing scope for local movements of species populations within the cell in response to climatic change, i.e. allowing persistence within the climatic envelope of a species by a vertical migration that requires only modest horizontal range shifts; and (vi) high mountains may act as hard barriers preventing the migration of species populations in response to climatic change. Another important variable at this scale of analysis may be surface water distribution, e.g. tree-lined rivers flowing through desert regions, raising tree richness above the zonally determined levels. In the other direction, reductions in richness might be anticipated from excess levels of water produced by run-off from large drainage basins. It seems likely that soil nutrient status can be considered largely (but not wholly) climatically determined in studies at coarse scales or across large extents (e.g. Clinebell *et al.* 1995). Once satisfactory models have been developed describing the macroscale, it will become possible to develop nested models incorporating those factors that display measurable heterogeneity of importance to modelling species richness at finer scales across landscapes.

Conclusions

The editors of the volume identified 'area', 'history' and 'energy' as distinct starting points for debate on diversity gradients. That area is important is a truism, and it is for this reason that we advocate holding area constant in analysing geographical gradients in species richness. That history is important, likewise: contemporary patterns have not emerged from thin air in the last few centuries. Indeed, the grand cline in diversity has persisted for millions of years (Crane & Lidgard 1989) pointing, once more, to the fundamental importance of climatic gradients across the surface of the globe. Although we regard the best starting point to be in a dynamic, climate–energetic model, we also argue that the paradigmatic barriers between these three headers must be broken down if we are to develop a thorough understanding of spatial patterns in diversity (Willis & Whittaker 2002). Thus, although we may appear to be taking a diametrically opposed view to advocates of area (e.g. Rosenzweig 1995; Rosenzweig & Sandlin 1997) and history (e.g. Latham & Ricklefs 1993; Dynesius & Jansson 2000), in fact we differ only as to the degree of importance to attach to these phenomena and how to deal with them in model building, analysis and development of theory.

What is needed is the reconciliation of these divergent perspectives in an overarching, consilient body of diversity theory (Maurer 2000). The panacea offered herein is scale. There appears to be a great deal of scale dependency in many 'ecological' phenomena, and yet many ecologists commonly review information with scant or incomplete regard to the scale of observation. Perhaps the safest approach to scale is to view ecological phenomena as scale dependent or scale delimited until proven otherwise. However, a comprehensive theory of diversity must be transcalar, and have the properties of a hierarchical theory, addressing multiple phenomena of richness and differentiation diversity and the linkages between these patterns and other macroecological patterns, e.g. those in range size, body size, abundance and metabolism (Brown 1995; Blackburn & Gaston 2001; Cardillo 2002). The development of improved general models of richness and diversity remains an important goal for ecology, with potential relevance to palaeoecology and conservation biology. A diversity of approaches to this goal should be encouraged to provide independent means of checking findings and to understand the sensitivity of model outcomes to changing assumptions and parameters.

Acknowledgements

R.J.W. thanks Tim Blackburn and Kevin Gaston for the invitation to participate in the Macroecology symposium, and the British Ecological Society for supporting his participation, and thanks the following for helpful responses to a variety of queries and/or for discussion in the preparation of this manuscript: David Bowman, Terry Dawson, Brian Enquist, Brad Hawkins, Eileen O'Brien and Carsten Rahbek.

References

Aava, B. (2001) Can resource use be the link between productivity and species richness in mammals? *Biodiversity and Conservation* **10**, 2011–2022.

Abrams, P.A. (1995) Monotonic or unimodal diversity–productivity gradients: what does competition theory predict? *Ecology* **76**, 2019–2027.

Adams, J.M. & Woodward, F.I. (1989) Patterns in tree species richness as a test of the glacial extinction hypothesis. *Nature* **339**, 699–701.

Allen, T.F.H. & Starr, T.B. (1982) *Hierarchy: Perspectives for Ecological Complexity.* Chicago University Press, Chicago.

Amthor, J.S. and members of the Ecosystem Working Group (1998) *Terrestrial Ecosystem Responses to Global Change: a Research Strategy.* ORNL Technical Memorandum 1988/27, Oak Ridge National Laboratory, Oak Ridge, Tennessee, 37pp.

Andrews, P. & O'Brien, E.M. (2000) Climate, vegetation, and predictable gradients in mammal species richness in southern Africa. *Journal of Zoology, London* **251**, 205–231.

Balmford, A., Moore, J.L., Brooks, T., Burgess, N., Hansen, L.A., Williams, P. & Rahbek, C. (2001) Conservation conflicts across Africa. *Science* **291**, 2616–2619.

Barthlott, W., Lauer, W. & Placke, A. (1996) Global distribution of species diversity in vascular plants: towards a world map of phytodiversity. *Erdkunde* **50**, 317–327.

Blackburn, T.M. & Gaston, K.J. (2001) Linking patterns in macroecology. *Journal of Animal Ecology* **70**, 338–352.

Boone, R.B. & Krohn, W.B. (2000) Partitioning sources of variation in vertebrate species richness. *Journal of Biogeography* **27**, 457–470.

Brooks, T., Balmford, A., Burgess, N., *et al.* (2001) Toward a blueprint for conservation in Africa. *BioScience* **51**, 613–624.

Brown, J. H. (1981) Two decades of homage to Santa Rosalia: toward a general theory of diversity. *American Zoologist* **21**, 877–888.

Brown, J.H. (1995) *Macroecology.* Chicago University Press, Chicago.

Brown, J.H. & Lomolino, M.V. (1998) *Biogeography,* 2nd edn. Sinauer, Sunderland, MA.

Brown, J.H. & West, G.B. (eds) (2000) *Scaling in Biology.* Oxford University Press, Oxford.

Brown, J.H., Lomolino, M.V. & Sax, D. (eds) (2003) *Foundations of Biogeography.* Chicago University Press, Chicago.

Brown, N.D. & Jennings, S. (1998) Gap-size niche differentiation by tropical rainforest trees: a testable hypothesis or a broken-down bandwagon? In: *Dynamics of Tropical Communities* (eds D.M. Newbery, H.H.T. Prins & N.D. Brown), pp. 79–94. Blackwell Science, Oxford.

Burgess, N.D., Clarke, G.P. & Rodgers, W.A. (1998) Coastal forests of eastern Africa: status, endemism patterns and their potential causes. *Biological Journal of the Linnean Society* **64**, 337–367.

Bush, M.B. (1994) Amazonian speciation: a necessarily complex model. *Journal of Biogeography* **21**, 5–17.

Bush, M.B. & Rivera, R. (2001) Reproductive ecology and pollen representation among neotropical trees. *Global Ecology and Biogeography* **10**, 359–367.

Cardillo, M. (2002) Body size and latitudinal gradients in regional diversity of New World birds. *Global Ecology and Biogeography* **11**, 59–65.

Chase, J.M. & Leibold, M.A. (2002) Spatial scale dictates the productivity–biodiversity relationship. *Nature* **416**, 427–430.

Clarke, A. & Lidgard, S. (2000) Spatial patterns of diversity in the sea: bryozoan species richness in the North Atlantic. *Journal of Animal Ecology* **69**, 799–814.

Clinebell, R.R. II, Phillips, O.L., Gentry, A.H., Starks, N. & Zuuring, H. (1995) Predictions of neotropical tree and liana species richness from soil and climatic data. *Biodiversity and Conservation* **4**, 56–90.

Colinvaux, P.A., Irion, G., Räsänen, M.E., Bush, M.B. & Nunes de Mello, J.A.S. (2001) A paradigm to be discarded: geological and paleoecological data falsify the Haffer & Prance refuge hypothesis of Amazonian speciation. *Amazoniana* **16**, 609–646.

Condit, R., Ashton, P.S., Baker, P., *et al.* (2000) Spatial patterns in the distribution of tropical tree species. *Science* **288**, 1414–1418.

Condit, R., Pitman, N., Leigh, E.G. Jr., *et al.* (2002) Beta-diversity in tropical forest trees. *Science* **295**, 666–669.

Connell, J.H. (1978) Diversity in tropical rain forests and coral reefs. *Science* **199**, 1302–1310.

Connell, J.H. & Orias, E. (1964) The ecological regulation of species diversity. *American Naturalist* **93**, 399–414.

Cornell, H. V. & Karlson, R.H. (1996) Species richness of reef-building corals determined by local and regional processes. *Journal of Animal Ecology* **65**, 233–241.

Crane, P.R. & Lidgard, S. (1989) Angiosperm diversification and paleolatitudinal gradients in Cretaceous floristic diversity. *Science* **246**, 675–678.

Currie, D.J. (1991) Energy and large-scale patterns of animal- and plant-species richness. *American Naturalist* **137**, 27–49.

Currie, D. J. & Paquin, V. (1987) Large-scale biogeographical patterns of species richness of trees. *Nature* **329**, 326–327.

Dynesius, M. & Jansson, R. (2000) Evolutionary consequences of changes in species' geographical distributions driven by Milankovitch climate oscillations. *Proceedings of the National Academy of Sciences, USA* **97**, 9115–9120.

Enquist, B. J., West, G. B., Charnov, E. L. & Brown, J. H. (1999) Allometric scaling of production and life-history variation in vascular plants. *Nature* **401**, 907–911.

Esser, G. (1998) *NPP Multi-biome: Global Osnabruck Data, 1937-1981.* Available on-line [http://www.daac.ornl.gov/] from Oak Ridge National Laboratory Distributed Active Archive Center, Oak Ridge, TN.

Fraser, R.H. (1998) Vertebrate species richness at the mesoscale: relative roles of energy and heterogeneity. *Global Ecology and Biogeography Letters* **7**, 215–220.

Givnish, T.J. (1999) On the causes of gradients in tropical tree diversity. *Journal of Ecology* **87**, 193–210.

Grace, J.B. (2001) The roles of community biomass and species pools in the regulation of plant diversity. *Oikos* **92**, 193–207.

Harms, K.E., Wright, S.J., Calderón, O., Hernández, A. & Herre, E.A. (2000) Pervasive density-dependent recruitment enhances seedling diversity in a tropical forest. *Nature* **404**, 493–495.

Hawkins, B.A. (2001) Ecology's oldest pattern? *Trends in Ecology and Evolution* **16**, 470.

Hubbell, S. P. (2001) *The Unified Neutral Theory of Biodiversity and Biogeography.* Princeton University Press, Princeton.

Huntley, B.J. (1996) Biodiversity conservation in the new South Africa. In: *Biodiversity, Science and Development: Towards a New Partnership* (eds F. di Castri & T. Younès), pp. 282–303. CABI, Wallingford.

Huston, M.A. (2001) People and biodiversity in Africa. *Science* **293**, 1591.

Hutchinson, G.E. (1959) Homage to Santa Rosalia, or why are there so many kinds of animals. *American Naturalist* **93**, 145–159.

Janzen, D.H. (1970) Herbivores and the number of tree species in tropical forests. *American Naturalist* **104**, 501–508.

Kay, R. F., Madden, R.H., Van Schaik, C. & Higdon, D. (1997) Primate species richness is determined by plant productivity: implications for conservation. *Proceedings of the National Academy of Sciences, USA* **94**, 13023–13027.

Koleff, P. & Gaston, K.J. (2002) The relationship between local and regional species richness and spatial turnover. *Global Ecology and Biogeography* **11**, 363–375.

Latham, R.E. & Ricklefs, R.E. (1993) Global patterns of tree species richness in moist forests: energy-diversity theory does not account for variation in species richness. *Oikos* **67**, 325–333.

Lennon, J.J., Koleff, P., Greenwood, J.J.D. & Gaston, K.J. (2001) The geographical structure of British bird distributions: diversity, spatial turnover and scale. *Journal of Animal Ecology* **70**, 966–979.

Lieth, H. & Whittaker, R.H. (1975) *The Primary Productivity of the Biosphere.* Springer-Verlag, New York.

Linder, H.P. (2001) Plant diversity and endemism in sub-Saharan tropical Africa. *Journal of Biogeography* **28**, 169–182.

Liu, J., Chen, J.M., Cihlar, J. & Chen, W. (2002) Net primary productivity mapped for Canada at 1-km resolution. *Global Ecology and Biogeography* **11**, 115–129.

Lomolino, M.V. (2001) Elevation gradients of species-density: historical and prospective views. *Global Ecology and Biogeography* **10**, 3–13.

MacArthur, R.H. (1972) *Geographical Ecology.* Harper & Row, New York.

Maslin, M.A. & Burns, S.J. (2000) Reconstruction of the Amazon Basin effective moisture availability

over the past 14,000 years. *Science* **290**, 2285–2287.
Maurer, B.A. (2000) Macroecology and consilience. *Global Ecology and Biogeography* **9**, 275–280.
Mittelbach, G.G., Steiner, C.F., Scheiner, S.M., *et al.* (2001) What is the observed relationship between species richness and productivity? *Ecology* **82**, 2381–2396.
Molino, J.-F. & Sabatier, D. (2001) Tree diversity in tropical rain forests: a validation of the intermediate disturbance hypothesis. *Science* **294**, 1702–1704.
Mouquet, N., Moore, J.L. & Loreau, M. (2002) Plant species richness and community productivity: why the mechanism that promotes coexistence matters. *Ecology Letters* **5**, 56–65.
Nekola, J.C. & White, P.S. (1999) The distance decay of similarity in biogeography and ecology. *Journal of Biogeography* **26**, 867–878.
Nichols, J.D., Agyeman, V.K., Agurgo, F.B., Wagner, M.R. & Cobbinah, J.R. (1999) Patterns of seedling survival in the tropical African tree *Milicia excelsa*. *Journal of Tropical Ecology* **15**, 451–461.
Niklas, K.J. & Enquist, B.J. (2001) Invariant scaling relationships for interspecific plant biomass production rates and body size. *Proceedings of the National Academy of Sciences, USA* **98**, 2922–2927.
O'Brien, E.M. (1993) Climatic gradients in woody plant species richness: towards an explanation based on an analysis of Southern Africa's woody flora. *Journal of Biogeography* **20**, 181–198.
O'Brien, E.M. (1998) Water–energy dynamics, climate, and prediction of woody plant species richness: an interim general model. *Journal of Biogeography* **25**, 379–398.
O'Brien, E.M., Whittaker, R. J. & Field, R. (1998) Climate and woody plant diversity in southern Africa: relationships at species, genus and family levels. *Ecography* **21**, 495–509.
O'Brien, E.M., Field, R. & Whittaker, R. J. (2000) Climatic gradients in woody plant (tree and shrub) diversity: water-energy dynamics, residual variation, and topography. *Oikos* **89**, 588–600.
Odum, E.P. (1983) *Basic Ecology.* Holt Saunders, New York.
Ohlemüller, R. & Wilson, J.B. (2000) Vascular plant species richness along latitudinal and altitudinal gradients: a contribution from New Zealand temperate rainforests. *Ecology Letters* **3**, 262–266.
Oksanen, J. (1996) Is the humped relationship between species richness and biomass an artefact due to plot size? *Journal of Ecology* **84**, 293–295.
Oksanen, J. (1997) The no-interaction model does not mean that interactions should not be studied. *Journal of Ecology* **85**, 101–102. (See preceding pages of issue for debate.)
Palmer, M.W. (1992) The coexistence of species in fractal landscapes. *American Naturalist* **139**, 375–397.
Palmer, M.W. & White, P.S. (1994) Scale dependence and the species-area relationship. *American Naturalist* **144**, 717–740.
Phillips, O.L., Hall, P., Gentry, A.H., Sawyer, S.A. & Vasquez, R. (1994) Dynamics and species richness of tropical rain forests. *Proceedings of the National Academy of Sciences, USA* **91**, 2805–2809.
Rahbek, C. (1995) The elevational gradient of species richness: a uniform pattern? *Ecography* **18**, 200–205.
Rahbek, C. (1997) The relationship among area, elevation, and regional species richness in Neotropical birds. *American Naturalist* **149**, 875–902.
Rahbek, C. & Graves, G.R. (2000) Detection of macro-ecological patterns in South American hummingbirds is affected by spatial scale. *Proceedings of the Royal Society, London, Series B* **267**, 2259–2265.
Rahbek, C. & Graves, G.R. (2001) Multiscale assessment of patterns of avian species richness. *Proceedings of the National Academy of Sciences, USA* **98**, 4534–4539.
Renner, S. (1998) Effects of habitat fragmentation on plant pollinator interactions in the tropics. In: *Dynamics of Tropical Communities* (eds D.M. Newbery, H.H.T. Prins & N.D. Brown), pp. 339–360. Blackwell Science, Oxford.
Richardson, J.E., Weitz, F.M., Fay, M.F., *et al.* (2001a) Rapid and recent origin of species richness in the Cape flora of South Africa. *Nature* **412**, 181–183.
Richardson, J.E., Pennington, R.T. Pennington, T.D. & Hollingsworth, P.M. (2001b) Rapid diversification of a species-rich genus of neotropical rain forest trees. *Science* **291**, 2242–2245.
Rosenzweig, M.C. (1995) *Species Diversity in*

Space and Time. Cambridge University Press, Cambridge.

Rosenzweig, M.C. & Sandlin, E.A. (1997) Species diversity and latitude: listening to area's signal. *Oikos* **80**, 172–176.

Smith, K.R. & Brown, J.H. (2002) Patterns of diversity, depth range and body size among pelagic fishes along a gradient of depth. *Global Ecology and Biogeography* **11**, 313–322.

Specht, R.L. & Specht, A. (1989a) Species richness of overstorey strata in Australian plant communities—the influence of overstorey growth rates. *Australian Journal of Botany* **37**, 321–326.

Specht, R.L. & Specht, A. (1989b) Species richness of schlerophyll (heath) plant communities in Australia—the influence of the overstorey cover. *Australian Journal of Botany* **37**, 337–350.

Svenning, J.-C. (2000) Small canopy gaps influence plant distribution in the rain forest understory. *Biotropica* **32**, 252–261.

Terborgh, J. (1986) Keystone plant resources in the tropical forest. In: *Conservation Biology: the Science of Scarcity and Diversity* (ed. M.E. Soulé), pp. 330–344. Sinauer, Sunderland, MA.

Thornthwaite, C. W. & Mather, J. R. (1962–1965) *Average Climatic Water Balance Data of the Continents*. Publications in Climatology 15–18, Laboratory of Climatology, Centerton, NJ.

Turner, J.R.G. (1991) Stochastic processes in populations: the horse behind the cart? In: *Genes in Ecology* (eds R.J. Berry, T.J. Crawford & G.M. Hewitt), pp. 199–220. Blackwell Scientific Publications, Oxford.

Turner, J.R.G., Lennon, J.J. & Lawrenson, J.A. (1988) British bird species distributions and the energy theory. *Nature* **335**, 539–541.

Turner, J.R.G., Lennon, J. & Greenwood, J.J.D. (1996) Does climate cause the global biodiversity gradient? In: *Aspects of the Genesis and Maintenance of Biological Diversity* (eds M. Hochberg, J. Clobert & R. Barbault), pp. 199–220. Oxford University Press, Oxford.

Vetaas, O.R. & Grytnes, J.-A. (2002) Distribution of vascular plant species richness and endemic richness along the Himalayan elevation gradient in Nepal. *Global Ecology and Biogeography* **11**, 291–301.

Webb, C.O. & Peart, D.R. (2001) High seed dispersal rates in faunally intact tropical rain forest: theoretical and conservation implications. *Ecology Letters* **4**, 491–499.

West, G.B., Brown, J.H. & Enquist, B.J. (1999) A general model for the structure and allometry of plant vascular systems. *Nature* **400**, 664–667.

Whittaker, R.H. (1977) Evolution of species diversity in land communities. In: *Evolutionary Biology* (eds M.K. Hecht, W.C. Steere & B. Wallace), Vol. 10, pp. 250–268. Plenum Press, New York.

Whittaker, R.J., Partomihardjo, T. & Jones, S.H. (1999) Interesting times on Krakatau: stand dynamics in the 1990s. *Philosophical Transactions of the Royal Society, London, Series B* **354**, 1857–1867.

Whittaker, R. J., Willis, K. J. & Field, R. (2001) Scale and species richness: towards a general, hierarchical theory of species diversity. *Journal of Biogeography* **28**, 453–470.

Williams, R. J., Duff, G.A., Bowman, D.M.J.S. & Cook, G.D. (1996) Variation in the composition and structure of tropical savannas as a function of rainfall and soil texture along a large-scale climatic gradient in the Northern Territory, Australia. *Journal of Biogeography* **23**, 747–756.

Willis, K. J. & Whittaker, R. J. (2000) The refugial debate. *Science* **287**, 1406–1407.

Willis, K.J. & Whittaker, R.J. (2002) Species diversity–scale matters. *Science* **295**, 1245–1248.

Wills, C., Condit, R., Foster, R.B. & Hubbell, S.P. (1997) Strong density- and diversity-related effects help to maintain tree species diversity in a neotropical forest. *Proceedings of the National Academy of Sciences, USA* **94**, 1252–1257.

Wright, D.H. (1983) Species–energy theory: an extension of species–area theory. *Oikos* **41**, 496–506.

Woodward, F.I., Smith, T.M. & Emanuel, W.R. (1995) A global primary productivity and phytogeography model. *Global Biogeochemical Cycles* **9**, 471–490.

Zheng, D.L., Prince S.D. and Wright R. (2001) *NPP Multi-Biome: Gridded Estimates for Selected Regions Worldwide, 1989–2001*. Available on-line [http://www.daac.ornl.gov/] from the Oak Ridge National Laboratory Distributed Active Archive Center, Oak Ridge, TN.

Chapter 8
The importance of historical processes in global patterns of diversity

Andrew Clarke and J. Alistair Crame*

Introduction

This review of the role of historical processes in generating large-scale patterns of diversity starts from the deceptively simple question of why the diversity of organisms varies from place to place, and specifically why there are so many species in the tropics. In ecology simple questions rarely have simple answers, and there remains no single consensus explanation as to what regulates biological diversity. In recent years, strong arguments have been advanced for the primacy of area (Rosenzweig 1995), heterogeneity (Huston 1994) and energy (Cousins 1989; Currie 1991; Currie & Fritz 1993). The problem we face is that it is very difficult to devise tests that distinguish convincingly between these competing explanations. We must also recognize that patterns of diversity may not be the result of a single driver but rather the outcome of several factors interacting (Tilman & Pacala 1993; Blackburn & Gaston 1996; Gaston & Blackburn 2000).

Global-scale patterns in diversity

The high diversity of tropical regions is part of a global pattern that at low resolution appears deceptively simple (Fig. 8.1a). A similar pattern of diversity is shown by some marine taxa (Fig. 8.1b), although here there are notable differences to the north and south of the Equator. This asymmetry has emerged recently as a common factor in the patterns of diversity of both marine and terrestrial taxa (Gaston & Spicer 1998; Crame 2000a, b). It suggests either that the main factor regulating large-scale patterns of diversity is itself asymmetric, or that the outcome is contingent upon other factors which differ between the two hemispheres.

Although the bold pattern evident at low spatial resolution would seem to imply a strong mechanism, the real world has two layers of additional complexity. The first is that the surface of the globe is partitioned unequally between land and sea, with a strong asymmetry in the relative proportions between the northern and southern hemispheres. The second is that both marine and terrestrial realms comprise a range

* *Correspondence address: accl@bas.ac.uk*

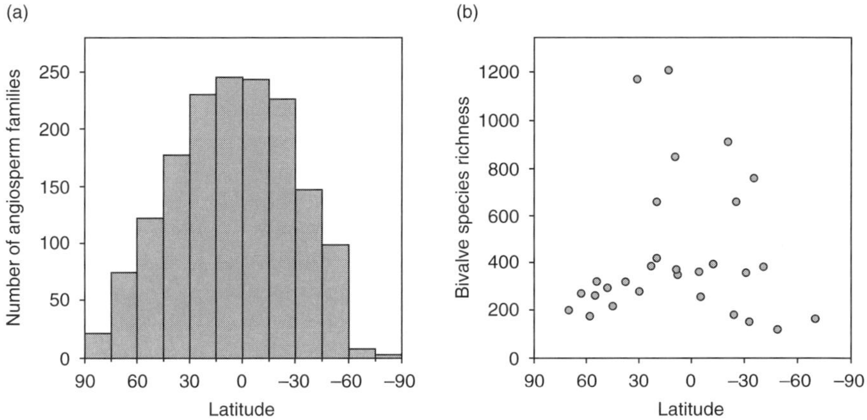

Figure 8.1 Two data sets showing simple patterns of diversity at low resolution (global scale); Southern Hemisphere latitudes are shown negative. (a) Angiosperms. Family richness pooled into bins of 15° latitude (redrawn from Woodward 1987). (b) Marine shallow-water (continental shelf) bivalve molluscs, showing a similar peak in the tropics and tailing off towards the poles; note subtly different, less symmetric pattern (from Crame 2000a).

of different habitats, themselves distributed unevenly. As a result, higher resolution patterns of diversity reveal a large amount of heterogeneity superimposed upon the large-scale trends, and the global distribution of plant species diversity exhibits areas of both high and low species richness at many latitudes (Barthlott *et al.* 1996; Groombridge & Jenkins 2000). The basic pattern that ecologists need to explain is thus one of a bold simple pattern at a global scale, with increasing heterogeneity apparent at higher resolution. An understanding of the role of spatial scale is thus integral to these patterns.

Spatial and temporal scale

The relationship between physical and biological processes is both subtle and complex. It has long been recognized that spatial and temporal scales of variability are closely linked in physical environmental processes (Steele 1985) and to a first approximation these two scales are directly linked. This is also true for biological processes, although here there are important differences between the marine and terrestrial environments (Fig. 8.2).

Marine organisms live in a fluid medium that is always in motion. Only the larger nekton can exhibit behaviour that is independent of this motion. Fundamental to the relationship between physical and biological processes in the sea are the length scales over which those processes of importance to organisms operate. Typical ocean basins are 10^7 m wide or more, and this is the biogeographical scale of whole marine communities. The Coriolis and gravitational forces define the Rossby internal de-

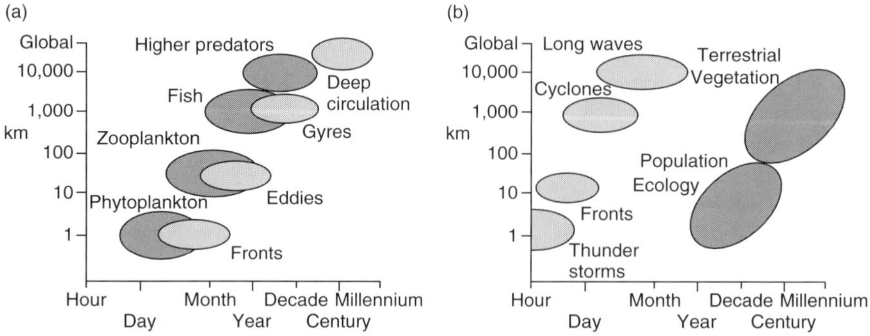

Figure 8.2 Spatial and temporal scales in ecology. (a) In the marine environment spatial and temporal scales are closely related, and the relationships for the physical environment and the biota are closely matched. In the physical environment energy is dissipated as it is transferred from larger to smaller scales; in the biota energy is transferred from smaller to larger scales. (b) In the terrestrial environment the relationships between spatial and temporal scale for the physical environment and for ecological processes are offset. This offset is related to the mechanisms by which energy is propagated through air and water (organisms being largely composed of water), although it also reflects the relatively static nature of organisms. The original formulation of this diagram was Stommel (1963), with many subsequent modifications.

formation scale; this varies with latitude and governs the width of oceanic currents, coastal upwelling regions and the size of oceanic eddies. Also important is the very much smaller scale at which viscosity balances the inertial forces of turbulent eddies, which marks an important change in locomotion and feeding mechanisms. For larger organisms (typically 1 cm or longer) nutrients and waste products are moved rapidly by turbulent diffusion, unaffected by viscosity. Small organisms must depend on molecular diffusion for these transfers.

An important difference between the marine and terrestrial environments lies in the relationship between physical and biological processes. In the marine environment these lie together in the time–space plot (Fig. 8.2a), whereas in the terrestrial environment they are offset (Fig. 8.2b). The difference is related to the mechanism and rate at which energy is propagated through air and water (organisms being composed primarily of water). Patterns of variability also vary between the two environments. If regular short-term cycles (diurnal, tidal, lunar and seasonal) are removed from environmental measures, there remains a large residual variability. For terrestrial temperature records this residual variability is essentially random up to timescales of about 50 years; over longer timescales the variability increases significantly (a pattern termed reddened, by analogy with light). What we know of the marine environment suggests this is very different. Here long-term temperature records from the deep sea indicate that variance increases continuously from hours to years, and inclusion of data from proxy records suggests that this relationship extends to very long timescales (Steele 1985).

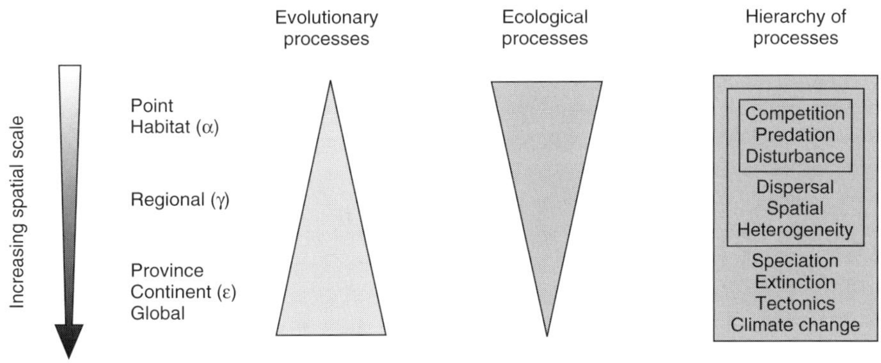

Figure 8.3 Schematic illustration of the balance between ecological and evolutionary timescales for diversity analysed at different spatial scales. The size of the triangular symbols is entirely arbitrary, although it is important that neither ecological nor evolutionary influences are zero at any spatial scale. The nested rectangles emphasize the hierarchical nature of the processes influencing diversity at different spatial scales.

This simple analysis of scale leads to important considerations of direct relevance to patterns of biological diversity. Differences between marine and terrestrial environments mean that even if we were to identify a single global factor responsible for patterns of diversity, we should expect that the outcome would differ between land and sea. Equally a factor that appears to work well at explaining patterns of diversity on land cannot be applied unmodified to the sea (and vice versa). The ecological consequences of the difference in variability between the marine and terrestrial environments are only now beginning to be explored (Halley 1996; Link 2002). Analysis of patterns of extinction in the fossil record have suggested differences between land and sea, indicating that these fundamental differences in environmental variability may also influence evolutionary processes (Solé *et al.* 1997, 1999; Hewzulla *et al.* 1999).

The importance of the relationship between spatial and temporal scales (Fig. 8.2) is that diversity at different spatial scales is regulated primarily by processes working over different timescales (Fig. 8.3) (Ricklefs 1989). In the simplest terms, if we are seeking explanations for patterns of diversity at the macroecological (global) spatial scale, then we need to look at processes operating over long temporal scales. Integration of palaeoecology into macroecological thinking is thus important (Jablonski *et al.*, this volume).

What is the relevant timescale for historical ecology?

The evolutionary history of the biosphere reveals an overall increase in marine diversity since the early Cambrian (Fig. 8.4). The range of basic body plans (*bauplane*, as expressed by the number of metazoan phyla) appears to have remained unaltered since the early Cambrian (Jablonski 2000) and the subsequent increase in diversity

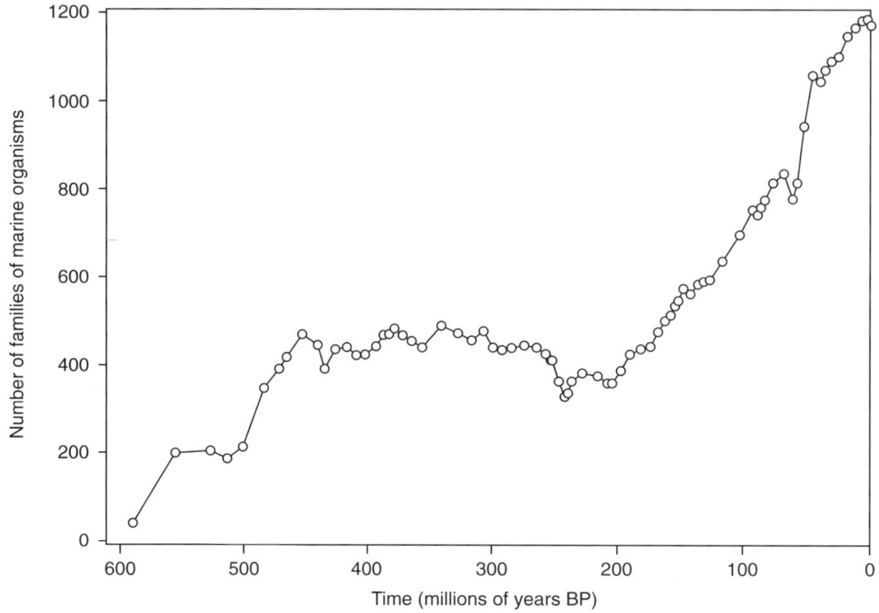

Figure 8.4 The evolutionary history of marine diversity. Data are for families of all known fossil marine organisms. Note the overall increase in diversity through geological time, interrupted by a series of mass extinction events, with a period of almost uninterrupted increase in diversity through the Cenozoic. (Data kindly provided by Mike Benton.)

has been at lower taxonomic levels (orders, families, genera and species). This is precisely what would be predicted from evolutionary diversification and a broadly similar pattern is seen for terrestrial habitats (Benton 1995, 1999).

This increase in total diversity has been neither steady nor monotonic. Important features in the marine record include the early Palaeozoic diversification, a plateau in diversity in the middle to late Palaeozoic, and the major late Mesozoic–Cenozoic radiation (Owen & Crame 2002, and references therein). Major features of the terrestrial record are the initial invasion of the land, the successive dominance of amphibians, reptiles, mammals and birds, and the angiosperm radiation (Benton 1999). Evidence suggests that there are more different kinds of organism alive in the sea and on land today than there have ever been.

The fossil record indicates that the increase in marine diversity has been particularly marked at all taxonomic levels through the Cenozoic. At the level of genus and species the diversification was the longest and most intense of the entire Phanerozoic (Signor 1990; Benton 1999; Foote 2000; but see Alroy *et al.* 2001). During this period groups such as neogastropods, heteroconch bivalves, cheilostome bryozoans, decapod crustaceans and teleost fish all underwent a widespread proliferation in shallow seas. At the same time, similar spectacular diversifications were taking place on land amongst amphibians, reptiles, birds, mammals, flowering plants and insects

(Hallam 1994). That so many different groups of organisms, both terrestrial and marine, should have diversified so strikingly and in parallel has suggested the possiblility of a single mechanism driving Cenozoic diversification (Signor 1990; Benton 1999). Crame (2001) has summarized evidence to suggest that a significant fraction of this diversification took place in lower latitude and tropical regions.

The Cenozoic: climatic change and tectonic activity

The Cenozoic has been characterized by three interrelated global-scale environmental processes: the continued fragmentation of Gondwana, changes in oceanic circulation, and overall climatic cooling (with associated glaciation at high latitudes). The fragmentation of Gondwana had been initiated in the mid-Mesozoic, with substantial areas of oceanic floor in the South Atlantic being created in the Jurassic and Cretaceous. The critical event in the establishment of the present-day oceanographic regime was the final separation of the Antarctic Peninsula from southern South America, with the opening of Drake Passage. The precise timing of this event is still uncertain: marine geophysical evidence indicates that sea-floor spreading began in the early Late Oligocene (28 Ma; where Ma represents million years ago) but full deep-water circulation may not have been established until 23–24 Ma (Barker *et al.* 1991). Palaeontological evidence, however, suggests that terrestrial faunal links between Antarctica and South America had been severed by the early Eocene (Woodburne & Case 1996). Other globally important Cenozoic tectonic events included the progressive northward movement of the Africa–Arabia landmass, constricting and eventually closing the Tethyan Ocean around 20 Ma, and the collision of the Australian Plate with Indonesia (20–15 Ma) (Crame & Rosen 2002).

These tectonic changes had two important implications for large-scale patterns of diversity. Fragmentation of Gondwana separated once continuous terrestrial floras and faunas, leading to isolation and separate evolutionary development. Classic examples of the latter include the lemuroid primate fauna of Madagascar, and the marsupial radiations in South America and Australia. The second consequence was a radical reorganization in the oceanic circulation, and particularly a shift from a predominantly equatorial to a strongly meridional (north–south) or gyral pattern. This was especially marked in the Pacific Ocean (Kennett *et al.* 1985; Grigg 1988). The enhanced transport of warm waters to high latitudes, with associated precipitation, is likely to have been a key factor in the building of continental ice-sheets.

Also important climatically was the gradual uplifting of the Central American Isthmus over the period 13–2 Ma. The closure of the Central American Seaway led to the deflection north of the major Atlantic western boundary current (the Gulf Stream), and in turn to the formation of cold saline North Atlantic Deep Water. The latter was a key factor both in the thermal isolation of the Arctic basin and the initiation of the modern thermohaline circulation (Stanley 1995).

These tectonic and oceanographic changes were undoubtedly critical to the long-term climate change that is such a marked feature of the Cenozoic. In the late Cretaceous and Palaeocene, global seawater temperatures were generally warm,

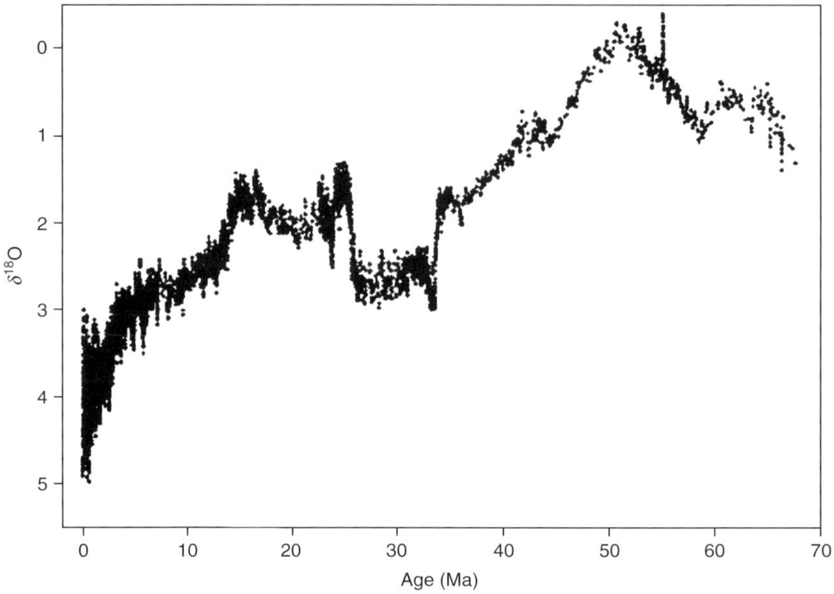

Figure 8.5 Oceanic palaeotemperature curve for the Cenozoic, based on oxygen isotopic composition for benthic foraminifera from 41 deep-sea cores. It thus represents an average picture for three oceanic basins. Pelagic temperatures are typically warmer, but show a similar overall trend. Data from Zachos *et al.* (2001).

with little latitudinal variation; terrestrial temperatures appear also to have been warm. Palaeotemperature records from the oxygen isotope composition of foraminiferal carbonate skeletons indicate that from a peak at the Palaeocene–Eocene boundary (57 Ma), high-latitude seawater temperatures have cooled to the present very low values. This long-term trend has intensified the cline in temperature between tropics and poles and narrowed the geographical distribution of the tropics. It has been estimated that during the Cenozoic cooling the tropical biome was roughly halved in area (Adams *et al.* 1990). The present latitudinal temperature cline is probably the most intense since late Carboniferous to early Permian times (Ziegler 1990).

The Cenozoic cooling was episodic with periods of very rapid cooling, as well as intermittent warming (Fig. 8.5). The first major global cooling event in the Cenozoic was at the Eocene–Oligocene boundary (37 Ma), when both surface and bottom waters cooled rapidly (Lazarus & Caulet 1993; Zachos *et al.* 2001a, b). This is widely interpreted as marking the onset of continental glaciation in East Antarctica. Further significant events were a middle Miocene warm period, an abrupt cooling in the late Miocene (17–11 Ma) and further cooling events at the Miocene–Pliocene boundary (6.2–4.8 Ma), later in the Pliocene (3.6–2.4 Ma) and the Pleistocene (Frakes *et al.* 1992; Barrett 1996, 2001; Zachos *et al.* 2001a, b).

Table 8.1 Milankovitch climatic variability. The three dominant orbital processes interact to produce a complex variation in the amount and distribution of solar energy received on Earth. These variations affect the timing, duration and intensity of the seasons, the latitudinal extent of the tropical and polar regions, and the size and extent of polar ice masses (Imbrie *et al.* 1993). Feedbacks within the climate system, mediated through biogeochemical processes, mean that some signals will lag the orbital cycles themselves (Shackleton 2000).

Period (years)	Process
100 000 (primary) 413 000 (secondary)	Eccentricity of the orbit, *e*, varying from 0 (perfectly circular) to 0.0607; $e = \sqrt{(a^2 - b^2)}/a$ where *a* is the semimajor axis and *b* the semiminor axis of elliptical orbit. The current value is $e = 0.016$
41 000	Axial tilt (obliquity), which varies between 22° and 24°30′. The current value is 23°27′
23 000 (primary) 19 000 (secondary)	Precession of the equinoxes

Higher frequency variability

It has been known since ancient times that Earth's relationship with the rest of the solar system is not constant. The orbital dynamics of Earth are affected predominantly by Venus (because it is close) and Jupiter (because it is so large). These interactions cause variations in the axial tilt of Earth, a precession in that tilt, and changes in the shape of the orbit. The predominant periods of variability are in the range 19 000 to >400 000 years and are together referred to as Milankovitch variability (Table 8.1).

Analyses of ice-core and sediment records also indicate significant climate variability on shorter timescales. This includes millennial-scale oscillations during the last glaciation (Dansgaard–Oeschger events with periodicities of 2000 to >5000 years), and major episodes of ice-rafting at intervals of 5000 to 15 000 years (Heinrich events). Evidence from high-resolution sediment cores and ice cores taken in Antarctica demonstrates variability on still finer scales (>1000 years) and the historical record reveals change on the timescale of centuries (e.g. the shift from the Medieval Warm Period to the Little Ice Age) and decades (the North Atlantic Oscillation and the El Niño–Southern Oscillation).

There also have been sudden events, which may be global in extent and catastrophic in nature, such as the Late Palaeocene Thermal Maximum. This was a sudden increase in global temperature that appears to have been the result of the sudden release of 2.5 Gt of carbon from sea-floor methane clathrates (Zachos *et al.* 1996). Events of this magnitude are genuinely rare in evolutionary history, but sudden small shifts in climate are emerging as an important feature of the climate record.

What is the importance of climatic variability to evolution, extinction and macroecology?

Climate has rarely if ever been stable, and this variability is expressed over a wide range of temporal scales. On evolutionary timescales (>10^5 years), the dominant variability is at Milankovitch and tectonic (>10^7) frequencies (Shackleton & Imbrie 1990). Organisms have therefore faced the evolutionary challenge of adapting their physiology to a continuously moving climatic target.

Very long wavelength variation in mean climate (Fig. 8.5) appears to have been slow enough for organisms to adapt evolutionarily. Periods of more rapid change, and especially cooling, are often correlated with enhanced regional or global extinction. This would seem intuitively reasonable, although even here mean rates of climate change are orders of magnitude slower than those with which many organisms living today are able to cope. This would suggest that the causes of such extinctions involve ecological, community-based processes and not simply the thermal death of individuals (Clarke 1993). It is possible, however, that in terms of the extinction component of evolutionary dynamics the critical features were rapid climate fluctuations. These represent changes in climate of a rapidity and extent quite outside human experience.

The fossil record reveals a range of responses to climate change, with no clear pattern. The best data are for the most recent switch from glacial to interglacial. Here the Northern Hemisphere signal is dominated by shifts in distribution, with relatively little evidence for either extinction or speciation. On longer timescales, encompassing a number of Milankovitch driven climate cycles, speciation and shifts in distribution become important. Extinction is always a feature (often termed *background extinction)* but neither the ultimate cause nor the proximate mechanisms for periods of enhanced extinction are clear. Patterns of variability in extinction rate in the marine and terrestrial biota are different, and internal dynamics may be an important factor (Halley 1996; Solé *et al.* 1997, 1999; Hewzulla *et al.* 1999).

Provinciality

One striking pattern in the modern biota is the existence of distinct biogeographical provinces. These can be defined operationally as areas where the biota are more similar to one another than they are to neighbouring regions. Often the borders between such provinces are marked by rapid changes in biotic composition, typically associated with changes in the physical environment (Brown & Lomolino 1998; Longhurst 1998). Rosenzweig (1995) has provided an evolutionary definition of a province, as an area where the composition of the biota is dictated by speciation and extinction alone. This is useful in theoretical terms, but it is difficult to recognize any such province with a biota that is unaffected by dispersal of species to and from neighbouring provinces.

Palaeobiologists have long noted that provinciality is higher when the latitudinal cline in climate is more intense (such as at present). This is shown very clearly in shallow-water benthic marine communities (Table 8.2). Valentine (1968) generalized this observation into a simple conceptual model for evolutionary dynamics whereby the overall marine diversity of the globe was determined by the strength

Table 8.2 Estimated number of provinces for shallow-water (continental shelf) marine faunas throughout the past 120 million years. From Crame (2000b).

Period	Estimated number of provinces
Late Cenozoic to Recent	31
Early Cenozoic	6
Late Cretaceous	6
Mid-Cretaceous	5
Early Cretaceous	4
Late Jurassic	5

of the latitudinal gradient in climate through an increase in provinciality (Valentine *et al.* 1978). It is likely that the degree of provinciality is greater at present than at any time since the Permian (see Crame (2001) for a recent review), but the influence of provinciality on global diversification has long been a contentious issue (Bambach 1977, 1990; Jablonski *et al.* 1985).

Case studies

The historical sciences and macroecology are both reliant on correlational approaches to understand the processes that underlie the observed patterns. Macroecology is concerned with spatial scales on which it would be unethical (and currently impractical) to undertake formal experimentation (Gaston & Blackburn 1999). Instead we must examine specific cases and attempt to generalize to basic principles. We will therefore examine a small number of specific studies, chosen for the light they can throw on general processes underpinning macroevolutionary patterns, and then attempt to draw some general conclusions.

Polar marine faunas

Although the Arctic and Antarctic marine environments are both cold, highly seasonal and heavily influenced by ice, they differ in many ways. The Arctic is an enclosed basin, deep in the centre but with restricted exchange with lower latitudes (especially with the North Pacific), and a large input of freshwater and sediment from the surrounding land. In contrast, the Antarctic continent is surrounded on all sides by a narrow and exceptionally deep continental shelf, and is contiguous with all three ocean basins.

The two polar regions also differ significantly in their history. The Arctic basin was covered by thick permanent ice at the height of the last glaciation, and all the fauna was almost certainly exterminated (Dunton 1992). Recolonization is currently underway and it is likely that the maximum age of the current fauna is only about 3–5 Ma (Vermeij 1991; Briggs 1995). In contrast there had been a shallow-water or continental-shelf fauna around Gondwana since the mid-Palaeozoic, and subsequent fragmentation of the supercontinent has provided increased habitat since the Early Cretaceous (Clarke & Crame 1989; Crame 1992).

The two polar faunas differ strikingly in their diversity, with the Arctic basin fauna being depauperate (Dunton 1992) and the Southern Ocean fauna quite rich and diverse (Arntz *et al.* 1997; Clarke & Johnston, in press). Although the Arctic basin is subject to intense physical and biological disturbance, which may restrict overall diversity (Dayton 1990), the major factor underpinning this difference in diversity is time. The continental shelf fauna of Antarctica has a long history of evolution *in situ* (Lipps & Hickman 1982; Clarke & Crame 1989) whereas that of the Arctic is relatively young, reflecting recent recolonization (Vermeij 1991; Dunton 1992; Briggs 1995).

Although the Southern Ocean fauna can be traced back deep into history, its composition has been affected by climatic processes during the Cenozoic. The Eocene fauna was typical of warm-temperate shallow waters at that time, and contained many cosmopolitan groups that are now absent (Feldmann & Tshudy 1989; Stilwell & Zinsmeister 1992; Eastman 1993). It is not known when these taxa died out, but there are indications that groups such as the decapods persisted long after the onset of widespread glaciation (Feldmann & Quilty 1997; Feldmann & Crame 1998). It does seem likely that the loss of many shallow-water habitats, as the continental ice-sheet extended on to the continental shelf together with the associated drop in seawater temperatures, was an important factor, although over what timescale is unclear.

The extinction of the majority of the shallow-water fish fauna was followed by a major radiation of a single clade of perciforme fish, the notothenioids (Eastman 1993). The low seawater temperature required the evolution of a glycopeptide antifreeze. The topography of the cladogram for notothenioids indicates that antifreeze evolved only once in this clade, and molecular evidence suggests a date of 13–11 Ma, coincident with the late Miocene cooling (Cheng & Chen 1999). The evolution of the notothenioids exhibits all the hallmarks of radiation into empty ecological niches left vacant by the extinction of the previous fauna. It is tempting to link similar radiations of amphipods and isopods in the Southern Ocean to the extinction of the previous decapod fauna, but this may be an oversimplification. Many of these taxa are characteristic of deeper water habitats (outer shelf, continental slope or abyssal) and this pattern also can be interpreted as the product of extensive onshore–offshore radiations over evolutionary timescales (Jablonski *et al.* 1985). A second key factor in the evolutionary history is undoubtedly the variability in the size and extent of the ice sheet, driven primarily by Milankovitch orbital cyclicity. Variations in extent undoubtedly generated great variability in the area of habitat for continental shelf benthos. This will have resulted in some taxa shifting distribution to deeper water on the continental slope (Brey *et al.* 1996), and also enhanced speciation and extinction through cycles of fragmentation of distribution and allopatric speciation (Clarke & Crame 1989).

Comparison of the two polar marine faunas emphasizes the role of time in allowing diversity to increase, the asymmetry in the glacial and evolutionary history of the

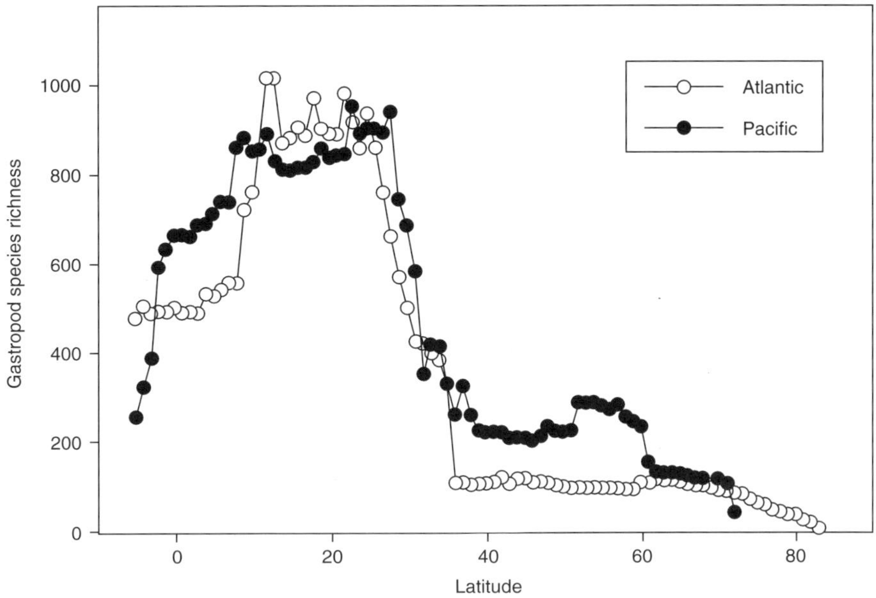

Figure 8.6 Latitudinal clines in diversity of shallow-water gastropod molluscs from west Atlantic and east Pacific coasts of North America. Redrawn from Roy *et al.* (1998).

two hemispheres, and the important role played by Milankovitch orbital cyclicity in evolutionary processes.

Latitudinal diversity clines in marine bivalves and gastropods

Gastropod and bivalve molluscs have long provided important insights into biogeography and evolutionary biology (Jablonski *et al.*, this volume), and the shallow-water continental shelf molluscan fauna of the Atlantic and Pacific coasts of North America have been particularly well studied. Despite the very different evolutionary histories of these two ocean basins the patterns of gastropod diversity show remarkable similarities (Fig. 8.6). Diversity is very high in the tropical latitudes and drops sharply around latitude 20°N. Further north diversity is more even, tailing off towards the highest latitudes.

This pattern points to both evolutionary and ecological processes at work. The sharp change in diversity at similar latitudes on opposite sides of the continent points to ecological control and Valentine *et al.* (2002) have suggested an important role for variability in community energy flow through oceanographic control (and specifically upwelling). The markedly high diversity in tropical regions may also have a historical component, as the Pleistocene climate fluctuations will have driven shifts in range, faunal turnover and speciation (Roy *et al.* 1996; Jackson & Overpeck 2000; Roy 2001).

A clear role for historical processes emerges when patterns of diversity are traced through time. Crame (2000a, b, 2001) has shown that in bivalve molluscs the steepest latitudinal clines in diversity are shown by the younger clades, and particularly the heteroconchs. Latitudinal clines in diversity can be traced back through the Cenozoic, but they intensify markedly in the mid-Miocene, about the time that the climatic cline also intensifies. This suggests that the diversity cline is driven by a burst of diversification of heteroconchs in tropical regions, with a slow spread into higher latitudes.

Corals and tropical marine faunas

It is now recognized that the Indo-West Pacific (IWP) region is exceptionally species-rich for many benthic marine taxa (Briggs 1995, 1996). This richness is a relatively recent development. In the Eocene and lower Oligocene the diversity of hermatypic corals in the IWP was lower than either the Mediterranean or Caribbean regions. In the early Miocene there was a marked increase in coral genera in the IWP, at a time when global seawater temperatures were cooling, but when the area of shallow water increased vastly as a result of tectonic activity (Crame & Rosen 2002).

Rosen (1984, 1988) has argued that the diversification was driven by an intensification of glacio-eustatic activity. The basic model combines speciation in the outer islands of the West Pacific and East Indian Ocean at low sea-level stands, with spread back into the central IWP at high stands. These cycles of sea-level thus pump species that originate allopatrically into secondary sympatry, with many cycles within the last 10–12 million years. The main driver for this sea-level change is Milankovitch orbital variability, mediated through the volume of high-latitude continental ice. Although the direct impact of Milankovitch variability on biogeographical range is expected to be greater at high latitudes, it would appear that diversification in the IWP is driven indirectly by polar glacial processes.

Northern Hemisphere floras and glaciation

The northward shift in the distribution of plants following the last glacial maximum has been exceptionally well documented in North America and Eurasia, predominantly from pollen records. The key feature has been a migration of individual taxa at different rates, rather than a movement of assemblages *per se* (Pielou 1991; Gates 1993). Colonization of newly deglaciated terrain, or invasion of established assemblages, thus depends on suitable source populations that have survived in refugia, dispersal ability, colonization and establishment ecology and factors such as invasibility of the previously established assemblage.

Many features of the environment would seem to us, as poorly insulated endotherms, to be harsher towards the poles. The lower richness of any assemblages at higher latitudes thus means that most measures of diversity will be strongly correlated with various measures of climate. This seems intuitively reasonable and there are undeniably ecological factors influencing local assemblage composition at high latitudes (e.g. the need to withstand freezing temperatures: Farrell *et al.* 1992; Latham & Ricklefs 1993); nevertheless the most important first-order explanation for the latitudinal cline in plant diversity towards high northern latitudes remains time. The last glacial maximum is simply too recent for every species that could evolve to live at northern latitudes to do so. Ecology remains important, however, in governing which extant taxa recolonize newly deglaciated terrain, and how quickly.

What general principles can we infer?

Current knowledge of the broad patterns of diversity together with tectonic and climatic history during the Cenozoic allow us to draw some general conclusions concerning the role of time in diversity.

The fossil record demonstrates an overall increase in diversity throughout the Phanerozoic (the past 450 million years), albeit with periods of increase, periods where overall diversity remained stable, and reductions in diversity caused by mass extinction events (Fig. 8.4). The basic pattern has remained robust to improving knowledge of the fossil record (Sepkoski 1993). Recently Peters & Foote (2001) have suggested that much of the short-term variation in marine diversity is largely a function of the volume of rock available for study, raising the intriguing possibility that taxonomic diversity may not have increased significantly since the early Palaeozoic. Nevertheless, there is currently general agreement that improving knowledge will change the detail but is unlikely to significantly alter the basic shape of the Phanerozoic diversity curve.

There is also general agreement that global diversity increased substantially in the Cenozoic. Many clades diversified spectacularly, notably angiosperms, insects, mammals and marine molluscs. This increase in global diversity occurred at a time when there was a substantial cooling of Earth's climate (both on land and in the sea), accompanied by the onset of a major glaciation. Although periods of rapid warming and cooling both probably stimulated short bursts of above-background extinction, the first-order pattern is for increased diversity and decreased mean temperature. This indicates that if there is any relationship between diversity and received solar energy or a climatic variable such as mean temperature it is either inverse, or the effect is masked by other factors.

An important factor in the increase in diversity throughout the Cenozoic has undoubtedly been the increase in provinciality, driven by the intensification of the latitudinal climatic gradient and the fragmentation of Gondwana. Because the tropics were always present, the key factor was the cooling outside the tropical regions, especially over the past 15 million years. For shallow-water marine faunas there may also have been a contribution from the increased area of habitat resulting from continental breakup. It is, however, striking that an earlier period of enhanced diversification in the marine benthos, in the Ordovician, was also a period of continental fragmentation and climate change (Miller & Foote 1996; Owen & Crame 2002).

For the later parts of the Cenozoic, when the increase in diversity was particularly marked, a second key factor has been the impact of climatic change driven by orbital variability on Milankovitch scales. Clarke & Crame (1989) suggested that variations in ice-sheet extent driven by Milankovitch orbital variability was a key factor underlying the high diversity of many taxa in the Southern Ocean shallow-water fauna. It is also now clear that these variations in ice-sheet volume have been important in driving speciation in some tropical marine areas through their control of sea-level (Valentine & Jablonski 1991; Palumbi 1997; Crame & Rosen 2002). The movement of species' range and distributions in response to Milankovitch climate variability has recently been termed orbitally-forced species' range dynamics (Chown & Gaston 2000; Dynesius & Jansson 2000).

The most recent manifestation of these range dynamics has been the warming of climate since the last glacial maximum. This has led to a major change in distribution of many organisms, with the recolonization of previously glaciated terrain in North America and Eurasia, although it is unlikely that the most recent cycle has alone been responsible for significant speciation or extinction (Roy *et al.* 1996). Although it is possible that after so many glacial cycles we are today studying the extinction-resistant flora and fauna (the susceptible taxa having gone extinct at the start), the history of the Southern Ocean and Indo-West Pacific marine fauna indicates clearly that other factors are involved. In particular a key feature has been the topography; long linear coastlines arranged N–S, or large areas of topographically uniform continent will allow for simple shifts in range. Where barriers intervene (e.g. mountain ranges) or the topography is complex (e.g. the pattern of islands in the IWP region), then the same cycles will produce contraction to refugia and/or vicariant speciation, followed by secondary sympatry. Consideration of the impact of the most recent cycle thus poses two important macroecological questions: whether any assemblage anywhere on Earth can be regarded as saturated or at equilibrium, and whether rates of speciation and extinction vary with latitude.

Is diversity at equilibrium?

The time-course of diversity throughout the Cenozoic shows no sign of having reached a plateau. Even allowing for the biases introduced to the fossil record (Raup, 1972, 1979; Alroy *et al.* 2001; Peters & Foote 2001) we appear to inhabit a world harbouring greater biological diversity than ever before.

The pattern of recolonization of deglaciated terrain since the last glacial maximum also argues strongly against high northern latitude assemblages being anywhere near saturated. Plant and animal taxa are still recolonizing, with migration rates varying widely between taxa (Bennett (1997) provides a valuable review). It is clear that at the current stage of recolonization, assemblages are quite different from the Clementsian view of saturated, equilibrium interactive communities.

The above view of equilibrium is concerned explicitly with very long spans of time. It implies that if we wait long enough we would observe a steady rise in the diversity and richness of these high-latitude assemblages. The timescale involved is, however, much greater than is allowed by Milankovitch-driven climatic variability. As such variability has greatest impact at higher latitudes, it is quite possible that assemblages here never reached long-term equilibrium since the intensification of the latitudinal climatic gradient and high-latitude glacial cycles in the Miocene. The obvious question is whether, given sufficient time in the absence of climatic variability, diversity at high latitudes would ever reach tropical levels? The evidence from assemblages where ecological factors are thought to prevail (e.g. pelagic copepods (Woodd-Walker *et al.* 2002), or the deep-sea (Levin *et al.* 2001; Lambshead *et al.* 2002)) suggests that this would not happen. The detail in the latitudinal cline in shallow-water molluscs (Fig. 8.6) shows a more or less constant diversity north of about 35°N, tailing off north of 60°N. One possible interpretation of this pattern is that the tail shows the increasing imprint of the glacial history of the Arctic, whereas the region from 35°N to 60°N gives an indication of equilibrium diversity for non-tropical shallow-water gastropod assemblages.

In recent years there has been considerable interest in the question of whether any assemblages are saturated, and particularly the relationship between local and regional diversity. Simple conceptual models developed by Cornell & Lawton (1992) drew a contrast between local assemblages that tracked regional pools in diversity and those that saturated. Although real data suggest that at least some local assemblages are saturated, these analyses beg the question of whether regional diversity dictates local richness, or vice versa (Cornell 1993; Ricklefs & Schluter 1993).

For examination of local/regional diversity relationships it would be most instructive for the regional scale to equate with an evolutionary province as defined by Rosenzweig (1995). It is clear that where a province has reached equilibrium then the steady-state diversity will be dictated by the balance between speciation and extinction processes. Under these circumstances the area effect will play an important role in determining equilibrium diversity (Rosenzweig 1995). The critical question is thus whether any area on Earth is at such equilibrium. Clearly high latitude areas influenced by orbital range dynamics are not, and it is now becoming clear that at least some lower latitude areas of high diversity are also relatively recent (Qian & Ricklefs 2000; Richardson *et al.* 2001).

Tropical Cenozoic diversification

The asymmetry between Northern and Southern Hemisphere latitudinal gradients reflects largely the difference in glacial impact and the consequent range dynamics in mid- to high-latitude organisms. The intense diversification in some tropical areas, and the associated longitudinal clines in diversity, indicate that tropical diversification also has been important in producing the global pattern of diversity we see today.

A key feature of tropical Cenozoic diversification has been the dominance of certain superclades, such as heteroconch bivalves, neogastropods and some angiosperm families. A second important feature has been the interplay between climate processes and topography, and a third the role played by the evolution of reefs. The Cenozoic diversification of hermatypic corals in the IWP region has provided new habitat for other taxa, and it is likely that the periodic evolution of substantial reef faunas throughout the Cenozoic has been an important factor in the overall pattern of marine diversity in the Phanerozoic (Wood 1999). Overall the Cenozoic reveals the complex interplay between climate change and tectonics (which set the scene), the internal evolutionary dynamics of some clades, and the increase in local scale within-province habitat diversity. One obvious remaining question is whether latitude or climate influences the rate of evolution itself.

Does evolutionary rate vary with latitude?

It has long been traditional to view the tropics as the birthplace of diversity, with the temperate and polar regions acting as the recipients of taxa that evolved elsewhere. In some cases this appears to be true: the recently evolved heteroconch bivalves exhibit their highest diversity in tropical regions and are spreading to lower latitudes (Crame 2000a,b, 2001). On the other hand there are also plentiful examples of taxa that have clearly evolved or radiated in polar regions; examples include isopod and amphipod crustaceans, pycnogonids, some predatory gastropods and notothenioid fishes (Clarke & Johnston, in press).

A critical factor in the rate of evolution is generation time. This tends to be faster in organisms living at warmer temperatures. Measures of the rate of molecular evolution, however, have suggested that rates in polar taxa are similar to those from warmer waters (Held 2000, 2001). Speciation rates in polar and non-polar molluscan taxa are very variable but mean rates are also similar (Crame & Clarke 1997). In contrast, Flessa & Jablonski (1996) demonstrated a difference in the dynamics of faunal turnover between tropical and non-tropical molluscs. These three measures are, however, very different metrics of evolutionary dynamics. Although conclusive evidence is currently lacking we are left with the impression from those superclades that have radiated so spectacularly in tropical regions that overall rates of speciation will prove to be higher in warmer climates.

The other important factor in evolutionary dynamics is extinction rate, but here we are entirely lacking in data to compare organisms living in different latitudes or temperatures.

Concluding remarks

The cyclical high-frequency climatic changes that have characterized the most recent history of Earth have had a particularly marked impact on those areas affected by glaciation and sea-level. In general these will have caused either or both large latitudinal shifts in range and fragmentation of ranges into refugia. There is currently some debate as to the extent to which such processes may have driven speciation (Roy *et al.* 1996). There is, however, strong evidence that they have been instrumental in generating key features of current macroecological patterns, including the latitudinal cline in marine and terrestrial diversity, the richness of the Indo-West Pacific marine province, the Mediterranean and fynbos floras, and the richness of some tropical forests (Qian & Ricklefs 1999, 2000; Richardson *et al.* 2001).

At the largest (global) scale historical processes dominate, and these are the most important in the origin of the present latitudinal cline in diversity. However, historical processes such as tectonics, glaciation and climate change simply set the scene. Speciation and extinction take place on local scales, and here ecological processes and contingency predominate. Ecological processes are critical to understanding how taxa originate and become extinct. But it is the historical processes that dictate where this happens, and hence underpin the large-scale patterns. Understanding both is essential to macroecology.

Acknowledgements

We thank Mike Benton and Jim Zachos for provision of data used in compiling figures, and Dave Jablonski and Peter Convey for preprints of recent publications. Our work is supported by the British Antarctic Survey.

References

Adams, C.G., Lee, D.E. & Rosen, B.R. (1990) Conflicting isotope and biotic evidence for tropical sea-surface temperatures during the Tertiary. *Palaeogeography, Palaeoclimatology, Palaeoecology* **77**, 289–313.

Alroy, J., Marshall, C.R., Bambach, R.K. *et al.* (2001) Effects of sampling standardization on estimates of Phanerozoic marine diversification. *Proceedings of the National Academy of Sciences, USA* **98**, 6261–6266.

Arntz, W.E., Gutt, J. & Klages, M. (1997) Antarctic marine biodiversity: an overview. In: *Antarctic Communities: Species, Structure and Survival* (eds B. Battaglia, J. Valencia & D.W.H. Walton), pp. 3–14. Cambridge University Press, Cambridge.

Bambach, R.K. (1977) Species richness in marine habitats through the Phanerozoic. *Paleobiology* **3**, 152–167.

Bambach, R.K. (1990) Late Paleozoic provinciality in the marine realm. In: *Palaeozoic Palaeogeography and Biogeography* (eds W.S. McKerrow & C.R. Scotese), pp. 307–323. The Geological Society, London.

Barker, P.F., Dalziel, I.W.D. & Storey, B.C. (1991) Tectonic development of the Scotia arc region. In: *The Geology of Antarctica* (ed. R.J. Tingey), pp. 215–248. Clarendon Press, Oxford.

Barrett, P.J. (1996) Antarctic palaeoenvironment through Cenozoic times. A review. *Terra Antartica* **3**, 103–119.

Barrett, P.J. (2001) Climate change—an Antarctic perspective. *New Zealand Science Review* **58**, 18–23.

Barthlott, W., Lauer, W. & Placke, A. (1996) Global distribution of species diversity in vascular plants: towards a world map of phytodiversity. *Erdkunde* **50**, 317–327.

Bennett, K.D. (1997) *Evolution and Ecology: the Pace of Life.* Cambridge University Press, Cambridge.

Benton, M.J. (1995) Diversification and extinction in the history of life. *Science* **268**, 52–58.

Benton, M.J. (1999) The history of life: large databases in palaeontology. In: *Numerical Palaeobiology: Computer-based Modelling and Analysis of Fossils and their Distributions* (ed. D.A.T. Harper), pp. 249–283. Wiley, Chichester.

Blackburn, T.M. & Gaston, K.J. (1996) A sideways look at patterns in species richness, or why there are so few species outside the tropics. *Biodiversity Letters* **3**, 44–53.

Brey, T., Dahm, C., Gorny, M., *et al.* (1996) Do Antarctic benthic invertebrates show an extended level of eurybathy? *Antarctic Science* **8**, 3–6.

Briggs, J.C. (1995) *Global Biogeography.* Elsevier, Amsterdam.

Briggs, J.C. (1996) Tropical diversity and conservation. *Conservation Biology* **10**, 713–718.

Brown, J.H. & Lomolino, M.V. (1998) *Biogeography*, 2nd edn. Sinauer, Sunderland, MA.

Cheng, C.H.C. & Chen, L.B. (1999) Evolution of an antifreeze glycoprotein. *Nature* **401**, 443–444.

Chown, S.L. & Gaston, K.J. (2000) Areas, cradles and museums: the latitudinal gradient in species richness. *Trends in Ecology and Evolution* **15**, 311–315.

Clarke, A. (1993) Temperature and extinction in the sea: a physiologist's view. *Paleobiology* **19**, 499–518.

Clarke, A. & Crame, J.A. (1989) The origin of the Southern Ocean marine fauna. In: *Origins and*

Evolution of the Antarctic Biota (ed. J.A. Crame), pp. 253–268. The Geological Society, London.
Clarke, A. & Johnston, N.M. (in press) Antarctic marine benthic diversity. *Oceanography and Marine Biology: an Annual Review.*
Cornell, H.V. (1993) Unsaturated patterns in species assemblages; the role of regional processes in setting local species richness. In: *Species Diversity in Ecological Communities: Historical and Geographical Perspectives* (eds R.E. Ricklefs & D. Schluter), pp. 243–252. University of Chicago Press, Chicago.
Cornell, H.V. & Lawton, J.H. (1992) Species interactions, local and regional processes, and limits to the richness of ecological communities: a theoretical perspective. *Journal of Animal Ecology* **61**, 1–12.
Cousins, S.H. (1989) Species richness and the energy theory. *Nature* **340**, 350–351.
Crame, J.A. (1992) Latitudinal range fluctuations in the marine realm through geological time. *Trends in Ecology and Evolution* **8**, 162–166.
Crame, J.A. (2000a) Evolution of taxonomic diversity gradients in the marine realm: evidence from the composition of Recent bivalve faunas. *Paleobiology* **26**, 188–214.
Crame, J.A. (2000b) The nature and origin of taxonomic diversity gradients in marine bivalves. In: *The Evolutionary Biology of the Bivalvia* (eds E.M. Harper, J.D. Taylor & J. A. Crame), pp. 347–360. The Geological Society, London.
Crame, J.A. (2001) Taxonomic diversity gradients through geological time. *Diversity and Distributions* **7**, 175–189.
Crame, J.A. & Clarke, A. (1997) The historical component of taxonomic diversity gradients. In: *Marine Biodiversity: Causes and Consequences* (eds R.F.G. Ormond, J.D. Gage & M.V. Angel), pp. 258–273. Cambridge University Press, Cambridge.
Crame, J.A. & Rosen, B.R. (2002) Cenozoic palaeogeography and the rise of modern biodiversity patterns. In: *Palaeobiogeography and Biodiversity Change: the Ordovician and Mesozoic–Cenozoic Radiations* (eds J.A. Crame & R.A. Owen), pp. 153–168. Special Publication 194, Geological Society Publishing House, Bath.
Currie, D.J. (1991) Energy and large-scale patterns of animal and plant species richness. *American Naturalist* **137**, 27–49.
Currie, D.J. & Fritz, J.T. (1993) Global patterns of animal abundance and species energy use. *Oikos* **67**, 56–68.
Dayton, P.K. (1990) Polar benthos. In: *Polar Oceanography, Part B: Chemistry, Biology and Geology* (ed. W.O. Smith), pp. 631–685. Academic Press, San Diego.
Dunton, K. (1992) Arctic biogeography: the paradox of the marine benthic fauna and flora. *Trends in Ecology and Evolution* **7**, 183–189.
Dynesius, M. & Jansson, R. (2000) Evolutionary consequences of changes in species' geographical distributions driven by Milankovitch climate oscillations. *Proceedings of the National Academy of Sciences, USA* **97**, 9115–9120.
Eastman, J.T. (1993) *Antarctic Fish Biology: Evolution in a Unique Environment.* Academic Press, San Diego, CA.
Farrell, B.D., Mitter, C. & Futuyma, D.J. (1992) Diversification at the insect–plant interface. *BioScience* **42**, 34–42.
Feldmann, R.M. & Crame, J.A. (1998) The significance of a new nephropid lobster from the Miocene of Antarctica. *Palaeontology* **41**, 807–814.
Feldmann, R.M. & Quilty, P.G. (1997) First Pliocene decapod crustacean (Malacostraca: Palinuridae) from the Antarctic. *Antarctic Science* **9**, 56–60.
Feldmann, R.M. & Tshudy, D.M. (1989) Evolutionary patterns in macrurous decapod crustaceans from Cretaceous to early Cenozoic rocks of the James Ross Island region, Antarctica. In: *Origins and Evolution of the Antarctic Biota* (ed. J.A. Crame), pp. 183–195. The Geological Society, London.
Flessa, K.W. & Jablonski, D. (1996) The geography of evolutionary turnover: a global analysis of extant bivalves. In: *Evolutionary Paleobiology*, pp. 376–397. University of Chicago Press, Chicago, IL.
Foote, M. (2000) Origination and extinction components of taxonomic diversity: Paleozoic and post-Paleozoic dynamics. *Paleobiology* **26**, 578–605.
Frakes, L.A., Francis, J.E. & Syktus, J.I. (1992) *Climate Modes of the Phanerozoic.* Cambridge University Press, Cambridge.
Gaston, K.J. & Blackburn, T.M. (1999) A critique for macroecology. *Oikos* **84**, 353–368.

Gaston, K.J. & Blackburn, T.M. (2000) *Pattern and Process in Macroecology*. Blackwell Science, Oxford.

Gaston, K.J. & Spicer, J.I. (1998) *Biodiversity: an Introduction*. Blackwell Science, Oxford.

Gates, D.M. (1993) *Climate Change and its Biological Consequences*. Sinauer, Sunderland, MA.

Grigg, R.W. (1988) Paleoceanography of coral reefs in the Hawaiian-Emperor Chain. *Science* **240**, 1737–1743.

Groombridge, B. & Jenkins, M.D. (2000) *Global Biodiversity: Earth's Living Resources in the 21st Century*. World Conservation Press, Cambridge.

Hallam, A. (1994) *An Outline of Phanerozoic Biogeography*. Oxford University Press, Oxford.

Halley, J.M. (1996) Ecology, evolution and 1/*f*-noise. *Trends in Ecology and Evolution* **11**, 33–37.

Held, C. (2000) Phylogeny and biogeography of serolid isopods (Crustacea, Isopoda, Serolidae) and the use of ribosomal expansion segments in molecular systematics. *Molecular Phylogenetics and Evolution* **15**, 165–178.

Held, C. (2001) No evidence for slow-down of molecular substitution rates at subzero temperatures in Antarctic serolid isopods (Crustacea, Isopoda, Serolidae). *Polar Biology* **24**, 497–501.

Hewzulla, D., Boulter, M.C., Benton, M.J. & Halley, J.M. (1999) Evolutionary patterns from mass originations and mass extinctions. *Philosophical Transactions of the Royal Society, London, Series B* **354**, 463–469.

Huston, M.A. (1994) *Biological Diversity: the Coexistence of Species on Changing Landscapes*. Cambridge University Press, Cambridge.

Imbrie, J., Berger, A., Boyle, E.A., *et al.* (1993) On the structure and origin of major glaciation cycles. **2**. The 100,000-year cycle. *Paleoceanography* **8**, 699–735.

Jablonski, D. (2000) Micro- and macroevolution: scale and hierarchy in evolutionary biology and paleobiology. In: *Deep Time: Paleobiology's Perspective. Supplement to Paleobiology*, Vol. 26(4) (eds D.H. Erwin & S.L. Wing), pp. 15–52. The Paleontological Society, Lawrence.

Jablonski, D., Flessa, K.W. & Valentine, J.W. (1985) Biogeography and paleobiology. *Paleobiology* **11**, 75–90.

Jackson, S.T. & Overpeck, J.T. (2000) Responses of plant populations and communities to environmental changes of the late Quaternary. In: *Deep Time: Paleobiology's Perspective. Supplement to Paleobiology*, Vol. 26(4) (eds D.H. Erwin & S.L. Wing), pp. 194–220. The Paleontological Society, Lawrence.

Kennett, J.P., Keller, G. & Srinivasan, M.S. (1985) Miocene planktonic foraminiferal biogeography and paleoceanographic development of the Indo-Pacific region. In: *The Miocene Ocean: Paleoceanography and Biogeography* (ed. J.P. Kennett), pp. 197–236. Geological Society of America Boulder, CO.

Lambshead, P.J.D., Brown, C.J., Ferrero, T.J., *et al.* (2002) Latitudinal diversity gradients of deep-sea marine nematodes and organic fluxes: a test from the central equatorial Pacific. *Marine Ecology Progress Series* **236**, 129–135.

Latham, R.E. & Ricklefs, R.E. (1993) Continental comparisons of temperate-zone tree species diversity. In: *Species Diversity in Ecological Communities: Historical and Geographical Perspectives* (eds R.E. Ricklefs & D. Schluter), pp. 294–314. University of Chicago Press, Chicago.

Lazarus, D. & Caulet, J.P. (1993) Cenozoic Southern Ocean reconstructions from sedimentologic, radiolarian and other microfossil data. In: *The Antarctic Paleoenvironment: a Perspective on Global Change*, Part Two (eds. J.P. Kennett & D.A. Warnke), pp. 145–174. American Geophysical Union, Washington, DC.

Levin, L.A., Etter, R.J., Rex, M.A., *et al.* (2001) Environmental influences on regional deep-sea species diversity. *Annual Review of Ecology and Systematics* **32**, 51–93.

Link, J. (2002) Does food web theory work for marine ecosystems? *Marine Ecology Progress Series* **230**, 1–9.

Lipps, J.H. & Hickman, C.S. (1982) Origin, age and evolution of Antarctic and deep-sea faunas. In: *Environment of the Deep Sea* (eds W.G. Ernst & J.G. Morris), pp. 324–256. Prentice Hall, Englewood Cliffs, NJ.

Longhurst, A. (1998) *Ecological Geography of the Sea*. Academic Press, San Diego.

Miller, A.I. & Foote, M. (1996) Calibrating the Ordovician radiation of marine life: implications for Phanerozoic diversity. *Paleobiology* **22**, 304–309.

Owen, A.W. & Crame, J.A. (2002) Palaeobiogeography and the Ordovician and Mesozoic–Cenozoic radiations. In: *Palaeobiogeography and Biodiversity Change: the Ordovician and Mesozoic–Cenozoic Radiations* (eds J.A. Crame & A.W. Owen), pp. 1–11. Special Publication 194, Geological Society Publishing House, Bath.

Palumbi, S.R. (1997) Molecular biogeography of the Pacific. *Coral Reefs* **16**, S47–S52.

Peters, S.E. & Foote, M. (2001) Biodiversity in the Phanerozoic: a reinterpretation. *Paleobiology* **27**, 583–601.

Pielou, E.C. (1991) *After the Ice Age: the Return of Life to Glaciated North America.* University of Chicago Press, Chicago, IL.

Qian, H. & Ricklefs, R.E. (1999) A comparison of the taxonomic richness of vascular plants in China and the United States. *American Naturalist* **154**, 160–181.

Qian, H. & Ricklefs, R.E. (2000) Large-scale processes and the Asian bias in species diversity of temperate plants. *Nature* **407**, 180–182.

Raup, D.M. (1972) Taxonomic diversity during the Phanerozoic. *Science* **177**, 1965–1071.

Raup, D.M. (1979) Biases in the fossil record of species and genera. *Bulletin of the Carnegie Museum of Natural History* **13**, 85–91.

Richardson, J.E., Weitz, F.M., Fay, M.F., Cronk, Q.C.B., Linder, H.P., Reeves, G. & Chase, M.W. (2001) Rapid and recent origin of species richness in the Cape flora of South Africa. *Nature* **412**, 181–183.

Ricklefs, R.E. (1989) Speciation and diversity: integration of local and regional processes. In: *Speciation and its Consequences* (eds D. Otte & J. Endler), pp. 199–622. Sinauer, Sunderland, MA.

Ricklefs, R.E. & Schluter, D. (1993) Species diversity: regional and historical influences. In: *Species Diversity in Ecological Communities: Historical and Geographical Perspectives* (eds R.E. Ricklefs & D. Schluter), pp. 350–364. University of Chicago Press, Chicago.

Rosen, B.R. (1984) Reef coral biogeography and climate through the late Cenozoic: just islands in the sun or a critical pattern of islands? In: *Fossils and Climate* (ed. P.J. Brenchley), pp. 210–262. Wiley, Chichester.

Rosen, B.R. (1988) Progress, problems and patterns in the biogeography of reef corals and other tropical marine organisms. *Helgoländer Meeresuntersuchungen* **42**, 269–301.

Rosenzweig, M.L. (1995) *Species Diversity in Space and Time.* Cambridge University Press, Cambridge.

Roy, K. (2001) Analyzing temporal trends in regional diversity: a biogeographic perspective. *Paleobiology* **27**, 631–645.

Roy, K., Valentine, J.W., Jablonski, D. & Kidwell, S.M. (1996) Scales of climatic variability and time averaging in Pleistocene biotas: implications for ecology and evolution. *Trends in Ecology and Evolution* **11**, 458–463.

Roy, K., Jablonski, D. & Valentine, J.W. (1998) Marine latitudinal diversity gradients: tests of causal hypotheses. *Proceedings of the National Academy of Sciences, USA* **95**, 3699–3702.

Sepkoski, J.J. (1993) Ten years in the library: new data confirm paleontological patterns. *Paleobiology* **19**, 43–51.

Shackleton, N.J. (2000) The 100 000-year ice-age cycle identified and found to lag temperature, carbon dioxide, and orbital eccentricity. *Science* **289**, 1897–1902.

Shackleton, N.J. & Imbrie, J. (1990) The $\delta^{18}O$ spectrum of oceanic deep water over a five-decade band. *Climatic Change* **16**, 217–230.

Signor, P.W. (1990) The geologic history of diversity. *Annual Review of Ecology and Systematics* **21**, 509–539.

Solé, R.V., Manrubia, S.C., Benton, M.J. & Bak, P. (1997) Self-similarity of extinction rates in the fossil record. *Nature* **388**, 764–767.

Solé, R.V., Manrubia, S.C., Benton, M., Kauffman, S. & Bak, P. (1999) Criticality and scaling in evolutionary ecology. *Trends in Ecology and Evolution* **14**, 156–160.

Stanley, S.M. (1995) New horizons for palaeontology, with two examples: the rise and fall of the Cretaceous Supertethys and the cause of the modern ice age. *Journal of Paleontology* **69**, 999–1007.

Steele, J.H. (1985) A comparison of terrestrial and marine ecological systems. *Nature* **313**, 355–358.

Stilwell, J.D. & Zinsmeister, W.J. (1992) *Molluscan Systematics and Biostratigraphy. Lower Tertiary La Mesata Formation, Seymour Island, Antarctic Peninsula.* American Geophysical Union, Washington, DC.

Stommel, H. (1963) Varieties of oceanographic experience. *Science* **139**, 572–576.

Tilman, D. & Pacala, S.W. (1993) The maintenance of species richness in plant communities. In: *Species Diversity in Ecological Communities: Historical and Geographical Perspectives* (eds R.E. Ricklefs & D. Schluter), pp. 13–25. University of Chicago Press, Chicago.

Valentine, J.W. (1968) Climatic regulation of species diversification and extinction. *Geological Society of America Bulletin* **79**, 273–275.

Valentine, J.W. & Jablonski, D. (1991) Biotic effects of sea level change: the Pleistocene change. *Journal of Geophysical Research* **96**, 6873–6878.

Valentine, J.W., Foin, T.C. & Peart, D. (1978) A provincial model of Phanerozoic marine diversity. *Paleobiology* **4**, 55–66.

Valentine, J.W., Roy, K. & Jablonski, D. (2002) Carnivore/non-carnivore ratios in northeastern Pacific marine gastropods. *Marine Ecology Progress Series* **228**, 153–163.

Vermeij, G.J. (1991) Anatomy of an invasion: the trans-Arctic interchange. *Paleobiology* **17**, 281–307.

Wood, R. (1999) *Reef Evolution*. Oxford University Press, Oxford.

Woodburne, M.O. & Case, J.A. (1996) Dispersal, vicariance, and Late Cretaceous to Early Tertiary land mammal biogeography from South America to Australia. *Journal of Mammalian Evolution* **3**, 121–161.

Woodd-Walker, R.S., Ward, P. & Clarke, A. (2002) Large scale patterns in diversity and community structure of surface water copepods from the Atlantic Ocean. *Marine Ecology Progress Series* **236**, 189–203.

Woodward, F.I. (1987) *Climate and Plant Distribution*. Cambridge University Press, Cambridge.

Zachos, J.C., Quinn, T.M. & Salamy, K.A. (1996) High-resolution deep-sea foraminiferal stable isotope records of the Eocene–Oligocene climate transition. *Paleooceanography* **11**, 256–266.

Zachos, J., Pagani, M., Sloan, L., Thomas, E. & Billups, K. (2001a) Trends, rhythms, and aberrations in global climate 65 Ma to present. *Science* **292**, 686–693.

Zachos, J.C., Shackleton, N.J., Revenaugh, J.S., Pälike, H. & Flower, B. P. (2001b) Climate response to orbital forcing across the Oligocene–Miocene boundary. *Science* **292**, 274–278.

Ziegler, A.M. (1990) Phytogeographic patterns and continental configurations during the Permian period. In: *Palaeozoic Palaeogeography and Biogeography* (eds W.S. McKerrow & C.R. Scotese), pp. 363–379. The Geological Society, London.

Part 4
Why are more species small-bodied?

Chapter 9

Why are most species small-bodied? A phylogenetic view

Andy Purvis, C. David L. Orme and Konrad Dolphin*

Introduction

Most animal species are small. More specifically, histograms of log(body size) of extant species are strongly right-skewed (e.g. May 1978; Dial & Marzluff 1988; Brown *et al.* 1993; Gaston & Blackburn 2000: see also Fig. 9.1). After considering biases in the data, we review non-phylogenetic comparative analyses of the relationship between body size and diversity (species richness), which typically support the hypothesis that net rates of diversification (speciation minus extinction) have been higher in small-bodied lineages. We detail two problems with non-phylogenetic analyses, before outlining two phylogenetic methods that have been used to investigate how size and diversity are related.

We sketch five models proposed to explain size distributions (Hutchinson & MacArthur 1959; Morse *et al.* 1985; McKinney 1990; Brown *et al.* 1993; Kozłowski & Weiner 1997), and outline predictions they make that can be tested using phylogenetic comparative methods on size data from extant species. (They also make plenty of predictions that can be tested in other ways, but we do not consider them here: see Gaston & Blackburn (2000) for the most comprehensive review.) We compare these predictions with comparative evidence to see which if any models receive strong support. We also attempt to tease apart the effects of body size on speciation rate and extinction rate, although we recognize many difficulties that such attempts must face, and conclude with a scenario that is compatible with all of the comparative evidence for mammals at least.

To avoid widespread use of such awkward phrases as 'species-richness' and 'large-bodied', we henceforth use 'large' and 'small' to describe body size only (not numbers of species in taxa), and use 'diversity' (not 'size') to indicate species-richnesses of taxa. Thus, elephants are a large clade, but not a diverse one.

* *Correspondence address: a.purvis@imperial.ac.uk*

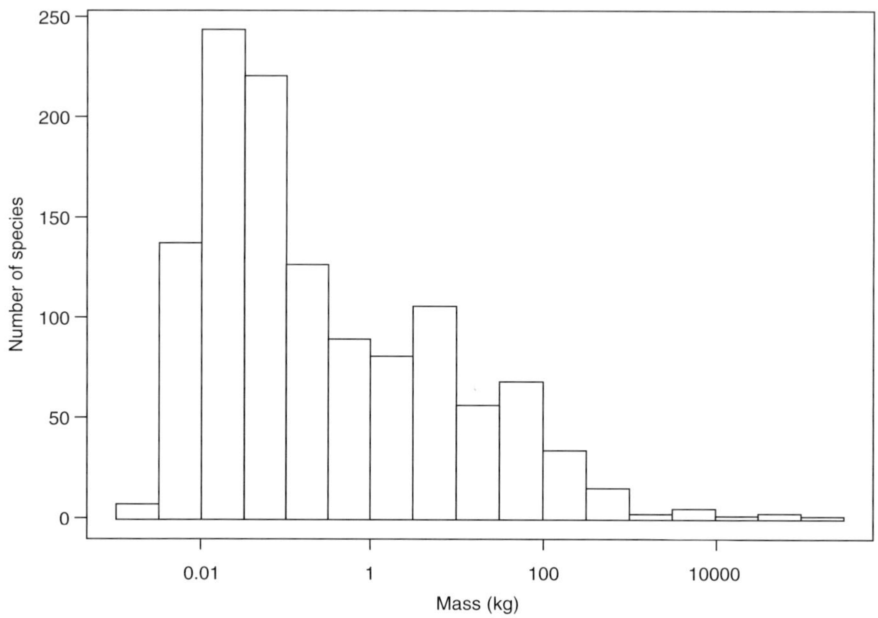

Figure 9.1 Histogram of mammalian adult female body mass. Data from Silva & Downing (1995).

The distribution of body sizes

When the sizes of terrestrial animal species are plotted as a histogram, the modal size class is towards the small end, even when size is first logarithmically transformed. Because a log-normal distribution might be a null expectation if species' sizes were determined by many factors combining multiplicatively (May 1986), the right skew of the log(size) distribution is a striking phenomenon needing explanation. May (1978, 1986) showed that numbers of described terrestrial animal species increase with decreasing body length, at least down to a length of about 0.5–1 cm. Birds and mammals (Fig. 9.1) both show size distributions with the same basic shape but with higher modes (Van Valen 1973; Caughley 1987; Maurer & Brown 1988).

May's (1978, 1986) data were, as he fully recognized, very incomplete: probably most extant species have not been formally described and named. This incompleteness matters if the probability of description depends on size, as May argued it does. A negative correlation between a species' body size and its date of description has been demonstrated subsequently in several taxa in various regions (Gaston 1991; Gaston & Blackburn 1994; Patterson 1994; Blackburn & Gaston 1995; Gaston *et al.* 1995; Allsopp 1997; Reed & Boback 2002). Several authors report high variance, however, or only a 'weak but significant' relationship, and exceptions are known (Reed & Boback 2002). For the analyses in the papers cited above, studies with more species tend to be more significant (regression of log(p-value) against log(number of species): $n = 12$; $p < 0.05$), suggesting that the pattern may be general.

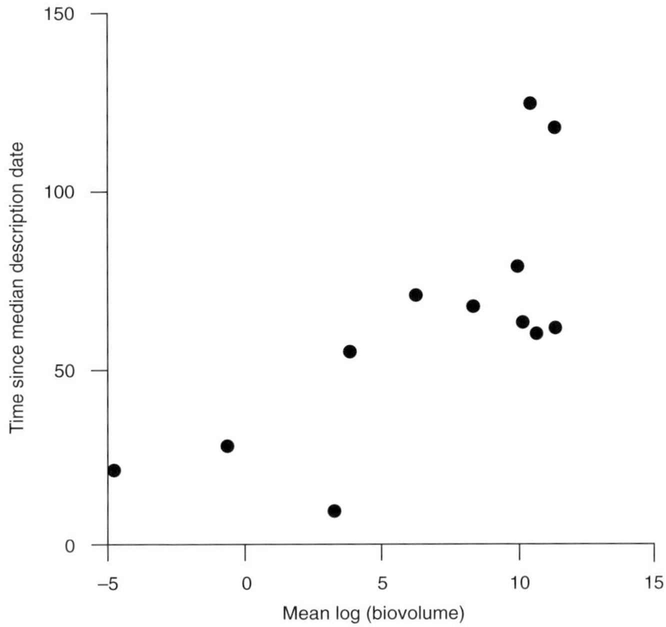

Figure 9.2 Time since the median description date of species in a range of taxa increases with mean biovolume. Data from May (1990) and Orme *et al.* (2002b). The taxa represented, in ascending order of time since median description (years), are: non-insect arthropods, Protozoa, 'Vermes', Insecta, Amphibia, Pisces, Echinodermata, Tunicata, Mollusca, Reptilia, Mammalia and Aves.

Figure 9.2 plots the median time since species were described against size across some very different higher taxa. The significant relationship (regression of log(median time) against log(size): $n = 12$; $r^2 = 0.55$, $p < 0.01$) indicates that larger bodied groups reach given levels of description earlier. Figure 9.3 shows that the same relationship holds even for such a well-studied group as mammals; it further holds when taxon diversity is controlled for (not shown) and, as in other groups (Gaston 1991; Allsopp 1997), is strongest in the earlier stages of discovery (not shown).

Direct causes of the relationship between size and description rate include an increased probability of recognizing and splitting large-bodied species because of (i) their greater level of taxonomic scrutiny (May 1988), (ii) our sensory bias in emphasizing differences between larger bodied lineages, and (iii) the higher chance of observing and recording larger species (Gaston 1991). The pattern has also been interpreted as an indirect effect of a correlation between body size and species' geographical range size, which would also increase the probability of description (Gaston & Blackburn 1994). Blackburn & Gaston (1995) suggest that this indirect effect might apply to taxa of larger absolute size, such as birds, rather than invertebrates; indeed, for very small species it seems that the body-size–range-size relationship is reversed (Hillebrand *et al.* 2001; Wilkinson 2001).

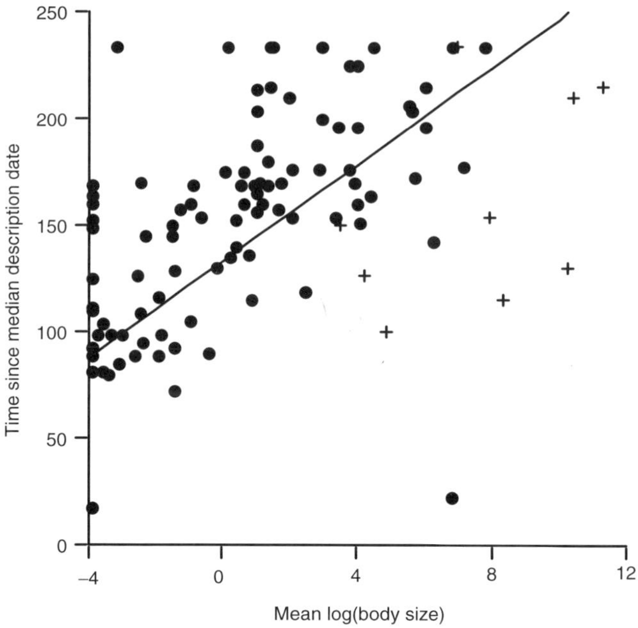

Figure 9.3 Regression of mean ln(body size in kilograms) against years since the median species description date for mammalian families, weighted by number of species in a family ($n = 95$; $p < 0.0001$). Crosses represent cetaceans, which were described late for their size. Data from Wilson & Reeder (1993) and N.J.B. Isaac (pers. comm.).

The underlying causes for the size–description relationship are likely to vary along the size axis. However, the pattern seems general and its effects on our shifting perspective of body size are undeniable. The first formally accepted taxonomic work of Linnaeus (1758) catalogued 2102 arthropod and 1334 vertebrate species—a ratio of roughly 1.6 to 1. That ratio 244 years later stands at approximately 20 : 1 and the true figure might be ten times higher. In all that follows, then, we must bear in mind that our view of the current size distribution is incomplete and biased. The problems are ameliorated, but not eliminated, by focusing on well-studied and/or large groups. The analyses reported in this chapter are all from taxa that are very much better studied than is typical, being the subject of either unusually high levels of general research interest or recent taxonomic revisions.

A non-phylogenetic view of diversity

Various authors (e.g. Van Valen 1973; Kochmer & Wagner 1988; Martin 1992) have plotted the numbers of species in taxa of a given rank against their typical body size (usually mean log body size), without considering phylogeny. Such plots typically show a negative relationship; Fig. 9.4 shows it for mammalian taxonomic families. A related approach was adopted by Dial & Marzluff (1988), who ranked the subtaxa

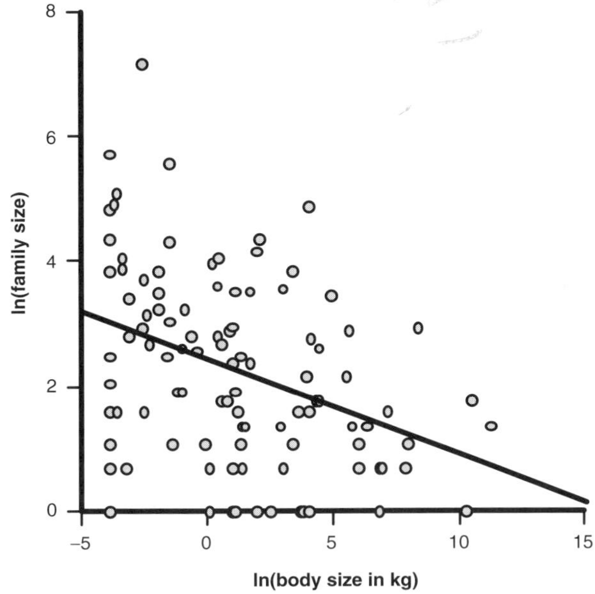

Figure 9.4 Non-phylogenetic view of the relationship between body size and species numbers across mammalian taxonomic families recognized by Wilson & Reeder (1993). Data from N.J.B. Isaac (pers. comm.).

within a taxon by their mean body size, and saw where in that ranking the most diverse subtaxon tended to lie. Their finding that it typically lay towards but not at the small-bodied end of the spectrum gels particularly well with the overall right-skewed size distribution. Such results have been taken as explicit support for models in which small or smallish species have the highest rates of diversification (Dial & Marzluff 1988; Kochmer & Wagner 1988; Martin 1992).

Three observations, then, seem to indicate a crucial role for body size in setting rates of diversification. First, most species are small. Second, small taxa of a given rank are (on average) the most diverse. Third, size correlates with huge numbers of other traits, many of which (e.g. fecundity, longevity, evolutionary rates, abundance and geographical ranges) might plausibly affect net rates of diversification. Why, then, do we need to bother with taking a phylogenetic view?

There are two reasons. The first is that taxa of a given rank can, and often do, vary greatly in age; this is true both at high (Avise & Johns 1999) and low (Purvis *et al.* 1995) taxonomic levels. This matters particularly if taxon age is correlated with body size, as is certainly true at some level (consider, e.g. mammals versus insects). Timescales are necessary to make sure taxa are the same age. The second reason applies if one wishes to test hypotheses about why the observed pattern arose. Closely related taxa may be pseudoreplicates in statistical tests (Nee *et al.* 1992; Barraclough *et al.* 1998; Gittleman & Purvis 1998; Dodd *et al.* 1999). Consider this purely illustrative and hypothetical example. Ten mammalian taxonomic families are linked by

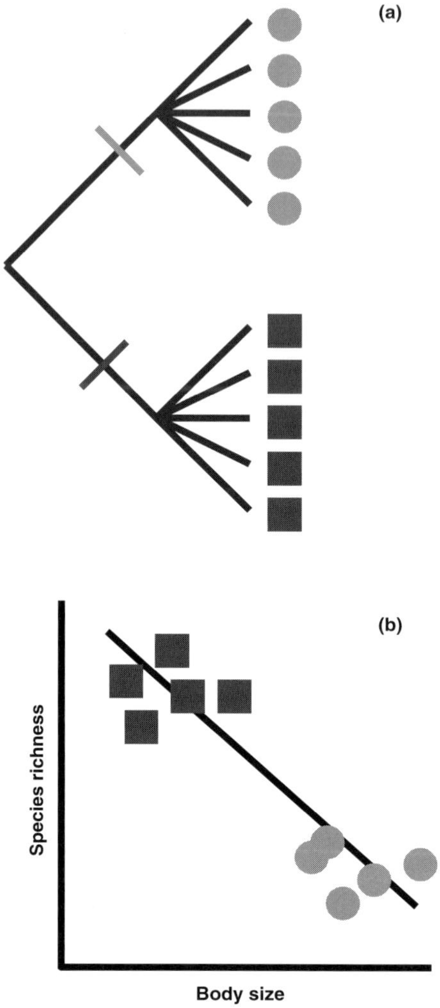

Figure 9.5 Illustration of how pseudoreplication can mislead non-phylogenetic tests. See text for explanation.

the phylogeny in Fig. 9.5a. The common ancestor of one of the five-family clades happened to evolve a grazing lifestyle (indicated by the light grey checkmark), in response to, say, climate change. Larger grazers are more efficient, so the ancestor became large; it has now left five families of descendants (light grey circles), all of them also large grazers. The common ancestor of the other clade happened to evolve a mating system in which female choice exerts strong sexual selection (indicated by the dark grey checkmark). Sexual selection accelerates speciation. Descendants of this family (dark grey squares) share the same mating system and so are diverse.

When diversity is now plotted as a function of size, a negative correlation is found (Fig. 9.5b). We know, however, what the non-phylogenetic statistical test cannot tell us: there is no causal link between size and diversity in this example. Note that this problem does not go away if all taxa are the same age. Some authors (e.g. Ricklefs & Renner 2000) have argued that results of phylogenetic and non-phylogenetic approaches are typically the same, and so phylogeny is an unnecessary complication. The results of the phylogenetic tests, described below, provide a test of this position. We do not argue that non-phylogenetic studies are useless whereas phylogenetic studies are perfect. Rather, we contend that non-phylogenetic analyses are even more seriously flawed than they need be, because they are open to more alternative interpretations than necessary.

Looking phylogenetically: some possible viewpoints

Even very coarse phylogenetic information can give insight into size distributions. The approach that makes fewest demands on the phylogeny is based on the subclades test proposed by McShea (1994). He noted that proposed processes behind a clade's log-skewed size distribution differed in their predictions for the shapes of the distributions in component subclades. As we shall see, some models predict that it should depend upon the subclade's position along the size axis. Testing this prediction requires only minimal phylogenetic information — just monophyly of a few subclades. The size distribution of subclades is useful in two further ways. First, some models predict all subclade distributions to have not only the same shape, but the same mode as well (see below). Second, inspection of subclades could show the overall distribution to be a simple composite of two very different distributions — one of a large clade, and one of a small clade, for instance — suggesting that the overall distribution may not be a sensible focus for research.

Where detailed phylogenetic information is available, we can gain a more precise view. If we predict that smaller species have (relative to larger species) higher net rates of diversification, then we expect to find smaller clades to be more diverse than larger clades, other things being equal. Other things can be made equal by comparing sister clades, i.e. lineages that are each other's closest relatives (e.g. Cracraft 1981; Felsenstein 1985; Mitter *et al.* 1988; Harvey & Pagel 1991). As well as automatically being the same age, sister clades often share a great deal of evolutionary history and so are likely to be similar in respect of many attributes that might otherwise confound comparisons.

Comparisons can be made between sister clades in the phylogeny. Is the more diverse clade of the pair typically larger or smaller? Under the null hypothesis that size is irrelevant for diversification, one would expect the proportion of comparisons in which small size is associated with higher diversity not to differ significantly from 0.5. More powerful statistical testing, with regression through the origin (Garland *et al.* 1992), is made possible by the use of a quantitative measure of diversity differences, rather than just 'more' or 'fewer'. Simulations (Isaac *et al.* 2003) indicate that a measure termed 'relative rate difference' (RRD) performs well under a wide range of scenarios. The RRD measure is defined as the natural logarithm of S_b/S_l, where S_b is the species richness of the larger-bodied clade and S_l the richness of the smaller-bodied. The RRD has an expectation of 0 at all body sizes under the null hypothesis. These comparisons also can be used to test whether net rates of diversification are highest at some intermediate size, as predicted by some models (see below): if they are, then a regression (not through the origin) of RRD against the mean body size of the clades being compared should have a significantly negative slope, and should cross the *x* axis at the optimum body size (Gittleman & Purvis 1998). MacroCAIC (Agapow & Isaac 2002), which we have used for all our own analyses, automates all these tests. Many other variants of the sister-clade approach are possible and also have been used.

Predictions from models of body size evolution

Plenty of models have been proposed to explain log-skewed size distributions. Here we very briefly sketch five in order to develop their predictions for phylogenetic comparisons (for more detail, and derivation of many of these predictions, see Gardezi & da Silva 1999; Gaston & Blackburn 2000; Kozłowski & Gawelczyk 2002).

Mosaic elements model (Hutchinson & MacArthur 1959)

In this model, the world is made up of many randomly distributed patches of a smaller number of patch types. An organism's view of the world depends upon its size. The world is largely homogeneous for both very small and extremely large organisms, but for different reasons: small organisms spend their whole lives in one patch, whereas the largest animals frequently pass through all patch types. In between, organisms will experience a subset of patch types, depending upon exactly where they are. If we further assume that different subsets of patches represent different niches, that no two species can share a niche, and that larger bodied species travel further so have larger subsets of patches, then the model predicts that the size distribution will be right-log-skewed overall, with the skew being present only in small subclades. When sister clades are compared, the result should depend upon whether they are large or small: among very small taxa, the larger clade is expected to be more diverse; near the mode there should be no systematic difference; among slightly larger taxa, smaller lineages should be more diverse; and among the largest clades the relationship flattens out leading once more to a prediction of no effect.

Fractal habitats model (Morse *et al.* 1985)

This model too relies on smaller organisms perceiving their environment in a more fine-grained way than larger organisms do. In its strictest form, it predicts the smallest size class to be most diverse. A size bias to discovery might lead to an intermediate mode (May 1986), but there must at some stage be a limit to either the fractality of the habitat or the size organisms can be. If such a limit is permitted, then the predictions of this model are basically the same as for the previous one. Otherwise, sister-clade comparisons predict that, where discovery is more or less complete, smaller size will be associated with higher diversity all along the size axis.

Passive diffusion with a lower bound (McKinney 1990)
This explains the size distribution as the result of a passive process and a constraint. Suppose there is a limit to how small species in a clade can be (e.g. the size limit to endothermy for mammals). Aside from this constraint, suppose that log(body size) is evolving as a random walk. (The log transformation is sensible because it is unlikely that absolute rates of body size evolution would be the same in, for instance, mice and elephants.) Following a clade's history, we would observe a size distribution starting as a spike at the ancestral size and then diffusing away from it, as lineages multiplied and size changed along them. The distribution of size would be log-normal, until and unless the lower limit is encountered. If ancestors are commonly quite small (Stanley 1973), the limit is encountered quite early on, and forces the size distribution to be right log-skewed. The skew is predicted only in small subclades (McShea 1994). In sister-clade comparisons, the model predicts, if anything, a weak positive correlation between rate and size among small taxa (depending upon whether the boundary is reflecting or absorbing), but no correlation at larger sizes.

Global optimum body size (Brown *et al.* 1993)
This model is based upon natural selection acting within mammalian populations to optimize reproductive power, which is the overall rate at which energy in the environment (potential food) is turned into offspring. This is viewed as a two-step process, with the rates of each step depending upon body size. First, energy must be assimilated; larger organisms are better than small ones at doing this (the rate equals $k_1 \times \text{mass}^{0.75}$). Second, the assimilated energy must be converted into offspring, and small animals are better than big ones at doing this (the rate equals $k_2 \times \text{mass}^{-0.25}$). Estimates of k_1 and k_2 from various empirical data suggest that reproductive power is maximized at a body mass of around 100 g; further, the distribution of reproductive power has a very similar shape to the log(body size) distribution. The model strictly predicts all species to evolve towards the same optimal size; species interactions are invoked to explain the species spreading out along the size axis in a similar way to reproductive power.

Predictions about subclades depend upon additional assumptions about importance of incumbency and rates of size evolution. If rate of change is not limiting, and incumbency unimportant, then all subclades should not only have the same shape as the overall distribution but also the same mode. Otherwise, there is no prediction about subclades. What is more, there is no longer any prediction for the overall size distribution of the clade: if incumbency and competition among mammalian clades prevent a subclade tracking the optimum size, then competition between mammals and other groups could prevent mammals tracking the fitness surface.

Moving to sister-clade comparisons, the predictions are again unclear. The model is about selection within populations. If body size affects species' persistence times

Table 9.1 Predictions of the models for phylogenetic tests: n.a., not applicable; 0, no correlation; parentheses indicate that predictions are dependent upon additional assumptions (see text).

Model	Right skew in subclades?			Sign of correlation between diversity contrasts and body size contrasts, when mean size is		
	Small size	Others	All same	Smaller than mode	Modal	Larger than mode
Mosaic	✓	✗	✗	+	0	–/0
Fractal habitat	✓	✓	✗	–	–	–/0
Diffusion	✓	✗	✗	+	0	0
Global optimum	n.a.	n.a.	✓	(+)	(0)	(–)
Specific optimum	n.a.	n.a.	(✓)	(+)	(0)	(–)

and/or rates of speciation, then lineages with near-optimal body sizes should tend to contain more species than sister lineages further from the optimum.

Species optimum body size (Kozłowski & Weiner 1997)
This model also invokes within-population optimality and has at its core a simple trade-off. The quantity that is optimized is lifetime reproductive success, and the trade-off arises because individuals delaying maturity are larger as adults, so can invest more in reproduction, but also have a greater chance of dying before maturity than do individuals that mature at smaller sizes. A population's optimal life history is determined by how the rates of assimilation, respiration and mortality scale with adult size among individuals within the species, and this optimal life history has a corresponding optimal body size for the population. However, unlike the previous model, the optimum differs among species. Furthermore, under many (but not all) circumstances, the distribution of optimal body sizes across many species is right-log-skewed (Kindlmann *et al.* 1999), provided that species of a given size do not competitively exclude one another.

As initially proposed, Kozłowski & Weiner's (1997) model predicts all subclades to have the same shape (mode and skew) as the overall distribution. However, the model could be extended to permit clades to specialize on different subsets of parameter space (Kozłowski & Gawelczyk 2002), in which case the model merely predicts most clades to show right-skew. As with Brown *et al.*'s (1993) model, this model makes predictions about sister clade comparisons only if clade selection is invoked, in which case lineages nearer to the overall modal size should be more diverse.

Table 9.1 summarizes each model's predictions about what should be revealed by a phylogenetic view of the relationship between size and diversity.

Looking phylogenetically: what do we see?

The subclades test has been used several times to examine body-size distributions. In

Table 9.2 Results of previously published phylogenetic tests, using independent contrasts, of the association between body size and species richness in sister taxa: level, taxonomic level of terminal taxa being compared; linear, sign of overall relationship between net rates of diversification and body size where significant; optimum, whether diversification rate was highest at some intermediate body size; n.s., not significant; n.t., not tested.

Group	Level	Linear	Optimum?	Reference
Primates	Species	n.s.	n.s.	Gittleman & Purvis 1998
Carnivores	Species	Negative	n.s.	Gittleman & Purvis 1998
Birds	Families	n.s.	n.t.	Nee *et al.* 1992; Owens *et al.* 1999
Hoverflies	Genera	n.s.	n.s.	Katzourakis *et al.* 2001
Metazoa	Phyla	n.s.	n.s.	Orme *et al.* 2002b
Mammals	Families	n.s.	n.s.	Gardezi & da Silva 1999

mammals, Gardezi & da Silva (1999) found that the smallest mammalian orders all show right-log-skew, with larger orders showing a mixture of patterns. In birds, orders show right-log-skew on average, with most orders (even the large ones) showing significant skew (Maurer 1998; Kozłowski & Gawelczyk 2002). Subclades do not generally have the same modal size. These results do not strongly support any model, but argue against the global optimum-body-size model.

Sister-clade comparisons also have been used to investigate the relationship between size and diversity, with many analytical details differing among studies. Table 9.2 shows the results; there is little evidence at best for any relationship—linear or humped—between size and net rates of diversification. Only one clade, Carnivora, shows a significant relationship between size and diversity; the relationship is patchy even within that clade (Gittleman & Purvis 1998). There are at least three possible artefactual reasons for the lack of a relationship. The first is that the analyses may lack power: simulations are needed to address this question. Second, studies finding a relationship might perhaps be viewed as uninteresting, given the widespread expectation that a relationship exists (the so-called 'desk drawer' problem). A corollary of this is that negative results should be easier to publish than significant ones; anecdotally, this has not been our experience. Finally, there may be a scale dependency. The taxa in Table 9.2 each include really rather different sorts of thing—there is much variation in traits other than body size, and perhaps this variation is more important in determining diversification. Perhaps looking at lower taxonomic levels, where confounding variables are presumably fewer, will give a clearer picture of how size relates to diversity.

These last two considerations suggest the need for a phylogenetic approach to be applied to many studies at low taxonomic level, in order to enable comparisons across a large sample of taxa with relatively small within-taxon variation in other life-history traits. Orme *et al.* (2002a) used sister-clade comparisons to investigate the size–diversity relationship in 38 complete species-level phylogenies of animal groups. Only one of the taxa, the dipteran genus *Bitheca*, exhibits a significant

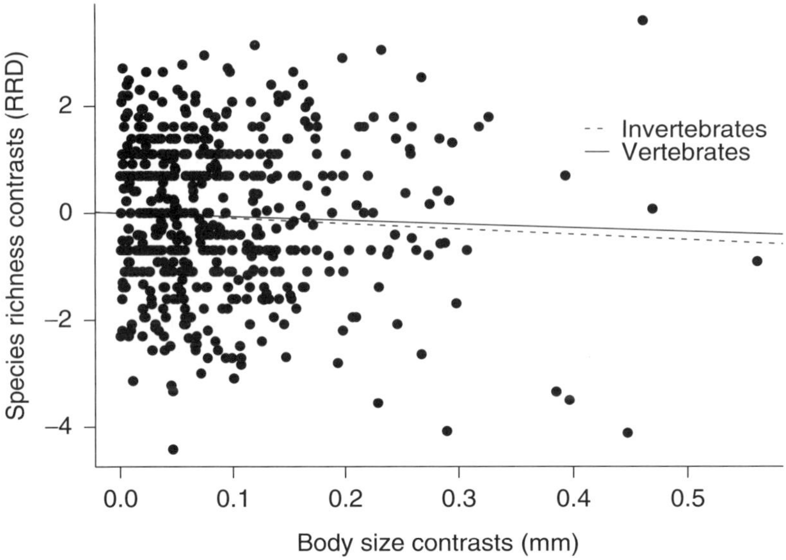

Figure 9.6 Pooled independent sister-clade comparisons from 38 species-level phylogenies reveal no linear relationship between RRD and body-size differences, in vertebrates, invertebrates or overall. See text.

relationship table-wide. Additionally, there is no overall trend across the taxa: the central tendency of the regression coefficients is not significantly different from zero (Wilcoxon signed-rank test: $V = 365$, $p = 0.94$). The lack of relationship persists if phylogenetic relationships among the 38 taxa are taken into account. Analysis of covariance of RRD against body-size differences pooled by clade (using the phylogeny relating the 38 taxa studied) can reveal both (i) whether any clades within the total set of taxa show a significant overall size–diversity relationship and (ii) whether subclades differ significantly in such relationships. Orme *et al.* (2002a) found no such differences, with the exception of comparing *Bitheca* to its sister taxa and comparing the clade of *Bitheca* and its sister taxa to their sister clade. The most basal comparison showed the relationship not to differ between invertebrate and vertebrate taxa ($F = 0.07$, $p = 0.79$; Fig. 9.6) and not to be significant overall ($b = -0.77$, $p = 0.13$). Finally, Orme *et al.* (2002a) found no evidence for any intermediate optimum body size, using linear regression, across all 38 taxa, of RRD against mean body size ($a = 0.02$, $b = -0.025$, $p = 0.32$; Fig. 9.7). This study strongly suggests that small size does not increase rates of net diversification even when differences among other life-history traits are minimal. These results relate best to the mosaic habitat and diffusion/boundary models, but are by no means a perfect fit.

A phylogenetic view, then, shows that body size is an atrocious predictor of diversity differences between sister clades. However, all the tests described above take a rather static view, focusing on the present-day distribution of body sizes. A clade might be diverse for its age because of high speciation rates, low extinction rates, or a

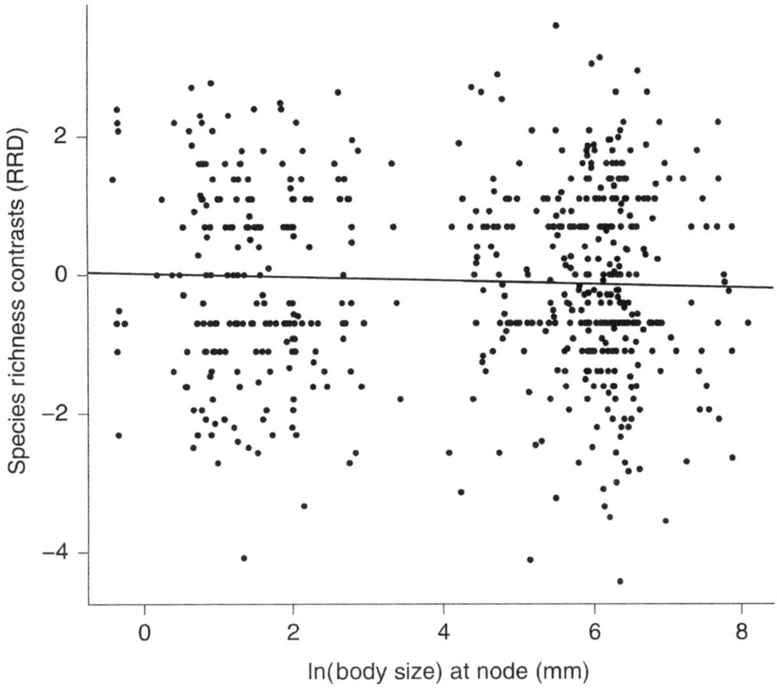

Figure 9.7 Pooled independent sister-clade comparisons from 38 species-level phylogenies reveal no tendency for net rates of diversification to be highest at an intermediate body size. See text.

combination of the two. Perhaps body size might play a role in at least one of these processes?

The fossil record is generally the only direct source of a long-term temporal dimension on macroecological phenomena. A perfect record would permit us to model directly how rates of speciation and extinction correlated with size. However, fossil records are incomplete in a wide range of ways (Kemp 1999), and a further problem is that species concepts typically applied to extant organisms cannot be applied straightforwardly to organisms from the past, leading some (e.g. Smith 1994) to suggest avoiding the use of the s-word altogether for fossil taxa.

A less direct approach is to try to reconstruct the past from information on extant species and their relationships. Methods are available for estimating ancestral character states from data on descendants (for recent reviews, see Cunningham *et al.* 1998; Pagel 1999; Webster & Purvis 2002a), and for identifying lineages with unusually high net rates of diversification (Slowinski & Guyer 1989; Nee *et al.* 1992; Purvis *et al.* 1995; Nee *et al.* 1996; Sanderson & Donoghue 1996; Paradis 1998; Mayhew 2002). Unfortunately, estimates of ancestral characteristics can be spectacularly inaccurate when the traits being reconstructed have shown evolutionary trends (Oakley & Cunningham 2000; Webster & Purvis 2002b), which is likely if they markedly affect speciation and extinction rates. Therefore, although there have been attempts explicitly to reconstruct the characteristics of lineages that have radiated (e.g. Schluter 2000), we do not follow that course of action here. Instead, we will use two even more indirect approaches, after a word of caution.

Literature-based analyses of speciation and extinction rates face a further problem. Most new lineages that form probably do not persist for long enough, or become widespread enough, to register in the literature. This is another veil line for macroecology to deal with, behind which new peripheral (or sympatric) isolates form and typically become extinct, with both processes quite possibly having very high rates (Hubbell 2001). Species tend to be described only if they reach some degree of abundance and/or geographical distribution (at which their probability of extinction has probably been much reduced); otherwise, they will typically either not be sampled or not be recognized as a distinct species. Rather than study rates of speciation and extinction, therefore, we are really studying the rates of species *establishment* and extinction of established species. In very well-studied groups and regions, the veil may have been pushed back a long way, minimizing the difference between what is seen and what happened, but the difference is likely to be marked in most groups.

Body size and establishment rate

One approach to evaluating how size affects establishment rate is to focus on young taxa, because extinction is likely to have had less of a chance to operate on them (Harvey *et al.* 1994). As an approximation, we have reanalysed data from the 29 clades in Orme *et al.*'s (2002a) study that are at or below the genus level. The regression coefficients for this restricted data set still do not differ significantly from zero (Wilcoxon signed-rank test: $V = 191$, $p = 0.58$), nor is there any evidence from pooled data for either an overall relationship between RRD and body-size differences (linear regression through the origin: $b = -0.53$, $p = 0.36$) or an intermediate optimum body size (linear regression: $a = 0.037$, $b = -0.032$, $p = 0.26$). This cannot, however, be taken as a strong rejection of a relationship between body size and establishment rate: the power of this approach is untested and taxonomic rank is a poor surrogate for clade age.

Body size and extinction rate

A second approach is to examine whether body size is associated with rates or probabilities of extinction of established species in the current extinction crisis. Large size often has been hypothesized to confer high rates or probabilities of extinction among present-day animal species, and might operate by any of several mechanisms (reviewed by McKinney 1997; Purvis *et al.* 2000b). Small size also has been proposed to increase extinction rates under certain circumstances (Pimm *et al.* 1988).

Table 9.3 lists the results of phylogenetic tests of association between body size and proneness to global species extinction. Large size is apparently a common correlate

Table 9.3 Results of phylogenetic comparative tests of the association between body size and global species extinction or extinction risk: n.s., not significant.

Study group	Index of extinction-proneness	Correlation	Reference
Primates	IUCN Red List assessments	+	Purvis *et al.* 2000a; Purvis 2001
Carnivores	IUCN Red List assessments	n.s.	Purvis *et al.* 2000a; Purvis 2001
Bats	IUCN Red List assessments	n.s.	Jones *et al.* in press
Australian mammals	IUCN Red List assessments	n.s.	Cardillo & Bromham 2001
Birds	Prevalence of threat in family in global survey	+	Bennett & Owens 1997
Birds	Prevalence of threat caused by habitat loss in family in IUCN Red List	n.s.	Owens & Bennett 2000
Birds	Prevalence of threat caused by overexploitation in family in IUCN Red List	+	Owens & Bennett 2000
Birds	Prevalence of threat in global survey	+	Gaston & Blackburn 1995
Malagasy primates	Subfossils compared with closest extant relatives	+	Walker 1967
New Zealand birds	Historically extinct versus extant	+	Cassey 2001

of present extinction-proneness in many groups. However, the (human) processes behind these extinctions must be borne in mind. Overexploitation—responsible for a high proportion of known historical extinctions and current severe threats (Mace & Balmford 2000)—is particularly likely to have an impact on large species, through the double disadvantage of their being obvious and reproducing slowly (Owens & Bennett 2000). There is no basis for expecting a similar process to have been responsible for much extinction before humans arrived. Consequently, although the correlation between large size and extinction-proneness has some generality across taxa, it may have no generality through time. Indeed, palaeontological comparative studies of the geological (pre-human) past tend to find no tendency for large body size to confer high extinction rates (Norris 1992; Jablonski & Raup 1995; McRoberts & Newton 1995; Jablonski 1996), although there are exceptions (Norris 1991).

Discussion and conclusion

In contrast to non-phylogenetic results, a phylogenetic view produces vanishingly little evidence that small species have higher net rates of diversification. Phylogeny can matter greatly when analysing diversity patterns and, given the statistical problems encountered by alternative approaches, we suggest that it must be considered. Phylogenetic comparative analysis of extant organisms is an important tool for explaining macroecological patterns, but the use of only extant species limits the view of process. Palaeontology alone can provide direct information about patterns in the past (Jablonski *et al.*, this volume). Unfortunately, the groups with the best fossil records are not the best known in today's fauna: macroecology has, for perfectly good reasons, focused mostly on birds and mammals, whereas the palaeontological studies most relevant to macroecology have largely been on foraminiferans and bivalve molluscs. Although not much can be done to help the fossil record of birds, the macroecology of taxa with good fossil records (e.g. Roy *et al.* 2000) must be an avenue worth exploring further.

Finally, why *are* most species small-bodied? We suggest the following scenario, building on ideas of Stanley (1973) and others, which seems to fit all the comparative evidence from mammals. Cretaceous mammals were typically small-bodied. It is increasingly evident that many mammalian species, perhaps placeable within many extant orders, survived the end-Cretaceous extinction (Bromham *et al.* 1999), but all such species were fairly small. In the Tertiary, mammals radiated into a much wider range of niches and evolved a much broader spectrum of body size (Alroy 1999). Something—perhaps the lower limit imposed by endothermy—prevented much exploration of smaller body sizes. Even if size had no influence on speciation or extinction rates in the Tertiary, the body-size distribution is expected to resemble that observed: more descendants are at the smaller end of the scale simply because more ancestors were. The non-phylogenetic correlation between small size and high diversity arises as an artefact of taxonomic practice, if small-bodied taxa of a given rank tend to originate and diversify earlier. There is no systematic relationship in sister-clade pairs between diversity and body size under this scenario, making it unsurprising that phylogenetic tests find no general evidence for one. This explanation has no trouble accommodating the fact of a general trend towards larger body size in mammalian lineages (Alroy 1998). It makes few key assumptions; none of them is unlikely, and none has been rejected by testing.

Acknowledgements

We thank Tim Blackburn, Kevin Gaston and the BES for inviting us to contribute to this book and the meeting; Kate Jones and Nick Isaac for help and data; Kevin Gaston and an anonymous referee for comments; and John Reynolds for his projector lead. The Natural Environment Research Council (UK) supported this work through grants GR3/11526 and GR3/13072, and through studentship GT04/98/MS/154.

References

Agapow, P.-M. & Isaac, N.J.B. (2002) MacroCAIC: revealing correlates of species richness by comparative analysis. *Diversity and Distributions* **8**, 41–43.

Allsopp, P.G. (1997) Probability of describing an Australian scarab beetle: influence of body size and distribution. *Journal of Biogeography* **24**, 717–724.

Alroy, J. (1998) Cope's Rule and the dynamics of body mass evolution in North American fossil mammals. *Science* **280**, 731–734.

Alroy, J. (1999) The fossil record of North American mammals: evidence for a Paleocene evolutionary radiation. *Systematic Biology* **48**, 107–118.

Avise, J.C. & Johns, G.C. (1999) Proposal for a standardized temporal scheme of biological classification for extant species. *Proceedings of the National Academy of Sciences, USA* **96**, 7358–7363.

Barraclough, T.G., Nee, S. & Harvey, P.H. (1998) Sister-group analysis in identifying correlates of diversification — comment. *Evolutionary Ecology* **12**, 751–754.

Bennett, P.M. & Owens, I.P.F. (1997) Variation in extinction risk among birds: chance or evolutionary predisposition? *Proceedings of the Royal Society, London, Series B* **264**, 401–408.

Blackburn, T.M. & Gaston, K.J. (1995) What determines the probability of discovering a new species?: a study of South American oscine passerine birds. *Journal of Biogeography* **22**, 7–14.

Bromham, L., Phillips, M.J. & Penny, D. (1999) Growing up with dinosaurs: molecular dates and the mammalian radiation. *Trends in Ecology and Evolution* **14**, 113–118.

Brown, J.H., Marquet, P.A. & Taper, M.L. (1993) Evolution of body size: consequences of an energetic definition of fitness. *American Naturalist* **142**, 573–584.

Cardillo, M. & Bromham, L. (2001) Body size and risk of extinction in Australian mammals. *Conservation Biology* **15**, 1435–1440.

Cassey, P. (2001) Determining variation in the success of New Zealand land birds. *Global Ecology and Biogeography* **10**, 161–172.

Caughley, G. (1987) The distributon of eutherian body weights. *Oecologia* **74**, 319–320.

Cracraft, J. (1981) Pattern and process in paleobiology: the role of cladistic analysis in systematic paleontology. *Paleobiology* **7**, 456–458.

Cunningham, C.W., Omland, K.W. & Oakley, T.H. (1998) Reconstructing ancestral character states: a critical reappraisal. *Trends in Ecology and Evolution* **13**, 361–366.

Dial, K.P. & Marzluff, J.M. (1988) Are the smallest organisms the most diverse? *Ecology* **69**, 1620–1624.

Dodd, M.E., Silvertown, J. & Chase, M.W. (1999) Phylogenetic analysis of trait evolution and species diversity variation among angiosperm families. *Evolution* **53**, 732–744.

Felsenstein, J. (1985) Phylogenies and the comparative method. *American Naturalist* **125**, 1–15.

Gardezi, T.F. & da Silva, J. (1999) Diversity in relation to body size in mammals: a comparative study. *American Naturalist* **153**, 110–123.

Garland, T.J., Harvey, P.H. & Ives, A.R. (1992) Procedures for the analysis of comparative data using phylogenetically independent contrasts. *Systematic Biology* **41**, 18–32.

Gaston, K.J. (1991) Body size and probability of description: the beetle fauna of Britain. *Ecological Entomology* **16**, 505–508.

Gaston, K.J. & Blackburn, T.M. (1994) Are newly described bird species small-bodied? *Biodiversity Letters* **2**, 16–20.

Gaston, K.J. & Blackburn, T.M. (1995) Birds, body size, and the threat of extinction. *Philosophical Transactions of the Royal Society, London, Series B* **347**, 205–212.

Gaston, K.J. & Blackburn, T.M. (2000) *Pattern and Process in Macroecology*. Blackwell Science, Oxford.

Gaston, K.J., Blackburn, T.M. & Loder, N. (1995) Which species are described first — the case of North American butterflies. *Biodiversity and Conservation* **4**, 119–127.

Gittleman, J.L. & Purvis, A. (1998) Body size and species richness in primates and carnivores. *Proceedings of the Royal Society, London, Series B* **265**, 113–119.

Harvey, P.H., May, R.M. & Nee, S. (1994) Phylogenies without fossils. *Evolution* **48**, 523–529.

Harvey, P.H. & Pagel, M.D. (1991) *The Comparative Method in Evolutionary Biology*. Oxford University Press, Oxford.

Hillebrand, H., Watermann, F., Karez, R. & Berninger, U.-G. (2001) Differences in species richness patterns between unicellular and multicellular organisms. *Oecologia* **126**, 114–124.

Hubbell, S.P. (2001) *The Unified Neutral Theory of Biodiversity and Biogeography*. Princeton University Press, Princeton, NJ.

Hutchinson, G.E. & MacArthur, R.H. (1959) A theoretical ecological model of size distribution among species of animals. *American Naturalist* **93**, 117–125.

Isaac, N.J.B., Agapow, P.-M., Harvey, P.H. & Purvis, A. (2003) Phylogenetically nested comparisons for testing correlates of species-richness: a simulation study of continuous variables. *Evolution* **57**, 18–26.

Jablonski, D. (1996) Body size and macroevolution. In: *Evolutionary Paleobiology* (eds D. Jablonski, D.H. Erwin & J.H. Lipps), pp. 256–289. University of Chicago Press, Chicago, IL.

Jablonski, D. & Raup, D.M. (1995) Selectivity of

end-Cretaceous marine bivalve extinctions. *Science* **268**, 389–391.

Jones, K.E., Purvis, A. & Gittleman, J.L. (in press) Biological correlates of extinction risk in bats. *American Naturalist.*

Katzourakis, A., Purvis, A., Azmeh, S., Rotheray, G. & Gilbert, F. (2001) Macroevolution of hoverflies (Diptera: Syrphidae): the effect of using higher-level taxa in studies of biodiversity, and correlates of species richness. *Journal of Evolutionary Biology* **14**, 219–227.

Kemp, T.S. (1999) *Fossils and Evolution.* Oxford University Press, Oxford.

Kindlmann, P., Dixon, A.F.G. & Dostalkova, I. (1999) Does body size optimization result in skewed body size distribution on a logarithmic scale? *American Naturalist* **153**, 445–447.

Kochmer, J.P. & Wagner, R.H. (1988) Why are there so many kinds of passerine birds? Because they are small. A reply to Raikow. *Systematic Zoology* **37**, 68–69.

Kozłowski, J. & Gawelczyk, A.T. (2002) Why are species body size distributions usually skewed to the right? *Functional Ecology* **16**, 419–432.

Kozłowski, J. & Weiner, J. (1997) Interspecific allometries are byproducts of body size optimization. *American Naturalist* **149**, 352–380.

Linnaeus, C. (1758) *Systema Naturae.* Stockholm.

Mace, G.M. & Balmford, A. (2000) Patterns and processes in contemporary mammalian extinction. In: *Future Priorities for the Conservation of Mammalian Diversity* (eds A. Entwhistle & N. Dunstone) pp. 27–52. Cambridge University Press, Cambridge.

Martin, R.A. (1992) Generic species richness and body mass in North American mammals: support for the inverse relationship of body size and speciation rate. *Historical Biology* **6**, 73–90.

Maurer, B.A. (1998) The evolution of body size in birds. I. Evidence for non–random diversification. *Evolutionary Ecology* **12**, 925–934.

Maurer, B.A. & Brown, J.H. (1988) Distribution of energy use and biomass among species of North American terrestrial birds. *Ecology* **69**, 1923–1932.

May, R.M. (1978) The dynamics and diversity of insect faunas. In: *Diversity of Insect Faunas* (eds L.A. Mound & N. Waloff), pp. 188–204. Blackwell, Oxford.

May, R.M. (1986) The search for patterns in the balance of nature: advances and retreats. *Ecology* **67**, 1115–1126.

May, R.M. (1988) How many species are there on earth? *Science* **241**, 1441–1449.

May, R.M. (1990) How many species? *Philosophical Transactions of the Royal Society, London, Series B* **330**, 292–304.

Mayhew, P.J. (2002) Shifts in hexapod diversification and what Haldane could have said. *Proceedings of the Royal Society, London, Series B* **269**, 969–974.

McKinney, M.L. (1990) Trends in body size evolution. In: *Evolutionary Trends* (eds K.J. McNamara), pp. 75–117. University of Arizona Press, Tucson.

McKinney, M.L. (1997) Extinction vulnerability and selectivity: combining ecological and paleontological views. *Annual Review in Ecology and Systematics* **28**, 495–516.

McRoberts, C.A. & Newton, C.R. (1995) Selective extinction among end-Triassic European bivalves. *Geology* **23**, 102–104.

McShea, D.W. (1994) Mechanisms of large-scale evolutionary trends. *Evolution* **48**, 1747–1763.

Mitter, C., Farrell, B. & Wiegmann, B. (1988) The phylogenetic study of adaptive zones: has phytophagy promoted insect diversification? *American Naturalist* **132**, 107–128.

Morse, D.R., Lawton, J.H., Dodson, M.M. & Williamson, M. (1985) Fractal dimension of vegetation and the distribution of arthropod body lengths. *Nature* **314**, 731–733.

Nee, S., Barraclough, T.G. & Harvey, P.H. (1996) Temporal changes in biodiversity: detecting patterns and identifying causes. In: *Biodiversity: a Biology of Numbers and Difference* (ed. K.J. Gaston), pp. 230–252. Blackwell Science, Oxford.

Nee, S., Mooers, A.Ø. & Harvey, P.H. (1992) The tempo and mode of evolution revealed from molecular phylogenies. *Proceedings of the National Academy of Sciences, USA* **89**, 8322–8326.

Norris, R.D. (1991) Biased extinction and evolutionary trends. *Paleobiology* **17**, 388–399.

Norris, R.D. (1992) Extinction selectivity and ecology in planktonic Foraminifera. *Palaeogeography, Palaeoclimatology, Palaeoecology* **95**, 1–17.

Oakley, T.H. & Cunningham, C.W. (2000) Independent contrasts succeed where ancestor reconstruction fails in a known bacteriophage phylogeny. *Evolution* **54**, 397–405.

Orme, C.D.L., Isaac, N.J.B. & Purvis, A. (2002a) Are most species small bodied? Not within species-level phylogenies. *Proceedings of the Royal Society, London, Series B* **269**, 1279–1287.

Orme, C.D.L., Quicke, D.L.J., Cook, J. & Purvis, A. (2002b) Body size does not predict species richness among the metazoan phyla. *Journal of Evolutionary Biology* **15**, 235–247.

Owens, I.P.F. & Bennett, P.M. (2000) Ecological basis of extinction risk in birds: Habitat loss versus human persecution and introduced predators. *Proceedings of the National Academy of Sciences, USA* **97**, 12144–12148.

Owens, I.P.F., Bennett, P.M. & Harvey, P.H. (1999) Species richness among birds: body size, life history, sexual selection or ecology? *Proceedings of the Royal Society, London, Series B* **266**, 933–939.

Pagel, M. (1999) Inferring the historical patterns of biological evolution. *Nature* **401**, 877–884.

Paradis, E. (1998) Detecting shifts in diversification rates without fossils. *American Naturalist* **152**, 176–187.

Patterson, B.D. (1994) Accumulating knowledge on the dimensions of biodiversity: systematic perspectives on neotropical mammals. *Biodiversity Letters* **2**, 79–86.

Pimm, S.L., Jones, H.L. & Diamond, J.M. (1988) On the risk of extinction. *American Naturalist* **132**, 757–785.

Purvis, A. (2001) Mammalian life histories and responses of populations to exploitation. In: *Exploited Species* (eds J.D. Reynolds, G.M. Mace, K.H. Redford & J.G. Robinson) pp. 169–181. Cambridge University Press, Cambridge.

Purvis, A., Nee, S. & Harvey, P.H. (1995) Macroevolutionary inferences from primate phylogeny. *Proceedings of the Royal Society, London, Series B* **260**, 329–333.

Purvis, A., Gittleman, J.L., Cowlishaw, G. & Mace, G.M. (2000a) Predicting extinction risk in declining species. *Proceedings of the Royal Society, London, Series B* **267**, 1947–1952.

Purvis, A., Jones, K.E. & Mace, G.M. (2000b) Extinction. *BioEssays* **22**, 1123–1133.

Reed, R.N. & Boback, S.M. (2002) Does body size predict dates of species description among North American and Australian reptiles and amphibians? *Global Ecology and Biogeography* **11**, 41–47.

Ricklefs, R.E. & Renner, S.S. (2000) Evolutionary flexibility and flowering plant familial diversity: a comment on Dodd, Silverton and Chase. *Evolution* **54**, 1061–1065.

Roy, K., Jablonski, D. & Martien, K.K. (2000) Invariant size–frequency distributions along a latitudinal gradient in marine bivalves. *Proceedings of the National Academy of Sciences, USA* **97**, 13150–13155.

Sanderson, M.J. & Donoghue, M.J. (1996) Reconstructing shifts in diversification rates on phylogenetic trees. *Trends in Ecology and Evolution* **11**, 15–20.

Schluter, D. (2000) *The Ecology of Adaptive Radiation.* Oxford University Press, Oxford.

Silva, M. & Downing, J.A. (1995) *CRC Handbook of Mammalian Body Masses.* CRC Press, Boca Raton, FL.

Slowinski, J.B. & Guyer, C. (1989) Testing the stochasticity of patterns of organismal diversity: an improved null model. *American Naturalist* **134**, 907–921.

Smith, A.B. (1994) *Systematics and the Fossil Record.* Blackwell Scientific, Oxford.

Stanley, S.M. (1973) An explanation for Cope's Rule. *Evolution* **27**, 1–26.

Van Valen, L. (1973) Body size and numbers of plants and animals. *Evolution* **27**, 27–35.

Walker, A.C. (1967) Patterns of extinctions among the subfossil Madagascan lemuroids. In: *Pleistocene Extinctions* (eds P.S. Martin & H.E.J. Wright), pp. 425–432. Yale University Press, New Haven, CT.

Webster, A.J. & Purvis, A. (2002a) Ancestral states and evolutionary rates of continuous characters. In: *Morphology, Shape and Phylogenetics* (eds N. MacLeod & P. Forey), pp. 247–268. Taylor & Francis, London.

Webster, A.J. & Purvis, A. (2002b) Testing the accuracy of methods for reconstructing ancestral states of continuous characters. *Proceedings of the Royal Society, London, Series B* **214**, 143–149.

Wilkinson, D.M. (2001) What is the upper size limit for cosmopolitan distribution in free-living microorganisms. *Journal of Biogeography* **28**, 285–291.

Wilson, D.E. & Reeder, D.-E. (eds) (1993) *Mammal Species of the World.* Smithsonian Institute Press, Washington, DC.

Chapter 10

Adaptive diversification of body size: the roles of physical constraint, energetics and natural selection

*Brian A. Maurer**

Introduction

Species vary considerably in body size within and among taxa. As body size has such a pervasive effect on the way that an organism interacts with its environment (Peters 1983; Calder 1984; Brown 1995), this variation in body size among species represents a substantial proportion of the diversity of modes by which living organisms have adapted to life on Earth. One of the fundamental properties of the universe is that there are many more small objects than there are large ones. This appears to be true for species as well (May 1978, 1986; Bonner 1988) to a certain degree. The pattern holds up well on a relatively coarse taxonomic scale. For example, there are more insects than there are mammals. However, a more careful analysis yields the insight that there are typically fewer of the smallest species within a taxon than there are moderately sized ones (Bonner 1988). Furthermore, there appears to be a more rapid decrease in the number of species below the modal size than the decrease in species above the modal size (Hutchinson & MacArthur 1959; May 1978; Bonner 1988; Maurer & Brown 1988; Maurer *et al.* 1992). This typically produces a positive skew in the frequency distribution of log-transformed body sizes, although there are exceptions (Fig. 10.1).

The variety of explanations for these empirical patterns have been examined elsewhere (Brown 1995; Chown & Gaston 1997; Gardezi & da Silva 1999; Maurer 1999; Kozłowski & Gawelczyk 2002). Here I integrate the insights from these explanations into a single, inclusive framework. This framework is based on the assumption that diversification of body sizes within taxa is the result of three sets of factors. The first set involves physical constraints that are implicit in making organisms of different sizes. There are two sets of physical constraints that must be considered: (i) external physical constraints dealing with the structure of the resources used by an organism; and (ii) internal physical constraints that emerge from the necessity of an organism to distribute nutrients to and collect waste products from its constituent cells. The

* *Correspondence address: maurerb@msu.edu*

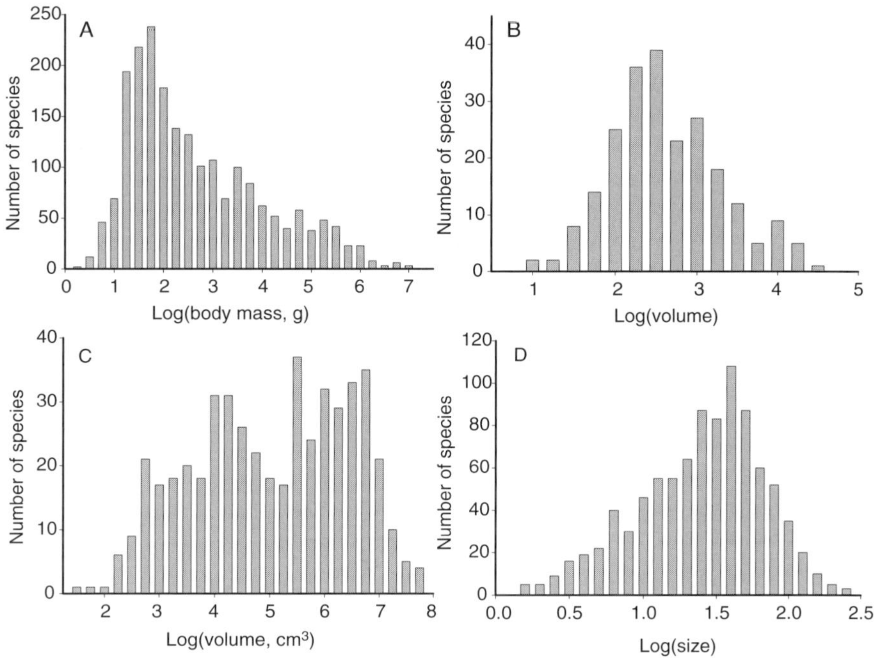

Figure 10.1 Frequency distributions of log transformed body sizes for different groups of species: (A). Late Pleistocene (202 species) and extant (1902 species) non-flying mammals from North and South America, Africa, and Australia; (B) 226 species of unicellular organisms; (C) 487 species of palms; (D) 916 species of clams (Roy *et al.* 2000).

second set of factors deals with the energetic trade-offs inherent in organismal maintenance and production, and how these trade-offs affect vital rates of populations of different sized organisms. The third component of the framework is the energetically constrained vital rates of populations that determine the outcome of natural selection on body size and other organismal attributes. These three sets of factors are incorporated into a cost-benefit analysis of the physical limits on energy flows that define a range of body masses for which the net benefits are positive for a given taxon. Within this range of body masses, there are fewer constraining factors for body masses of intermediate sizes, although constraints become increasingly more severe for organisms that are larger or smaller than the modal body size for a taxon. The evolutionary significance of these constraints is that the creation of new species, or the adaptive evolution of existing species within the range of possible body sizes for a taxon, has a lower probability of occurring for species near either size extreme.

A framework for understanding the diversification of body sizes

External physical constraints

One of the first attempts to develop a theoretical basis for the existence of a variety of body sizes among species was built on the assumption that the resources used by species could be divided into discrete 'mosaic elements' that represented the different needs of individual organisms (Hutchinson & MacArthur 1959). The biological insight was that small organisms are limited in the degree that they can move about in geographical space, and therefore are limited to a subset of possible resources available to a larger organism using the same resources. The method of quantifying this insight, however, required the use of a somewhat obscure probability argument. The parameters needed to develop this model were ill-defined biologically, and no objective method for statistical evaluation of the parameters was able to be developed, hence, it was of limited usefulness.

May (1978, 1986) revisited this problem, and suggested there was an apparent scaling law for organism sizes across multicellular terrestrial animals. As terrestrial organisms use space in essentially a two-dimensional fashion, May (1978, 1986) suggested that the scaling of species numbers should be approximately L^{-2}, where L is organism size. Morse *et al.* (1985) analysed the spatial structure of leaves eaten by insects in trees, and suggested that leaves presented a fractal-like spatial arrangement of resources for insects of different sizes. If, in general, resources exist in fractal-like spatial structures for many different kinds of species, then the scaling of species number with body size should be negative, assuming that the resource is infinitely divisible. Clearly resources are not infinitely divisible, so there must be some minimum size below which no resources are available. Within the range of resources used by a taxon, the spatial structure of resources suggests that there should be many more organisms of smaller size than larger size. If numbers of organisms of different size are proportional to numbers of species, then the modal body size of species within a taxon should be the same as the size of the smallest species. As can be seen from Fig. 10.1, this is not the case within a variety of taxa. Thus, we are led to conclude that although the spatial arrangement of resources on Earth might limit the number of large species within a taxon that can use them, the number of the smallest species within a taxon must be limited by something else.

Another limitation of a model of external physical constraint on body masses is that there is no explicit or implicit form that the scaling of diversity with body mass need follow. That is, the fractal dimension of the non-biological environment is arbitrary with regard to fundamental organismal attributes such as survival and reproduction. It is therefore necessary to ask whether diversification of body masses might be controlled or constrained by properties of the physical structure of organisms themselves rather than by the structure of their environments.

Internal physical constraints

The fundamental problem of an organism is to take resources available in the environment and use them to produce and maintain tissues. Resources are composed of a small set of relatively simple molecules. These molecules are extracted from the environment and assembled into more complex molecules that are the building blocks of tissues. During this process of extraction and assembly, combinations of molecules are formed that cannot be used in tissue growth and maintenance. If it is assumed that evolutionary processes have maximized the rate at which resource molecules are extracted from and waste molecules are excreted to the environment, then the most efficient spatial arrangement of tissues to accomplish this is fractal-like space-filling networks in which the smallest branches are invariant in size (West *et al.* 1997, 1999).

Autotrophs and heterotrophs extract resource molecules from the environment in different ways. Autotrophs typically extract resource molecules directly from the environment via passive diffusion across organismal boundaries. For multicellular autotrophs, the most efficient way to do this is to maximize the surface area across which these passive exchanges take place. Hence, plant tissues are typically highly branched networks that maximally fill environmental volumes with external exchange surfaces. Rigid structures in these organisms are typically constructed to allow the organism to locate exchange surfaces in environmental space to minimize the amount of overlap with other organisms (Horn 2000).

Heterotrophs extract resource molecules either from aggregated tissues ingested within the internal structure of the organism or directly across internal exchange surfaces via passive diffusion. Ingested tissues are typically broken down physically and then disassembled biochemically into resource molecules. These resource molecules are then distributed to exchange surfaces within the organism and transported via active or passive diffusion. These internal exchange surfaces have the same geometric properties that the external exchange surfaces of autotrophs have (West *et al.* 1997).

In large heterotrophic organisms, there is an additional necessity to distribute resource molecules among widely dispersed tissues and collect waste molecules from those tissues. The physical problem is the same for these internal circulatory systems as it is for external exchange networks, hence they tend to have fractal-like network structures that are maximally space filling in the same manner as external exchange surfaces (Li 2000).

In some taxa, coordination of the behaviour of the variety of tissues that make up the structure of the organism requires the collection and distribution of information in a manner similar to the distribution and collection of molecules. Moreover, the movement of this information often must be extremely rapid relative to the rates of diffusion of molecules. Some coordination of tissues can be accomplished via endocrine secretions into internal circulatory systems, but this is often too slow to respond to threats or opportunities to the whole organism in its environment. Information is circulated and coordinated through fractal-like networks of nerve tissue that have similar space-filling properties to circulatory and exchange networks.

The necessity of organisms to be constructed of fractal-like spatial structures that maximize the exchange of matter, energy and information within the organism and with its environment has profound consequences for the variation in sizes among organisms. In these fractal networks, the surface area across which exchanges are made is proportional to the organism's volume raised to the 3/4 power (West *et al.* 1999). As the flow of matter and energy into the organism is proportional to the surface area of exchange and the biomass of the organism is proportional to its volume, the rate of energy flow into an organism is proportional to its mass raised to the 3/4 power. Hence, the scaling of whole organism metabolic rate (B) with body mass (M) observed across many taxa is $B = aM^{3/4}$ (Peters 1983; Calder 1984). As a corollary to this scaling, per unit biological rates scale as $M^{-1/4}$ and biological transit times as $M^{1/4}$ (West *et al.* 1997, 1999). I turn now to the consequences of these fundamental physical constraints for the biological processes that determine variation among species in their body sizes.

Trade-offs in organismal maintenance and production

The incoming flow of energy and nutrients in an organism must be apportioned into the maintenance of existing tissues and the production of new tissues. As organisms are constructed of fractal-like networks, the rate of incoming energy scales as $M^{3/4}$ but the number of cells that must be provided by this inflowing energy scales as M^1. Consequently, for multicellular organisms, growth of new tissues in organisms will increase rapidly up to a given size, but then will begin to decline with time (West *et al.* 2001).

A fraction of the incoming energy devoted to production is allocated to the growth of reproductive tissues (Kozłowski 1992; Charnov 1993) and hence the rate of growth of non-reproductive tissues must slow (for indeterminant growth) or go to zero (for determinant growth). Over the lifetime of an organism, there is a trade-off between production of additional somatic tissues and production of reproductive tissues. All multicellular organisms require a finite length of time from the origin of the organism (birth or hatching) to the time at which it begins to reproduce (Fig. 10.2). The difference between total production and the biomass allocated to non-reproductive tissues when the organism begins to reproduce represents the amount of energy available for reproduction. The success of an organism in reproducing new organisms is a function of the amount of this energy relative to the requirements of producing offspring that survive.

There are two important allometric consequences of this trade-off between growth and reproduction within an organism. First, age at first reproduction can be shown from first principles to be proportional to $M^{1/4}$ (West *et al.* 2001), a value that is well substantiated empirically (Peters 1983; Charnov 1993). Thus, larger organisms spend a longer absolute time accumulating resources and maturing tissues than do smaller organisms. Second, if an organism of a given species produces a mass of offspring (R) that is some fraction of its adult mass, say λ, then total production, P, will be

$$P = M + R = M + \lambda M \tag{10.1}$$

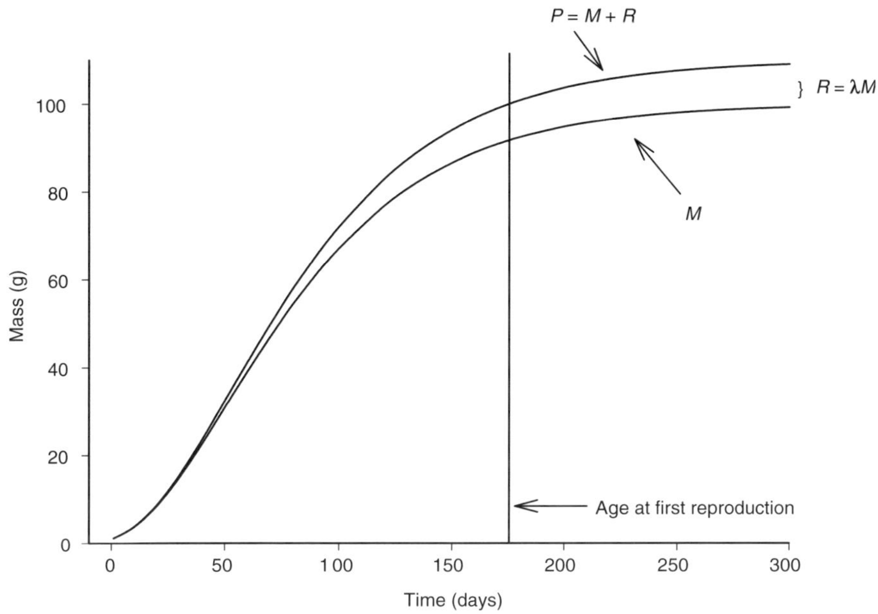

Figure 10.2 Generalized production curve for an organism, where *P* is total production and *M* is the biomass of somatic tissues. The mass put into reproduction, *R*, is assumed to be some fraction of *M*. This is assumed to be relatively constant for a given individual. It may vary among individuals or species.

Empirically, *R* has been shown to scale approximately as $M^{3/4}$ (Peters 1983), and from first principles it can be argued that when reproduction begins, the amount of reproductive biomass accumulated will be proportional to the amount of energy intake (West *et al.* 2001). Hence, the proportion of an organism's body mass that its offspring represent at birth will be proportional to $M^{-1/4}$. That is, an individual organism of small species produces a mass of offspring that represents a larger fraction of its adult mass than an individual of a larger species. Thus, if we compare large and small species within a taxon, in general, we would expect that reproduction in the largest species will be constrained by the longer times it takes to reach reproductive maturity, and the proportionately smaller mass of offspring that can be produced.

It is important to recognize that these two allometric constraints on reproduction arise from the energetic processes that underlie organismal growth, given the organism is constructed of tissues organized into fractal-like networks for the distribution of energy and matter. This, I believe, was the fundamental insight implicit in the controversial concept of reproductive power (Brown & Maurer 1987; Brown *et al.* 1993; Kozłowski 1996; Chown & Gaston 1997; Maurer 1998; Bokma 2001). Reproductive power was defined as the rate of production of reproductive biomass (Brown *et al.* 1993). Much confusion has been generated in attempting to interpret reproductive power as an energetic measure of fitness. Fitness, which I examine in the next section, is itself a concept that has generated much confusion, because most definitions of fitness do not differentiate between the fitness of individual organisms, and the average fitness of a population of organisms. The trade-offs between growth and reproduction discussed above are component processes that occur within organisms, and it is the interaction of these organismal processes with a particular set of environmental circumstances that determines the fitness of an organism.

Fitness and the adaptiveness of body mass

Our understanding of the diversification of body sizes among species depends on how much we understand about the evolutionary dynamics of the body sizes of individual species. Adaptation is the outcome of an evolutionary process by which some phenotypic character, in this case body mass, 'solves' an environmental problem. The process itself is sometimes called adaptation, although this can generate some confusion that can be construed in some senses to be circular (Peters 1976). The present problem is to understand whether the average body sizes of individual species within a taxon are adaptations. To do this, we need to answer three questions: (i) what the environmental 'problem' is; (ii) how the body size, or range of body sizes, of a species 'solves' the problem; and (iii) the population mechanisms by which the solution is obtained.

The third question is the most straightforward, so we begin there. The mechanism by which an adaptive body size arises within a species is natural selection. If natural selection occurs within a species, then there must be heritable variation among individual organisms in body size that corresponds to differences in individual production of offspring. Generally, the rate of production of new offspring is equated with fitness. The body size, or range of body sizes, that allow individuals to produce the most offspring will be perpetuated in the population over time.

It is more difficult to define the problem that a fitness-maximizing body size solves, because there may be a number of different phenotypic consequences of body mass. For example, an individual that begins to reproduce at a smaller body mass may be able to produce offspring slightly faster than a slightly larger individual of the same species, but may not be able to use certain kinds of resources as efficiently as the larger individual. One way around this is to recast the problem in terms of life histories (Roff 1992; Stearns 1992; Charnov 1993; Kozłowski 1999). From a life-history perspective, the fundamental problem to be solved is to balance mortality with future reproduction. If an organism begins to reproduce at a smaller size it may not be able to produce as large a quantity of offspring as it would if it waited until it was larger, but in waiting it faces a higher likelihood of dying before it reproduces. Differences in specific adaptive consequences of body size such as foraging efficiency or competitive ability can then be modelled in terms of their effects on this trade-off between mortality and future reproduction.

The answer to the question of how a particular body size maximizes fitness depends on how one defines fitness. From the perspective of an individual organism, its body size will result in it producing a specific number of offspring over the course of its reproductive life, each of which will have a suite of body-size-determining genes that is similar to its parent. In this sense, fitness is a property of the individual organism, as Darwin envisioned. However, fitness also can be construed to have a statistical meaning. In the statistical sense, a population of phenotypes (in this case body sizes) is considered, and the rate of change of a specific phenotype within that population is taken as the fitness of that phenotype (Fisher 1958; Yodzis 1989). The phenotype that grows most rapidly within the population is then construed to be the best adapted phenotype.

Fitness measures used in life-history studies often can be construed to be an average fitness, where the average is calculated over a specified population of organisms, often a single cohort. To illustrate the relationship between individual fitness and average fitness in a life-history context, consider the following argument. Let the fitness (R_i) of individual i, in a population of N organisms, be the total number of offspring produced by an organism over its reproductive life. If t_0 is the first breeding season for that individual and t is the last breeding season, then

$$R_i = \sum_{x=t_0}^{t} b_{i,x} \tag{10.2}$$

where $b_{i,x}$ is the number of offspring individual i produces at age x. Following the convention of life-history theory to consider only a single cohort of individuals that begin reproducing at t_0, we can calculate the average fitness of the population R_0 as

$$R_0 = \frac{\sum_{i=1}^{N} R_i}{N} = \sum_{x=t_0}^{t} \frac{n_x}{N} b_x \tag{10.3}$$

where n_x is the number of individuals alive at time x, $b_x = \Sigma_i\, b_{i,x}/n_x$ is the average number of offspring produced by individuals alive at time x, and $N = \Sigma_x\, n_x$. Taking $l_x = n_x/N$ as the proportion of individuals in the cohort alive at age x gives the conventional definition of R_0. Thus, R_0 is not a measure of individual fitness. It is related to the rate of change of a population of individuals (Yodzis 1989). The significance of R_0 in life-history theory is that it can be used to determine the evolutionary stable strategy (ESS) of a life history when population growth is regulated (Yodzis 1989; Charnov 1993).

In what follows I develop a relatively simple model for the size of the optimal phenotype in a population based on the assumption that a phenotype's life history is governed by the production and growth constraints described in the previous sections. A number of authors have developed more complicated arguments that share the same underlying structure (Roff 1992; Stearns 1992; Charnov 1993; Kozłowski & Weiner 1997). What I wish to emphasize here is that the optimal size of an organism within a species living in a particular environment is a direct consequence of not only the environmental circumstances that determine 'selection pressures' and the specific combinations of genes in a gene pool that provide the variation upon which selection acts, but also is in a fundamental way determined by the general physical constraints that phenotypes must exhibit as a consequence of their fractal-like physical structure.

In order to incorporate the insights obtained from the previous discussion on trade-offs between growth and production of individual organisms, I make some simplifying assumptions that make the following argument a bit more transparent, but that still capture the essence of the optimization problem. I first assume that survivorship, l_x, and average number of births, b_x, are both constant from t_0 to t. Equation (10.3) can then be written

$$R_0 = (t - t_0)lb \tag{10.4}$$

where l and b represent constant survivorship and births, respectively. This gives three life-history parameters: $t - t_0$, l and b, each of which might be functions of body size. As $t - t_0$ represents a biological time we can write $t - t_0 = cM^{1/4}$. For many types of organisms (e.g. birds and mammals), the number of births typically declines with body size across species (Peters 1983). It is not clear from first principles whether this should be true within a species. Here, for purposes of developing the argument, I assume that the average number of births per year will scale with body mass within a species in the same manner as it does among species. Using this assumption, we can write $b = kM^{-1/2}$ (e.g. Peters 1983). The constants c and k are arbitrary, and I assume here are constant among individual phenotypes within the population undergoing selection.

In order to incorporate environmental factors, I assume that survivorship is a unimodal function of body size, say $l(M)$, so that at some intermediate body size in the population, survivorship is maximized. This might be the case, for example, if resources in the environment were most efficiently usable by a phenotype of intermediate size. The important point is that, for this model, the survivorship function is determined by the particular environment in which selection is occurring, and not by any internal physical constraint. The particular parameters that determine what this function is I assume to be set by the environment in which selection is occurring. Thus, equation (10.4) becomes

$$R_0 = zl(M)M^{-1/4} \tag{10.5}$$

where $z = ck$.

The ESS for body size, M^*, under the assumption of population stability (Yodzis 1989) is

$$\frac{l(M^*)}{l'(M^*)} = 4M^* \tag{10.6}$$

where $l'(M)$ is the derivative of $l(M)$ with respect to M. The solution for the ESS is illustrated in Fig. 10.3 under the assumption of a quadratic relationship between l and M. More complicated models might be constructed using different assumptions about how natural selection operates on body size. The important point I wish to

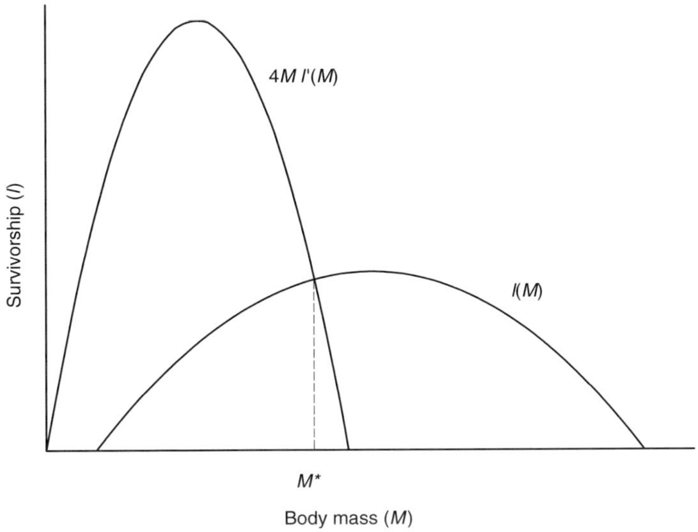

Figure 10.3 Evolutionary stable strategy under the assumption of a quadratic relationship between survivorship and body size. In this solution, population stability is assumed. Survivorship and reproduction are assumed to be constant in adults. Reproductive lifespan is assumed to scale with body mass as $M^{1/4}$ and offspring production is assumed to scale with body mass as $M^{-1/2}$.

emphasize here is that the form of the ESS for body size is determined by the assumption that the organisms within a population are constructed of fractal-like networks that transport matter, energy and information within the organism, and between the organism and its environment.

The argument up to this point can be generalized as follows. Equation (10.6) assumes that natural selection operates primarily on body size via adult mortality. This may be unsatisfactory if selection is operating on some other character. Suppose we assume that survivorship is affected by some quantitative character X that is correlated with body size. Let $l(X)$ and $M(X)$ describe the dependency of survivorship and body size on X, respectively. The ESS can then be obtained from

$$\frac{l(X^*)}{l'(X^*)} = \frac{4M(X^*)}{M'(X^*)} \tag{10.7}$$

where X^* is the ESS. As before, the form of the ESS is determined by constraints inherent in the nature of the physical structure of organisms.

The preceding argument was based on some simplifying assumptions regarding fitness. However, the method outlined above can be applied to other, more complicated models of natural selection. The fundamental insight that comes from assuming a fractal-like physical structure for organisms is that body size will profoundly influence the outcome of natural selection on *any* phenotypic character (Maurer 1998), and this insight is robust to the details of the particular model used to analyse a particular life history.

Before continuing, it should be pointed out that each species within a taxon will have its own, unique ESS. This is because the environment in which the ESS of a given species evolves is complex. Therefore, any two environments chosen at random will have a low probability of being identical. It also should be noted that the environment for any given species will contain many other species, such as predators, competitors and prey. In the simple model developed above, the environment was included in the function that related survivorship to body size without specifying how individuals of other species might affect survivorship. In a more general model, the effects of other species on the fitness of an evolving species should be considered (Taper & Case 1985, 1992; Case & Taper 2000). The distribution of body sizes of a given taxon might be considered to represent a joint, multispecies ESS if each species within the taxon affects fitness of other species. In actuality, it is unlikely that each species in a taxon will affect the ESSs of all other species in that taxon. The multispecies ESS may often include only a limited number of cross-species effects, with many ESSs being relatively independent of others. It is to this multispecies ESS that I now turn my attention.

A cost-benefit assessment of the diversification problem

What I have shown up to this point is that it is possible to determine the effects of body size on the adaptive evolution of a population of organisms by assuming a rather general physical structure that most organisms approximate in one way or another. The framework I discussed does not specify how new species with different adaptations originate, nor how species with certain sets of adaptations eventually become extinct. A complete theory of adaptive diversification requires not only the description of the process of adaptation within single species, it also requires a description of how new sets of adaptations are produced and how existing sets of adaptations become obsolete, rendering the species that possess them unviable in evolutionary time.

Here I consider the problem of how the pervasive effects of body size on energetic trade-offs within organisms, and consequently on patterns of natural selection within populations, might constrain the outcome of adaptive diversification among populations. I will not discuss the mechanics of speciation or extinction, nor how these processes might be related to body size (see Purvis *et al.*, this volume). I assume that whatever mechanisms generate new species will produce species that are sufficiently adapted to their environment so that they are able to persist over evolutionary time. Thus, the body size of each species that is able to persist, or suite of traits functionally correlated with body size, will be assumed to represent an ESS for that species.

There are two somewhat opposite views that have been proposed to explain the adaptive diversification of body sizes. At one extreme, individual species can be considered to be random samples from some statistical population of potential species (Kozłowski & Weiner 1997). Under this view, the statistical distribution of body sizes of species within a taxon is a posterior distribution defined by prior probability distributions for each life-history parameter that defines the ESS for a taxon. Individual species are considered to have their life-history parameters drawn randomly from the prior distributions; the resulting posterior distribution of body sizes can show many similarities to empirical distributions (Kozłowski & Weiner 1997). At the other extreme, all species can be considered to be realizations of a single, global ESS that is modified for a particular species by competition with other, similar sized species (Kelt 1997; Kelt & Brown 1998). Under this view, species are added sequentially, and the ESSs for each species are parameterized identically except for a reduction in fitness caused by similarity in body size with other, existing species. The latter alternative has been dismissed because the original formulation of the global ESS (Brown *et al.* 1993) has been questioned on technical grounds (Brown *et al.* 1996; Kozłowski 1996; Chown & Gaston 1997; Bokma 2001). However, there is no reason to suspect that an appropriate formulation cannot be developed. The question is whether either of these two alternatives are sufficient descriptions of the diversification process.

Both alternatives discussed in the preceding paragraph share the assumption that body size is the primary character upon which natural selection works. This is clearly a limitation if there is significant adaptive evolutionary change driven by other characters that might be correlated with body size. Changes in average body sizes as new species evolve need not be adaptive. Neither model will provide an adequate description of body-size diversification under such a scenario. A more general model that defines ESSs in terms of constraints on life history owing to body size, as discussed above, would overcome this limitation.

A second concern is that both views of body size evolution, especially the first alternative, do not incorporate the idea that there are strong genetic constraints on what combination of characters can be exhibited by a new species. New species cannot simply be random draws from a set of prior probability distributions that define all possible values a life-history parameter might take on. Rather, new species must be constrained to be similar to the species from which they arose because they share a large number of genes with the parent species. This means that prior probability distributions for life-history parameters will have exceedingly low variances at speciation. Furthermore, prior probability distributions for life-history parameters are likely to be different each time a speciation event occurs, owing to accumulated mutations in the gene pool from which individuals of an emerging species are drawn.

The fundamental problem with both models is that they take too simplistic a view of the process by which new, adaptive characters arise, and how new species are generated. The set of processes responsible for the generation of new species is undoubtedly highly complex, perhaps even too complex to model adequately. Thus, a somewhat different approach is called for. Instead of developing a model that predicts each and every one of the ESSs of a taxon, I will discuss an approach to identifying what kinds of ESSs cannot exist.

Given the potential for a pervasive influence of body size on the adaptive process via its role in organismal function and consequently fitness, it is possible to postulate that the kinds of species that arise within a taxon might be constrained by body size as well. Consider the following argument. The energy ingested by an organism that can be used for reproduction is a measure of the 'benefit' that the organism, as an energy processing system, yields as a function of its 'design', where design is used here to denote the functionality of its phenotypic characteristics. The goal is to compare designs that differ in size and ask what limitations those designs might have. Such limitations can be conceived of as 'costs', such that the energy available for reproduction must be greater than these costs in order for a viable design to exist over evolutionary time. Within a species, natural selection will result in a single viable design that constitutes the ESS given the specific genetic potential of the species and the environment within which that species resides. The question that needs to be addressed is: what constrains the distribution of ESSs among taxa?

One way to answer this question is to compare the costs and benefits associated with different ESS designs as a function of their size. Here I examine a fairly simple way to make this assessment. I will assume that the benefit of a design, in terms of the reproductive work that can be generated by that design, will scale with the body size of the ESS as $M^{3/4}$. The precedent for this follows from the discussion on the geometric structure of organisms developed above. If a single phenotype, say phenotype *i*, of species *j* is constrained by its geometry to generate energy at a rate proportional to $M_{ij}^{3/4}$, then phenotype *k*, of the same species that shares the same design but differs in size, will generate energy for its mass at a rate proportional to $M_{kj}^{3/4}$. The rate of energy generated by the average sized phenotype for the species is proportional to $M_j^{3/4}$, where M_j is the body size of the average individual (Welsh *et al.* 1988); note that even if the proportionality constant for all individuals is the same, the proportionality constant for the average individual will be different because it includes a term for variability among individuals. The scaling of rate of energy production for reproductive work will remain the same across species as well, although there may be considerable variation in proportionality constants among species. However, we can define the maximum possible benefit as a single scaling relationship. The proportionality constant for this maximum benefit function is related to the physical limits on the rate of energy production within an organism of a given size.

Given the maximum power output scaling for a set of ESS designs, the next problem is to decide what sets the maximum and minimum sizes. We first must assume that, in addition to an energetic benefit, any set of designs will also carry energetic costs for somatic growth and maintenance, where maintenance is construed to be any activity by the organism required for its survival. These costs can be modelled in the same way used to describe benefits. We are interested in describing how minimal costs might scale with organism size. Viable designs are those designs where the

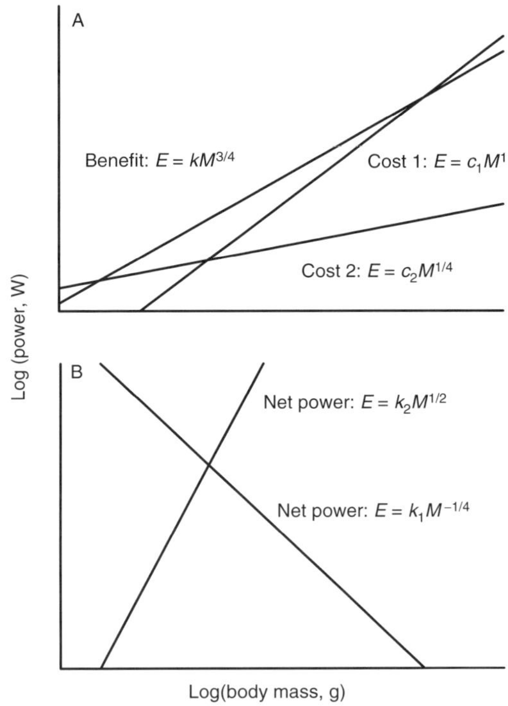

Figure 10.4 (A) Scaling of maximum net benefit (reproduction) and two minimum costs. Cost 1 limits the largest possible size and cost 2 limits the smallest possible size. The scaling exponent of cost 1 must exceed 3/4, while the scaling exponent of cost 2 must be less than 3/4. (B) Net power available for reproduction is greatest for intermediate body masses. The benefit/cost ratio of a particular evolutionary design must fall in the area beneath the two net benefit curves in order to be viable in evolutionary time. If species diversity for a given body mass is proportional to the height of the lowest curve, then the location and shape of the frequency distribution of species body masses will be determined by the exponents of the two net benefit curves. Note that for a given body size, the net power curve that gives the smallest value of net power represents the maximum net power achievable for that size. Thus, for species smaller than the body size at which the curves intersect, maximum power achievable will be limited by cost 2. Conversely, for species larger than the body size at which curves intersect, maximum power achievable will be limited by cost 1.

maximum benefit exceeds the minimum cost (Fig. 10.4). In order for there to be limits on species size, at least two allometrically scaling costs are needed: one to set the upper size, and one to set the lower size (Fig. 10.4A). Furthermore, the cost that sets the maximum size must have a scaling exponent greater than 3/4, and the cost that sets the minimum size must have a scaling exponent less than 3/4 (Fig. 10.4A). As an example, consider what kinds of costs might set the size of the largest possible organism of a given design. For a wide variety of locomotor activities, the total cost of moving an organism a fixed distance is proportional to about $M^{3/4}$ (Peters 1983). The rate of movement scales as roughly $M^{1/4}$ (Peters 1983). Note that both of these relationships also can be derived from the fractal-structure argument discussed above (see West *et al.* 1999). Hence, the power output for transportation costs is the total cost multiplied by the rate of movement, which scales as M^1. As the power output for reproduction scales as $M^{3/4}$, the largest possible organism of a given design would be determined by the body size at which power output needed for reproduction approximately equals the power needed to move the organism about in space. In other words, organisms larger than the largest size would have no surplus energy for reproduction because what surplus they generated above maintenance costs would be spent for locomotion costs.

It should be noted that allometric arguments, such as the one made above, are simplifications of more complex phenomena that emphasize certain trade-offs implicit in changes in size. Used in this sense they can be thought of as theoretical models that describe general properties that organisms are expected to have. Typically, empirical data show varying degrees of variability about expected allometric relationships. Some of this variability can be attributed to other deterministic factors that affect the 'dependent' variable in the allometric relationship, such as phylogeny, temperature, etc. Although the present argument assumes such factors to be constant among species, such variability may be an important source of information to build more general models.

The cost-benefit analysis just described suggests that the range of viable designs within a taxon is determined in a fundamental way by size. As each design is an ESS, each design will have a net benefit ratio (benefit/cost) that is >1. The range of viable designs will be greatest at some intermediate size, because this is where the maximum benefit exceeds the minimum cost by the greatest amount (Fig. 10.4B). Another way of viewing this result is to visualize each species as a unique ESS design that produces a unique cost and benefit. For a species of a given size, the minimum possible cost is set by the greater of the two allometric costs in Fig. 10.4A, and the maximum possible benefit is determined by the allometric reproductive power requirement. There are more possible cost-benefit combinations available to a species of an intermediate body size than one with a body size near either extreme.

In order to relate the cost-benefit model to empirical distributions such as those in Fig. 10.1, we must assume that the process of diversification maps individual species into the design space visualized in Fig. 10.4B. The shape of the empirical distribution will then be determined by the shape of the design space. The most common body size will correspond to that collection of ESSs where the cost/benefit ratio spans the largest range of viable designs. Design constraints implicit in the underlying biological and ecological structure of the ESS will limit the range of body sizes that produce viable designs. In addition to determining the modal body size, allometric costs will also determine the shape of the body size distribution. To see this, let the maximum net power that sets the lower bound on body size be $E = k_2 M^{3/4 - p}$ and the maximum net power associated with the upper bound on body size as $E = k_1 M^{q - 3/4}$. The body size distribution resulting from these constraints will be positively skewed if $(3/4 - p) > | q - 3/4 |$. A left-skewed distribution will be produced when $(3/4 - p) < | q - 3/4 |$.

The discussion up to this point has been qualitative, primarily because not enough is known about how the costs and benefits associated with the viability of adaptive designs vary with body size. Conceivably, one might be able to predict the location and shape of a body size distribution if one could derive from first principles the underlying allometric costs and benefits of body size for a group of species sharing similar designs. Alternatively, with appropriate measures of net benefit as a function of body size, it may be possible to fit alternative models that incorporate different types of allometric costs, and therefore test specific hypotheses about what factors are associated with the diversification of body size within a taxon.

Summary

The problem of understanding the adaptive process whereby body sizes diversify within an evolving lineage depends on our understanding of the inherent geometric structure of individual organisms, how this structure constrains individual productivity, and how individual productivity relates to differences in fitness among different sized organisms. The constraints on life histories that arise from the nature of their physical structure result in specific forms that the ESS of any character correlated with body size may exhibit. The diversification of designs (i.e. of ESSs) among species is most likely not random, rather it is the outcome of a set of complex processes that are not (or perhaps cannot be) described adequately given our current state of knowledge about the ecology and biology of different taxa. However, the outcomes of this complex process might be amenable to cost/benefit analyses, where costs and benefits are defined allometrically. This approach does not predict the specific set of ESSs that should evolve within a taxon, rather, it predicts which ESS cannot physically exist. This analysis suggests that more species of intermediate size exist within a taxon because there are more severe physical constraints experienced as species depart from that intermediate size.

Acknowledgements

The ideas for this chapter have stemmed from numerous discussions with J. Brown, B. Enquist, and members of the National Center for Ecological Analysis and Synthesis Working Group on Body Size in Ecology and Paleoecology. The data used in Fig. 10.1 were compiled by the members of this working group.

References

Bokma, F. (2001) Evolution of body size: limitations of an energetic definition of fitness. *Functional Ecology* **15**, 696–699.

Bonner, J.T. (1988) *The Evolution of Complexity by Means of Natural Selection*. Princeton University Press, Princeton, NJ.

Brown, J.H. (1995) *Macroecology*. University of Chicago Press, Chicago, IL.

Brown, J.H. & Maurer, B.A. (1987) Evolution of species assemblages: effects of energetic constraints and species dynamics on the diversification of the North America avifauna. *American Naturalist* **130**, 1–17.

Brown, J.H., Marquet, P.A. & Taper, M.L. (1993) Evolution of body-size—consequences of an energetic definition of fitness. *American Naturalist* **142**, 573–584.

Brown, J.H., Taper, M.L. & Marquet, P.A. (1996) Darwinian fitness and reproductive power: reply. *American Naturalist* **147**, 1092–1097.

Calder, W.A. (1984) *Size, Function, and Life History.* Harvard University Press, Cambridge, MA.

Case, T.J. & Taper, M.L. (2000) Interspecific competition, environmental gradients, gene flow, and the coevolution of species' borders. *American Naturalist* **155**, 583–605.

Charnov, E.L. (1993) *Life History Invariants: Some Explorations of Symmetry in Evolutionary Ecology.* Oxford University Press, Oxford.

Chown, S.L. & Gaston, K.J. (1997) The species–body size distribution: energy, fitness and optimality. *Functional Ecology* **11**, 365–375.

Fisher, R.A. (1958) *The Genetical Theory of Natural Selection.* Dover Publications, New York.

Gardezi, T. & da Silva, J. (1999) Diversity in relation to body size in mammals: a comparative study. *American Naturalist* **153**, 110–123.

Horn, H.S. (2000) Twigs, trees, and the dynamics of carbon in the landscape. In: *Scaling in Biology* (eds J.H. Brown & G.B. West), pp. 199–220. Oxford University Press, Oxford.

Hutchinson, G.E. & MacArthur, R.H. (1959) A theoretical ecological model of size distributions among species. *American Naturalist* **93**, 117–125.

Kelt, D.A. (1997) Assembly of local communities: consequences of an optimal body size for the organization of competitively structured communities. *Biological Journal of the Linnean Society* **62**, 15–37.

Kelt, D.A. & Brown, J.H. (1998) Diversification of body sizes: patterns and processes in the assembly of terrestrial mammal faunas. In: *Biodiversity Dynamics: Turnover of Populations, Taxa, and Communities* (eds M.L. McKinney & J.A. Drake), pp. 110–131. Columbia University Press, New York.

Kozłowski, J. (1992) Optimal allocation of resources to growth and reproduction: implications for age and size at maturity. *Trends in Ecology and Evolution* **7**, 15–19.

Kozłowski, J. (1996) Energetic definition of fitness? Yes, but not that one. *American Naturalist* **147**, 1087–1091.

Kozłowski, J. (1999) Adaptation: a life history perspective. *Oikos*, **86**, 185–194.

Kozłowski, J. & Gawelczyk, A.T. (2002) Why are species body size distributions usually skewed to the right? *Functional Ecology* **16**, 419–432.

Kozłowski, J. & Weiner, J. (1997) Interspecific allometries are by-products of body size optimization. *American Naturalist* **149**, 352–380.

Li, J.J.-K. (2002). Scaling invariance in cardiovascular biology. In: *Scaling in Biology* (eds J.H. Brown & G.B. West), pp. 113–128. Oxford University Press, Oxford.

Maurer, B.A. (1998) The evolution of body size in birds. II. The role of reproductive power. *Evolutionary Ecology*, **12**, 935–944.

Maurer, B.A. (1999) *Untangling Ecological Complexity: the Macroscopic Perspective.* University of Chicago Press, Chicago, IL.

Maurer, B.A. & Brown, J.H. (1988). Distribution of energy use and biomass among species of North American terrestrial birds. *Ecology* **69**: 1923–1932.

Maurer, B.A. Brown, J.H. & Rusler, R.D. (1992) The micro and macro in body size evolution. *Evolution* **46**, 939–953.

May, R.M. (1978). The dynamics and diversity of insect faunas. In: *Diversity of Insect Faunas* (eds L.A. Mound & N. Waloff), pp. 188–204.Blackwell Scientific, Oxford.

May, R.M. (1986) The search for patterns in the balance of nature – advances and retreats. *Ecology* 67, 1115–1126.

Morse, D.R., Lawton, J.H., Dodson, M.M. & Williamson, M.H. (1985) Fractal dimension of vegetation and the distribution of arthropod body lengths. *Nature* **314**, 731–733.

Peters, R.H. (1976) Tautology in ecology and evolution. *American Naturalist* **110**, 1–12.

Peters, R.H. (1983) *The Ecological Implications of Body Size.* Cambridge University Press, Cambridge.

Roff, D.A. (1992) *The Evolution of Life Histories: Theory and Analysis.* Chapman & Hall, New York.

Roy, K., Jablonski, D. & Martien, K.K. (2000) Invariant size-frequency distributions along a latitudinal gradient in marine bivalves. *Proceedings of the National Academy of Sciences, USA* **97**, 13150–13155.

Stearns, S.C. (1992) *The Evolution of Life Histories.* Oxford University Press, Oxford.

Taper, M.L. & Case, T.J. (1985) Quantitative genetic models for the coevolution of character displacement. *Ecology* **66**, 353–371.

Taper, M.L. & Case, T.J. (1992) Models of character displacement and the theoretical robustness of taxon cycles. *Evolution* **46**, 317–333.

Welsh, A.H., Peterson, A.T., & Altmann, S.A. (1988) The fallacy of averages. *American Naturalist* **132**, 277–288.

West, G.B., Brown, J.H. & Enquist, B.J. (1997) A general model for the origin of allometric scaling laws in biology. *Science* **276**, 122–126.

West, G.B., Brown, J.H. & Enquist, B.J. (1999) The fourth dimension of life: fractal geometry and allometric scaling of organisms. *Science* **284**, 1677–1679.

West, G.B., Brown, J.H. & Enquist, B.J. (2001) A general model for ontogenetic growth. *Nature* **413**, 628–631.

Yodzis, P. (1989) *Introduction to Theoretical Ecology.* Harper and Row, New York.

Part 5

Why are some species more likely to become extinct?

Chapter 11
Life histories and extinction risk

*John D. Reynolds**

Introduction

Conservation biologists always wish they had more information. Which populations are most vulnerable to extinction? What processes threaten them? What do they need to recover? Who will be affected by management plans, and how can we enlist their advice and support? In the vast majority of cases we will never be able to answer such questions with full investigations into population ecology. For every well-characterized population of deer or oystercatcher, there are hundreds of thousands of species for which we cannot even count their numbers. But if we could develop a better understanding of the biology of vulnerability, at least we might be better able to prioritize our efforts by predicting which species are apt to be in most trouble.

In this chapter I focus on correlations between life-history traits and population vulnerability to extinction. I will review theory and evidence related to traits such as fecundity, age at maturity, generation time, lifespan and body size. The focus will be primarily on human-induced changes in populations, especially due to direct mortality, although I will also consider habitat loss. Sources of mortality include exploitation such as hunting and fishing, as well as impacts of introduced predators and diseases. Together with habitat loss, these human-caused impacts are responsible for the vast majority of modern extinctions and listings of threatened status of the world's plants and animals (Hilton-Taylor 2000).

This review has been motivated not only by a desire to inform conservation biology, but also to inform our understanding of macroecology by linking life histories with population dynamics. Macroecology is a very ambitious field, attempting to explain large-scale biotic patterns in space and time. I am somewhat less ambitious, restricting myself primarily to temporal patterns of change in abundances of populations, rather than examining range size (but see Gaston & Blackburn 2000). By setting my sights on this scale, I hope to have a better chance of uncovering the processes that lie behind the patterns.

* *Correspondence address: Reynolds@uea.ac.uk*

Theory

Population dynamics

A simple way to consider correlations between life histories and responses of populations to changes in habitats or mortality rates is to model population dynamics using the logistic equation for deterministic growth:

$$N_t = \frac{r}{a + \left(\frac{r}{N_0} - a\right)e^{-rt}} \tag{11.1}$$

where N_t is the number of individuals at time *t*, *a* is the per capita reduction in population growth with increasing time and *r* is the intrinsic rate of natural increase, i.e. the maximum rate of growth that a population would achieve at small population sizes where there is no density dependence (Fig. 11.1). This model has an equilibrium or carrying capacity of $K = r/a$:

$$N_t = \frac{K}{1 + \left(\frac{K}{N_0} - 1\right)e^{-rt}} \tag{11.2}$$

This simple model ignores stochasticity, which is reviewed extensively in the next chapter (Sæther & Engen, this volume). Briefly, environmental stochasticity is caused by fluctuations in external forces such as climate, and demographic stochasticity is due to random successes or failures of individuals to survive and reproduce. The latter can be important in small populations. Both forms of stochasticity may cause strong reductions in long-run population sizes, and hence time to extinction (Sæther & Engen, this volume).

All of the parameters of equations (11.1) and (11.2) are extremely difficult to measure in the wild. For example, by definition *r* must be measured at population sizes that are sufficiently low that density-dependence will not alter per capita birth and death rates. Arguably, populations that suffer from direct removal of individuals offer the best chance of measuring this, and indeed there has been progress in measuring surrogates for *r* in commercially exploited fish populations by examining maximum population growth rates from time series (Myers *et al.* 1999). But measurements of *r* also require stable age distributions, which are also rather hard to come by in any population, including those of fish species, which are perpetually fluctuating, declining or (sometimes) recovering (Hutchings 2000; Jennings *et al.* 2001; Reynolds *et al.* 2002).

Another complication is Allee effects, known as depensation in most fisheries literature, in which there is a positive relationship between population growth and population size. There are many examples of components of fitness declining as numbers of individuals decline, and if these translate into effects on demography, they may cause the extinction of small populations (Courchamp *et al.* 1999; Stephens & Sutherland 1999; Petersen & Levitan 2001).

We will return to these formulations to illustrate effects of various kinds of threats

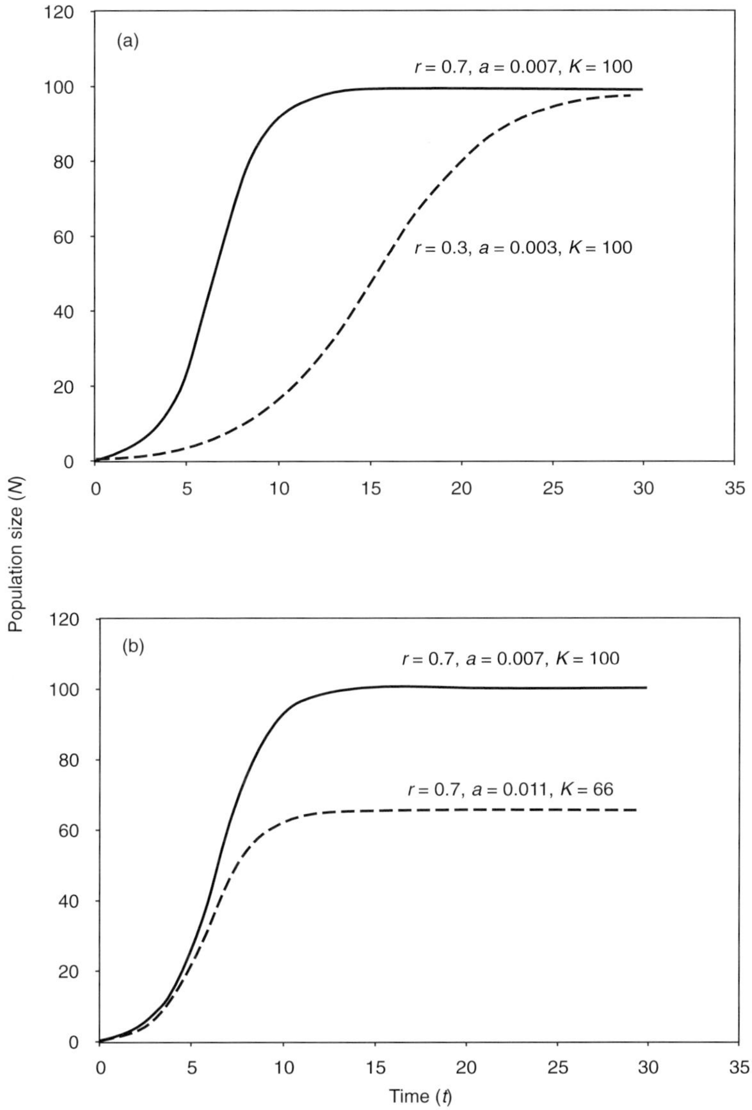

Figure 11.1 Logistic models of population growth. (a) A high intrinsic rate of increase, r, increases the rate of growth. The equilibrium population size or carrying capacity, K, is unaffected by the differences in r if the high value of r is matched by a high per capita reduction in growth, a ($K = r/a$). (b) Strong density dependence, a, reduces K if r is held constant.

Table 11.1 Fast–slow continuum of life-history traits matched to predicted maximum population growth rates, abundance and range size.

Biological characteristic	Fast	Slow
Life-history trait		
Maturity	Early	Late
Generation time	Short	Long
Fecundity	High	Low
Offspring size	Small	Large
Lifespan	Short	Long
Body size	Small	Large
Population trait		
Maximum growth rate	High	Low
Abundance	Common	Rare
Range size	Small–large	Large

to populations, and the manner in which they respond according to life-history characteristics of individuals.

Links to life histories

Life histories describe individual lifetime schedules of growth, maturation and reproduction (Roff 1992, 2002; Stearns 1992; Charnov 1993). Natural selection leads to the evolution of combinations of these traits that maximize lifetime reproductive success, subject to constraints imposed by trade-offs between time and energy devoted to each component. Responses to selection imposed by particular environments, for example, according to productivity or predation pressure, commonly lead to coevolution of suites of life-history traits. These can be characterized crudely according to a 'fast–slow' continuum between species that produce large numbers of small offspring at an early age, and those that delay maturity until they reach a larger body size, often producing fewer and larger offspring over a longer period of time (Table 11.1). A third type of life history, bet-hedging, involves both high productivity and high adult survivorship, which can evolve in response to high annual variation in survivorship of offspring. Both fast life histories and bet-hedging typically have high potential population growth rates (r or λ in the discrete version of the logistic model). Species with fast life histories often live at high densities, owing to their small body sizes, requiring less food and space per individual (reviewed by Gaston & Blackburn 2000). Conversely, they often have smaller geographical range sizes. However, species with large body sizes, including those with slow life histories and bet-hedgers, tend to have larger ranges. All such patterns have exceptions, such as the tendency for large-bodied species of ectotherms to have higher fecundity. They also depend on the breadth of the taxonomic scales of comparison. These crude characterizations therefore must be applied carefully, on a taxon by taxon basis.

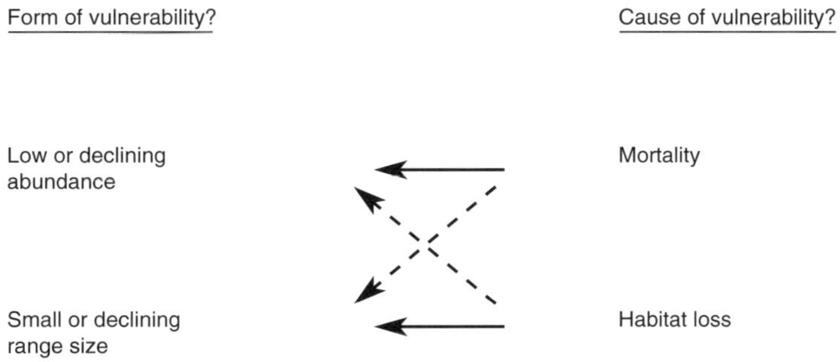

Figure 11.2 Forms of rarity and their causes. Mortality operates directly on abundance, and may reduce range sizes. Habitat loss may have a more direct effect on range sizes, and usually also reduces abundance.

Population vulnerability and its causes

To understand the biological correlates of extinction risk, we should ask two questions (Fig. 11.2): what form of vulnerability are we interested in, and what is causing it? Vulnerable populations are usually in some sense rare or becoming so, although not all rare species are necessarily vulnerable, if we leave them and their habitats alone. Vulnerability can be defined by either (i) low or declining abundance or (ii) small or declining geographical range size (Fig. 11.2). These distinctions are important because they do not necessarily correlate in the same direction with life histories. For example, large-bodied animals often have low densities but large geographical ranges (Gaston & Blackburn 2000). Such species are rare in one way and not in another.

The answer to the second question, 'what is causing vulnerability?', will also affect our predictions about correlates with life histories. When humans threaten population viability they usually do so either by removing individuals directly from the population, or by destroying their habitats (Fig. 11.2). Elevated mortality is often caused by exploitation, which is the second-most important cause of threatened status of animal species, after habitat loss (reviewed in Reynolds *et al.* 2001b). Mortality also can be caused by predation by introduced species, as well as diseases. Although such mortality clearly reduces population sizes, it also can reduce geographical range sizes through local extinctions or responses of individuals to disturbance. Habitat loss and degradation obviously reduce range sizes, at least on a local scale. Furthermore, the activities that cause it, such as forest clearance or water pollution, often entail direct mortality as well.

Predictions for elevated mortality

If we imagine a one-time removal of individuals from the population shown in Fig. 11.1a, we can see that species with high *r* will bounce back more quickly. If the elevated mortality is continuous, for example, as a result of hunting or fishing, the number of individuals that can be taken continuously from a population depends

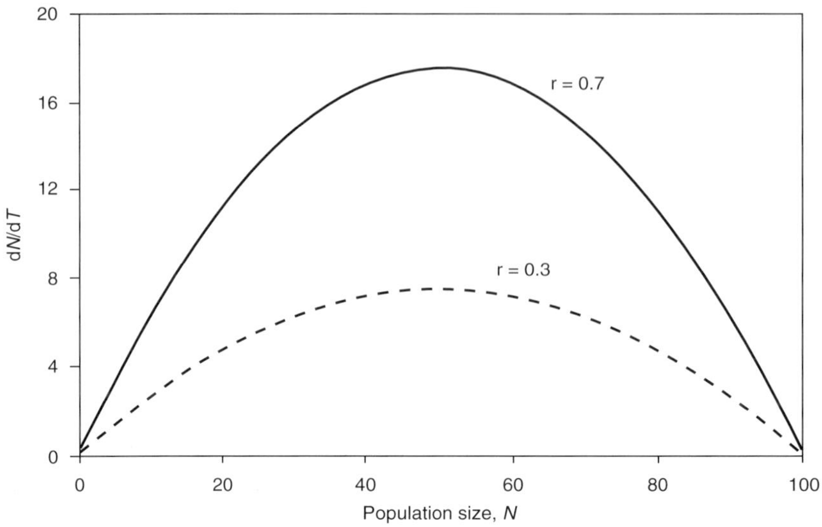

Figure 11.3 Rates of change in population sizes per unit time from the logistic model, showing the theoretical maximum sustainable yield at half of the equilibrium population size. Species with high intrinsic rates of increase, *r*, have higher maximum growth rates, and hence produce greater yields to hunting or fishing. In the top curve, the density-dependent parameter $a = 0.007$, and in the bottom curve $a = 0.003$, hence equilibrium population sizes $K (= r/a)$ are the same.

on the rate of growth of the population per unit time. For the logistic equation this, in turn, depends on *r*, *a* and the population size:

$$\frac{dN}{dt} = N(r - aN) \tag{11.3}$$

or with $K = r/a$,

$$\frac{dN}{dt} = rN\left(1 - \frac{N}{K}\right) \tag{11.4}$$

Figure 11.3 shows that populations characterized by high values of *r* grow more quickly, and can therefore provide greater yields. The absolute difference in growth rates between populations with different values of *r* is highest at intermediate population sizes, which correspond to the so-called maximum sustainable yield (MSY). Note that in fisheries, where MSY made its biggest splash, it has become clear that it is impossible to estimate this parameter accurately, and that it is not sufficiently precautionary. Therefore, it has now been downgraded to a theoretical reference point in population models, rather than a management target (Punt & Smith 2001).

We can link population responses to life histories through the fact that if mortality is the problem, then other things being equal (are they ever?), fast life histories (high *r*) ought to confer greater resilience than slow life histories (Brown 1971;

Diamond 1984; Pimm *et al.* 1988). One approach is to use life tables to examine sensitivities and elasticities of population growth rates to individual life-history components (Benton & Grant 1999; Grant & Benton 2000; Kokko *et al.* 2001). For example, Rochet (2000) found that population growth of North Sea plaice, *Pleuronectes platessa*, and Greenland halibut, *Reinhardtius hippoglossoides*, from coastal Labrador were most sensitive to age at maturity, lifetime fecundity and an index of natural mortality. Heppell *et al.* (1999) found that for North Sea haddock, *Melanogrammus aeglefinus*, survival from the first year to maturity contributes approximately 60% to the population growth rate, whereas survival in the first year of life contributed only about half as much.

Another approach is to calculate how much additional mortality populations can withstand based on natural mortality rates. Kirkwood *et al.* (1994) used an age-structured population model to suggest that fishing mortality should be proportional to natural mortality rates. Pope *et al.* (2000) combined natural mortality with body growth rates to show that high values of either parameter, alone or together, led to higher levels of sustainable fishing mortality rates.

Predictions for habitat loss

If habitat loss causes high densities of surviving individuals, this will raise density-dependent competition. In the logistic model this can be represented by an increase in the value of the density-dependent parameter a (eqn 11.1). We have already shown this in the bottom curve of Fig. 11.1b.

But why should life histories be correlated with impacts of habitat loss on populations? If we clear-cut a forest or drain a wetland, the inhabitants must surely disappear regardless of their age at maturity or fecundity. Here, ecological specialization might be the key (Brown 1971; Diamond 1984). We could still bring in some life-history correlates of extinction risk, although the tie-in to populations may be an extra step further removed than in the case of direct mortality. For example, if the form of habitat loss is fragmentation, then species with fast life histories could be more resilient if they have better dispersal characteristics. The devil is in the details. For example, the sizes of the fragments may be critical, with threshold amounts of habitat required for persistence. Furthermore, there may be non-linear relationships between the amount of habitat available and population sizes. If habitat fragments are large enough, then species with good dispersal characteristics may benefit from 'rescue effects' in metapopulations, overcoming isolation and problems with small population sizes, such as environmental and demographic stochasticity (Hanski 1999). Conversely, this beneficial effect of fast life histories could be counteracted by small natural range sizes if small-bodied species have more restricted distributions (Gaston & Blackburn 1996).

Two case studies

A study of extinction risk in birds addressed the second question in Fig. 11.2, concerning causes of vulnerability, but not the first one (Owens & Bennett 2000). This study analysed data from BirdLife International on threat status of the world's birds,

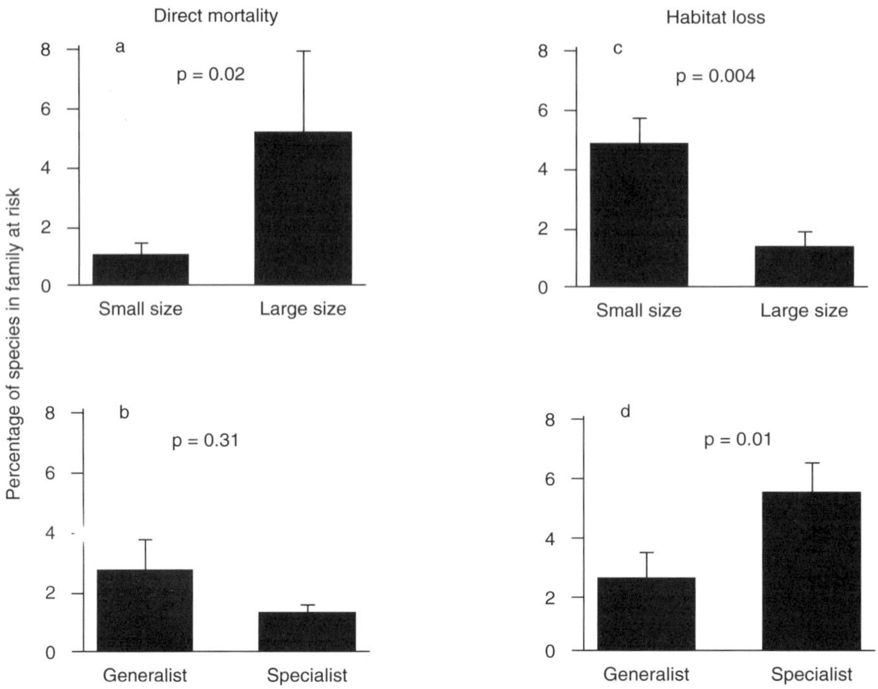

Figure 11.4 Relationships between life-history traits and extinction risk in birds. Data show means and standard errors of proportions of species in families at risk owing to mortality, in terms of body size (a) and habitat specialization (b), and those at risk owing to habitat loss, in terms of body size (c) and habitat specialization (d). (After Owens & Bennett 2000.)

which had been evaluated using criteria of the World Conservation Union (IUCN). The objectives were similar to an earlier study by Gaston & Blackburn (1995), which showed that larger bodied birds were under greater threat. The main advance of the Owens & Bennett (2000) study was that it distinguished between categories of threat, and examined two additional variables: generation time controlled for body size, and degree of habitat specialization. The authors noted that 43% of birds were threatened solely by habitat loss and 12% were threatened by mortality as a result of human persecution (primarily hunting) and introduced predators. When the threat was due to mortality, large-bodied species (> 1 kg) were most at risk (Fig 11.4a). The same was true for generation time, when controlled for body size (not shown). The authors also found a non-significant trend for habitat generalists to be more at risk (Fig. 11.4b). In contrast, when habitat loss was the problem the opposite patterns emerged, with smaller bodied habitat specialists being more at risk (Figs 11.4c and d). This study provides a clear demonstration of why we need to distinguish between causes of vulnerability if we are to advance our understanding of links between life histories and extinction risk. I suspect the body-size relationships will become even more pronounced if the data are broken down according to the first question in Fig.

11.2, concerning the form of vulnerability. For example, the theory outlined in an earlier section predicts that large-bodied species should decline most strongly as a result of hunting, not only because they will suffer greater mortality, but also because of lower *r*. They should be less apt to be listed as vulnerable owing to small range sizes.

A comparative study of extinction risk in primates and carnivores complemented the avian study by addressing the first question in Fig. 11.2, by providing analyses that focused on one form of vulnerability—population decline—in addition to overall analyses of all forms of threat combined (Purvis *et al.* 2000). However, they did not focus on the second question, concerning the nature of the threat, except anecdotally. Regression models confirmed that large-bodied species were more at risk of extinction in the two orders, but the relationship was not strong, and only came through when effects of geographical range size were factored out. When all species were examined in a multiple regression, nearly half of the variation in IUCN level of threat status was explained by geographical range, body size, gestation length, age at sexual maturity, trophic level and population density. The authors were then able to examine the second question in Fig. 11.2 anecdotally, by asking whether outliers could be explained by habitat loss (or lack thereof). This often proved to be the case. Species for which threat was underpredicted included primates that live in areas suffering from severe deforestation, such as Java and Madagascar. Conversely, species that were less threatened than predicted included species living in relatively undisturbed habitats, or that were able to live in secondary forests.

The literature since the mid-1990s has been preparing us for the sorts of complicated patterns found by Owens & Bennett (2000) and Purvis *et al.* (2000). Indeed, the editors of this volume have been prominent in showing that understanding correlations between body size, ecology and abundance is something of a minefield (e.g. Blackburn *et al.* 1993; Blackburn & Lawton 1994; Blackburn *et al.* 1994; Gaston 1994; Gaston & Blackburn 1995,1996; Blackburn & Gaston 1997). Much to my relief, this work has been reviewed recently (Gaston & Blackburn 2000). But I will reiterate three points that are particularly important here, concerning the phylogeny of rarity and links between body size, abundance and geographical range size. First, rarity, however it is measured, is non-randomly distributed among phylogenetic groups: a parrot or a primate should be a lot more worried than a finch or a mouse. This means that nearly all studies use phylogenetically based analyses where possible. Second, although on the scale of bacteria to whales large-bodied species occur at lower densities than smaller ones, this does not necessarily hold within narrower phylogenetic groups. Indeed, within some old avian lineages, large-bodied species may be more abundant (Cotgreave & Harvey 1991; Nee *et al.* 1991; Blackburn *et al.* 1994). Finally, although large-bodied species often have large geographic range sizes, small ones may have a wide variety of range sizes from the smallest to the largest (reviewed by Gaston & Blackburn 1996). So the scatter of body-size–range-size relationships is often messy, and approximately triangular.

More studies need to pick up where the avian and mammal studies have left off, by addressing both questions in Fig. 11.2 simultaneously, thereby distinguishing between the forms of vulnerability as well as their causes. If rates of habitat loss and the scale of hunting can be quantified, we could then determine the importance of each process, and the sensitivity of relationships between life histories and ecological specialization to each. A large amount of ecology stands between habitat loss, population growth rates, abundance and life histories. This presents a considerable challenge to those attempting to provide quantitative estimates from population models such as those presented earlier.

Comparative studies

Table 11.2 lists examples of life-history components that have been tested empirically against some aspect of population declines or low abundance. I have not included studies that focused solely on range size, although this aspect of rarity is included in some of the studies that examine IUCN levels of threat. The list is not exhaustive, but I hope it is fairly representative for animal species.

Invertebrates

Considering the enormous taxonomic diversity of invertebrates, it is unfortunate that so few studies have examined rarity and extinction of these animals in comparison with vertebrates (Reynolds *et al.* 2003). A general review of patterns of extinction of British invertebrates did not examine life-history attributes *per se* (as listed in Table 11.1), but it did examine habitat types and specialization (Thomas & Morris 1995). This study found that species at greatest risk usually occupied either extremely early successional habitats, or very late successional stages, namely involving dead wood.

The Red Data Book status of hoverflies (Syrphidae) in Britain and three regions of Germany has been compared with aspects of life histories and number of habitats occupied (Sullivan *et al.* 2000). Phylogenetically based analyses found that increasing levels of threat were associated with short flight periods in all four regions, as well as with long wings. There was no association with number of broods produced per season, nor with the number of habitats occupied. Although this provides a good first indication of threat status based on life-history traits, interpretation of the results is hampered by the fact that the index of rarity based on Red Data Book status includes the vulnerability of the habitats in which the species occur. Therefore, we have the two forms of vulnerability identified by Fig. 11.2 (based on population size and range size) as well as one of the causes of vulnerability (habitat loss) all bound up together in the response variable. Such are the realities of working with compound threat classifications when this is the only information available. A study of British macro-moths, gelechiid micro-moths, beetles and tephritid flies also used a composite measure of rarity based on Red Data Book status, and found that rarity of these insects was correlated with the rarity of their host plants (Hopkins *et al.* 2002).

Studies of European butterflies have shed light on links between mobility and extinction–colonization dynamics. Thomas (2000) found that species with

Table 11.2 Examples of studies testing for links between life histories and population vulnerability, defined according to small populations and decline, or based on composite measures of risk that include such aspects of population status. Studies based solely on range size have been excluded.

Taxon	Form of vulnerability	Cause of vulnerability	Traits examined	Trait correlated with vulnerability	Comments	Reference
Hoverflies, Britain and Germany	Varied	Varied	Size, flight period, number broods, habitat specialization	Long wings, short flight periods	Based on Red Data Book status, including vulnerability of habitat	Sullivan *et al.* 2000
Butterflies, Europe	Population extinction	Habitat fragmentation	Mobility, minimum area required for persistence	Intermediate mobility, require larger areas than sedentary	Mobility not confounded with losses of specific habitats	Thomas 2000
Butterflies, Britain	Population decline	Habitat and climate change	Mobility, habitat specialization	Lack of mobility, habitat specialization	Species with opposite traits increased in response to warming	Warren *et al.* 2001
Fishes, North America	Varied	Varied	51 aspects of life histories, ecology, behaviour	Not size, but late maturity, various ecologies	Unclear whether phylogenetic biases	Parent & Schriml 1995
Fishes, northeast Atlantic	Population decline	Mortality (fishing)	Size, growth, fecundity, age mature	Large size, fast growth, late maturity	Controlled for differences among taxa in mortality	Jennings *et al.* 1998
Skates and rays, western Britain	Population decline	Mortality (fishing)	Size (various measures), growth	Large maximum size	Crude growth data and some control for differences in mortality	Dulvy *et al.* 2000
Frogs, Australian wet tropics	Population decline	Unknown	Size, fecundity, reproductive mode, ecology, behaviour	Low fecundity, stream reproduction, specialized habitat	Based on guild classification	Williams & Hero 1998
Frogs, world	Population fluctuation	Varied	Size, development (direct versus aquatic stage)	Development involving aquatic larvae	Could not distinguish development mode from family characteristics	Marsh 2001
Reptiles, eastern Mediterranean	Inferred extinction	Island habitats	Size, longevity, habitat specialization	Large size, long lifespan, specialized habitat	Only habitat specialization significant in multiple regressions	Foufopoulos & Ives 1999

Table 11.2 *continued*

Taxon	Form of vulnerability	Cause of vulnerability	Traits examined	Trait correlated with vulnerability	Comments	Reference
Elapid snakes, Australia	Varied	Varied	19 aspects of life histories, ecology, behaviour	Ambush predation, lack of male competition	Lack of male-male competition related to large female : male size	Reed & Shine 2002
Birds, neotropical islands	Abundance	Island habitats	Size	None significant	Large-bodied species tended to be over-represented on islands	Gotelli & Graves 1990
Birds, Indonesian islands	Low abundance	Island habitats	Size, taxonomic distinctness, ecology, range size	Lack of taxonomic distinctness, some ecological traits	Weak or inconsistent trends with aspects of ecology	Jones *et al.* 2001
Birds, world	Varied	Varied	Size, island habitat	Large size, island habitat	Both traits independently correlated with threat status	Gaston & Blackburn 1995
Birds, world	Varied	Mortality (various)	Size, residual generation time, habitat specialization	Large size, long generation time	Species categorized into two size groups	Owens & Bennett 2000
Birds, world	Varied	Habitat loss	Size, residual generation time, habitat specialization	Small size, specialized habitat	Species categorized into two size groups	Owens & Bennett 2000
Primates, world	Population decline	Varied	Size, gestation length, age mature, others	Large size	Based on IUCN listings	Purvis *et al.* 2000
Carnivores, world	Population decline	Varied	Size, gestation length, age mature, others	Long gestation time, high trophic level	Based on IUCN listings	Purvis *et al.* 2000
Mammals, western North America parks	Extinction	Varied	Size, age mature, ecological specialization	Late maturity	Multiple regressions, extinctions restricted to 'natural' cases	Newmark 1995
Mammals, world	Extinction risk	Population fluctuation	Size, fecundity, age maturity	Combination of all three	Extinction risk was theoretical, based on demographic models	Fagan *et al.* 2001
Mammals, southeast Asia	Extinction	Island habitats	Size, age mature, interbirth interval, diet, range traits	Large size, large home range, low maximum latitude	Compared species restricted to large islands with those also on small	Harcourt & Schwartz 2001
Birds, reptiles, mammals, Amazon	Population decline	Mortality (hunting)	Size	Size	Hunting was size-selective	Peres 2000

intermediate levels of mobility have lost more of their populations than have species that are more sedentary. This was explained by mortality of individuals dispersing through fragmented landscapes, as well as the need for larger minimum areas for population persistence. Changes in population status should also depend on habitat specificity. This was confirmed by a study of changes in abundance and range size of 46 species of British butterflies over the past 30 years, in relation to climate change and habitat loss (Warren *et al.* 2001). This study found that half of the species that were mobile and habitat generalists have increased their abundance and range sizes, whereas other generalists and 89% of specialists have declined in response to habitat loss. This updates an earlier study by Hodgson (1993), which found that when highly mobile butterflies were excluded from the analyses, rare species tended to be large, produced a single brood per year and had long-lived larvae, with specialized host plant requirements. However, this study was based exclusively on range size, and the author was forced to wrestle with a large number of intercorrelated variables.

Fish species

A study by Parent & Shriml (1995) of the Great Lakes–St Lawrence basin of North America found several aspects of ecology and behaviour of freshwater fish to be correlated with threatened status, although body size was not one of them. Similarly, Angermeier (1995) found no relationship between body length and threat status among the 197 species native to Virginia, and neither did Duncan & Lockwood (2001) in a study based on IUCN Red List classifications. Duncan & Lockwood suggested that freshwater habitats may be impacted so heavily that biological characteristics of fish species may be over-ridden by extrinsic factors. This highlights the importance of making the distinction illustrated in Fig. 11.2 between forms of rarity (population sizes versus range sizes) and between forms of threat.

Marine commercial fisheries provide the best long-term, large-scale data sets for testing for the role of life histories in mediating population responses to elevated mortality (Reynolds *et al.* 2001a, 2002; Jennings *et al.*, in press). Although humans are certainly having profound impacts on marine habitats, especially in coastal zones, there is little dispute that for the vast majority of exploited marine fish species, these effects are dwarfed by the impacts of fishing. Many, but not all, of the theoretical predictions about slow life-history traits being linked to population declines have been confirmed for fish species in both temperate and tropical seas. For example, a phylogenetically based study of fish populations in the northeast Atlantic showed that late maturity and large body size were correlated with population declines, but there was no correlation with fecundity, and body growth rate was correlated in the opposite direction (Jennings *et al.* 1998). These analyses had been corrected for differences among populations in fishing mortality, suggesting that the population trends were driven by links between life histories and demographics, rather than by higher mortality on large-bodied species. Similarly, large-bodied parrotfish species were shown to be more susceptible in a study in Fiji, although this study could not control for differential fishing mortality (Jennings *et al.* 1999). Large-bodied skates and rays have declined significantly in many regions (Casey & Myers 1998; Walker &

Hislop 1998; Dulvy *et al.* 2000; Dulvy & Reynolds 2002), and an index of rebound potential of Pacific sharks proved to be correlated with body size and especially with maximum age (Smith *et al.* 1998).

It is worth considering fecundity in a little more detail, because some biologists have claimed that marine fishes are so much more fecund than birds and mammals that they will have much larger potential rates of population growth. This growth potential should make them much better able to bounce back from low numbers (e.g. Musick 1999). Indeed, it has been argued that this stronger resilience means that the IUCN's criteria for designating threat status of other animals do not work for marine fish species. This has led organizations such as the American Fisheries Society, the Convention on International Trade in Endangered Species (CITES), and the Committee on the Status of Endangered Wildlife in Canada (COSEWIC) to consider other criteria for threat listing. These criteria either explicitly (American Fisheries Society) or implicitly exempt 'highly productive' fish species from being listed as threatened with extinction. As with IUCN criteria, they ignore details of population growth, such as the potential existence of Allee effects.

But is it true that high fecundity is related to ability to sustain elevated mortality, and ability to bounce back from low numbers? Sadovy (2001) and Dulvy *et al.* (in press) have examined the evidence for this belief, and found little support for it. In addition to the lack of relationship between fecundity and sustainability in northeast Atlantic fishes mentioned above (Jennings *et al.* 1998), Sadovy lists many highly fecund fish species for which numbers have declined by more than 90%. Many populations of groupers (Epinephalinae), which are highly fecund, have become extinct, in both the Caribbean and in the Indo-Pacific (Sadovy & Eklund 1999; Morris *et al.* 2000; Sadovy 2001; Reynolds *et al.* 2002). The totoaba (*Totoaba macdonaldi*) is a very large and very fecund species of croaker (Scienidae) that is now restricted to the Gulf of California. Its numbers declined very quickly as a result of fisheries, and it continues to suffer from bycatches (Cisneros-Mata *et al.* 1995). These facts suggest that, in practice, highly fecund fishes do not win the lottery of reproductive success with their millions of tiny eggs, and indeed on theoretical grounds I am not sure why we would have ever expected them to. They are outflanked by other aspects of their life histories, such as late maturity and large body size.

Amphibians

It is widely recognized that many populations of amphibians are undergoing sharp population declines in many parts of the world (Alford & Richards 1999; Houlahan *et al.* 2000; Gardner 2001). Various causes have been blamed, singly or in combination, including pathogens, exposure to UV-B radiation, susceptibility to introduced predators, pollution, habitat loss and climate change. I do not know of any phylogenetically based comparative analyses that have tested for life-history correlates of decline and extinction risk of amphibians. However, a study of guilds of Australian rainforest frogs used multidimensional scaling to indicate that species that have disappeared or declined sharply are associated with stream habitats and have low fecundity (Williams & Hero 1998). This conclusion was based on two guilds of seven species in three genera that have undergone recent declines. A comprehensive comparative analysis of the Australian frog fauna would be very interesting. To date we have an analysis of 178 species of Australian endemics, which has related population abundance (but not declines or extinction risk) to body size (Murray *et al.* 1998). This study found no relationship between size and abundance, using cross-species comparisons and analyses that took phylogenetic relationships into account. (They did find a body-size–range-size relationship.) The authors were able to rule out various methodological artefacts that might have masked the predicted negative relationship, but they could not explain why their results differed from the usual trends.

Reptiles

Two studies of reptiles are listed in Table 11.2. Foufopoulos & Ives (1999) examined presence or absence of 35 species of turtles, snakes and lizards on 87 land-bridge islands in the Aegean and Ionian Seas. All of these islands had been connected to each other and most had been connected to the mainland during the last ice age. The authors inferred, therefore, that all of the reptiles had occurred at one time on all of the islands, and the absence today of any species indicates an extinction event. They also provided several lines of argument suggesting that any subsequent recolonization events were rare. Their analyses showed that the species that were most extinction-prone were large-bodied, long-lived, and had specialized habitat requirements. The authors also provided abundance estimates derived from 1400 h of surveys, and found that species with low abundances were also extinction-prone. Indeed, this proved to be the strongest effect, and only this trait as well as habitat specialization were significantly correlated with extinctions in multiple regressions. So although life histories provide a crude guide to extinction risk, the next step would be to ask about what, exactly, is causing low abundance, such as natural or human-caused rarity of habitats.

A comparative study of threat status of 70 species of Australian elapid snakes examined 19 aspects of life histories, ecology and behaviour (Reed & Shine 2002). Threatened species were less likely to display male–male combat than non-threatened species. This may be because females are larger than males in such species, and appear to be more vulnerable to humans. Ambush foraging was also correlated with threat, probably because this foraging mode is correlated with low rates of growth and reproduction. The authors used a linear discriminant function analysis to classify threat status of species according to the variables they had examined, and achieved a 90% success rate in assigning species to the correct category. Interestingly, as with the mammal study by Purvis *et al.* (2000), misclassification was ascribed to benefits or problems that species encountered when dealing with changes made by humans to habitats. Such changes were not in the model. Thus, forest clearance has benefited two large species, whereas the one threatened species that was incorrectly classified as non-threatened on the basis of its life history and ecology, the dwarf copperhead *Austrelaps labialis*, was suffering from a restricted range that was vulnerable to threatening processes (Reed & Shine 2002).

Birds

Many studies have examined population demography and vulnerability of birds in relation to life histories and ecology (Table 11.2; see also Diamond 1984; Pimm *et al.* 1988; Tracy & George 1992; Sæther *et al.* 1996; Fagan *et al.* 2001). I have already discussed two of these in the case studies presented in an earlier section (Gaston & Blackburn 1995; Owens & Bennett 2000). Most of the avian studies have found that large-bodied species tend to be more vulnerable, although as noted earlier various measures of rarity and forms of threat have been used, sometimes yielding complex and contradictory results. For example, Gotelli & Graves (1990) tested the hypothesis that for birds living on islands, large-bodied species would be particularly extinction-prone. They compared size distributions of assemblages of birds living on seven islands in the southern Caribbean with those on adjacent mainland. Contrary to expectation, there was a slight excess of large-bodied species on islands. The authors speculated that this could be a result of larger-bodied species being less specific in their ecological requirements, being better able to fly between islands and the mainland, or better able to forego breeding in bad years. Once again, the message is that we cannot ignore ecology and behaviour when predicting links between life histories, population dynamics and vulnerability.

Another study of birds on islands by Jones *et al.* (2001) shows that we should not ignore evolutionary history either. This study found that body size had little relationship to abundance of forest birds on Sumba and Buru in Indonesia. Rather, a strong correlate of abundance within each island was an index of taxonomic distinctness, measured on a seven-point scale from endemic to the island to widespread elsewhere. The authors speculated that taxonomically distinct species could be common owing to increases in abundance as species evolved adaptations to specific islands (and hence diverged from others), or it could be because successful colonists were pre-adapted to forest life, with low dispersal and therefore greater divergence from those on other islands.

Mammals

Mammals have also been studied extensively in tests for biological correlates of rarity and vulnerability to extinction. Examples in addition to those in Table 11.2 include a pioneering study of 15 species of mammals living on mountaintops in the Sierra Nevada and Rocky Mountains of the USA (Brown 1971), 16 species of rainforest mammals in north-east Queensland, Australia (Laurence 1991), a demographic study of 86 species (Fagan *et al.* 2001), a comparative study of 52 North American carnivores (Ferguson & Larivière 2002), and a study of various rarity indices in primates (Harcourt *et al.* 2002). The results, with respect to life histories, are rather mixed. For example, in the study by Brown (1971), small population sizes appeared to be correlated with large body size, high trophic level and specialized habitat requirements. However, the relative importance of each of these traits was not clear. A study of primates examined three indices of rarity, all of which included geographical range size, and found that only ecological specialization was correlated with rarity (Harcourt *et al.* 2002). Laurence's (1991) study of Australian rainforest mammals found that extinction-proneness in relation to forest fragmentation was overwhelmingly correlated with natural abundances in control areas. Once this was factored out, traits such as body size, longevity, fecundity, trophic level and dietary specialization did not contribute significantly to vulnerability. Even in comprehensive studies such as the carnivore and primate comparative analyses by Purvis *et al.* (2000), comparisons of threatened status of carnivores and primates with ecology and life histories showed that different traits were important in each of the orders, although the results fit with the general expectation of 'slow' life histories being correlated with extinction risk (Table 11.2). This finding also emerged from a study of extinctions of lagomorphs, carnivores and artiodactyls in 14 western North American parks, where species with late maturity were more extinction-prone (Newmark 1995). However, that relationship was secondary to the importance of initial population size, mirroring the result obtained by Laurence (1991) for Australian mammals and by Foufopoulos & Ives (1999) for Mediterranean reptiles. Within all of these mammal studies, there were multiple threats to the species.

Mixed communities

A study of the impacts of a single threat—subsistence hunting—on vertebrate communities in Amazonian forests provides a clear link between mortality and declines of large-bodied species (Peres 2000). Peres compared densities and biomasses of various birds, mammals and tortoises at 25 sites that differed in hunting pressure, and found a clear trend toward a decrease in percentage biomass of large-bodied species that was matched by an increase in small-bodied species (Fig. 11.5). Intermediate-sized species (weighing 1–5 kg) showed no relationship between level of hunting pressure and percentage density, but their percentage biomass also increased with level of hunting (not shown). These changes were a result of the size-selectivity of hunting, perhaps in combination with an inability of species with associated life histories (e.g. long generation time and low fecundity) to respond quickly enough demographically to sustain the pressure.

What do the comparative studies tell us?

The prominence of large body size as a predictor of vulnerability in many (but by no means all) of the studies described in Table 11.2 and in the previous sections probably occurs for three reasons. First, as a variable, size must benefit from a sampling and publication bias. Size is always the easiest thing to measure, so it is hardly surprising that more studies would examine body size (and with more precision) than age at maturity or fecundity. Second, large-bodied animals are often the most valuable, providing the most food, and looking most impressive when stuffed. As John Lawton put it (1995, p. 155): 'Being big is dangerous in a world dominated by *Homo sapiens*.' So, large-bodied animals typically face higher mortality, and it is hardly surprising that their populations would suffer more as a result. Again, we have to moderate this generalization in the face of potential complications with the animals' habitat characteristics. In fact, Reed & Shine (2002) note that the Australian taipan

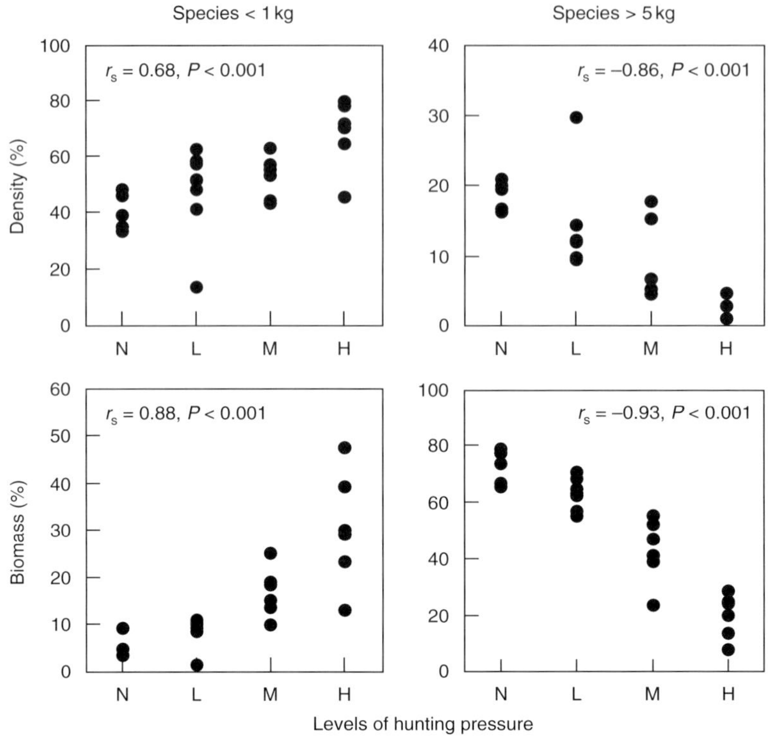

Figure 11.5 Relationships between the impact of hunting pressure (N, none; L, light; M, medium; H, high) and percentage biomass of Amazonian birds, mammals and tortoises that weigh less than 1 kg, and those that weigh more than 5 kg. (After Peres 2000.)

Oxyuranus scutellatus and eastern brown snake *Pseudonaja textiles* have benefited from forest clearance, and concluded that 'Whereas big, fierce animals are generally rare, big fierce Australian elapids are rarely rare'. Third, size is often correlated strongly with various life-history traits (Table 11.1). Size therefore can be thought of as a composite of many other aspects of life histories, as well as ecology (e.g. trophic position) and behaviour (e.g. home range size) (Peters 1983). So when we measure size we are also probably measuring age at maturity, lifespan and fecundity to a much greater extent than would be the case if we chose one of those variables and related it to the others. In the 1970s and 1980s this might have seemed a blessing to those hoping for simple correlates of extinction risk. But it has been a curse to those trying to disentangle whether size itself, or some correlate of size, explains patterns. Distinguishing the importance of interdependent predictor variables remains a major challenge today.

One of the messages that emerges from the studies reviewed here is that for many taxa there is little point in focusing solely on body size and its correlates in the quest to predict population dynamics and vulnerability. This point was brought out by a study of 10 species of large carnivores (Woodroffe & Ginsberg 1998). The authors found that home range size was significantly correlated with the size of reserves needed to allow population persistence. This was a result of the mortality suffered by animals that ranged outside reserve boundaries. Although potential confounding effects of body size were not reported in the study, it is clear from inspection of the data that body size was not related to home range size for the 10 species studied. Another reason to be concerned about what is happening outside of reserves is the fact that small reserves tend to be located in areas of high human density (Harcourt *et al.* 2001; Parks & Harcourt 2002). Therefore, small reserves face the double jeopardy not only of their small size, but also of their tendency to occur in hostile surroundings. In these cases body size clearly is not everything, but an interaction between behavioural ecology and human activities might be.

Conclusions

Some studies have reported success rates of 60–90% in predicting population trends and threat status of animals based on a small number of life-history and ecological variables. Yet I detected a sense of frustration running through most of the studies, caused by lack of data for the species being examined, which certainly mirrors my own experience. But this brings us back to one of the original motivations of these studies and of this review: to find short-cuts to predicting threatened status of species in a world in which conservation threats are pressing, time and money are in short supply, and critical ecological and demographic information is largely absent. As a means of prioritizing species for conservation assessments, I think we should be encouraged by the successes of many of these comparative studies.

To advance this field further, we need more studies to distinguish between forms of vulnerability and their causes. Is vulnerability based on low abundance or small range sizes? What is the threatening process? An inability to distinguish between threatening processes and the use of composite threat indices is hampering our ability to find consensus among some taxa, and it may be partly responsible for different findings in studies of similar taxa (e.g. primates studied by Purvis *et al.* (2001) and Harcourt *et al.* (2002)). Furthermore, although nearly all studies had access to body-size data, they differed widely in other variables examined, many of which were correlated with each other and with body size. So the manner in which different studies have handled interrelated variables must be important. Of course different taxa in different places are bound to have real differences in their biology of extinction risk. But at this rate, it will take a long time to find out. A difficult but potentially worthwhile project would be to assemble as many of the original data sets as possible and use standardized analyses of common variables, to obtain a better picture of where real similarities and differences among taxa lie. So far, the studies reviewed here have repeatedly discovered that when life histories have failed to explain vulnerability, ecology, behaviour and habitat loss usually did.

Acknowledgements

I thank Rob Freckleton, Jeremy Greenwood and Nick Isaacs for helpful comments, and Toby Gardner, Graham Hopkins and Owen Lewis for helping with literature suggestions.

References

Alford, R.A. & Richards, S.J. (1999) Global amphibian declines: a problem in applied ecology. *Annual Review of Ecology and Systematics* **30**, 133–165.

Angermeier, P.L. (1995) Ecological attributes of extinction-prone species: loss of freshwater fishes of Virginia. *Conservation Biology* **9**, 143–158.

Benton, T.G. & Grant, A. (1999) Elasticity analysis as an important tool in evolutionary and population ecology. *Trends in Ecology and Evolution* **14**, 467–471.

Blackburn, T.M. & Gaston, K.J. (1997) A critical assessment of the form of the interspecific relationship between abundance and body size in animals. *Journal of Animal Ecology* **66**, 233–249.

Blackburn, T.M. & Lawton, J.H. (1994) Population abundance and body size in animal assemblages. *Philosophical Transactions of the Royal Society, London, Series B* **343**, 33–39.

Blackburn, T.M., Brown, V.K., Doube, B.M., *et al.* (1993) The relationship between body size and abundance in natural animal assemblages. *Journal of Animal Ecology* **62**, 519–528.

Blackburn, T.M., Gates, S., Lawton, J.H. & Greenwood, J.J.D. (1994) Relations between body size, abundance and taxonomy of birds wintering in Britain and Ireland. *Philosophical Transactions of the Royal Society, London, Series B* **343**, 135–144.

Brown, J.H. (1971) Mammals on mountaintops: nonequilibrium insular biogeography. *American Naturalist* **105**, 467–478.

Casey, J.M. & Myers, R.A. (1998) Near extinction of a large, widely distributed fish. *Science* **281**, 690–692.

Charnov, E.L. (1993) *Life History Invariants.* Oxford University Press, Oxford.

Cisneros-Mata, M.A., Montemayor-Lopez, G. & Roman-Rodriguez, M.J. (1995) Life-history and conservation of *Totoaba macdonaldi. Conservation Biology* **9**, 806–814.

Cotgreave, P. & Harvey, P.H. (1991) Bird community structure. *Nature* **353**, 123.

Courchamp, F., Clutton-Brock, T. & Grenfell, B. (1999) Inverse density dependence and the Allee effect. *Trends in Ecology and Evolution* **14**, 405–410.

Diamond, J.M. (1984) 'Normal' extinctions of isolated populations. In: *Extinctions* (ed. M.H. Nitecki), pp. 191–246. University of Chicago Press, Chicago.

Dulvy, N.K. & Reynolds, J.D. (2002) Predicting extinction vulnerability in skates. *Conservation Biology* **16**, 440–450.

Dulvy, N.K., Metcalfe, J.D., Glanville, J., Pawson, M.K. & Reynolds, J.D. (2000) Fishery stability, local extinctions, and shifts in community structure in skates. *Conservation Biology* **14**, 283–293.

Dulvy, N.K., Sadovy, Y. & Reynolds, J.D. (in press) Extinction vulnerability in marine populations. *Fish and Fisheries.*

Duncan, J.R. & Lockwood, J.L. (2001) Extinction in a field of bullets: a search for causes in the decline of the world's freshwater fishes. *Biological Conservation* **102**, 97–105.

Fagan, W.F., Meir, E., Prendergast, J., Folarin, A. & Karieva, P. (2001) Characterizing population vulnerability for 758 species. *Ecology Letters* **4**, 132–138.

Ferguson, S.H. & Larivière, S. (2002) Can comparing life histories help conserve carnivores? *Animal Conservation* **5**, 1–12.

Foufopoulos, J. & Ives, A.R. (1999) Reptile extinctions on land-bridge islands: life-history attributes and vulnerability to extinction. *American Naturalist* **153**, 1–25.

Gardner, T. (2001) Declining amphibian populations: a global phenomenon in conservation biology. *Animal Biodiversity and Conservation* **24.2**, 25–44.

Gaston, K.J. (1994) *Rarity.* Chapman & Hall, London.

Gaston, K.J. & Blackburn. T.M. (1995) Birds, body size and the threat of extinction. *Philosophical Transactions of the Royal Society, London, Series B* **347**, 205–212.

Gaston, K.J. & Blackburn, T.M. (1996) Conservation implications of geographic range size–body size relationships. *Conservation Biology* **10**, 638–646.

Gaston, K.J. & Blackburn, T.M. (2000) *Pattern and Process in Macroecology*. Blackwell Science, Oxford.

Gotelli, N.J. & Graves, G.R. (1990) Body size and the occurrence of avian species on land-bridge islands. *Journal of Biogeography* **17**, 315–325.

Grant, A. & Benton, T.G. (2000) Elasticity analysis for density-dependent populations in stochastic environments. *Ecology* **81**, 680–693.

Hanski, I. (1999) *Metapopulation Ecology*. Oxford University Press, Oxford.

Harcourt, A.H. & Schwartz, M.W. (2001) Primate evolution: a biology of Holocene extinction and survival on the southeast Asian Sunda Shelf islands. *American Journal of Physical Anthropology* **114**, 4–17.

Harcourt, A.H., Parks, S.A. & Woodroffe, R. (2001) Human density as an influence on species/area relationships: double jeopardy for small African reserves? *Biodiversity and Conservation* **10**, 1011–1026.

Harcourt, A.H., Coppeto, S.A. & Parks, S.A. (2002) Rarity, specialization and extinction in primates. *Journal of Biogeography* **29**, 445–456.

Heppell, S.S., Crowder, L.B. & Menzel, T.R. (1999) Life table analysis of long-lived marine species with implications for conservation and management. In: *Life in the Slow Lane: Ecology and Conservation of Long-lived Marine Animals* (ed. J.A. Musick), pp. 137–147. American Fisheries Society Symposium **23**, Bethesda, MD.

Hilton-Taylor, C. (2000) *The IUCN Red List of Threatened Species*. IUCN, Gland, Switzerland.

Hodgson, J.G. (1993) Commonness and rarity in British butterflies. *Journal of Applied Ecology* **30**, 407–427.

Hopkins, G.W., Thacker, J.I., Dixon, A.F.G., Waring, P. & Telfer, M.G. (2002) Identifying rarity in insects: the importance of host plant range. *Biological Conservation* **105**, 293–307.

Houlahan, J.E., Findlay, C.S., Schmidt, B.R., Meyer, A.H. & Kuzmin, S.L. (2000) Quantitative evidence for global amphibian population declines. *Nature* **404**, 752–755.

Hutchings, J.A. (2000) Collapse and recovery of marine fishes. *Nature* **406**, 882–885.

Jennings, S., Reynolds, J.D. & Mills, S.C. (1998) Life history correlates of responses to fisheries exploitation. *Proceedings of the Royal Society of London, Series B* **265**, 333–339.

Jennings, S., Reynolds, J.D. & Polunin, N.V.C. (1999) Predicting the vulnerability of tropical reef fishes to exploitation with phylogenies and life histories. *Conservation Biology* **13**, 1466–1475.

Jennings, S., Kaiser, M.J. & Reynolds, J.D. (2001) *Marine Fisheries Ecology*. Blackwell Science, Oxford.

Jennings, S., Reynolds, J.D. & Greenstreet, S.P.R. (in press) Fish community responses to selective exploitation. In: *Fisheries-induced Adaptive Change* (eds U. Diekmann, O.R. Godoe, M. Heino & J. Mork). Cambridge University Press, Cambridge.

Jones, M.J., Sullivan, M.S., Marsden, S.J. & Linsley, M.D. (2001) Correlates of extinction risk of birds from two Indonesian islands. *Biological Journal of the Linnean Society* **73**, 65–79.

Kirkwood, G.P., Beddington, J.R. & Rossouw, J.A. (1994) Harvesting species of different lifespans. In: *Large-scale Ecology and Conservation Biology* (eds P.J. Edwards, R.M. May & N.R. Webb), pp. 199–227. Blackwell Science, Oxford.

Kokko, H., Lindström, J. & Ranta, E. (2001) Life histories and sustainable harvesting. In: *Conservation of Exploited Species* (eds J.D. Reynolds, G.M. Mace, K.H. Redford & J.G. Robinson), pp. 301–322. Cambridge University Press, Cambridge.

Laurence, W.F. (1991) Ecological correlates of extinction proneness in Australian tropical rain forest mammals. *Conservation Biology* **5**, 79–89.

Lawton, J.H. (1995) Population dynamic principles. In: *Extinction Rates* (eds J.H. Lawton & R.M. May), pp. 147–163. Oxford University Press, Oxford.

Marsh, D.M. (2001) Fluctuations in amphibian populations: a meta-analysis. *Biological Conservation* **101**, 327–335.

Morris, A.V., Roberts, C.M. & Hawkins, J.P. (2000) The threatened status of groupers (Epinephelinae). *Biodiversity and Conservation* **9**, 919–942.

Murray, B.R., Fonseca, C.R. & Westoby, M. (1998) The macroecology of Australian frogs. *Journal of Animal Ecology* **67**, 567–579.
Musick, J.A. (1999) Criteria to define extinction risk in marine fishes. *Fisheries* **24**, 6–14.
Myers, R.A., Bowen, K.G. & Barrowman, N.J. (1999) Maximum reproductive rate of fish at low population sizes. *Canadian Journal of Fisheries and Aquatic Sciences* **56**, 2404–2419.
Nee, S., Read, A.F., Greenwood, J.J.D. & Harvey, P.H. (1991) The relationship between abundance and body size in British birds. *Nature* **325**, 430–432.
Newmark, W.D. (1995) Extinction of mammal populations in western North American national parks. *Conservation Biology* **9**, 512–526.
Owens, I.P.F. & Bennett, P.M. (2000) Ecological basis of extinction risk in birds: habitat loss versus human persecution and introduced predators. *Proceedings of the National Academy of Sciences, USA* **97**, 12144–12148.
Parent, S. & Schriml, L.M. (1995) A model for the determination of fish species at risk based upon life-history traits and ecological data. *Canadian Journal of Fisheries and Aquatic Sciences* **52**, 1768–1781.
Parks, S.A. & Harcourt, A.H. (2002) Reserve size, local human density, and mammalian extinctions in U.S. protected areas. *Conservation Biology* **16**, 800–808.
Peres, C. (2000) Effects of subsistence hunting on vertebrate community structure in Amazonian forests. *Conservation Biology* **14**, 240–253.
Peters, R.H. (1983) *The Ecological Implications of Body Size.* Cambridge University Press, Cambridge.
Petersen, C.W. & Levitan, D.R. (2001) The Allee effect: a barrier to recovery by exploited species. In: *Conservation of Exploited Species* (eds J.D. Reynolds, G.M. Mace, K.H. Redford & J.G. Robinson), pp. 282–300. Cambridge University Press, Cambridge.
Pimm, S.L., Jones, H.L. & Diamond, J. (1988) On the risk of extinction. *American Naturalist* **132**, 757–785.
Pope, J.G, MacDonald, D.S., Daan, N., Reynolds, J.D., & Jennings, S. (2000) Gauging the impact of fishing mortality on non-target species. *ICES Journal of Marine Science* **57**, 689–696.
Punt, A.E. & Smith, A.D.M. (2001) The gospel of maximum sustainable yield in fisheries management: birth, crucifixion and reincarnation. In: *Conservation of Exploited Species* (eds J.D. Reynolds, G.M. Mace, K.H. Redford & J.G. Robinson), pp. 41–66. Cambridge University Press, Cambridge.
Purvis, A., Gittleman, J.L., Cowlishaw, G. & Mace, G.M. (2000) Predicting extinction risk in declining species. *Proceedings of the Royal Society, London, Series B* **267**, 1947–1952.
Reed, R.N. & Shine, R. (2002) Lying in wait for extinction: ecological correlates of conservation status among Australian elapid snakes. *Conservation Biology* **16**, 451–461.
Reynolds, J.D., Bruford, M.W., Gittleman, J.L. & Wayne, R.K. (2003) Editorial: the first five years. *Animal Conservation* **6**, 1–2.
Reynolds, J.D., Jennings, S. & Dulvy, N.K. (2001a) Life histories of fishes and population responses to exploitation. In: *Conservation of Exploited Species* (eds J.D. Reynolds, G.M. Mace, K.H. Redford & J.G. Robinson), pp. 148–168. Cambridge University Press, Cambridge.
Reynolds, J.D., Mace, G.M., Redford, K.H. & Robinson, J.G. (eds) (2001b) *Conservation of Exploited Species.* Cambridge University Press, Cambridge.
Reynolds, J.D., Dulvy, N.K. & Roberts, C.R. (2002) Exploitation and other threats to fish conservation. In: *Handbook of Fish Biology and Fisheries*: Vol. 2, *Fisheries* (eds P.J.B. Hart & J.D. Reynolds), pp. 319–341. Blackwell Publishing, Oxford.
Rochet, M.-J. (2000) May life history traits be used as indices of population viability? *Journal of Sea Research* **44**, 145–157.
Roff, D.A. (1992) *The Evolution of Life Histories: Theory and Analysis.* Chapman & Hall, New York.
Roff, D.A. (2002) *Life History Evolution.* Sinauer, Sunderland, MA.
Sadovy, Y. (2001) The threat of fishing to highly fecund fishes. *Journal of Fish Biology Supplement A* **59**, 90–108.
Sadovy, Y. & Eklund, A. M. (1999) Synopsis of the biological data on the Nassau grouper, *Epinephelus striatus* (Bloch, 1972), and the Jewfish, *E. itajara* (Lichtenstein, 1822). NOAA Technical Report NMFS 146. 65 pp.
Sæther, B.-E., Ringsby, T.H. & Røskaft, E. (1996) Life history variation, population processes and priorities in species conservation: towards a reunion of research paradigms. *Oikos* **77**, 217–226.

Smith, S.E., Au, D.W. & Show, C. (1998) Intrinsic rebound potentials of 26 species of Pacific sharks. *Marine and Freshwater Research* **49**, 663–678.

Stearns, S.C. (1992) *The Evolution of Life Histories.* Oxford University Press, Oxford.

Stephens, P.A. & Sutherland, W.J. (1999) Consequences of the Allee effect for behaviour, ecology and conservation. *Trends in Ecology and Evolution* **14**, 401–405.

Sullivan, M.S., Gilbert, F., Rotheray, G., Croasdale, S. & Jones, M. (2000) Comparative analyses of correlates of Red Data Book status: a case study using European hoverflies (Diptera: Syrphidae). *Animal Conservation* **3**, 91–95.

Thomas, C.D. (2000) Dispersal and extinction in fragmented landscapes. *Proceedings of the Royal Society, London, Series B* **267**, 139–145.

Thomas, J.A. & Morris, M.G. (1995) Rates and patterns of extinction among British invertebrates. In: *Extinction Rates* (eds J.H. Lawton & R.M. May), pp. 111–130. Oxford University Press, Oxford.

Tracy, C. & George, T.L. (1992) On the determinants of extinction. *American Naturalist* **139**, 102–122.

Walker, P.A. & Hislop, J.R.G. (1998) Sensitive skates or resilient rays? Spatial and temporal shifts in ray species composition in the central and northwestern North Sea between 1930 and the present day. *ICES Journal of Marine Science* **55**, 392–402.

Warren, M.S., Hill, J.K., Thomas, J.A., *et al.* (2001) Rapid responses of British butterflies to opposing forces of climate and habitat change. *Nature* **414**, 65–69.

Williams, S.E. & Hero, J.-M. (1998) Rainforest frogs of the Australian wet tropics: guild classification and the ecological similarity of declining species. *Proceedings of the Royal Society, London, Series B* **265**, 597–602.

Woodroffe, R. & Ginsberg, J.R. (1998) Edge effects and the extinction of populations inside protected areas. *Science* **280**, 2124–2128.

Chapter 12
Routes to extinction

Bernt-Erik Sæther and Steinar Engen*

Introduction

All species will inevitably become extinct. The longevity of most species in the fossil record is less than a couple of million years (Jablonski 1994) so the majority of the species that have existed during the past two billion years are now extinct. Although many of those species disappeared during distinct periods with mass extinctions because of rapid habitat alterations or degradation (Eldredge 1999), or other catastrophic events (Shapton *et al.* 1993), the majority of all extinctions in the fossil record still occurred at a nearly constant rate, giving an exponential distribution of taxon longevity (Van Valen 1973; Sepkoski 1993).

Humans are now causing a new period of mass extinctions, probably the sixth one in Earth's history. About one-eighth of bird species and one-quarter of mammalian species of the world are now considered threatened as a result of human activities (Groombridge 1992; Hilton-Taylor 2000). Most of the declines are the result of habitat loss or fragmentation, overexploitation, introduction of alien species or pollution. As a consequence, the current extinction rate is probably about three to four orders of magnitude greater than the rate in the normal background extinctions (May *et al.* 1995; Pimm *et al.* 1995).

Basically, there are two, not mutually exclusive, routes to extinction. Populations may become extinct because of a negative mean population growth rate or decreased carrying capacity owing to habitat destruction. However, populations with a positive growth rate also may become extinct through the operation of demographic or environmental stochasticity affecting the population growth rate or fluctuations in the carrying capacity. Caughley (1994) contrasted these two major focuses in the study of extinction into a 'declining species' versus 'a small population' paradigm.

Despite the huge importance of extinction in determining the composition of biological diversity, we still lack an understanding of how different population characteristics affect the chances that a population will become extinct. This has prevented the development of testable predictions for how extinction will act as a process. Here

* *Correspondence address: Bernt-Erik.Sather@bio.ntnu.no*

we will try to review some of the insight that has been gained from theoretical and empirical analyses of extinction, and try to suggest some general patterns that characterize extinction as a process. Our empirical focus is strongly biased in favour of birds, but we hope some of the findings may have relevance for a wider range of taxa.

Stochastic components in population dynamics

In a fluctuating environment, the chance of population extinction is influenced by two kinds of stochastic effects, demographic and environmental. Thus, we consider catastrophes as extreme effects of environmental stochasticity. To illustrate the stochastic effects on population dynamics we consider a simple model describing density-independent growth in a random environment $N(t+1) = \lambda(t)N(t)$ (Lewontin & Cohen 1969). Here $N(t)$ is population size in year t, and $\lambda(t)$, the population growth rate in year t, is assumed to have the same probability distribution each year with mean $\bar{\lambda}$ and variance σ_λ^2. At large population sizes, the *environmental stochasticity* is $\sigma_e^2 = \sigma_\lambda^2 = \mathrm{var}\left(\frac{\Delta N}{N}\middle|N\right)$. On a logarithmic scale $X(t+1) = X(t) + r(t)$, where $X(t) = \ln N(t)$ and $r(t) = \ln \lambda(t)$. The growth rate during a period of time t, $(X(t) - X(0))/t$, has mean $s = \mathrm{E}\bar{r}$ and variance $\mathrm{var}(r)/t = \sigma_r^2/t$, where E denotes expectation. For long time intervals the mean slope of the trajectories on log-scale all approach the constant s as t increases, which is called the stochastic population growth rate. The first-order approximation of the mean and variance are $s \approx \ln \lambda - \frac{\sigma_r^2}{2}$ and $\sigma_r^2 \approx \frac{\sigma_\lambda^2}{\bar{\lambda}^2} = \frac{\sigma_e^2}{\bar{\lambda}^2}$, respectively.

Demographic stochasticity is caused by random fluctuations in individual fitness that are independent among individuals, giving $\mathrm{var}\left(\frac{\Delta N}{N}\middle|N\right) = \sigma_e^2 + \sigma_d^2/N$ (Engen *et al.* 1998), where σ_d^2 is called the demographic variance. It produces a similar reduction in the stochastic growth rate s as the environmental stochasticity σ_e^2, given by (Lande 1998)

$$s \approx \ln \bar{\lambda} - \frac{\sigma_e^2}{2} - \frac{\sigma_d^2}{2N} \tag{12.1}$$

We see from this that to obtain a proper understanding of the risk of extinction stochastic effects must be modelled and estimated in addition to the deterministic components.

Density-independent populations

Tragically, an increasing number of populations are now found only in small numbers, far below their carrying capacity. From equation (12.1) it is evident that the population is doomed to extinction when the deterministic growth rate r is zero or less. Ignoring stochastic effects, the time to reach extinction being at population size N will in this case be $T(N) = \ln N/\ln \bar{\lambda}$.

A population also may become extinct even though the deterministic growth rate is positive because of stochastic effects on the population dynamics. Thus, the demographic and environmental stochasticity have a twofold effect on the risk of extinction through a reduction of the long-run growth rate s as well as through random fluctuations in population numbers (Lande 2002).

Our next step will be to examine quantitatively the relative contribution of the different components of equation (12.1) on the time to extinction. Assuming all individuals to be equal ($\sigma_d^2 = 0$), time to extinction at $N = 1$ has an inverse Gaussian distribution (Cox & Miller 1970). This distribution has relatively well understood statistical properties (Seshadri 1999). When $s < 0$ extinction is certain because the cumulative probability of extinction (see Dennis *et al.* 1991, equation 16) approaches 1 as the time increases towards infinity. However, when $s > 0$, extinction occurs with probability e^{-2X_0S/σ_r^2}, where X_0 is the logarithm of the initial population size. Engen & Sæther (2000) showed that these results are particularly useful for predicting the time to quasi-extinction, i.e. the time to reach a barrier that is sufficiently large to ignore σ_d^2 (Ginzburg *et al.* 1982).

When both demographic and environmental stochasticity are present, no analytic expression for the time to extinction is known, and we have to compute the quasi-stationary distribution by stochastic simulations. In general, the distribution of time to extinction has an initial period with low probability and a long right-hand tail (Fig. 12.1). This means that even for negative growth rates a time will elapse before extinction will occur. We also see that the time to extinction is strongly influenced by σ_d^2 (Fig. 12.1) because demographic stochasticity operates more strongly at small population sizes (eqn 12.1), increasing the downward trend dramatically. Thus, reliable predictions of time to extinction cannot be obtained without knowledge of σ_d^2.

Risk of extinction of small populations far below carrying capacity is also strongly influenced by the environmental stochasticity. From equation (12.1), we find, using general results from diffusion theory (Karlin & Taylor 1981), that the probability of ultimate extinction at $N = 1$ starting at initial population size $N = N_0$ is 1 when $s \le 0$, and $\left(\dfrac{\sigma_d^2 + \sigma_e^2}{\sigma_d^2 + \sigma_e^2 N_0}\right)^{2s/\sigma_e^2}$ when $s > 0$. Thus, the probability that a population goes extinct decreases with increasing initial population size, but even larger populations may have a high chance for ultimate extinction for larger values of σ_d^2 (Fig. 12.2).

In a seminal paper, Lande (1998) was able to analyse the relative contribution of σ_d^2 and σ_e^2 to the decline of small populations by applying a scale transformation that keeps the infinitesimal variance constant. Demographic stochasticity may create an unstable equilibrium on the transformed scale. Below this equilibrium population size, the population is likely to decline to extinction. This effect of demographic stochasticity closely resembles the operation of a traditional Allee effect (Allee *et al.* 1949).

In practice, demographic stochasticity is estimated from data on individual variation between females in reproductive success and survival (Sæther *et al.* 1998,

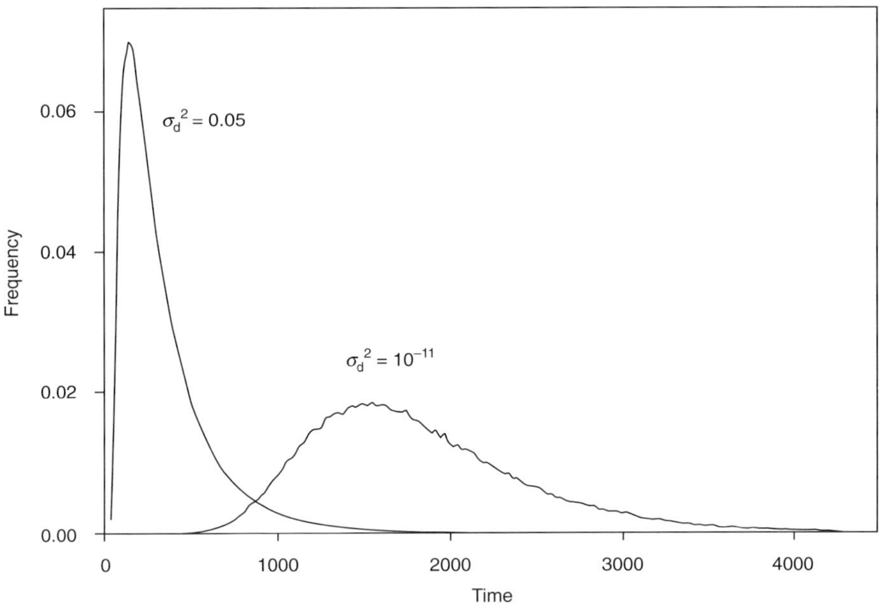

Figure 12.1 The distribution of time to extinction from an initial population size $N = 100$ of a declining population with no density dependence for two different levels of demographic stochasticity. Other parameters are $r = 0.01$ and $\sigma_d^2 = 0.05$.

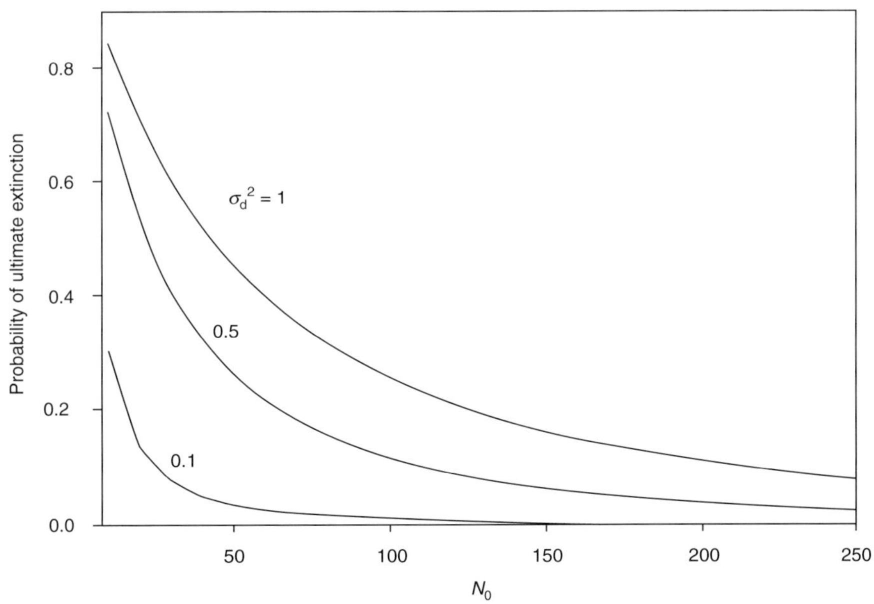

Figure 12.2 The probability of ultimate extinction of a density-independent population (see eqn 12.1) in relation to initial population size for different values of demographic variance σ_d^2. Other parameters are $r = 0.01$ and $\sigma_e^2 = 0.01$.

2000). The demographic variance is the among-years mean of the within-years variance in the contributions of the individuals to the next generation (Engen *et al.* 1998). The contribution is defined as the number of female offspring that is recruited into the next generation plus 1 if the female herself survives (Sæther *et al.* 1998). Estimates of σ_d^2 are now available for several vertebrate populations (Tufto *et al.* 2000; Lande *et al.* 2003), and range typically from 0.15 to 0.65.

Dennis *et al.* (1991) presented a method for estimating *s* and σ_e^2 of strictly declining species from time-series data of abundances, using a diffusion approximation and assuming $\sigma_d^2 = 0$. Thus, estimates of these parameters are now available for a number of declining species (Dennis *et al.* 1991; Gaston & Nicholls 1995; Engen & Sæther 2000; Sæther & Engen 2002a). A problem with these parameter estimates is that they will be strongly influenced by sampling errors in the population estimates (Solow 1998). To reduce the impact of this problem, Sæther & Engen (2002a) analysed time series of 10 strictly declining bird populations where the population counts are quite accurate because they were based either on direct counting of nests or recapture of colour-ringed individuals. In this data set, σ_r^2 contributed significantly to *s* (correlation coefficient = −0.66, $P = 0.039$, $n = 10$). This shows that not only knowledge of the expected dynamics, but also quantification of stochastic effects are important for predicting the rate of decline in such decreasing populations.

Engen *et al.* (2001) extended the approach of Dennis *et al.* (1991) and developed a method for computing the risk of extinction of declining populations that included separate estimation of σ_d^2 and σ_e^2. This enabled us to examine factors affecting the time to extinction of a declining barn swallow *Hirundo rustica* population in an area of highly intensified agricultural practice in Denmark. Assuming no uncertainty in the parameter estimates, the distribution of the time to extinction has an initial period with a small probability of extinction and a peak at about 50 years (Fig. 12.3). This illustrates that even for rapidly declining populations there will be a time delay, dependent on the initial population size, before the consequences of a demographic change in terms of extinction or quasi-extinction can be recorded. In contrast to what is to be expected in more slowly declining populations (Fig. 12.1), the difference between the mean (52 years) and median (48 years) time to extinction was quite small.

So far, we have only considered the female part of the population. However, in a population with two sexes the demographic variance is strongly influenced by adult sex ratios and behavioural dynamics (e.g. mating systems). To illustrate this, let us first consider a polygamous mating system, where we assume all females will be mated. Engen *et al.* (in press) show that the demographic variance can be split into three components owing to a binomial distribution of male survival (σ_{dm}^2), stochastic variation among females in survival and reproduction (σ_{df}^2) and random fluctuations in the adult sex ratio (σ_{sex}^2), so that

$$\sigma_d^2 = (1-q)\sigma_{dm}^2 + q\sigma_{df}^2 + \sigma_{sex}^2 \qquad (12.2)$$

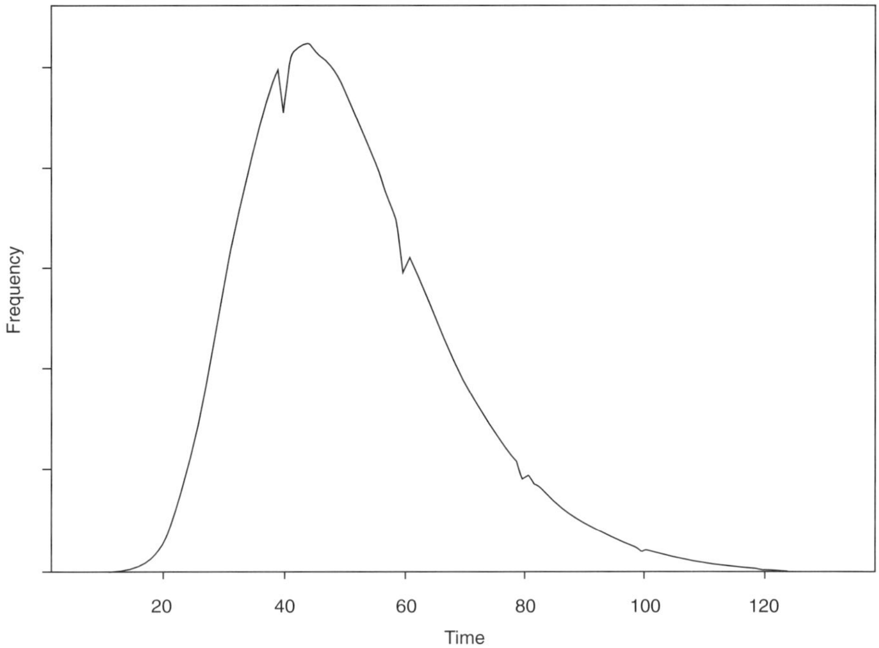

Figure 12.3 The distribution of the time to extinction of a declining population of barn swallow in Jutland, Denmark, from an initial size of 58 pairs recorded in 1999. The parameters were $r = -0.078$, $\sigma_e^2 = 0.024$ and $\sigma_d^2 = 0.18$. For further details, see Engen *et al.* (2001).

where q is the probability of being a female at birth. For small populations close to the extinction barrier

$$\sigma_{\text{sex}}^2 \approx \bar{f}(0)^2 q(1-q)(2T-1) \qquad (12.3)$$

where $\bar{f}(0)$ is the mean fecundity of mated females at very low densities and T is the generation time. This shows a very strong effect of T on σ_{sex}^2 and hence on σ_d^2, owing to temporal autocorrelations in the sex ratio fluctuations. We also see that for $q = 1$ we have a one-sex model where σ_{df}^2 is exactly equal to the demographic variance for a female population. In general, σ_d^2 is, however, larger than σ_{df}^2. Let us then assume a strictly monogamous system where the number of mated females is constrained by number of males or females in the population. This creates, as in the case with isotropic noise (Lande 1998), a stochastic Allee effect, with an unstable equilibrium at population sizes as large as 40–50 individuals (Fig. 12.4). However, demographic stochasticity is smaller in a monogamous than in a polygamous mating system (Fig. 12.5). These results strongly indicate the importance of recent attempts (e.g. Caro 1998; Gosling & Sutherland 2000) to examine behavioural changes that especially occur at small population sizes because they are likely to affect the components of demographic variance and the stochastic growth rate, and then in turn the probabil-

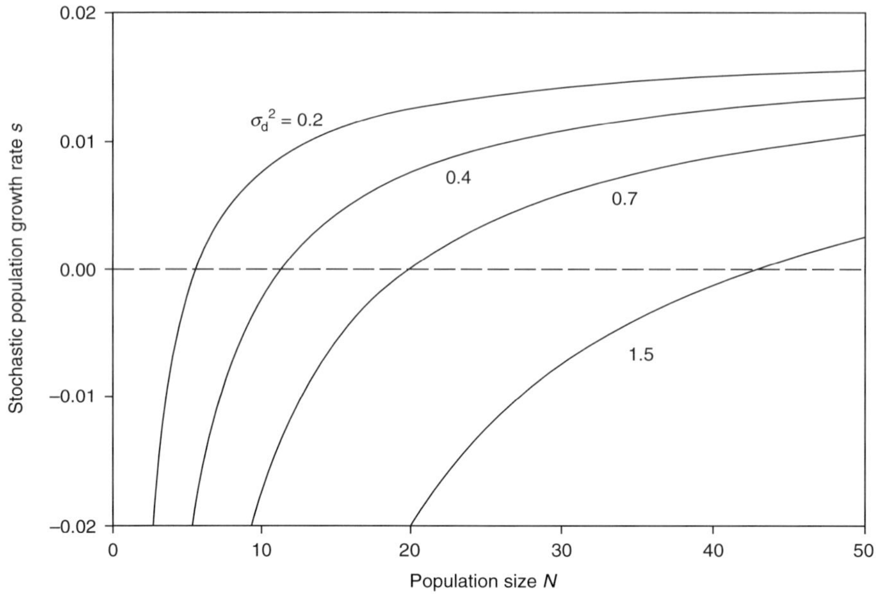

Figure 12.4 The stochastic population growth rate *s* as a function of the population size *N* for different values of the demographic variance σ_d^2. Other parameters were $r = 0.02$ and $\sigma_e^2 = 0.005$.

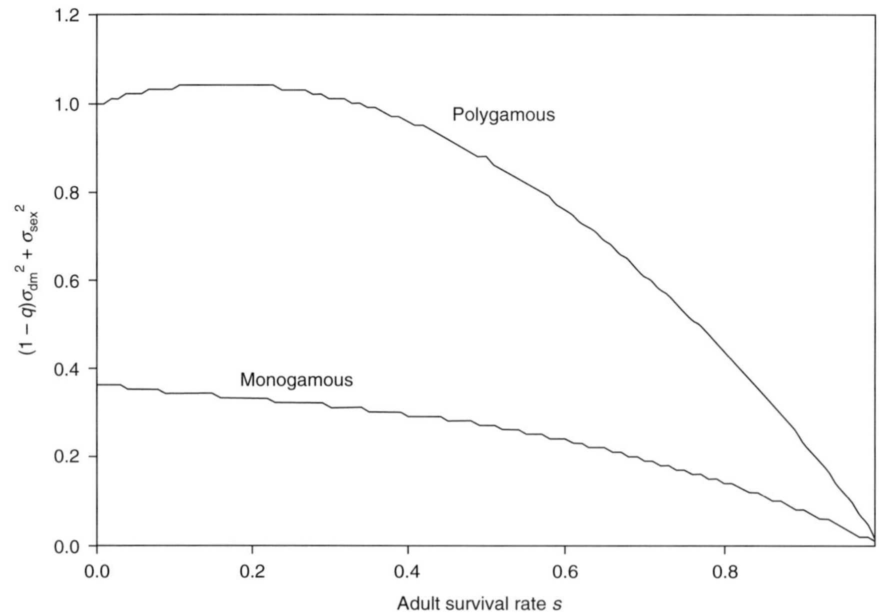

Figure 12.5 The contribution to the demographic variance σ_d^2 from the components of male survival σ_{dm}^2 and fluctuations in the sex ratio σ_{sex}^2, in relation to mating system (from Engen *et al.* in press).

ity of extinction. Accordingly, for a given number of introduced individuals, the extinction probability of introduced monogamous passerine bird species on New Zealand was higher than among polygamous species (Legendre *et al.* 1999).

Density-dependent populations

In density-dependent populations many of the same effects are found on time to extinction as in density-independent populations. This means that both variations in the expected dynamics as well as environmental and demographic stochastic effects will strongly affect the risk of extinction. In addition, we must also consider the effects of variation in the carrying capacity for the persistence of the population. For instance, habitat loss causing a decrease in K represents another type of deterministic decline to extinction. However, we cannot obtain any analytical results for the time to extinction, assuming an annual decrease of $\ln K$, but have to rely on simulations. In the initial phase of the loss of habitat, most sample paths lay above the actual carrying capacity (Fig. 12.6). After a period, the covariation of a reduction in K and the operation of the stochastic effects together bring the sample paths below K. Thus, for small values of K, time to extinction is speeded up compared with the deterministic case. This illustrates that there is a delay in time before we are able to record the effects of habitat loss on the population process and reveal the effects on the time to extinction.

Density-dependent populations with a positive long-run growth rate also can become extinct. Lande (1993) examined how different factors contributed to extinction, using a very simple ceiling-model of density dependence. He found that the time to extinction from the carrying capacity ($T(K)$) increased almost exponentially with K under the influence of demographic stochasticity alone. In contrast, $T(K)$ increased as a power function of K (dependent on the ratio $\bar{r}/\sigma_e^2$) when the population was influenced only by environmental stochasticity. Thus, demographic stochasticity affected the time to extinction only of small populations whereas environmental stochasticity influenced the persistence of even larger populations.

Similar patterns were also present when the joint effects of the variables were considered in more realistic density-dependent models. For instance, when we used the familiar logistic form of density dependence with a stochastic density-independent growth rate (Lande *et al.* 2003), $T(K)$ increased as a power function of K. However, at a given K, increased environmental stochasticity (σ_e^2) shortened $T(K)$ (Fig. 12.7a). Environmental stochasticity can also affect the carrying capacity. We see that under the influence of the variance in the carrying capacity ($\sigma_{\ln K}^2$), $T(K)$ increased as a power function of K (Fig. 12.7b). However, only a large increase in the value of $\sigma_{\ln K}^2$ decreased $T(K)$ significantly at a given K.

Differences in the expected dynamics will also affect the time to extinction. Elsewhere (Sæther *et al.* 2000, 2002) we have used the theta-logistic model (Gilpin & Ayala 1973; Gilpin *et al.* 1976; Diserud & Engen 2000) to quantify the effects of density dependence on the population growth rate, expressed as

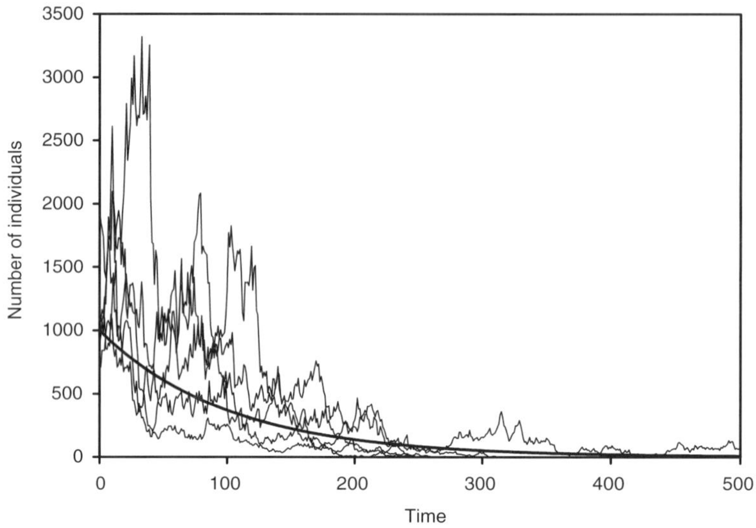

Figure 12.6 Simulated sample paths of a population with logistic density regulation at an initial size of $N = 1000$ (thin lines) in which the logarithm of the carrying capacity decreases annually at a constant rate α (solid line) so that $K(t) = N_e + (K(0) - N_e)e^{-\alpha t}$, where N_e is the extinction barrier and t is the time. The parameters were $K(0) = 1000$, $N_e = 1$, $\alpha = 0.01$, $r = 0.01$, $\sigma_d^2 = 1$ and $\sigma_e^2 = 0.02$.

$$E\left(\frac{\Delta N}{N}\middle| N\right) = r_0\left[1 - \left(\frac{N}{K}\right)^{\theta}\right] \tag{12.4}$$

where r_0 is the mean specific growth rate of the population at $N = 0$ and θ specifies the density regulation. Equation (12.4) also can be formulated as (Sæther *et al.* 2000)

$$E\left(\frac{\Delta N}{N}\middle| N\right) = r_1\left[1 - \frac{N^{\theta} - 1}{K^{\theta} - 1}\right] \tag{12.5}$$

where r_1 is the mean specific growth rate of the population at $N = 1$. We can then describe a wide variety of density regulation functions by varying only one parameter θ (Sæther *et al.* 1996, 2002). For instance, for $\theta = 0$ (taking the limit as θ approaches 0) we obtain the Gompertz form of density regulation, whereas $\theta = 1$ produces the logistic model. When the time to extinction is computed for different values of θ (Fig. 12.7c), we see that populations with larger values of θ become extinct sooner than populations with smaller values of θ.

Further insight into extinction as a process can be obtained from an examination of the distribution of the times to extinction in models with a Gompertz density regulation. After a period with a relatively small probability of extinction, we then have a period with a relatively high rate of extinction before we enter the final long phase where the distribution of time to extinction is asymptotically exponential

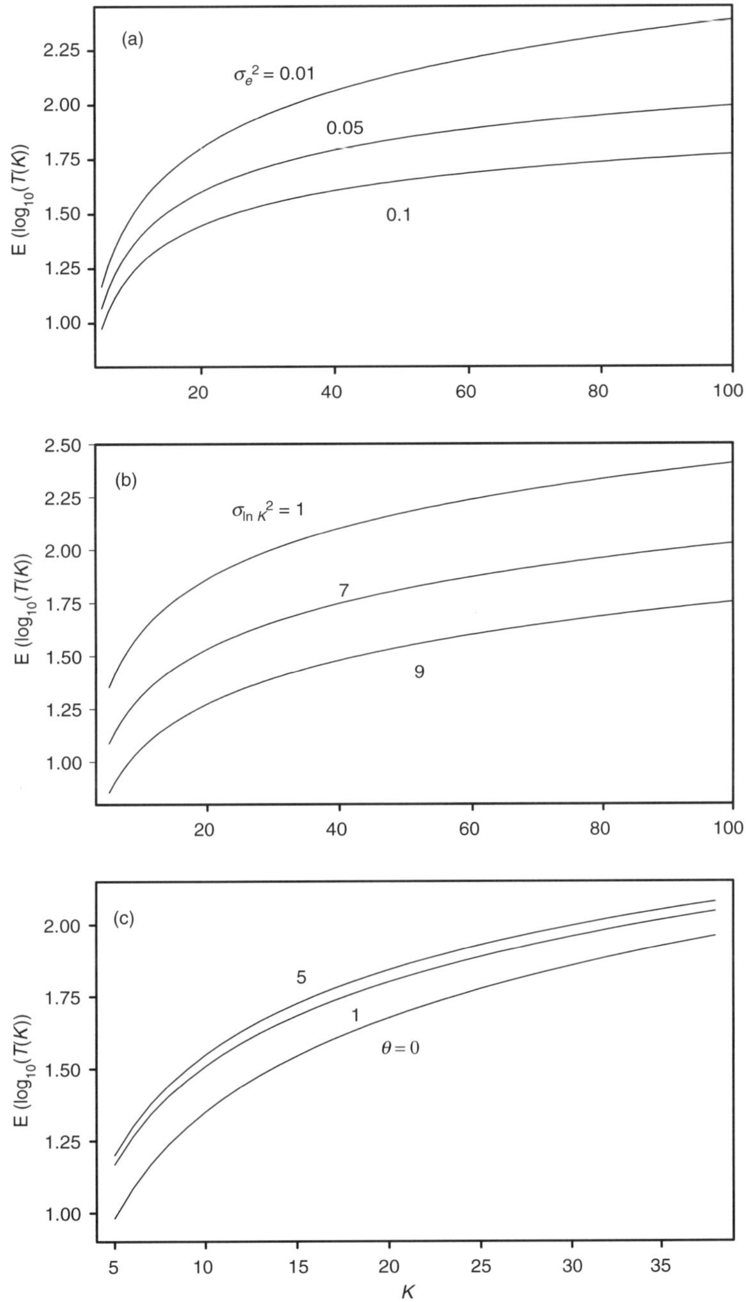

Figure 12.7 The logarithm of the expected time to extinction in relation to the carrying capacity K for different values of environmental stochasticity in $r\,\sigma_e^2$ (a), environmental stochasticity in $\ln K\sigma_{\ln K}^2$ (b), and form of the density regulation θ (c). Parameters were $r_1 = 0.01$, $\sigma_d^2 = 1$, $\sigma_e^2 = 0.01$ in (b) and (c), and $\theta = 1$ in (a) and (b).

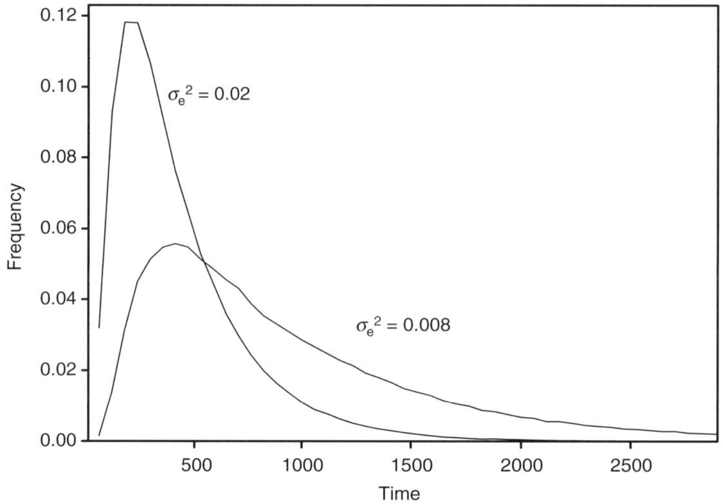

Figure 12.8 The distribution of time to extinction of a population with Gompertz density regulation ($\theta = 0$) initially being at the carrying capacity K for different values of the environmental stochasticity σ_e^2. Other parameters were $r_1 = 0.01$, $K = 1000$ and $\sigma_d^2 = 1$.

(Fig. 12.8). The length of the initial period is approximately on the order of the duration of the final decline from K to extinction (Lande *et al.* 2003). It increases with the initial population size (Lande *et al.* 2003) and decreasing environmental stochasticity (Fig. 12.8).

Thus, these analyses show that variation in most parameters in the theta-logistic model of density regulation (eqns 12.4 and 12.5) strongly affects the time to extinction. These effects will be further influenced by covariation between the different parameters. We examined the presence of such patterns by performing a comparative analysis of avian population fluctuations by using long-term time series of fluctuations in the abundance of populations where the population counts were assumed to be quite accurate. Two sets of data were analysed: one with approximate stationary fluctuations around K with r_1 estimated from demographic information (Sæther *et al.* 2002), and another data set where all parameters could be estimated by maximum likelihood methods and $\theta > 0$ (Sæther & Engen 2002b). Even though we find large uncertainties in the estimates, a pattern of covariation appeared among several parameters of the theta-logistic model. In both data sets, the environmental variance increased, and the logarithm of θ decreased, with r_1 (Fig. 12.9). This shows maximum density regulation to occur at lower relative (to K) population sizes in species with higher population growth rates. Furthermore, the interspecific variation in r_1 was also correlated to life-history differences. Lower population growth rates at small densities were found in species with higher survival rates (Sæther *et al.* 2002; Sæther & Engen 2002b) and smaller clutch sizes (Sæther & Engen 2002b). Thus, differences in demography (Sæther *et al.* 2002) and population-dynamics characteris-

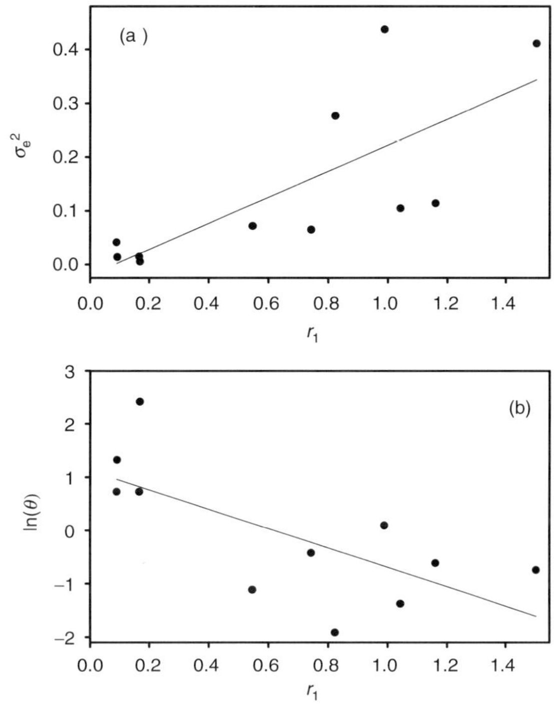

Figure 12.9 Interspecific variation among 11 bird species (for sources see Sæther & Engen 2002b) in the environmental variance σ_e^2 (a) and the logarithm of the form of the density regulation θ (b) in relation to the specific population growth rate r_1.

tics (Sæther *et al.* 2002; Sæther & Engen 2002b) may be related to the position of a species along a 'slow–fast continuum' of life-history adaptations (Sæther & Bakke 2000; Bennett & Owens 2002).

As an example of the impact of population-dynamics characteristics on the risk of extinction, we compare the time to extinction of the sparrowhawk *Accipiter nisus*, with a large θ and small values of r_1 and σ_e^2, with the time to extinction of the blue tit *Parus caeruleus*, with large values of r_1 and σ_e^2, but in which density regulation acts more strongly at lower population sizes (small θ). The time to extinction from a carrying capacity of 50 was much longer in the blue tit than in the sparrowhawk (Table 12.1), indicating that time to extinction of both species was strongly influenced by the demographic variance σ_d^2 (as expected: Figs 12.1 and 12.2).

Spatially structured populations

Most natural populations are interconnected with other populations through migration. First consider such a metapopulation where the local populations are found in discrete habitat patches. Changes in the habitat occupancy p will affect the local as well as metapopulation dynamics because an increase in p will increase local population size and thereby reduce the risk of extinction, the so-called 'rescue effect' (see discussion in Hanski 1999). Furthermore, the probability of successful establishment per migrant will also increase with p through increasing the number of immigrants in the first crucial stage in the establishment of the new population. Lande *et al.* (1998) called this the 'establishment effect'.

Including stochastic local dynamics effects in a Levins-type metapopulation (Levins 1969, 1970) strongly affected the mean time to extinction (Lande *et al.* 1998). In addition to a stable equilibrium of habitat occupancy, there also may be an unstable equilibrium at small habitat occupancies, below which the metapopulation is expected to decline to extinction (see also Hanski & Gyllenberg 1993). Positive density dependence in emigration and negative density dependence in immigration increase this stable equilibrium occupancy of habitat in a metapopulation (Sæther *et al.* 1999). Furthermore, the mean time to extinction of a metapopulation with stochastic local dynamics is shorter than in a classic metapopulation, assuming a stable equilibrium occupancy of the expected dynamics (Lande *et al.* 2003). This occurs because the increase in local extinction and decrease in colonization as habitat occupancy decreases are not accounted for in the classic model.

Comparing population fluctuations over larger regions we often find correlations among locations in the temporal variations in population size (Hanski & Woiwood 1993; Sutcliffe *et al.* 1996; Ranta *et al.* 1998; Stenseth *et al.* 1999; Paradis *et al.* 2000). Such synchronized population fluctuations can arise from the operation of a common environmental factor such as weather (Moran 1953; Grenfell *et al.* 1998) or predation (Ims & Andreassen 2000), or from dispersal of individuals (Lande *et al.* 1999). Differences in population synchrony strongly affect the risk of local or regional extinction (Allen *et al.* 1993; Bolker & Grenfell 1996; Heino *et al.* 1997; Ranta *et al.* 1997a; Palmqvist & Lundberg 1998). A quantitative evaluation of the relative contribution of different factors to the risk of extinction showed that with large, nonlinear population fluctuations with a log-normal stationary distribution (Engen 2001), dispersal increases mean local population size, but also increases the spatial scale of population synchrony (Engen *et al.* 2002a, b). Thus, increasing the population size will increase mean time to local as well as regional extinction (Engen *et al.* 2002b), whereas the increased synchrony will decrease it because local populations will fluctuate and become extinct together. However, the overall effect of dispersal is to decrease the risk of local extinction. These analyses (Engen *et al.* 2002b) also show that the spatial scale of regional extinction will be similar to the scale of population synchrony. This suggests that the size of protected areas ideally should be larger than the scale of population synchrony.

Discussion

This review of the routes to extinction shows that when the deterministic growth rate $r < 0$ extinction certainly will occur. The duration of the final decline from K to

Table 12.1 The logarithm of the expected time to extinction ($\log_{10}(\mathrm{E}\,T(K))$) from the carrying capacity $K = 50$ and the mean ($\overline{N}$) and standard deviation (SD (N)) of the quasi-stationary distribution for a 'survival-restricted species' the sparrowhawk and a 'recruitment-driven species' the blue tit (see Sæther *et al.* 2002) for different values of demographic variance σ_d^2.

	σ_d^2	$\log_{10}(\mathrm{E}\,T(K))$	$\overline{N}$	SD ($\log_{10}(T(K))$
Blue tit*	0.1	17.64	46.91	13.74
	0.5	10.56	46.66	14.22
Sparrowhawk†	0.1	9.57	45.67	10.77
	0.5	6.16	45.18	11.37

* $r_1 = 1.16$, $\theta = 0.54$, $\sigma_e^2 = 0.1137$ (Sæther & Engen 2002b).
† $r_1 = 0.39$, $\theta = 2.57$, $\sigma_e^2 = 0.090$ (Sæther *et al.* 2002).

the extinction barrier will be of the same order of magnitude as the deterministic exponential decline, nearly independent of the level of environmental stochasticity. This strongly emphasizes the importance of the focus Caughley (1994) placed on identifying the factors affecting long-term declines for the conservation of many endangered or threatened species. Similarly, a deterministic decline in the carrying capacity will, after a period, strongly reduce the time to extinction (Fig. 12.6). However, populations with a positive *r* also may become extinct under the influence of stochastic factors (Fig. 12.2). Especially, the demographic stochasticity strongly affects the risk of extinction of small populations (Fig. 12.1 and Table 12.1).

Several studies have recently tried to compare the current extinction rates in well-studied taxa such as birds and mammals against the background loss of species from the fossil record (e.g. May *et al.* 1995; Pimm *et al.* 1995; Pimm 1998). However, such comparisons may be difficult because the results presented here indicate that an important characteristic of extinction as a process is the presence of a time delay before extinction starts to occur as a consequence of a change in the population dynamics (Figs 12.1, 12.3, 12.6 and 12.8). Such effects are even present in deterministically declining populations (Figs 12.1 and 12.6). An implication of this is that a long time will pass before the actual impact of current human activities on the rate of species loss is evident (Rosenzweig 1995).

The theoretical analyses suggest that the risk of extinction should increase with lowered carrying capacity *K* (Fig. 12.7). These relationships suggest that the risk of extinction will be closely related to the size of areas with suitable habitats and thereby increase with the rate of habitat loss. Indeed, studies of several taxa have shown that the extinction rate does increase with decreasing area (Newmark 1987, 1996; Pimm & Askins 1995; Brashares *et al.* 2001). These relationships differ among taxa. For instance, the area-specific extinction rate of carnivores in Ghanian natural reserves is far larger for carnivorous mammals than for herbivores (Brashares *et al.* 2001). Similarly, the reduction in extinction risk with population size seems to be

Table 12.2 A summary of studies of factors affecting the risk of population extinction through time.

Taxa	Location	Extinction measures	Significant correlates	Source
Interspecific variation within communities				
Plants	Krakatau	Species loss	Correlated to dispersal mode, decreases with abundance	Whittaker *et al.* 2000
Plants	Auckland, New Zealand	Species loss	Decreases with abundance, correlated to growth form, taxonomic differences, habitat affinity	Duncan & Young 2000
Birds	British islands	Log (T)	Increases with mean number of nesting pairs and body size, correlated to migratory status	Tracy & George 1992
Orb spiders	Bahamas	$1/\bar{T}$	Increases with population variability, and decreases with N	Schoener & Spiller 1992
Experimental studies *Drosophila hydei*		T	Decreases with K and σ_e^2	Phillipi *et al.* 1987
Brine shrimp		T	Increases with K and initial N. Decreases with variability in N, but independent of variability in K	Belovsky *et al.* 1999

one of the few general patterns emerging from studies of variation in extinction risk within communities as well as from experimental studies (Table 12.2).

Time to extinction was strongly influenced by the demographic stochasticity (Figs 12.1 and 12.2, Table 12.1). The demographic variance σ_d^2 may be density dependent (Sæther *et al.* 1998), suggesting that behavioural changes affecting individual variation in reproductive success and survival at low densities may influence time to extinction, and even give rise to stochastic Allee effects (Lande 1998; Engen *et al.*, in press). This emphasizes an important role for behavioural ecology in conservation biology (see Caro 1998). An important task for behavioural studies will be to disentangle the mechanisms that affect the rate of interchange of individuals among subpopulations. This strongly influences fluctuations of populations locally as well as regionally (Ranta *et al.* 1997b; Lande *et al.* 1999; Engen 2001; Engen *et al.* 2002a). For instance, the pattern of density dependence in dispersal rate strongly affects the time to extinction of metapopulations (Sæther *et al.* 1999).

We have identified several factors that strongly affect time to extinction. Estimating many of those parameters will require precise long-term time series and data on individual fitness variation (see Lande *et al.* 2002; Sæther & Engen 2002a). Such data will be available for only a very limited set of species. Successful predictions of future changes in species composition will depend on relating variation in those parameters to more easily collected variables, such as life-history data or ecological characteristics (e.g. Jennings *et al.* 1999; Owens & Bennett 2000; Purvis *et al.* 2000; Sæther *et al.* 2002).

Acknowledgements

This study was financed by grants from the European Commission (project METABIRD) and the Research Council of Norway.

References

Allee, W.C., Emerson, A.E., Park, O., Park, T. & Schmidt, K.P. (1949) *Principles of Animal Ecology.* W.B. Saunders, Philadelphia.

Allen, J.C., Schaffer, W.M. & Rosto, D. (1993) Chaos reduces species extinction by amplifying local population noise. *Nature* **364**, 229–232.

Belovsky, G.E., Mellison, C., Larson, C. & Van Zandt, P.A. (1999) Experimental studies of extinction dynamics. *Science* **286**, 1175–1177.

Bennett, P.M. & Owens, I.P.F. (2002) *Evolutionary Ecology of Birds.* Oxford University Press, Oxford.

Bolker, B.M. & Grenfell, B.T. (1996) Impact of vaccination on the spatial correlation and the persistence of measles dynamics. *Proceedings of the National Academy of Sciences, USA* **93**, 12648–12653.

Brashares, J.S., Arcese, P. & Sam, M.K. (2001) Human demography and reserve size predict wildlife extinction in West Africa. *Proceedings of the Royal Society, London, Series B* **268**, 2473–2478.

Caro, T.M. (1998) *Behavioral Ecology* and *Conservation Biology.* Oxford University Press, New York.

Caughley, G. (1994) Directions in conservation biology. *Journal of Animal Ecology* **63**, 215–244.

Cox, D.R. & Miller, H.D. (1970) *The Theory of Stochastic Processes.* Chapman & Hall, London.

Dennis, B., Munholland, P. & Scott, J.M. (1991) Estimation of growth and extinction parameters for endangered species. *Ecological Monographs* **61**, 115–143.

Diserud, O. & Engen, S. (2000) A general and dynamic species abundance model, embracing the lognormal and the gamma models. *American Naturalist* **155**, 497–511.

Duncan, R.P. & Young, J.R. (2000) Determinants of plant extinction and rarity 145 years after European settlement of Auckland, New Zealand. *Ecology* **81**, 3048–3061.

Eldredge, N. (1999) Cretaceous meteor showers, the human ecological 'niche', and the sixth extinction. In: *Extinctions in Near Time* (ed. R.D.E. MacPhee), pp. 1–15. Kluwer Academic, New York.

Engen, S. (2001) A dynamic and spatial model with migration generating the log-Gaussian field of population densities. *Mathematical Biosciences* **173**, 85–102.

Engen, S. & Sæther, B.-E. (2000) Predicting the time to quasi-extinction for populations far below their carrying capacity. *Journal of theoretical Biology* **205**, 649–658.

Engen, S., Bakke, Ø. & Islam, A. (1998) Demographic and environmental stochasticity—concepts and definitions. *Biometrics* **54**, 840–846.

Engen, S., Sæther, B.-E. & Møller, A.P. (2001) Stochastic population dynamics and time to extinction of a declining population of barn swallows. *Journal of Animal Ecology* **70**, 789–797.

Engen, S., Lande, R. & Sæther, B.-E. (2002a) Migration and spatiotemporal variation in population dynamics in a hetereogeneous environment. *Ecology* **83**, 570–579.

Engen, S., Lande, R. & Sæther, B.-E. (2002b) The spatial scale of population fluctuations and quasi-extinction risk. *American Naturalist* **160**, 439–451.

Engen, S., Lande, R. & Sæther, B.-E. (in press) Demographic stochasticity and Allee effects in populations with two sexes. *Ecology.*

Gaston, K.J. & Nicholls, A.O. (1995) Probable times to extinction of some rare breeding bird species in the United Kingdom. *Proceedings of the Royal Society, London, Series B* **259**, 119–123.

Gilpin, M.E. & Ayala, F.J. (1973) Global models of growth and competition. *Proceedings of the National Academy of Sciences, USA* **70**, 3590–3593.

Gilpin, M.E., Case, T.J. & Ayala, F.J. (1976) *θ*-selection. *Mathematical Biosciences* **32**, 131–139.

Ginzburg, L.R., Slobodkin, L.B., Johnson, K. & Bindman, A.G. (1982) Quasiextinction probabilities as a measure of impact on population growth. *Risk Analysis* **21**, 171–181.

Gosling, M. & Sutherland, W.J. (2000) *Behaviour and Conservation.* Cambridge University Press, Cambridge.

Grenfell, B.T., Wilson, K., Finkenstadt, B.F., *et al.* (1998) Noise and determinism in synchronized sheep dynamics. *Nature* **394**, 674–677.

Groombridge, B. (1992) *Global Diversity. Status of the Earth's Living Resources.* Chapman & Hall, New York.

Hanski, I. (1999) *Metapopulation Ecology.* Oxford University Press, Oxford.

Hanski, I. & Gyllenberg, M. (1993) Two general metapopulation models and the core–satellite species hypothesis. *American Naturalist* **142**, 17–41.

Hanski, I. & Woiwood, I.P. (1993) Spatial synchrony in the dynamics of moth and aphid populations. *Journal of Animal Ecology* **62**, 656–668.

Heino, M., Kaitala, V., Ranta, E. & Lindström, J. (1997) Synchronous dynamics and rates of extinction in spatially structured populations. *Proceedings of the Royal Society, London, Series B* **259**, 119–123.

Hilton-Taylor, C. (2000). *2000 IUCN Red List of Threatened Species.* IUCN, Gland, Switzerland.

Ims, R.A. & Andreassen, H.P. (2000) Spatial synchronization of vole population dynamics by predatory birds. *Nature* **408**, 194–196.

Jablonski, D. (1994) Extinctions in the fossil record. *Philosophical Transactions of the Royal Society, London, Series B* **344**, 11–17.

Jennings, S., Reynolds, J.D. & Mills, S.C. (1999) Life history correlates of responses to fisheries exploitation. *Proceedings of the Royal Society, London, Series B* **265**, 333-339.

Karlin, S. & Taylor, H.M. (1981) *A Second Course in Stochastic Processes.* Academic Press, New York.

Lande, R. (1993) Risks of population extinction from demographic and environmental stochasticity and random catastrophes. *American Naturalist* **142**, 911-927.

Lande, R. (1998) Demographic stochasticity and Allee effect on a scale with isotropic noise. *Oikos* **83**, 353–358.

Lande, R. (2002) Incorporating stochasticity in population viability analysis. In: *Population Viability Analysis* (eds S.R. Beissinger & D.R. McCullough), pp. 18–40. University of Chicago Press, Chicago.

Lande, R., Engen, S. & Sæther, B.-E. (1998) Extinction times in finite metapopulation models with stochastic local dynamics. *Oikos* **83**, 383–389.

Lande, R., Engen, S. & Sæther, B.-E. (1999) Spatial scale of synchrony: environmental correlation versus dispersal and density regulation. *American Naturalist* **154**, 271–281.

Lande, R., Engen, S., Sæther, B.-E., *et al.* (2002) Estimating density dependence from population time series using demographic theory and life-history data. *American Naturalist* **159**, 321–337.

Lande, R., Engen, S. & Sæther, B.-E. (2003) *Stochastic Population Models in Ecology and Conservation.* Oxford University Press, Oxford.

Legendre, S., Clobert, J., Møller, A.P. & Sorci, G. (1999) Demographic stochasticity and social mating system in the process of extinction of small populations: the case of passerines introduced to New Zealand. *American Naturalist* **153**, 449–463.

Levins, R. (1969) Some demographic and genetic consequences of environmental heterogeneity for biological control. *Bulletin of the Entomological Society of America* **15**, 237–240.

Levins, R. (1970) Extinction. In: *Some Mathematical Problems in Biology* (ed. M. Gerstenhaber), pp. 75–107. American Mathematical Society, Providence, RI.

Lewontin, R.C. & Cohen, D. (1969) On population growth in a randomly fluctuating environment. *Proceedings of the National Academy of Sciences, USA* **62**, 1056–1060.

May, R.M., Lawton, J.H. & Stork, N.E. (1995) Assessing extinction rates. In: *Extinction Rates* (eds J.H. Lawton & R.M. May), pp. 1–24. Oxford University Press, Oxford.

Moran, P.A.P. (1953) The statistical analysis of the Canadian lynx cycle. II. Synchronization and meteorology. *Australian Journal of Zoology* **1**, 281–298.

Newmark, W.D. (1987) A land-bridge island perspective on mammalian extinctions in western North American parks. *Nature* **325**, 430–432.

Newmark, W.D. (1996) Insularization of Tanzanian parks and the local extinction of large mammals. *Conservation Biology* **10**, 1549–1556.

Owens, I.P.F. & Bennett, P.M. (2000) Ecological basis of extinction risk in birds: Habitat loss versus human persecution and introduced predators. *Proceedings of the National Academy of Sciences, USA* **97**, 12144–12148.

Palmqvist, E. & Lundberg, P. (1998) Population extinctions in correlated environments. *Oikos* **83**, 359–367.

Paradis, E., Baillie, S.R., Sutherland, W.J. & Gregory, R.D. (2000) Spatial synchrony in populations of birds: effects of habitat, population trend, and spatial scale. *Ecology* **81**, 2212–2125.

Phillipi, T.E., Carpenter, M.P., Case, T.J. & Gilpin, M.E. (1987) *Drosophila* population dynamics: chaos and extinction. *Ecology* **68**, 154–159.

Pimm, S.L. (1998) Extinction. In: *Conservation Science and Action* (ed. W.J. Sutherland), pp. 20–38. Blackwell Science, Oxford.

Pimm, S.L. & Askins, R.A. (1995) Forest losses predict bird extinctions in eastern North America. *Proceedings of the National Academy of Sciences, USA* **92**, 1056–1060.

Pimm, S.L., Russell, G.J., Gittleman, J.L. & Brooks, T.M. (1995) The future of biodiversity. *Science* **269**, 347–350.

Purvis, A.P., Gittleman, J.L., Cowlishaw, G. & Mace, G.M. (2000) Predicting extinction risk in declining species. *Proceedings of the Royal Society, London, Series B* **267**, 1947–1952.

Ranta, E., Kaitala, V. & Lundberg, P. (1997a) The spatial dimension in population fluctuations. *Science* **278**, 1621–1623.

Ranta, E., Lindström, J. & Helle, E. (1997b) The Moran effect and synchrony in population dynamics. *Oikos* **78**, 136–142.

Ranta, E., Kaitala, V. & Lindström, J. (1998) Spatial dynamics of populations. In: *Modeling Spatiotemporal Dynamics in Ecology* (eds J. Bascompte & R.V. Solé), pp. 47–62. Springer-Verlag, Berlin.

Rosenzweig, M.L. (1995) *Species Diversity in Space and Time.* Cambridge University Press, Cambridge.

Sæther, B.-E. & Bakke, Ø. (2000) Avian life history variation and contribution of demographic traits to the population growth rate. *Ecology* **81**, 642–653.

Sæther, B.-E. & Engen, S. (2002a) Including uncertainties in population viability analysis. In: *Population Viability Analysis* (eds S.R. Beissinger & D.R. McCullough), pp. 191–212. University of Chicago Press, Chicago.

Sæther, B.-E. & Engen, S. (2002b). Pattern of variation in avian population growth rates. *Philosophical Transactions of the Royal Society, London, Series B* **357**, 1185–1195.

Sæther, B.-E., Engen, S. & Lande, R. (1996) Density-dependence and optimal harvesting of fluctuating populations. *Oikos* **76**, 40–46.

Sæther, B.-E., Engen, S., Islam, A., McCleery, R.H. & Perrins, C. (1998) Environmental stochasticity and extinction risk in a population of a small songbird, the great tit. *American Naturalist* **151**, 441–450.

Sæther, B.-E., Engen, S. & Lande, R. (1999) Finite metapopulation models with density-dependent migration and stochastic local dynamics. *Proceedings of the Royal Society, London, Series B* **266**, 113–118.

Sæther, B.-E., Engen, S., Lande, R., Arcese, P. & Smith, J.N.M. (2000) Estimating time to extinction in an island population of song sparrows. *Proceedings of the Royal Society, London, Series B* **267**, 621–626.

Sæther, B.-E., Engen, S. & Matthysen, E. (2002) Demographic characteristics and population dynamical patterns of solitary birds. *Science* **295**, 2070–2073.

Schoener, T.W. & Spiller, D.A. (1992) Is extinction rate related to temporal variability in population size? An empirical answer for orb spiders. *American Naturalist* **139**, 1176–1207.

Sepkoski, J.J. Jr. (1993) Ten years in the library: new data confirms paleontological patterns. *Paleobiology* **19**, 43–51.

Seshadri, V. (1999) *The Inverse Gaussian Distribution. Statistical Theory and Applications.* Springer, New York.

Shapton, V.L., Burke, K., Camargozanoguera, A., *et al.* (1993) Chicxulub multiring impact basin-size and other characteristics derived from gravity analysis. *Science* **261**, 1564–1567.

Solow, A.R. (1998) On fitting a population model in the presence of observation error. *Ecology* **79**, 1463–1466.

Stenseth, N.C., Chan, K.S., Tong, H., *et al.* (1999) Common dynamic structure of Canada lynx populations with three climatic regions. *Science* **285**, 1071–1073.

Sutcliffe, O.L., Thomas, C.D. & Moss, D. (1996) Spatial synchrony and asynchrony in butterfly population dynamics. *Journal of Animal Ecology* **65**, 85–95.

Tracy, C.R. & George, T.L. (1992) On the determinants of extinction. *American Naturalist* **139**, 102–122.

Tufto, J., Sæther, B.-E., Engen, S., *et al.* (2000) Bayesian meta-analysis of demographic parameters in three small, temperate passerines. *Oikos* **88**, 273–281.

Van Valen, L. (1973) A new evolutionary law. *Evolutionary Theory* **1**, 1–30.

Whittaker, R. J., Field, R. & Partomihardjo, T. (2000) How to go extinct: lessons from the lost plants of Krakatau. *Journal of Biogeography* **27**, 1049–1064.

Part 6
Why are species not more widely distributed?

Chapter 13
Why are species not more widely distributed? Physiological and environmental limits

F. I. Woodward and C. K. Kelly*

Introduction

The notion that species are narrowly distributed can be readily concluded from maps of species distributions and diversity at the global scale (Heywood 1978; Williams *et al.* 1994). Willis (1922) was one of the first ecologists to quantify the effect for vascular plants and demonstrated that more species occupied narrow than wide geographical ranges. Ideally, the relationship between physiological and distributional limits would be ascertained by determining experimentally the physiological limits of a species, and then comparing that fundamental capacity with environmental correlates of its realized geographical distribution. Unfortunately, these sorts of data are few and far between and the potential confounding effects of phylogeny make those that do exist of little use in inferring general patterns. We have therefore used a sequence of 'model systems' in order to develop and apply hypotheses regarding interactions between physiological and distributional limits at the level of species.

Because our areas of expertise are primarily to do with vascular plants, we will deal largely, although not exclusively, with that group of organisms. We will first look at a group of more or less related species, some of which are extremely widespread, some of which are not, in order to determine if there are any obvious differences between those species that appear to have few or no limits on distribution and those that have such limits. We will discuss the possible implications of these differences, and investigate potentially confounding factors through comparison with another group of organisms with similar constraints. Some of these ideas are then applied to a single, particularly well-documented species, *Arabidopsis thaliana*. Finally, we investigate controls on species ranges within genera (*Pinus* and *Jatropha*). Here the aim is to avoid the possible confounding effects of phylogeny when comparing different species from different parts of the tree of life, and we have therefore used in-depth paired comparisons (*Pinus sylvestris* versus *P. radiata* and *Jatropha chamelensis*

* *Correspondence address: f.i.woodward@sheffield.ac.uk*

versus *J. standleyi*) and appropriate comparative methods using a fully resolved phylogeny of species in the genus *Pinus* in order to tease out determinants of range size.

Microorganisms

Free-living microbial organisms appear to provide clear examples of species that are globally very widespread. Wilkinson (2001) has collected data showing that many terrestrial protozoa are globally widespread, from pole to pole. In a compilation of recorded data on the sizes of different species of testate amoebae from the Arctic and the Antarctic (Fig. 13.1)—close to the absolute maximum possible geographical range for any species—Wilkinson found that 23% of the 127 species were classified as cosmopolitan, occurring at both poles. Contrary to patterns found in higher plants, where body size does not predict range size (Kelly 1996; Kelly & Woodward 1996), the more widespread species were smaller than species restricted to either the Arctic or Antarctic (Fig. 13.2). Wilkinson accounts for the global distribution of the smaller species by two processes. The first occurs through a sampling bias of the generally larger population sizes of the smaller species. The second occurs because the smaller species have smaller airborne terminal velocities and so are likely to be distributed much further by the wind. So differences in distributions of protozoans are a mobility/dispersibility argument, but the inferred effect of body size is reversed relative to mammals, for example, because differing scale brings in differing physical processes.

However, care must be taken in concluding that size confers a simple physical effect in protozoans in terms of population numbers and dispersibility. Size may also limit the number of morphological characters that can be used for distinguishing one species from another. If it were larger species that were more widespread, then an effect of size on identification would be less of an issue as one could assume that if small species could be readily distinguished then morphological characters are sufficient for differentiating between larger species. As it is, it may be that smaller species are not actually more widespread, merely less easily identified; what is seen as one widespread species may actually be several species each of more restricted distributions. This concern is exacerbated when it is noted that molecular fingerprinting of bacterial species shows greater species diversity than can be detected by morphological methods (Pace 1997). So for protozoans, it is probably appropriate to hold off on general theories of species distributions based on size until molecular phylogenies are more readily available.

Flowering plants—*Arabidopsis thaliana*

The investigations on protozoans outlined above raise the suggestion that dispersal may be a major limitation to the large-scale geographical spread of species. Fortunately, for *Arabidopsis thaliana* there is considerable genetic information about populations and ecotypes and the potentially confounding effects of morphological similarity noted above are virtually eliminated. Hoffmann (2002) found climate to

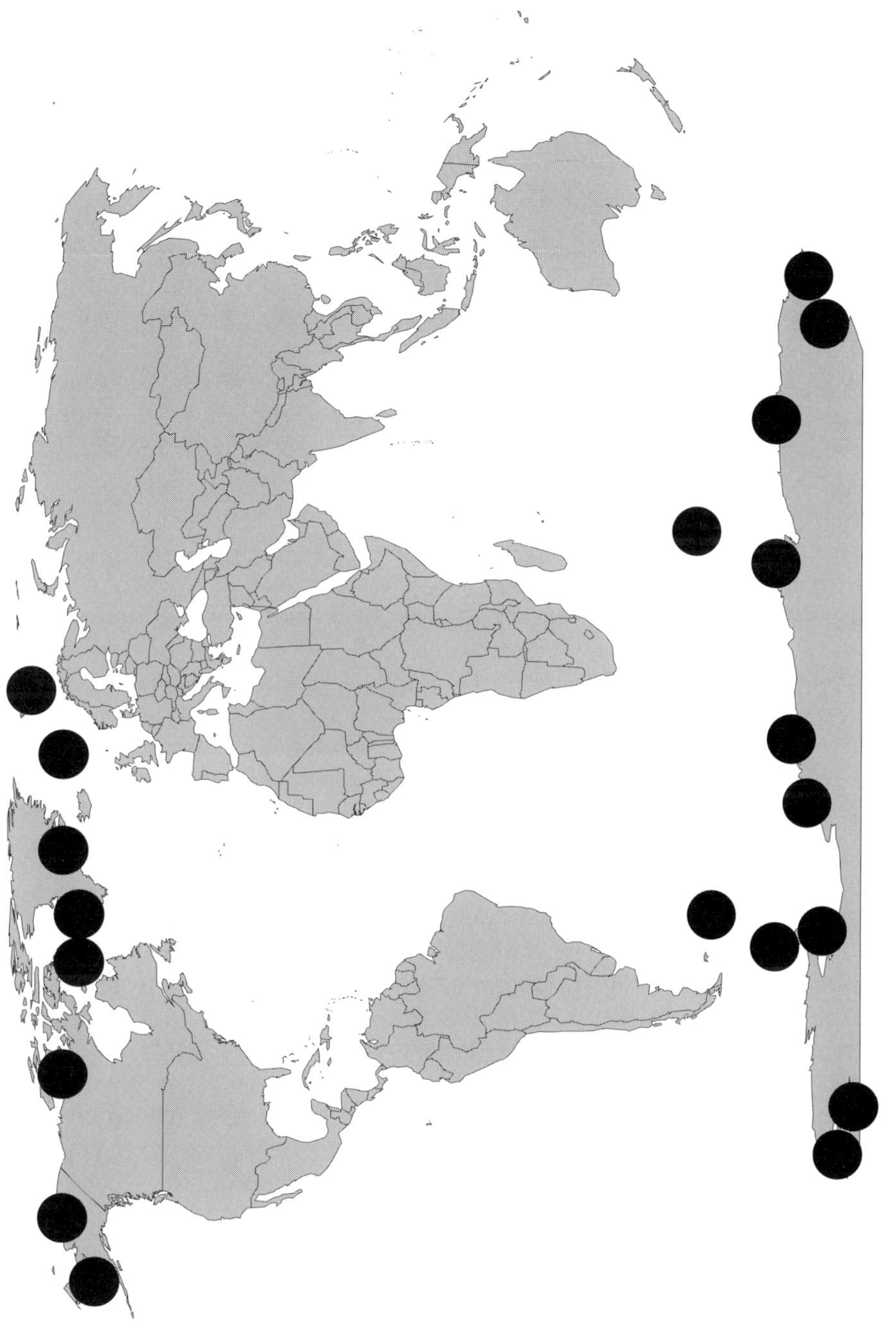

Figure 13.1 Arctic and Antarctic sample sites (circles) of testate amoebae. (Adapted from Wilkinson 2001.)

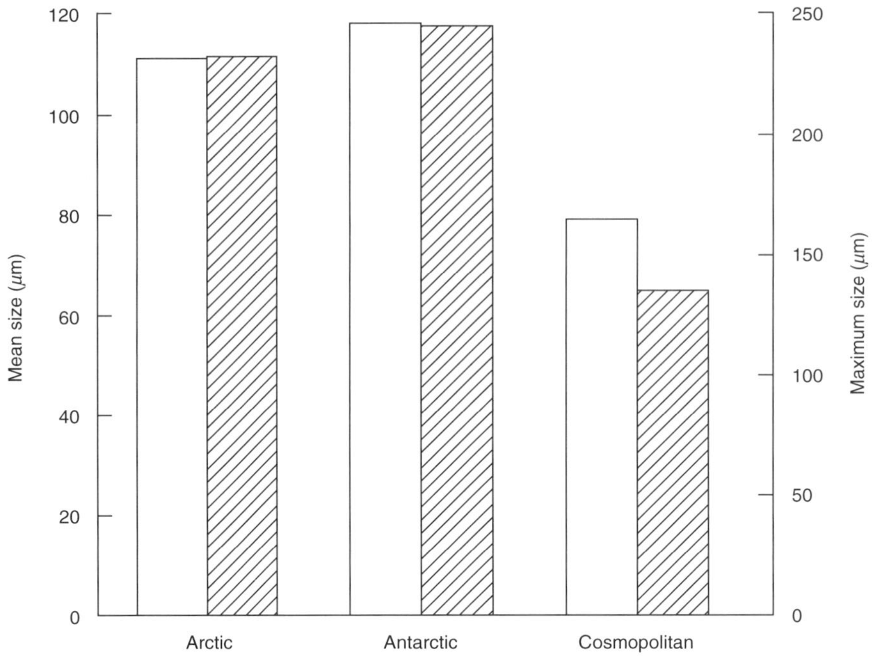

Figure 13.2 Mean size (open bar) and maximum size (hatched bar) of testate amoebae from the Arctic, Antarctic and with a cosmopolitan distribution. (Adapted from Wilkinson 2001.)

be the major correlate explaining the current distributional spread of *A. thaliana*, with three major temperature limits. Winter minima should be greater than –20°C, spring minima should exceed 0°C and summer temperatures should not exceed 25°C. These temperature limits clearly define the global extent of the species and exclude the species from the hot tropical and cold boreal climates. Low annual precipitation appears also to keep *A. thaliana* out of the dry but otherwise suitable temperatures of central Asia (Hoffmann 2002).

There is some independent quantitative evidence that broadly supports Hoffmann's determination of the climatic limits of the global-scale distribution of *A. thaliana*. Laboratory observations indicate freezing tolerance, in cold hardened plants, to an absolute minimum within the range of –8 to –10°C (Llorente *et al.* 2000; Tamminen *et al.* 2001). Pickett *et al.* (1996) demonstrated normal growth of *A. thaliana* at a temperature of 29°C, although a temperature of 21 to 22°C is generally considered optimal for the growth of all ecotypes of *A. thaliana*. At temperatures above 29°C plants develop with increasing quantities of heat shock proteins (Xiong *et al.* 1999), indicating negative impacts of high temperatures on growth and survival. These data have also demonstrated rather small differences between ecotypes in their temperature responses, a feature that might be expected from the genetic mixing of the species since the last glaciation.

Even so, the contemporary distribution of *A. thaliana* (Fig. 13.3) indicates that natural dispersal has also contributed to *A. thaliana* distribution. It is known that the native distribution of *A. thaliana* extends from Europe to central Asia (Hoffmann 2002). Yet when human actions disperse the seed, the species has survived and spread quite well in the previously unoccupied Southern Hemisphere as well as its native Northern Hemisphere. A parallel, notable feature is that the North American distribution has a small range of ecotypes with a relatively small geographical distribution overall, less than would be expected from European distributions. Thus it may be that the geographical range could be significantly extended if new ecotypes arrive. However, it is also possible that range expansion of these recent arrivals is still incomplete. The latter explanation appears more likely given, as noted above, that most ecotypes appear to grow optimally under very similar conditions (Hoffmann 2002).

Dispersal can also affect genetic structure, and the interpretation of historical processes. Genetic analyses of *A. thaliana* ecotypes (Fig. 13.4) indicate the highest diversity in central Europe, although diversity in central Asia has probably not been fully sampled. Analyses of amplified fragment length polymorphisms in 142 accessions of *A. thaliana* (Sharbel *et al.* 2000) indicated increasing genetic isolation with geographical distance of accessions from Asia and southern Europe. These data suggest that the post-glacial expansion of *A. thaliana* in central and northern Europe occurred both from Asia and the Iberian peninsula. As a consequence, the area of central Europe that has been most recently colonized since the last glaciation is the most diverse genetically (Fig. 13.4), representing the combined diversity from the two refugia. The observed low variability in environmental tolerances between different accessions, or ecotypes, may be a further consequence of this genetic combination.

Flowering plants—comparative analyses

The inferences of causation for any trait, physical, physiological or ecological, can be obscured by the effects of similarity owing to relatedness (Harvey & Pagel 1991; Kelly & Purvis 1993; Harvey 1996; Woodward & Kelly 1997). For example, when phylogenetic relatedness is appropriately taken into account, the classic 'functional type' of growth form is not a simple predictor of range size (Kelly 1996; Kelly & Woodward 1996). Therefore potentially confounding, extraneous factors will be removed, to as great a degree as possible, by comparing morphologically similar congeneric species.

Jatropha

Jatropha standleyi and *J. chamelensis* are closely related, deep forest trees (Dehgan & Schutzman 1994) and physically highly similar (Buckley & Kelly, in press). The two species do not differ in abundance or habitat use within the forest in which they co-exist and possess similar adult and juvenile growth rates, signifying a large degree of physiological similarity. Nonetheless, *J. standleyi* occurs throughout the tropical

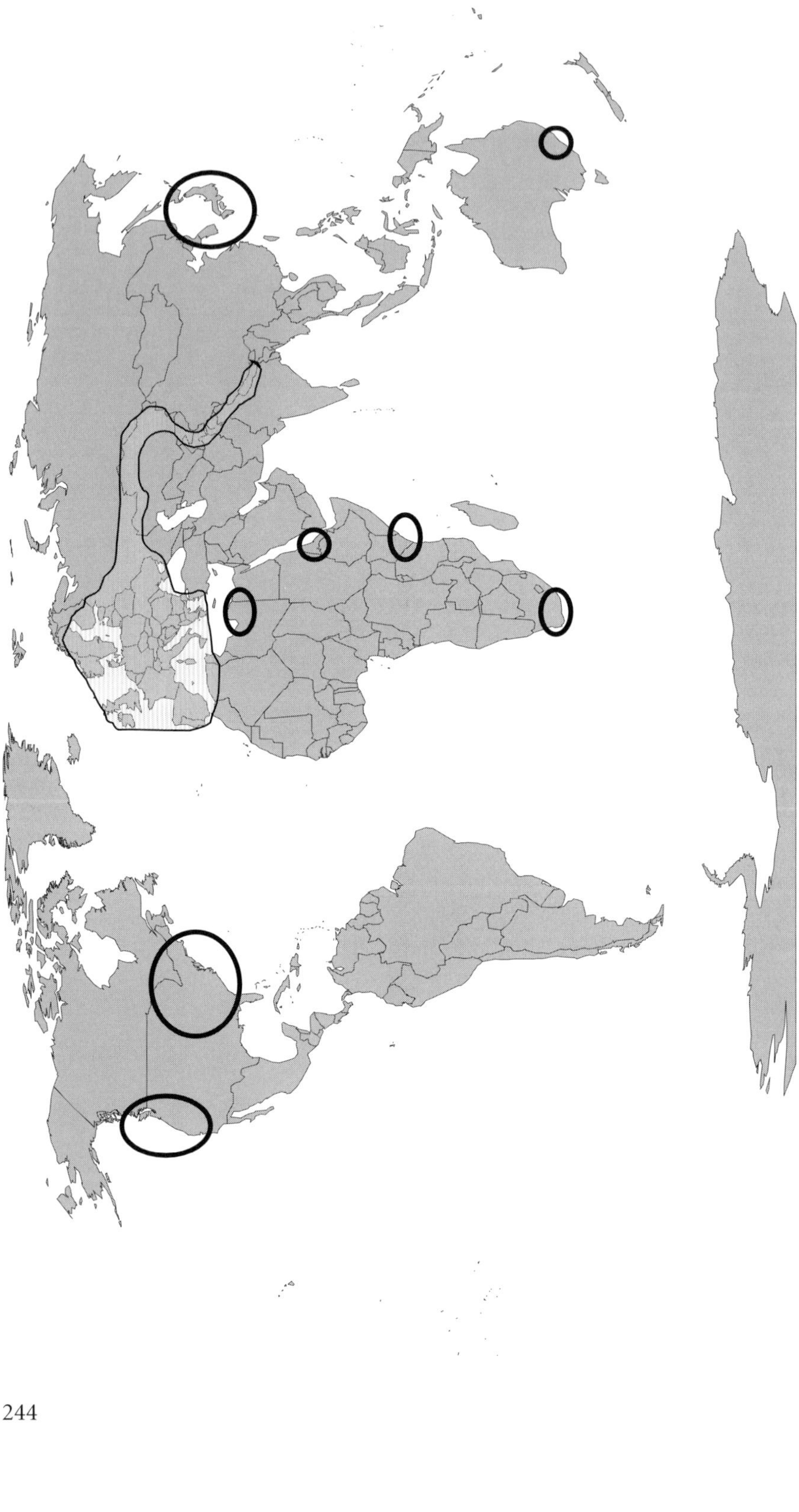

Figure 13.3 Distribution of *Arabidopsis thaliana* (adapted from Alonso-Blanco & Koornneef 2000; Hoffmann 2002). Rings, introductions; shaded area, native distribution.

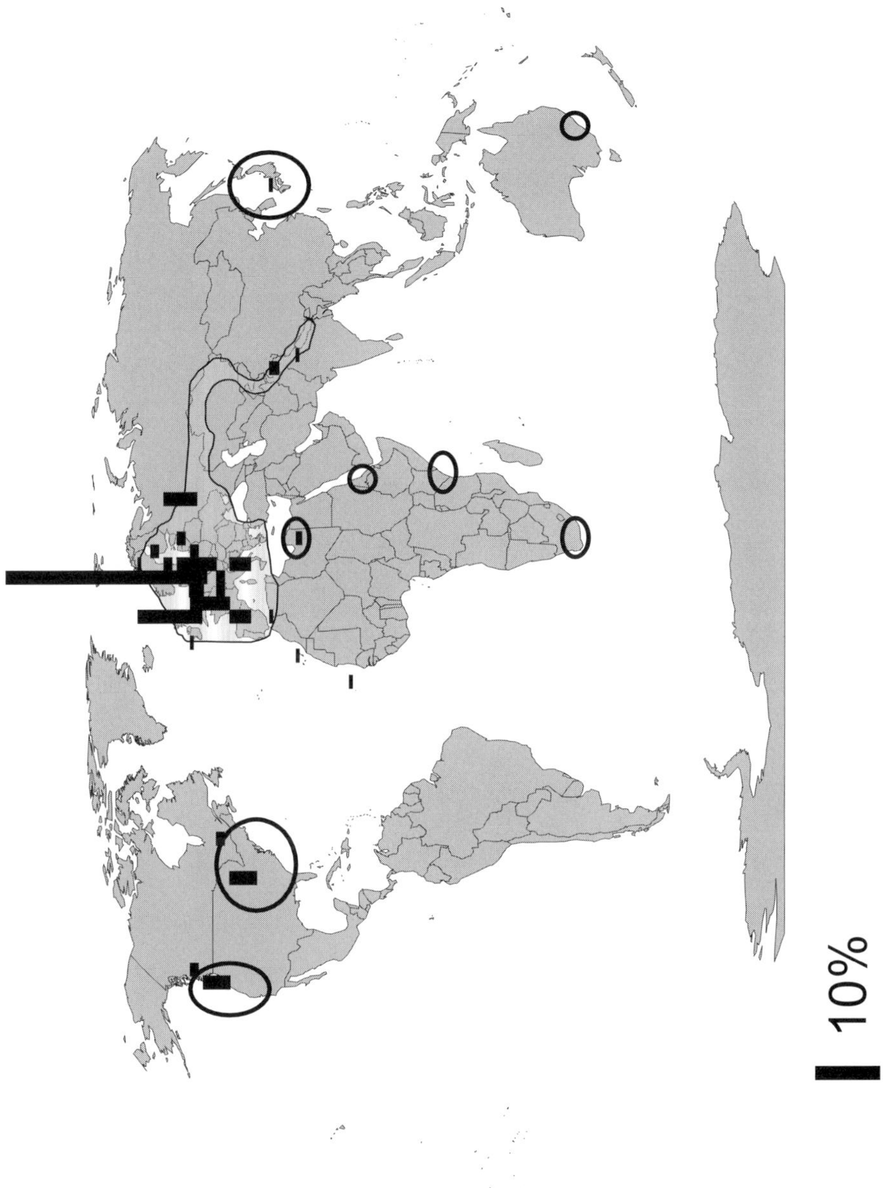

Figure 13.4 Percentage distribution of 224 *Arabidopsis thaliana* ecotypes (data from the Nottingham Arabidopsis Centre). Scale bar 10%.

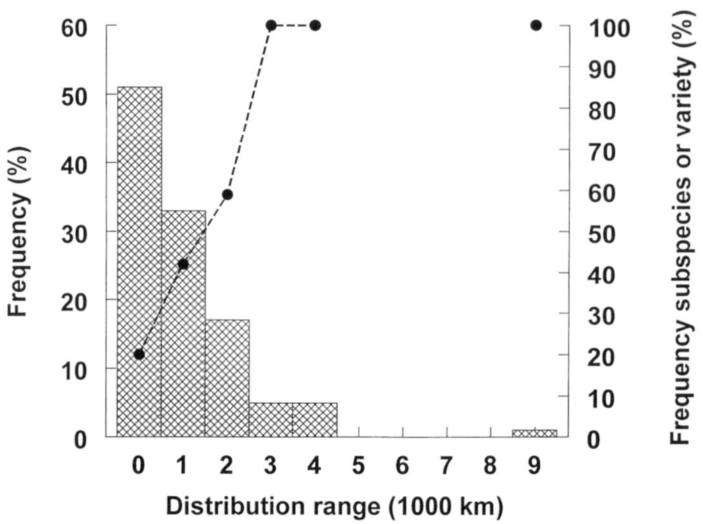

Figure 13.5 Distributional ranges of *Pinus* species (histogram) and frequency of species with subspecies or varieties (adapted from Vidakovic 1991; Price *et al.* 1998). The distributional range is calculated as the maximum distance between range boundaries on any axis.

deciduous forest of México whereas *J. chamelensis* is endemic to the Chamela Bay area (Buckley & Kelly, in press). After detailed study, Pirie *et al.* (2000) found no significant difference between the two species in dispersal mode; the highly similar fruits in both species fall to the ground immediately beneath the parent tree and are removed by rodents in similar proportions. However, they proposed that the observed distributional differences may yet be a function of dispersal. They suggested that the greater seed crop size of *J. standelyi* (approximately three times that of *J. chamelensis*) increased the probability of infrequent long-distance dispersal sufficiently to overcome the difficulties of population establishment, and found this pattern repeated in all the other such congeneric pairs available in the literature at the time. Murray *et al.* (2002) added their individual study to the data collated by Pirie *et al.* (2000) and found that in all five of the examples available to them, the congener with the larger seed crop was also the more widespread.

Pines

The genus *Pinus* contains over 100 species with well-characterized and widely differing geographical ranges (Fig. 13.5; Vidakovic 1991). Just over half of the species occur over a geographical range of less than 1000 km, whereas the most widespread species, *Pinus sylvestris*, has a range of nearly 10 000 km (maximum range on any compass axis). Within the *Pinus* genus, the most widespread species defines the relative term of wide and we therefore consider 10 000 km to be a wide geographical range. In this context, the majority of the species of pine are narrowly distributed.

Scots pine and Monterey pine

Both *Pinus sylvestris* (Scots pine) and *P. radiata* (Monterey or radiata pine) are well understood physiologically and historically and offer a useful paired comparison: the former species is widespread whereas the latter is of narrow distribution (Fig. 13.6) by the standard of the genus (Vidakovic 1991; Lavery & Mead 1998). Although in *A. thaliana* the wide range of genetic variability does not seem to interact with the environmental and physiological capacity of the species to dictate geographical distribution (Hoffmann 2002), the reverse may be true in conifers. Data in Vidakovic (1991) indicate that in conifers there is a strong correlation between the occurrence of recognizable subspecies or varieties and the geographical range of a species, indicating an apparently inevitable change in genetic characteristics across different environments (but see Gosler *et al.* (1994) on *Crataegus*). Consistent with this general finding, Scots pine contrasts strongly with Monterey pine, showing marked differences of growth in relation to both the latitude and longitude of origin (Cregg & Zhang 2001; Oleksyn *et al.* 2001). In addition the low temperature tolerance of Scots pine is known to exceed –90°C (Sakai 1983; Woodward 1987, 1988), contrasting with a much more limited tolerance (to about –13°C) in Monterey pine (Oohata & Sakai 1982). Such large differences in physiological tolerances will explain at least some of the large differences in the native distributions of the two pines (see below; Fig. 13.7). In addition, it is reported that Monterey pine is confined to areas of fog (Ornduff 1974), which will reduce the impact of drought during dry summers, and it has been noted that drought tolerance is limited in Monterey pine (Wakamiya *et al.* 1996).

It is also of interest to note that genetic diversity is much higher in Scots pine than in Monterey pine (Fig. 13.8). However, such genetic variability in itself may not be the factor determining the differences between these two species in range size. These data are for neutral markers and do not of necessity indicate the range of adaptive variability (Ledig 1998). Although genetic variation may not have a causal relationship with temperature or drought tolerance, one practical demonstration of the limited genetic variability in Monterey pine is the virtually complete susceptibility to pitch canker fungus in both native and commercial populations (Hodge & Dvorak 2000).

The genetic variability of Scots pine may have been crucial during the relatively frequent ice ages of the Quaternary Period, as indicated for *Arabidopsis thaliana* (Sharbel *et al.* 2000). Palynological evidence indicates that in its western distribution, Scots pine retreated southwards to refugia in the Balkans, the Alps and the Iberian Peninsula (Bennett *et al.* 1991) during the last ice age. Subsequently Scots pine migrated northwards, in parallel with warming at the termination of the ice age. These repeated migrations through different ice ages will allow some genetic exchange between newly adjacent populations, and this modified or enhanced genetic variability may enhance survival in new environments.

The southernmost populations of Scots pine in Europe occur in the Iberian Peninsula, at high altitudes (Soranzo *et al.* 2000). During an ice age these populations will have migrated to lower altitudes, tracking the survivable environmental

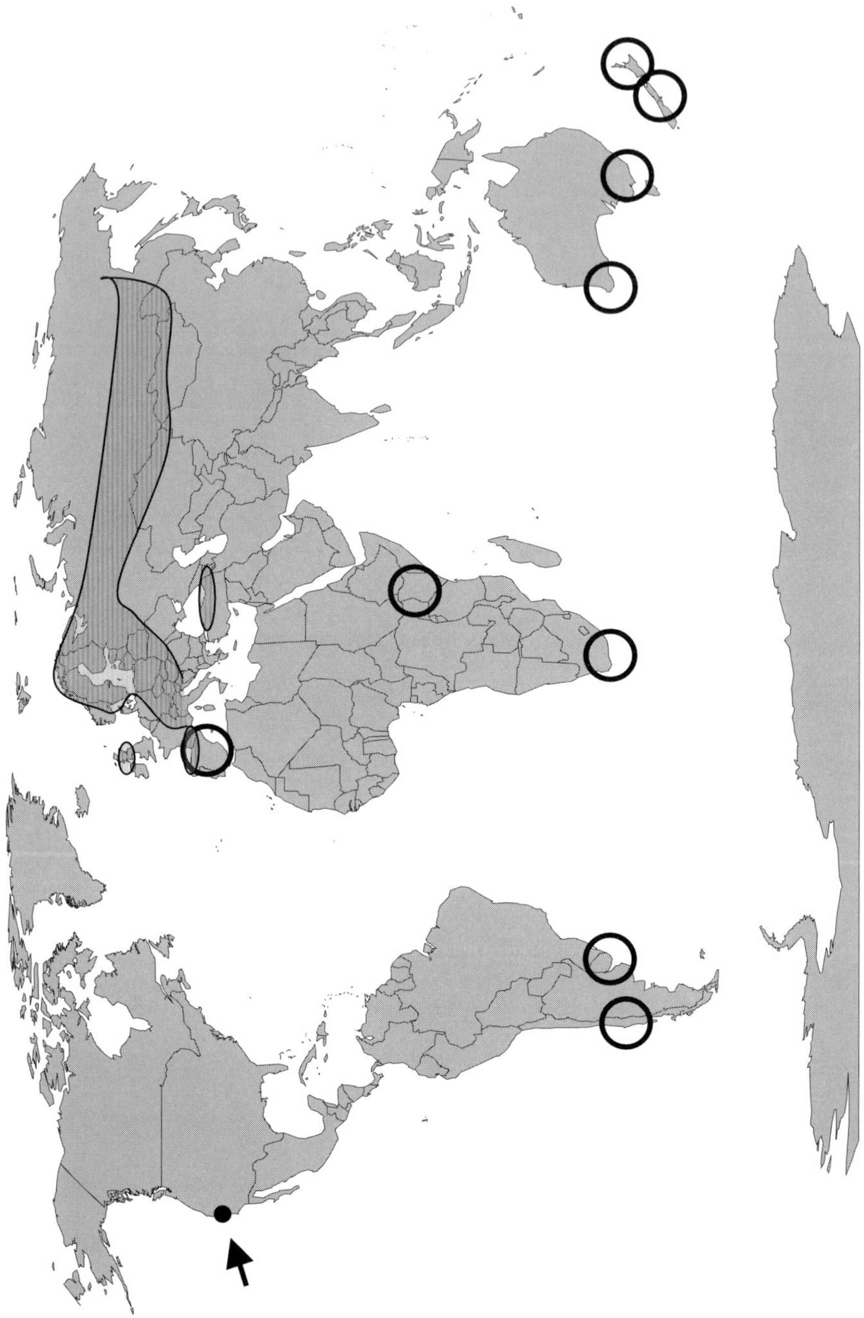

Figure 13.6 Native distributions of Scots pine (shaded) and Monterey pine (indicated by arrow) (adapted from Vidakovic 1991; Lavery & Mead 1998). Introductions of Monterey pine indicated by rings.

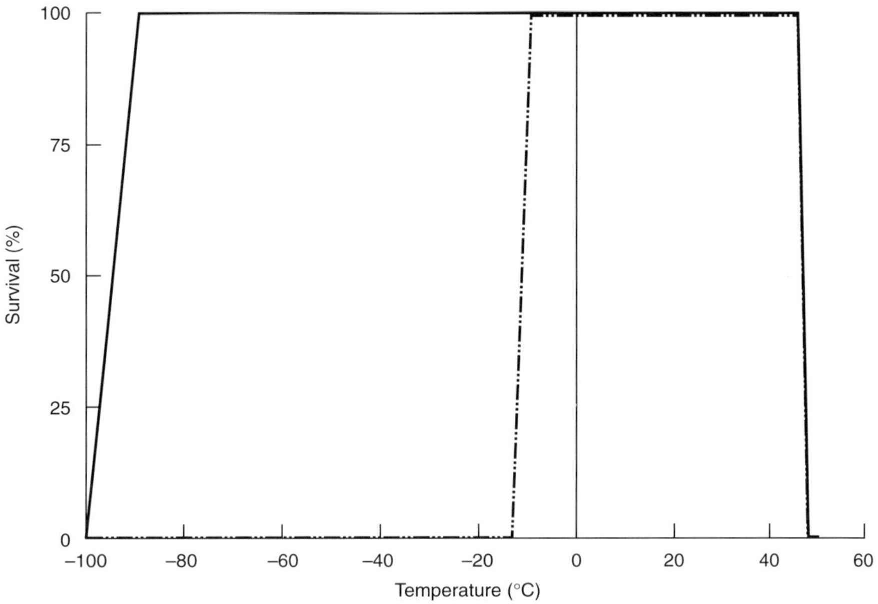

Figure 13.7 Temperature tolerances of Scots pine (continuous line) and Monterey pine (dash dot). (Adapted from Oohata & Sakai 1982; Sakai 1983; Woodward 1987; Woodward 1988; Booth & Jones 1998.)

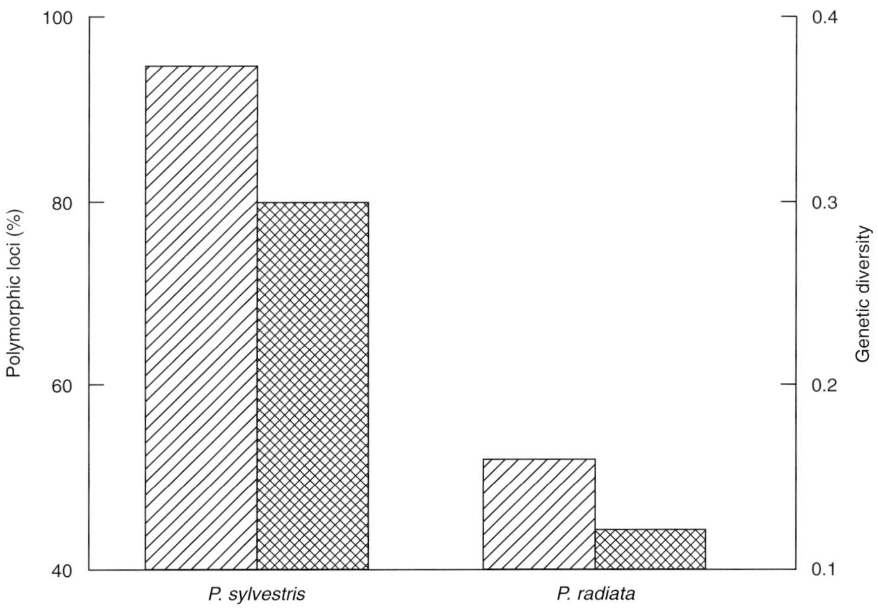

Figure 13.8 Frequency of polymorphic loci (hatched bar) and genetic diversity (cross hatched bar) for *Pinus sylvestris* (Scots pine) and *P. radiata* (Monterey pine). (Adapted from Ledig 1998.)

range of the species. It has been calculated that the movement from the current altitude to sea-level would allow a tolerance of greater than 7°C of cooling during the ice age. Survival at high altitudes depends, in part, on physiological tolerances of reduced partial pressures of carbon dioxide and structural tolerances of high wind speeds and avalanches (Woodward 1987). Tolerance of reduced carbon dioxide concentrations would also be applicable at low altitudes during ice ages. Populations of Scots pine from high latitude (>56°N) have similar growth rates to southern populations (Oleksyn *et al.* 2001), which could imply that populations from high latitude have the capacity to survive in high-altitude relict environments.

Evidence from maternally inherited mitochondrial markers, which can only migrate between populations by seed, indicates that current populations of Scots pine in northern Europe did not originate from the Iberian Peninsula and also did not experience significant genetic exchange with populations in the Peninsula (Soranzo *et al.* 2000), suggesting post-glacial origins from the Balkans. Such ice age changes, with a significant number of isolated mountain refugia, indicate how a widespread species can maintain and may even enhance genetic diversity as a consequence of wide physiological tolerances and a capacity to migrate over long distances. By contrast, there appears to be no evidence of a significantly more widespread distribution of Monterey pine over the past million years (Axelrod & Govean 1996), a feature that is consistent with its low environmental tolerance (Fig. 13.7).

We conclude by noting that even here, with such strong evidence of physiological determinants of distribution, dispersal limitations may still be seen to have played a role in the geographical distribution of Monterey pine. When initially planted into areas of mild winters and wet summers in Australia and New Zealand, the species spontaneously spread from its original managed area (Burdon & Chilvers 1977).

Freezing resistance and geographical range of pines

Although Scots pine possesses both a greater tolerance of freezing and a greater range size than does Monterey pine, without additional external evidence any inference of causality is weak. Evolutionary comparative methods have therefore been used in order to determine if a difference in cold tolerance between related species predicts a difference in range size, thus allowing inference of a general functional relationship between the two variables. A measure of freezing resistance (cold tolerance) is available for a number of *Pinus* species (Oohata & Sakai 1982) for which range sizes are also known (Vidakovic 1991). Evolutionary comparative and cross-species analyses have been applied to these data in order to assess the inference that increasing freezing tolerance is functionally related to increased range size in this genus. Further, the species for which data were available also allowed a comparison of the functional relationship for these two traits in 'northern' and 'southern' species of *Pinus*.

A fully resolved phylogenetic tree of 20 *Pinus* species was constructed using the molecular phylogenies of Liston *et al.* (1999) and Dvorak *et al.* (2000) (Fig. 13.9). Six of the 20 species were from a clade arising from a secondary radiation of the pines in which the species all occur into México and Central America (Price *et al.* 1998). In-

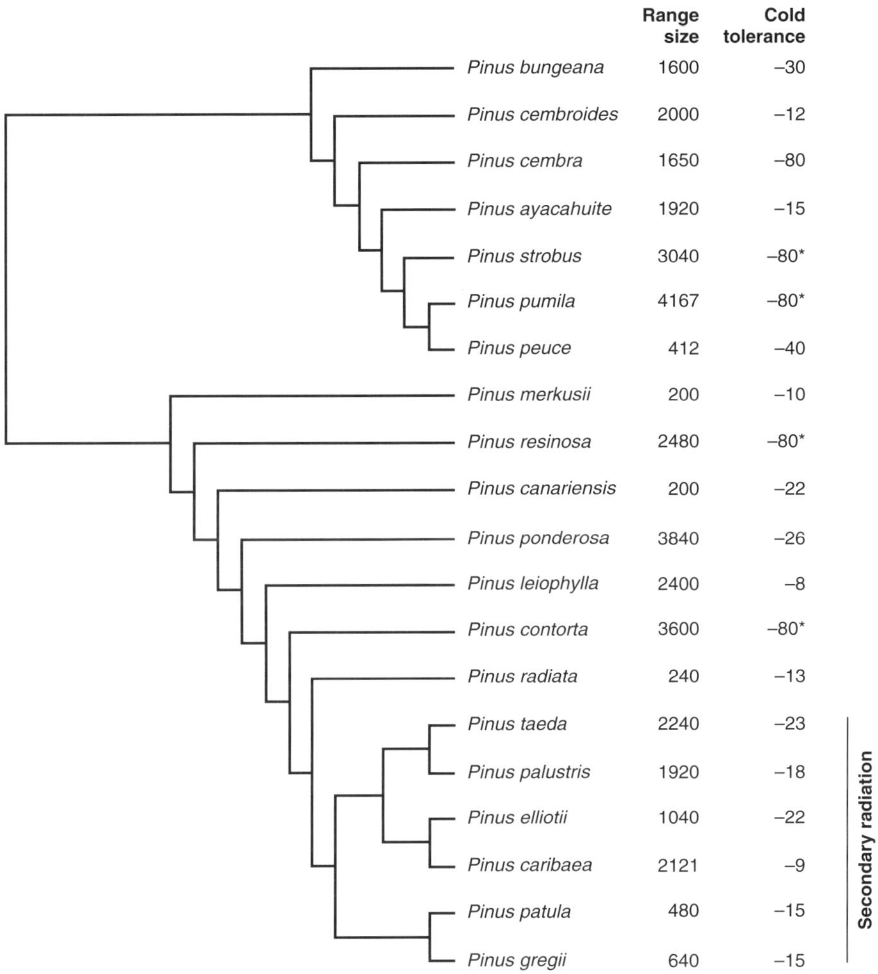

Figure 13.9 Phylogenetic tree for selected *Pinus* species. Values to the right of the species name indicate the range size and the temperature (°C) at which the species begins to suffer freezing damage; *, no mortality at the minimum treatment temperature.

dependent contrasts analysis of the two continuous variables followed Pagel (1992) based on Felsenstein (1985). Resistance was converted to degrees Kelvin and $\log_{10}$ transformed prior to analysis. For the evolutionary comparative analyses, the regression was forced through the origin (Garland *et al.* 1992).

A general functional relationship emerges between freezing tolerance and range size for the species in the primary radiation. In the independent contrasts, the greater the increase in cold tolerance between paired taxa, the greater the increase in range size ($p = 0.0359$; $r^2 = 0.32$; Fig. 13.10). Cross-species analysis also showed a correlation between cold tolerance and range size for the primary radiation

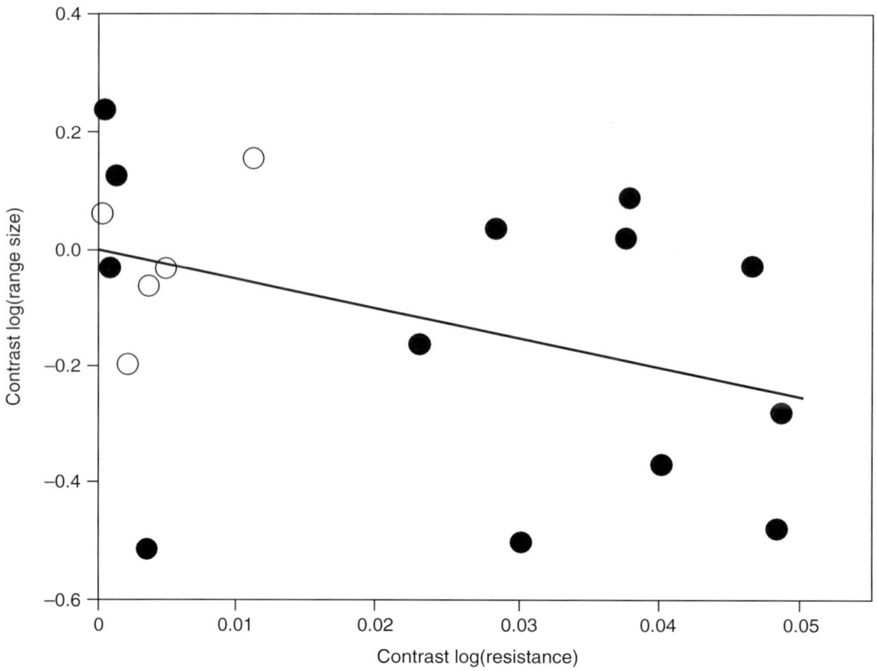

Figure 13.10 Relationship between contrasts of freezing resistance [$\log_{10}$(K)] and maximum range size [$\log_{10}$(km)] in the genus *Pinus*. Species data are presented in Fig. 13.9. Open circles represent species from the secondary radiation of southern clades; closed circles represent northern species from the primary radiation in the genus. The solid line shows the significant relationship (forced through the origin) among the northern species.

($p = 0.0567$; $r^2 = 0.255$), but even with $\log_{10}$ transformation, resistance values were significantly non-normal ($p < 0.0001$) making interpretation of the results problematic.

Overall, the species from the secondary radiation are not particularly freezing tolerant. This is not tremendously surprising—much the greatest proportion of land surface in México and Central American is not subject to prolonged freezing. Nonetheless, although cold mountain tops may not represent a large proportion of land available for colonization, the differences between lowlands and uplands are more dramatic in the tropics (Janzen 1967) and freezing tolerance may yet serve to overcome that potential barrier to dispersal. Thus the important point is whether what little variation there is in freezing resistance predicts any part of the markedly large differences in range size between closely related taxa, and it does not ($p = 0.590$; Fig. 13.10).

It is concluded for those *Pinus* species in the secondary radiation that freezing tolerance is not a functional determinant of range size. However, that may not be because freezing tolerance is not ultimately a limiting factor for these species. Willis

(1922) found a relationship between the age of a vascular plant genus and the largest range size to be found among the species of the genus. His interpretation of this pattern is that larger range sizes may be found in older genera because the species in that genus have had more time to invade their entire potential habitat. What may be occurring in the second, later radiation of *Pinus* is a pattern that is still a function of dispersal rather than physiology.

Conclusion

Although a species certainly will not exist outside the range of its physiological capacities, its physiological capacities will only partly determine its range. In particular, dispersal may play a role. The title of this paper has been addressed by the examination of several different sources of information. The analysis of the testate amoebae indicated that species can be extremely widely distributed—virtually the maximum possible at the global scale—but the more limited distributions of some species can be explained, at least in part, by dispersal. A dispersal limitation to range size is also true for *Arabidopsis thaliana*, for Monterey pine and probably the tropical species *Jatropha chamelensis*.

Data for *Arabidopsis* and for pines indicate that the limits of distributions, and therefore the width of their geographical distributions, can in some cases, at least, be explained by climatic tolerances, in particular of temperature. Species of pines from northern latitudes demonstrated clear correlations between freezing resistance and range size. However, no such relationship was observed for pines from the secondary radiation into México and Central America.

Acknowledgements

We are grateful to Alastair Jump and David Read for their comments on the manuscript.

References

Alonso-Blanco, C. & Koornneef, M. (2000) Naturally occurring variation in *Arabidopsis*: an underexploited resource for plant genetics. *Trends in Plant Science* **5**, 22–29.

Axelrod, D.I. & Govean, F. (1996) An early Pleistocene closed-cone pine forest at Costa Mesa, southern California. *International Journal of Plant Sciences* **157**, 323–329.

Bennett, K.D., Tzedakis, P.C. & Willis, K.J. (1991) Quaternary refugia of north European trees. *Journal of Biogeography* **18**, 103–115.

Booth, T.H. & Jones, P.G. (1998) Identifying climatically suitable areas for growing particular trees in Latin America. *Forest Ecology and Management* **108**, 167–173.

Buckley, Y.M. & Kelly, C.K. (in press) Comparison of population structures and ecology of a congeneric pair of common and rare neotropical tree species. *Plant Ecology*.

Burdon, J.J. & Chilvers, G.A. (1977) Preliminary studies on a native eucalypt forest invaded by exotic pines. *Oecologia* **31**, 1–12.

Cregg, B.M. & Zhang, J.W. (2001) Physiology and morphology of *Pinus sylvestris* seedlings from diverse sources under cyclic drought stress. *Forest Ecology and Management* **154**, 131–139.

Dehgan, B. & Schutzman, B. (1994) Contributions toward a monograph of neotropical *Jatropha*: phenetic and phylogenetic analyses. *Annals of the Missouri Botanical Garden* **81**, 349–367.

Dvorak, W.S., Jordan, A.P., Hodge, G.P., & Romero, J.L. (2000) Assessing evolutionary relationships of pines in the *Oocarpae* and *Australes* subsections using RAPD markers. *New Forests* **20**, 163–192.

Felsenstein, J. (1985) Phylogenies and the comparative method. *American Naturalist* **125**, 1–15.

Garland, T., Harvey, P.H. & Ives, A.R. (1992) Procedures for the analysis of comparative data using phylogenetically independent contrasts. *Systematic Biology* **41**, 18–32.

Gosler, A.G., Kelly, C.K. & Blakey, J.K. (1994) Phenotypic plasticity in leaf morphology of *Crataegus monogyna* (Rosaceae): an experimental study with taxonomic implications. *Botanical Journal of the Linnean Society* **115**, 211–219.

Harvey, P.H. (1996) Phylogenies for ecologists: the 1995 Tansley Lecture. *Journal of Animal Ecology* **65**, 255–263.

Harvey, P.H. & Pagel, M.D. (1991) *The Comparative Method in Evolutionary Biology*. Oxford University Press, Oxford.

Heywood, V.H. (1978) *Flowering Plants of the World*. Oxford University Press, Oxford.

Hodge, G.R. & Dvorak, W.S. (2000) Differential responses of Central American and Mexican pine species and *Pinus radiata* to infection by the pitch canker fungus. *New Forests* **19**, 241–258.

Hoffmann, M.H. (2002) Biogeography of *Arabidopsis thaliana* (L.) Heynh. *Journal of Biogeography* **29**, 125–134.

Janzen, D.H. (1967) Why mountain passes are higher in the tropics. *American Naturalist* **101**, 233–249.

Kelly, C.K. (1996) Identifying plant functional types using floristic data bases: ecological correlates of plant range size. *Journal of Vegetation Science* **7**, 417–424.

Kelly, C.K. & Purvis, A. (1993) Seed size and establishment conditions in tropical trees: on the use of taxonomic relatedness in determining ecological patterns. *Oecologia* **94**, 356–360.

Kelly, C.K. & Woodward, F.I. (1996) Ecological correlates of plant range size: taxonomies and phylogenies in the study of plant commonness and rarity in Great Britain. *Philosophical Transactions of the Royal Society, London, Series B* **351**, 1261–1269.

Lavery, P.B. & Mead, D.J. (1998). *Pinus radiata*: a narrow endemic from North America takes on the world. In: *Ecology and Biogeography of Pinus* (ed. D.M. Richardson), pp. 432–449. Cambridge University Press, Cambridge.

Ledig, F.T. (1998). Genetic variation in *Pinus*. In: *Ecology and Biogeography of Pinus* (ed. D.M. Richardson), pp. 251–280. Cambridge University Press, Cambridge.

Liston, A.W., Robinson, A., Piñero, D. & Alvarez-Buylla, E.R. (1999) Phylogenetics of *Pinus* (Pinaceae) based on nuclear ribosomal DNA internal transcribed spacer region sequences. *Molecular Phylogenetics and Evolution* **11**, 95–109.

Llorente, F., Oliveros, J.C., Martinez-Zapater, J.M. & Salinas, J. (2000) A freezing-sensitive mutant of *Arabidopsis*, frs1, is a new aba3 allele. *Planta* **211**, 648–655.

Murray, B.R., Thrall, P.H., Gill, A.M. & Nicotra, A.B. (2002) How plant life-history traits related to species rarity and commonness at varying spatial scales. *Australian Ecology* **27**, 291–310.

Oleksyn, J., Reich, P.B., Tjoelker, M.G., & Chalupka, W. (2001) Biogeographic differences in shoot elongation pattern among European Scots pine populations. *Forest Ecology and Management* **148**, 207–220.

Oohata, S. & Sakai, A. (1982). Freezing resistance and thermal indices with reference to distribution of the genus *Pinus*. In: *Plant Cold Hardiness and Freezing Stress: Mechanisms and Crop Implications* (eds P.H. Li & A. Sakai), pp. 437–446. Academic Press, New York.

Ornduff, R. (1974) *Introduction to California Plant Life*. University of California Press, Los Angeles.

Pace, N.R. (1997) A molecular view of microbial diversity and the biosphere. *Science* **276**, 734–740.

Pagel, M.D. (1992) A method for the analysis of comparative data. *Journal of Theoretical Biology* **156**, 431–442.

Pickett, F.B., Champagne, M.M. & Meeks Wagner, D.R. (1996) Temperature-sensitive mutatations that arrest *Arabidopsis* shoot development. *Development* **122**, 3799–3807.

Pirie, C.D., Walmsley, S., Ingle, R., *et al.* (2000) Investigations in commonness and rarity: a comparison of seed removal patterns in the widespread *Jatropha standleyi* and the endemic *J.*

chamelensis (Euphorbiaceae). *Biological Journal of the Linnean Society* **71**, 501–512.

Price, R.A., Liston, A. & Strauss, S.H. (1998). Phylogeny and systematics of *Pinus*. In: *Ecology and Biogeography of Pinus* (ed. D.M. Richardson), pp. 49–68. Cambridge University Press, Cambridge.

Sakai, A. (1983) Comparative studies on freezing resistance of conifers with special reference to cold adaptation and its evolutive aspects. *Canadian Journal of Botany* **61**, 2323–2332.

Sharbel, T.F., Haubold, B. & Mitchell-Olds, T. (2000) Genetic isolation by distance in *Arabidopsis thaliana*: biogeography and postglacial colonization of Europe. *Molecular Ecology* **9**, 2109–2118.

Soranzo, N., Alia, R., Provan, J. & Powell, W. (2000) Patterns of variation at a mitochondrial sequence-tagged-site locus provides new insights into the postglacial history of European *Pinus sylvestris* populations. *Molecular Ecology* **9**, 1205–1211.

Tamminen, I., Makela, P. Heino, P. & Palva, E.T. (2001) Ectopic expression of ABI3 enhances freezing tolerance in response to abscisic acid and low temperatures in *Arabidopsis thaliana*. *Plant Journal* **25**, 1–8.

Vidakovic, M. (1991) *Conifers: Morphology and Variation*. Graficki Zavod, Hrvatske.

Wakamiya, I., Price, H.J., Messina, M.G. & Newton, R.J. (1996) Pine genome size diversity and water relations. *Physiologia Plantarum* **96**, 13–20.

Wilkinson, D.M. (2001) What is the upper size limit for cosmopolitan distribution in free-living microorganisms? *Journal of Biogeography* **28**, 285–291.

Williams, P.H., Gaston, K.J. & Humphries, C.J. (1994) Do conservationists and molecular biologists value differences between organisms in the same way? *Biodiversity Letters* **2**, 67–78.

Willis, J.C. (1922) *Age and Area*. Cambridge University Press, Cambridge.

Woodward, F.I. (1987) *Climate and Plant Distribution*. Cambridge University Press, Cambridge.

Woodward, F.I. (1988) Temperature and the distribution of plant species. In: *Plants and Temperature*, Society for Experimental Biology Symposium, XXXXII (eds S.P. Long & F.I. Woodward), pp. 59–75. Company of Biologists, Cambridge.

Woodward, F.I. & Kelly, C.K. (1997) Plant functional types: towards a definition by environmental constraints. In: *Plant Functional Types* (eds T.M. Smith, H.H. Shugart & F.I. Woodward), pp. 47–65. Cambridge University Press, Cambridge.

Xiong, L.M., Ishitani, M. & Zhu, J.K. (1999) Interaction of osmotic stress, temperature, and abscisic acid in the regulation of gene expression in *Arabidopsis*. *Plant Physiology* **119**, 205–211.

Chapter 14

Macroecology and microecology: linking large-scale patterns of abundance to population processes

Andrew R. Watkinson, Jennifer A. Gill and Robert P. Freckleton*

Introduction

From a population perspective, answering the question of why species are not more widely distributed entails understanding what determines the number of individuals in a population as well as the number of potential sites a species can occupy. Essentially this involves quantifying the various density-dependent and density-independent processes that determine population size: the numbers of births, deaths, immigrants and emigrants. Only when the births and immigrants exceed the emigrants and deaths will a population persist. The potential range of the species is thus defined as the proportion of the habitat within which persistence is possible. Despite the apparent simplicity of this conceptual framework there have been relatively few studies that have examined, from a population perspective, why a species does not persist outside of its current range or even that have explained levels of incidence across the observed range.

Related to the question of why species are not more widely distributed is the question of why the range of a species is so often correlated with its abundance (Gaston & Blackburn 2000). Both positive and negative correlations have been documented between range size and abundance (both mean and maximum), the clear implication being that the same factors that determine abundance also determine range size. In this paper we examine the demographic processes that occur within populations and how they relate to range size and also the correlation between range size and abundance. In so doing we link the processes that occur within populations at a microecological level to the macroecological questions concerning range size and abundance.

* *Correspondence address: a.watkinson@uea.ac.uk*

Linking demographic processes to range limits

The influence of the finite rate of population increase

The majority of population studies have been concerned primarily with quantifying births and deaths at a local level and secondarily with the movement of individuals. Although such studies inform us about what determines the abundance of individuals within a population, they tell us rather little about what determines range limits. For example, consider a population with dynamics that can be captured by the equation

$$N_{t+1} = \lambda N_t f(N_t) \tag{14.1}$$

Here N_t and N_{t+1} are the numbers of individuals at time t and $t + 1$ respectively, λ is the finite rate of population increase summarizing the birth and death rates at low population densities, and $f(N_t)$ is a density-dependent feedback function that reduces the growth rate of the population as population size increases. This equation captures the population dynamics of a range of annual species, including both plants and animals. More specifically, the density-dependent feedback term in a range of animal and plant species can be described by $f(N_t) = (1 + aN_t)^{-b}$, where a and b are parameters that determine the form of the density-dependent feedback (Hassell 1975; Watkinson 1980). Substituting this expression into equation (14.1) gives

$$N_{t+1} = \lambda N_t (1 + aN_t)^{-b} \tag{14.2}$$

At equilibrium, population size can therefore be calculated from

$$\hat{N} = (\lambda^{1/b} - 1)/a \tag{14.3}$$

No assumptions are made here about the carrying capacity of the habitat. Rather population size is a function of λ and the form of the density-dependent feedback function. Considering an environmental gradient that results in a decline in the birth rate or an increase in the death rate such that λ declines along the gradient by a fraction c allows the population size along the gradient to be calculated from

$$\hat{N} = \left[(\lambda c)^{1/b} - 1\right]/a \tag{14.4}$$

In the specific case where $\lambda = 10$, $b = 1$ and $a = 0.001$, abundance along the gradient is as depicted in Fig. 14.1. The species is most abundant in those parts of the range where $c = 1$ ($\lambda c = 10$) and becomes extinct when $c < 0.1$ ($\lambda c < 1$). If, however, the value of λ was 2 at the core of the range then the range limit would be reached when $c = 0.5$ and the range therefore would be considerably smaller (Fig. 14.1). There are two important conclusions to be reached from this analysis (see Watkinson 1985; Holt *et al.* 1997): (i) range size and abundance are correlated (we explore this conclusion later); (ii) the range limits can be identified from a quantification of births and deaths (i.e. when $\lambda < 1$).

In plant populations, at least, it has been noted that there is often a positive correlation between λ and a. This results from the fact that the fecundity of isolated plants

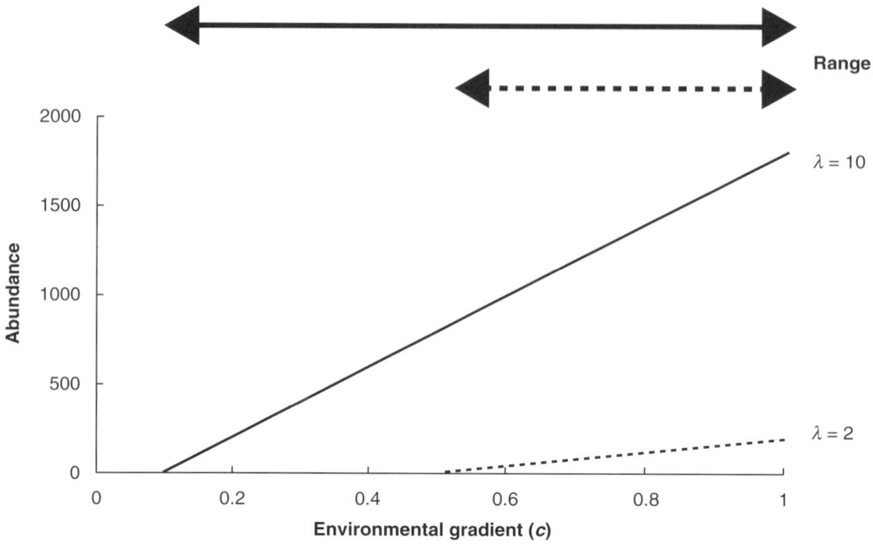

Figure 14.1 The relationship between abundance (see eqn 14.4) and range size along an environmental gradient (c) where $\lambda = 10$, $a = 0.001$ and $b = 1$ (solid line) and $\lambda = 2$, $a = 0.001$ and $b = 1$ (dashed line). See text for further details.

is correlated with the area of resources that are available to them, reflected in the values of the parameters λ and a respectively. Assuming that the two parameters are positively correlated (Watkinson 1984), then it can be shown that the range limit stays at the same position but that the pattern of abundance along the gradient changes. This can be seen by setting N to zero in equation (14.3); only the value of λ influences the solution at equilibrium if b does not vary along the gradient.

Although the range limit of a species in the above analysis is shown to depend critically on the value of λ, it might be expected that an Allee effect could influence the position of the range limit (Stephens & Sutherland 1999). The Allee effect can be defined as a positive relationship between any component of individual fitness and either numbers or density of conspecifics (Stephens *et al.* 1999) and may result from a number of mechanisms, including anti-predator vigilance, social thermoregulation, the reduction of inbreeding and the facilitation of reproduction. Incorporating the Allee effect into traditional models of population dynamics indicates that population growth rates will be depressed at low levels of abundance and that populations will have two equilibria, an upper stable equilibrium and a lower unstable equilibrium. The consequence of the former is that it is extremely difficult to measure the finite rate of population increase, whereas the latter means that it is highly likely that populations will become extinct below the lower equilibrium (Stephens & Sutherland 1999). As a result, the range limits of species may be truncated, even though suitable habitat is available, if populations fall below the unstable equilibrium point. Equation (14.2) can be modified to include an Allee effect by including a parameter θ:

$$N_{t+1} = \lambda(1-\theta N_t^{-1})N_t(1+aN_t)^{-b} \tag{14.5}$$

In this form of the model the population becomes extinct if N falls below $N = \theta + \sqrt{\theta(a^{-1}+\theta)}$. This is an important example of why it is necessary accurately to characterize the dynamics of populations across a range of densities, because the persistence of populations growing according to equation (14.5) does not depend upon λ, but instead on the parameters determining the strength of the Allee effect and the strength of density dependence.

There are remarkably few cases that allow us to test the role of such factors in determining range boundaries. This is in part because so many population studies are restricted to a single location, and in part because few studies estimate the finite rate of population increase or estimate population growth rates across a range of densities, in order to determine whether density-dependence is positive or negative as populations grow from low densities. Also very few studies have transplanted individuals outside of their current range to test whether a species does not occur at a site because it has not dispersed there or because conditions are unsuitable (Crawley *et al.* 1993). In an analysis of the distribution of the perennial shrub *Atriplex vesicaria* in relation to water points in South Australia, Hunt (2001) found that populations only had population growth rates greater than 1 at sites greater than 2200 m from water. The implication from the population matrix analysis was that sheep grazing of the palatable shrub close to watering points would eventually lead to local extinction of the plants within approximately 2 km of the watering point. The phenomenon of grazing around watering points resulting in zonation of vegetation has been well documented (James *et al.* 1999; Nash *et al.* 1999; Brits *et al.* 2002) and this example provides a clear indication of how the range limit at a local level is set by the impact of sheep grazing on the finite rate of population increase.

The role of Allee effects in limiting population establishment and rates of spread into new habitats is demonstrated by work on classical biocontrol (Memmott *et al.* 1998; Freckleton 2000; Shea & Possingham 2000). It is often not certain whether the frequent failure of the release of biocontrol agents is because of a low value of λ, stochasticity or Allee effects, and release programmes are now being designed in order to distinguish between these mechanisms of extinction.

The influence of dispersal

The above analyses indicate how the distribution of a species depends critically upon the finite rate of population increase in terms of the births and deaths that occur within a population. However, classical invasion theory also predicts, as a result of random diffusion, that populations grow in area in proportion to the square root of the finite rate of population increase and the diffusion coefficient (Fisher 1937; Skellam 1951; Okubo 1980). The reason why the rate of spread depends on the finite rate of increase results from the fact that spread is determined by processes at the fringe of the population where densities are low. Inevitably the Allee effect that reduces population growth at low densities will slow the rate of advance (Lewis & Kareiva 1993) and in a patchy landscape may cause an invasion to fail (Keitt *et al.* 2001).

Tests of the theory have proved remarkably difficult because it is problematic to estimate both the finite rate of population increase and the diffusion coefficient that provides a measure of the dispersal ability of a species (Williamson 1996). Both of these parameters also can be expected to vary with habitat and environment. Nevertheless, Williamson (1996) concluded from an analysis of invasive species that reaction–diffusion equations give a good first-order description of spread in many cases. In considering the spread of *Vulpia* species in Great Britain and Australia, Watkinson *et al.* (2000) concluded, on the basis of the higher finite rate of increase alone, that the spread of *Vulpia* in Australia should be much greater than in Great Britain; on the same time-scale they predicted that patch sizes should be an order of magnitude greater in Australia. Moreover they concluded that the highly restricted size of patches in Great Britain was largely a function of the limited dispersal of the species coupled with the low finite rate of population increase.

Much of our understanding of the role of dispersal in determining range boundaries comes from the analysis of invasive species (Williamson 1996; Rushton *et al.* 1997; Collingham *et al.* 2000; Wadsworth *et al.* 2000; Lurz *et al.* 2001). One of the most striking examples of the role of dispersal in determining the local distribution limits of a native species is described by Keddy (1981), who found that the abundance of the annual plant *Cakile edentula* along a sand dune gradient in Nova Scotia was typically greatest in the middle of the dune ridge and lower towards the seaward and landward ends. Along this gradient the levels of fecundity and mortality were such that the populations could be maintained (i.e. $\lambda > 1$) only at the seaward end of the gradient. On the middle and landward sites the level of mortality exceeded reproductive output ($\lambda < 1$), so that the populations would not have been able to persist in the absence of considerable seed dispersal inland from the seaward end of the gradient (Watkinson 1985). Essentially the seaward population acted as a source population, whereas the inward populations acted as sinks (Pulliam 1989). In such cases the range of a species is greater than would be expected from a consideration of the finite rate of population increase alone; the range boundary in this case depends on the source–sink dynamics, with individuals dispersing from the more productive centre of the range to the less productive extremes.

It is, however, extremely difficult to recognize sink populations (Watkinson & Sutherland 1995) as a result of the fact that true sinks, where deaths exceed births, are difficult to distinguish from pseudosinks. This is because the net movement of individuals into a population will, as a consequence of density-dependence, result in a population where the death rate will exceed the birth rate—a pseudosink. The exploration of abundance patterns across the landscape nevertheless depends critically on understanding the interplay of the variation between sites in mortality and fecundity, their relationship with density, and the pattern of movement between sites (Baillie *et al.* 2000).

Metapopulation processes

In considering the population dynamics of organisms, ecologists have traditionally concentrated on the processes that occur within a population. Inevitably the definition of a population was rather arbitrary in such studies and it is only recently that ecologists have focused more explicitly on how the population dynamics of organisms relate explicitly to the spatial distribution of individuals at a range of scales through the quantification of dispersal and the patterns of colonization and extinction. In this context metapopulation theory (Levins 1970; Levins & Culver 1971) has been particularly influential in that it focuses attention on the regional dynamics of species by considering the proportion of suitable habitat patches that are occupied (p) relative to the rates of colonization (m) and extinction (e). In its original formulation the rate of appearance of new occupied patches is given by the equation

$$\frac{dp}{dt} = mp(1-p) - ep \tag{14.6}$$

At equilibrium

$$\hat{p} = 1 - e/m \tag{14.7}$$

The important point about this result is the existence of a threshold; the migration rate must be large enough and the extinction rate low enough such that $e/m < 1$ otherwise the system of patches will decline to extinction.

This analysis leads to three important conclusions when considering the influence of metapopulation dynamics on the spatial distribution of species: (i) a species may not occur within a region where suitable habitat patches exist if the balance of immigration and extinction is beyond the threshold such that $e/m < 1$; (ii) a species may reach its range limit along an environmental gradient if the density of habitat patches becomes so low that the rate of colonization falls and $e/m < 1$; (iii) a species may reach its range limit along an environmental gradient if the area of habitat patches becomes smaller or they become less suitable such that the extinction rate increases and $e/m < 1$ (see also Holt & Keitt 2000). Moreover Lennon *et al.* (1997) found that relatively sharp boundaries could form along gradual environmental gradients separating the main geographical range from a zone of relatively sparse and ephemeral colonization.

A number of studies have shown within a metapopulation framework that the occupancy of sites depends critically on the proximity of suitable habitat patches (e.g. Sjögren 1991; Thomas *et al.* 1992) to other occupied patches. More critically in the context of the above argument, however, is the demonstration of a threshold of metapopulation persistence. Husband & Barrett (1998), for example, demonstrated that the persistence of metapopulations of *Eichhornia paniculata* along roads in Brazil depends on there being a density of patches greater than approximately $0.2\,km^{-1}$. Densities lower than this result in the probability of colonization being less than the probability of extinction.

The importance of the metapopulation concept lies not only in the fact that it takes account specifically of the processes of colonization and extinction but also in that it disassociates regional from local dynamics. If populations have a metapopulation structure then the regional dynamics cannot be predicted from studies of local dynamics; persistence depends critically on the amount and regional configuration of suitable habitat. This is an important result in terms of understanding range limits but only if populations show a metapopulation structure. For the regional population of a species to persist as a metapopulation, four conditions need to hold (Hanski 1997): (i) suitable habitat occurs in discrete patches that may be occupied by local reproducing populations; (ii) even the largest populations have a measurable risk of extinction; (iii) habitat patches must not be too isolated to prevent recolonization following local extinctions; and (iv) local populations do not have completely synchronous dynamics. Undoubtedly a number of species have a metapopulation structure (Hanski 1999), but the concept has been applied very loosely to a range of taxa for which there is only limited evidence of metapopulation structure (Elmhagen & Angerbjorn 2001; Freckleton & Watkinson 2002). It is therefore not clear to what extent metapopulation processes are important in restricting abundance.

Linking local abundance to range size and regional population numbers

It has been widely observed that the abundance and distribution of species tend to be positively correlated, such that species declining in abundance also tend to show declines in the number of sites they occupy, whereas species increasing in abundance tend also to be increasing in occupancy (Gaston *et al.* 2000). Nevertheless there are exceptions to this general rule (Gaston *et al.* 1999a). In theory, it is possible to explain a positive, negative or indeed no relationship between abundance and range size/occupancy (Holt *et al.* 1997). We consider here the relationship between local abundance (i.e. N, the number of individuals within local populations), occupancy (i.e. I, the fraction of habitat occupied), total population size (i.e. the sum of all local population sizes, i.e. $I \times N$) and range size (i.e. the spatial area encompassing these populations). We note that several measures are used in the literature, and that total and local population size are often not distinguished, and also that incidence is often used as a surrogate for range size. As we show below, alternative theories differ in their predictions about how these components of abundance may be related. The relationship between abundance or incidence and range size will depend on the nature of the scaling from local to regional dynamics. Thus an important issue is the degree to which populations show various forms of regional population structure (Freckleton & Watkinson 2002). Here we discuss three classes of mechanism that link local and regional abundance (Gaston *et al.* 2000) and contrast the predictions of simple models for large-scale population abundance.

Vital rates

It was emphasized above that the dynamics of populations can be viewed as a func-

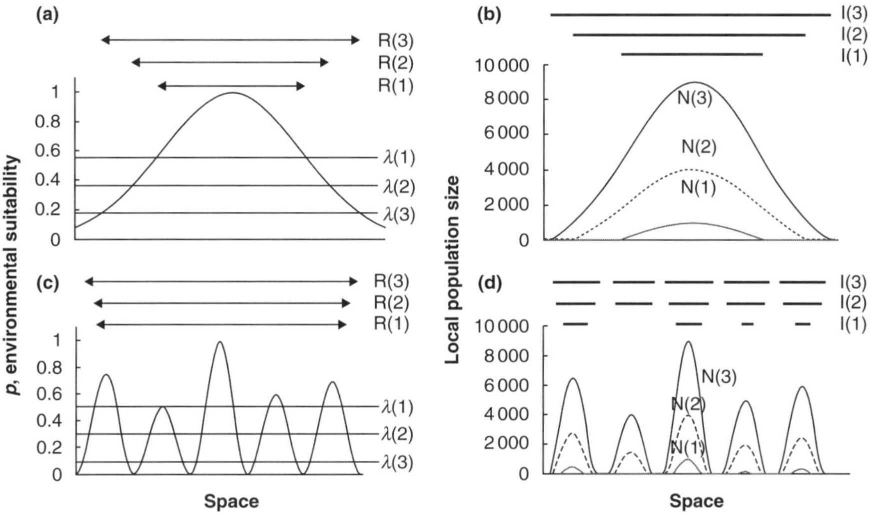

Figure 14.2 Schematic models for range–occupancy–abundance relationships: (a) and (b) follow patterns of range and occupancy expansion in an area of continuous habitat. (a) The curve represents habitat suitability (*p* is defined in the text and represents mortality). As the finite rate of increase (λ) is changed from a low value, $\lambda(1)$, through to a high value, $\lambda(3)$, the potential range (R) increases as shown. (b) Habitat is continuous, hence incidence (I) is the same as the total range. Increasing the finite rate of increase leads to an increase in population size, as increasing λ increases the difference between production of offspring and mortality. A second model is shown in (c) and (d) in which habitat is distributed patchily. (c) Habitat suitability is patchy rather than continuous as in (a). This means that as λ is varied the impacts on geographical range size (R) are rather small. (d) By contrast there is an effect of varying λ on the incidence of species within the habitat, as well as local population size. Note both models predict that the effect of increasing λ on incidence should be saturating.

tion of several key vital rates, and that the dynamics of species at a local scale can be readily modelled as a function of these variables. One class of answer to the question of why species are not more widely distributed at a very large scale may be framed in terms of large-scale variations in such rates, following on from the patterns described in Fig. 14.1. Figure 14.2 shows two informal models of how variation in vital rates may relate to variations in regional abundance (e.g. see Holt *et al.* 1997). In the first case (Fig. 14.2a and b), suitable habitat is distributed continuously across a bell-shaped gradient. Habitat suitability is greatest at the centre of the range and lowest at the edge of the range. Figure 14.2a shows how increasing λ, the finite rate of increase, will affect the potential range of the species. As λ is increased there is a steady increase in the potential geographical range size as more of the area is able to be occupied. Figure 14.2b shows the effect of changing λ on local population size and incidence. Accompanying the increase in range size there is an increase in abundance, so there is an initial positive relationship between range size and abundance. If λ were increased to the point that all suitable habitat were used up, then local population sizes could increase further, but range size would remain constant. In this model habitat is contiguous, and patterns of occupancy are therefore identical to range size, as represented by the incidence functions in Fig. 14.2b.

In many cases, however, suitable habitat may be patchy and not distributed contiguously as in Fig. 14.2a and b. Figure 14.2c and d show a patchy distribution of suitable habitat. In this case increasing λ has little impact on the total range of the species (Fig. 14.2c). This is because it is assumed that within the habitable area, good and bad patches are distributed at random so that there is no core area of most suitable habitat. As a consequence, in this model there is a large difference between geographical range size and incidence (Fig. 14.2d), with both local population size and incidence increasing together at low values of λ, as patches become initially occupied (Fig. 14.2d). However, when all patches become occupied at large values of λ, population size then increases independently of incidence as λ is increased further.

Both models predict that there should be an asymptotic relationship between occupancy and local population size. This prediction is outlined schematically in Fig. 14.3a, in which the relationship between incidence and local population size is represented by a rectangular hyperbola. At low local population densities occupancy is proportional to local population size (N), whereas at high population densities occupancy is independent of local population size. As shown in Fig. 14.3b, this pattern has important implications for the relationship between total abundance (i.e. the summed population size across local populations) and local abundance. Total population size is a function of local population size multiplied by occupancy. When N is low occupancy is proportional to N and hence total population size is proportional to N^2. However when N is large occupancy changes independently of N and hence total population size is proportional to N, rather than N^2. This leads to a form of relationship between total and local population size shown in Fig. 14.3b. On a log–log scale the slope of the relationship between total and local population size declines from 2 at low population sizes to 1 at high population sizes (Fig. 14.3c). This change represents an important shift in dynamics: at low densities increasing the average size of populations (through increasing λ, and hence the potential area that can be occupied) increases population size in two ways: the number of individuals increases, but the area or number of patches occupied also increases. By contrast, at high densities the habitat area is full and increasing density increases population size only and not the area or number of patches occupied.

Rescue effects and metapopulation processes

Most populations live in patchy environments, as in Fig. 14.2c and d. Thus the relationship between abundance and occupancy will depend on the configuration of the habitat and the ability of species to colonize new habitat as it becomes available. The processes of colonization and extinction that operate in metapopulations are therefore likely to be of key importance. Two aspects of metapopulation dynamics have been suggested as being important in generating relationships between distribution and abundance: rescue effects and decreased extinction in large local populations (Hanski 1999; Gaston *et al.* 2000).

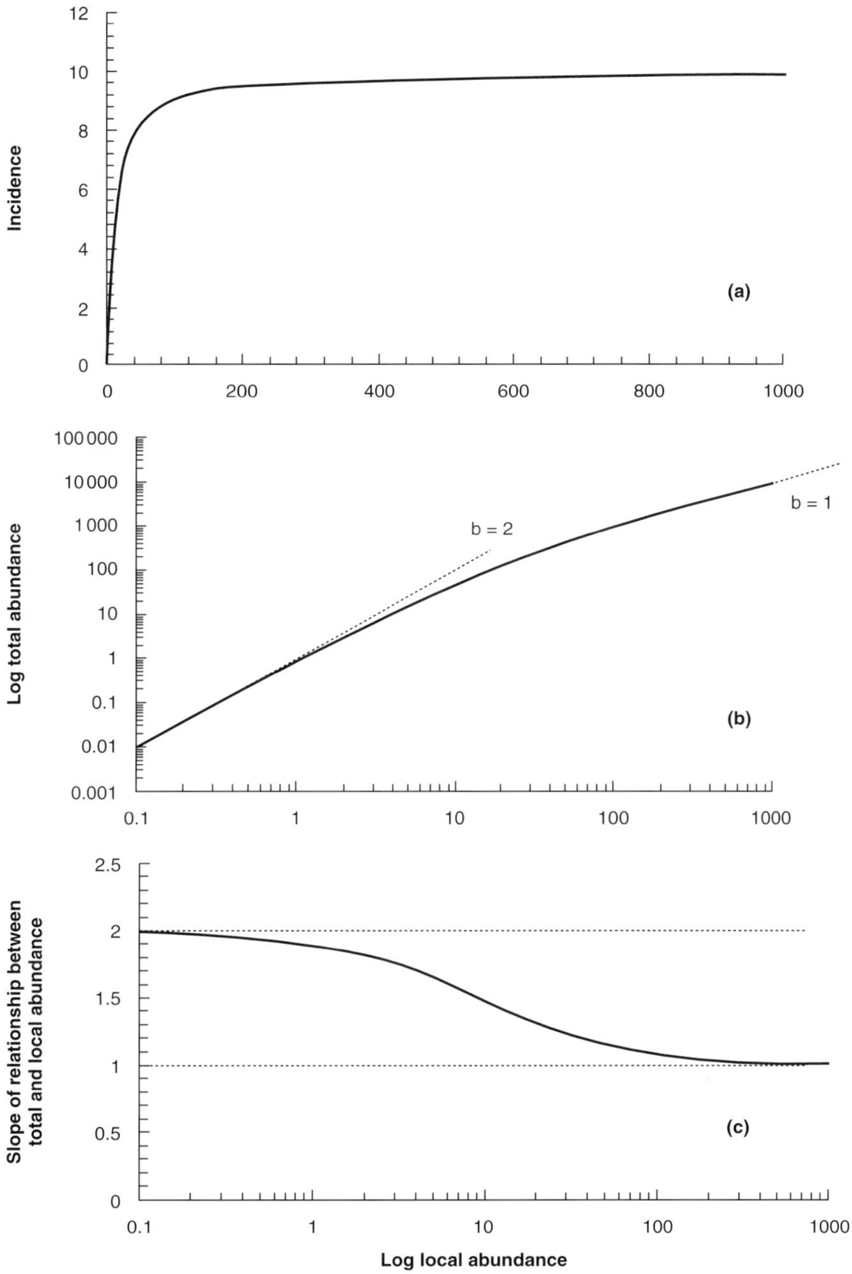

Figure 14.3 The models in Fig. 14.2 both predict an asymptotic limit to the relationship between incidence (proportion of habitat occupied) and local abundance, as represented by the rectangular hyperbola in (a). (b) The relationship between total population size (local abundance × incidence) and local abundance, derived from the curve in (a). At low local densities, log total density increases with slope 2 as log local abundance increases, whereas the slope of this relationship will be 1 at high local densities, as shown in (c).

The rescue effect hypothesis is based on the observation that (i) immigration into patches in a metapopulation tends to decrease the probability of a local patch becoming extinct, and (ii) the rate of immigration into patches increases as the proportion of occupied sites increases. As a consequence there tends to be a positive relationship between patch occupancy and abundance for metapopulation models (Hanski 1991; Hanski *et al.* 1995). There is some experimental evidence to support this contention (e.g. Gonzalez *et al.* 1998).

The rescue effect highlights the importance of immigration in generating relationships between abundance and occupancy. If immigration was rare and extinction of patches very frequent, then overall incidence would be lowered, and the relationship between local abundance and incidence would be shallower. In the extreme, if immigration was extremely rare then local populations persisting on local areas of suitable habitat could become very abundant, but isolated from other suitable patches and hence overall incidence would be low. Thus in populations in which there is little immigration, the pattern will break down and there may be no relationship between occupancy and abundance. There is, however, an exception to this. Under the scenario envisaged in Fig. 14.2a and b, populations are assumed to exist within relatively uniform areas of suitable habitat. Thus metapopulation dynamics are not relevant, because habitat is not patchily distributed. Under this model, however, there should be a close relationship between the proportion of habitat occupied (incidence, *I*) and geographical extent (range, *R*), whereas in models in which habitat is patchy, this will not be the case. This could form a diagnostic of whether habitat patchiness contributes to such relationships.

Hanski (1999) showed that a metapopulation model incorporating both local and regional dynamics would yield a positive relationship between occupancy and local abundance. In his 'carrying capacity' model increasing local population size decreases the probability of patch extinction and hence increases the proportion of patches occupied. In both the rescue effect model and the carrying capacity models it is assumed that population extinction is inevitable in the absence of immigration. In Fig. 14.2c and d, populations increase in density through an increase in the proportion of the habitat for which λ is greater than unity, whereas in the metapopulation models the probability of extinction of patches is decreased as local population sizes become larger. Thus one way to identify whether metapopulation processes are important in generating relationships between abundance and occupancy is to determine whether changes in the flux of existing populations are correlated with changes in abundance and occupancy.

The models of Hanski (1999) make similar predictions to the habitat filling model, but with subtle differences (Fig. 14.4). The relationship between the proportion of occupied sites (*P*) and local population size (*N*) is predicted to be asymptotic (Fig. 14.4a), and approximated by the following equation (see also He *et al.* 2002):

$$\log\left[\frac{P}{1-P}\right]=c+d\log N$$

$$\text{i.e. } P=\left(1+e^{-c}N^{-d}\right)^{-1} \qquad (14.8)$$

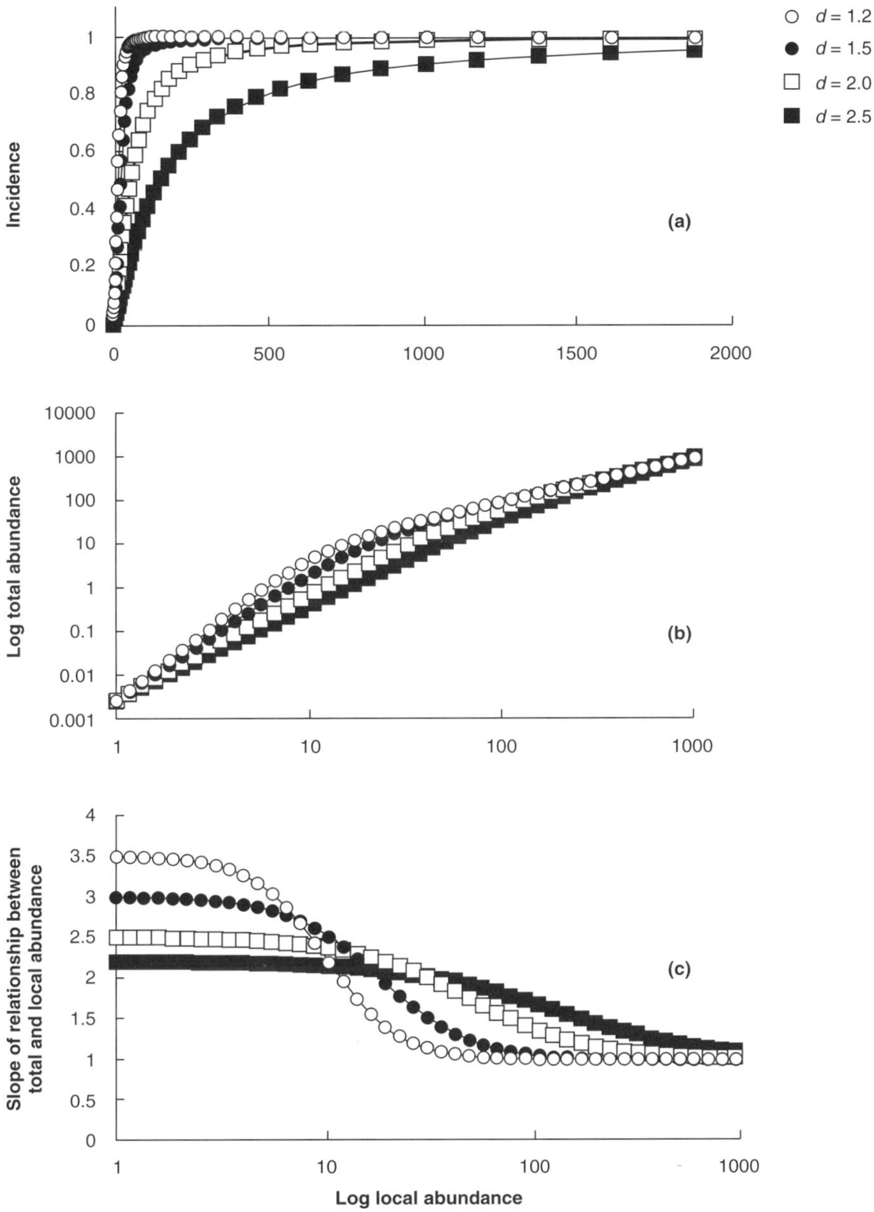

Figure 14.4 As Fig. 14.3, but plotted for the logistic model for patch occupancy as a function of local density in metapopulations (eqn 14.8). The different curves show increasing values of *d*, the slope of the relationship between logit occupancy and log local population size. Increasing values of *d* reflect increasing importance of metapopulation dynamics and rescue effects.

For metapopulations the slope parameter d should be greater than 1, and steepened further by the rescue effect (Hanski 1999). This parameter measures how occupancy increases with local population sizes when local populations are small. In metapopulations, when occupancy is low, the rate of change in occupancy with increasing density is more than proportionate, because increasing occupancy increases colonization and decreases extinction. In point of fact, equation (14.8) is only an approximation, as in genuine metapopulations occupancy will be zero if colonization rates become too low and extinction rates too high. As both of these (may) be a function of local density, this implies that occupancy will decline to zero below some critical local density, but equation (14.8) does not account for this. In the habitat filling models outlined in the previous section, species simply fill habitat as it becomes available and thus the parameter d is predicted to be equal to 1 for these models. A test of whether d is greater than unity could thus be used to test whether patterns of occupancy show evidence for metapopulation or rescue effects versus simple habitat filling.

The importance of metapopulation dynamics for the scaling from local to regional population sizes is shown in Fig. 14.4b and c. Because in metapopulations the value of d is greater than unity, the slope of the relationship between average local population size and total population size is increased at low local population sizes. This is because increasing habitat suitability not only increases the fraction of suitable habitat, but in metapopulations also increases the probability of patches remaining colonized in the face of local extinctions. Thus in small metapopulations, total population size is extremely sensitive to changes in habitat quality, leading to average reductions in local populations.

As noted above, suitable habitat is not always patchy and, for example, within well-defined habitat types we may expect to observe a relationship between abundance and occupancy, even if this were not evident at a wider scale. This may explain the results of Thompson *et al.* (1999) who found no relationship between abundance and occupancy for the British flora. Plants only rarely show metapopulation dynamics (Bullock *et al.* 2002; Freckleton & Watkinson 2002). More commonly plants exist as isolated populations with no or little migration between populations (Freckleton & Watkinson 2002), and hence would be expected to show no relationship between occupancy and abundance. However, many plant populations occur as either continuous or patchy populations within well-defined habitat types (Freckleton & Watkinson 2002), and Thompson *et al.* (1998) found positive relationships between occupancy and abundance within habitat types, as might be expected from the model in Fig. 14.2a and c.

Positive density-dependent habitat selection

For mobile species, intraspecific patterns of site occupancy may be based not only on dispersal abilities but also on individual habitat selection decisions influenced by levels of competition within habitats. In systems such as this, patterns of site occupancy will reflect habitat quality, with highest quality habitats being occupied first and animals only moving to poorer quality habitats at high density. This type of density-dependent habitat selection has been termed the buffer effect (Kluyver & Tinbergen 1953) because poor quality habitat effectively buffers good quality habitat from any changes in numbers when total population size fluctuates. Buffer effect processes can therefore clearly result in increases in range size with abundance, as individuals either expand into or contract out of poorer quality habitats as population size changes. A good example of this is the population of black-tailed godwits *Limosa limosa islandica*, which breeds in Iceland and winters mainly on coastal sites in Britain and Ireland. Long-term census data from the Wetland Bird Survey (Pollit *et al.* 2000) show that this population has been rapidly increasing since the 1970s but that the pattern of population increase has varied across wintering sites. Populations on traditionally used estuaries in the south of England have remained high and stable throughout the period of population increase but estuaries in the east of England have seen very rapid increases from very small initial populations (Gill *et al.* 2001). Detailed analyses of the interactions between godwits and their invertebrate prey on these sites revealed that birds wintering in the south had significantly higher prey intake rates at the end of the winter than those in the east. Birds wintering in the south also had higher annual survival rates and arrived earlier on the Icelandic breeding grounds than east-coast birds (Gill *et al.* 2001). Thus, the population increase in this species has resulted in an expansion in range as a result of buffer effect processes, with poorer quality sites becoming occupied only as numbers increase.

Buffer effect patterns (disproportionate rates of population change in habitats or locations of varying quality) are frequently recorded in a range of vertebrate taxa, both at small (e.g. Krebs 1970; Merkt 1981; Dhondt *et al.* 1992) and large (e.g. O'Connor 1982; Moser 1988; Ferrer & Donazar 1996) spatial scales. Density-dependent habitat selection thus may be a major cause of positive range-size–abundance relationships.

Although positive intraspecific range-size–abundance relationships are most frequent (Gaston *et al.* 1999b, 2000), negative relationships also have been recorded (Gaston *et al.* 1999a). Buffer effects can also generate negative range-size–abundance relationships if density is used as the measure of abundance rather than population size. For example, Gaston *et al.* (1999a) described a negative relationship between mean density and the proportion of 10-km squares in the UK occupied by sparrowhawks, *Accipiter nisus.* This species has been increasing in the UK as a result of the population recovering from the crash caused by organochlorine pesticides in the 1950s and 1960s (Newton 1986). The likely explanation for the decline in density with increasing range size was that sparrowhawks were expanding into poorer quality locations where territories had to be larger and, hence, average densities declined as population size increased (Gaston *et al.* 1999a). Figure 14.5 shows the relationships between range size and both population size and average density for Icelandic black-tailed godwits. Figure 14.5a shows that the population expansion has been mirrored by a range expansion, which, as described above, is the result of animals expanding into poorer quality wintering locations (Gill *et al.* 2001). However, the average density of godwits is not related to range size (Fig. 14.5b), because density does not increase linearly with population size; rather, the sequen-

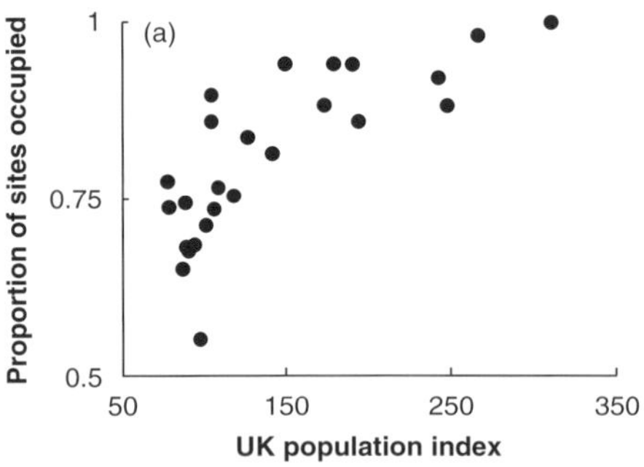

Figure 14.5 The relationships between the proportion of estuaries occupied by black-tailed godwits and (a) the UK black-tailed godwit population index and (b) mean density (godwits per site). Data from the Wetland Bird Survey (Pollit *et al.* 2000).

tial pattern of site occupancy results in average densities initially declining as new sites are occupied and then increasing as those sites are filled. Thus, in buffer effect situations the relationship between density and site occupancy can be far more complex than the relationship between population size and range size.

Concluding remarks

The links between large-scale abundance and local dynamics are of fundamental importance. If local processes dominate then large-scale population dynamics represent an extrapolation of local dynamics. By contrast if large-scale processes are dominant then large-scale dynamics are dominated by metapopulation processes, such as migration, patch-scale extinction and recolonization. This is evident in relationships between abundance and occupancy. In the previous section we reviewed three classes of theories that yield asymptotic relationships between range size and local abundance. The mechanisms underlying these are very different, however. The vital rates and density-dependent habitat selection hypotheses represent extrapolations of local-scale dynamics to a regional level, with increases in occupancy and local densities resulting from an increase in suitable habitat. The metapopulation hypotheses, however, rely on decreased patch extinction and flux of occupancy with increasing local densities. Although the differences between these mechanisms are profound, they share two important features. Firstly, both classes of mechanism assume that organisms are mobile, and are thus able to colonize suitable available habitat. Secondly, both mechanisms assume that an increase in habitat quality leads to an increase in local density. If this is not the case, as may be true in some examples of density-dependent habitat selection, then positive relationships between range size and local population size do not result.

Whatever the underlying mechanism, the scaling from local- to large-scale population sizes is highly dependent on the form of the abundance–occupancy relationship. The impacts of environmental changes observed on local populations cannot be linearly extrapolated to regional populations for species at low densities. Loss of patch occupancy together with reductions in density lead to disproportionate impacts of environmental degradation on total population sizes. For such reasons the question of how to scale from local-scale observations to regional-scale predictions is extremely important.

References

Baillie, S.R., Sutherland, W.J., Freeman, S.N., Gregory, R.D. & Paradis, E. (2000) Consequences of large-scale processes for the conservation of bird populations. *Journal of Applied Ecology* **37**, 88–102.

Brits, J., van Rooyen, M.W. & van Rooyen, N. (2002) Ecological impact of large herbivores on the woody vegetation at selected watering points on the eastern basaltic soils in the Kruger National Park. *African Journal of Ecology* **40**, 53–60.

Bullock, J.M., Moy, I.L., Pywell, R.F. *et al.* (2002) Plant dispersal and colonization processes at local and landscape scales. In: *Dispersal Ecology* (eds J.M. Bullock, R.E. Kenward & R. S. Hails), pp. 279–302. Blackwell Science, Oxford.

Collingham, Y.C., Wadsworth, R A., Huntley, B. & Hulme, P.E. (2000) Predicting the spatial distribution of non-indigenous riparian weeds: issues of spatial scale and extent. *Journal of Applied Ecology* **37**, 13–27.

Crawley, M.J., Hails, R.S., Rees, M., Kohn, D. & Buxton, J. (1993) Ecology of transgenic oilseed rape in natural habitats. *Nature* **363**, 620–623.

Dhondt, A.A., Kempenaers, B. & Adriaensen, F. (1992) Density-dependent clutch size caused by habitat heterogeneity. *Journal of Animal Ecology* **61**, 643–648.

Elmhagen, B.& Angerbjorn, A. (2001) The applicability of metapopulation theory to large mammals. *Oikos* **94**, 89–100.

Ferrer, M. & Donazar, J.A. (1996) Density-dependent fecundity by habitat heterogeneity in an increasing population of Spanish Imperial Eagles. *Ecology* **77**, 69–74.

Fisher, R.A. (1937) The wave of advance of advantageous genes. *Annals of Eugenics* **7**, 355–369.

Freckleton, R.P. (2000) Biological control as a learning process. *Trends in Ecology and Evolution* **15**, 263–264.

Freckleton, R.P. & Watkinson, A.R. (2002) Large-scale spatial dynamics of plants: metapopulations, regional ensembles and patchy populations. *Journal of Ecology* **90**, 419–434.

Gaston, K.J. & Blackburn, T.M. (2000). *Pattern and Process in Macroecology*. Blackwell Science, Oxford.

Gaston, K.J., Blackburn, T.M. & Gregory, R.D. (1999a) Intraspecific abundance–range size relationships: case studies of six bird species in Britain. *Diversity and Distributions* **5**, 197–212.

Gaston, K.J., Gregory, R.D. & Blackburn, T.M. (1999b) Intraspecific relationships between abundance and occupancy among species of Paridae and Sylviidae in Britain. *Ecoscience* **6**, 131–142.

Gaston, K.J., Blackburn, T.M., Greenwood, J.J.D., *et al.* (2000) Abundance–occupancy relationships. *Journal of Applied Ecology* **37**, 39–59.

Gill, J.A., Norris, K., Potts, P.M., *et al.* (2001) The buffer effect as a mechanism for large-scale population regulation of a migratory bird. *Nature* **412**, 436–438.

Gonzalez, A., Lawton, J.H., Gilbert, F.S., Blackburn, T.M. & Evans-Freke, I. (1998) Metapopulation dynamics, abundance, and distribution in a microecosystem. *Science* **281**, 2045–2047.

Hanski, I. (1991) Single-species metapopulation dynamics: concepts, models and observations. *Biological Journal of the Linnean Society* **42**, 17–38.

Hanski, I. (1997) Metapopulation dynamics: from concepts and observations to predictive models. In: *Metapopulation Biology: Ecology, Genetics and Evolution* (eds I. Hanski & M.E. Gilpin), pp. 5–26. Academic Press, London.

Hanski, I. (1999) *Metapopulation Ecology*. Oxford University Press, Oxford.

Hanski, I., Pakkala, T., Kuussaari, M. & Lei, G.C.

(1995) Metapopulation persistence of an endangered butterfly in a fragmented landscape. *Oikos* **72**, 21–28.

Hassell, M.P. (1975) Density-dependence in single-species populations. *Journal of Animal Ecology* **44**, 283–295.

He, F., Gaston, K.J. & Wu, J.G. (2002) On species occupancy–abundance models. *Ecoscience* **9**, 119–126.

Holt, R.D. & Keitt, T.H. (2000) Alternative causes for range limits: a metapopulation perspective. *Ecology Letters* **3**, 41–47.

Holt, R.D., Lawton, J.H., Gaston, K.J. & Blackburn, T.M. (1997) On the relationship between range size and local abundance: back to basics. *Oikos* **78**, 183–190.

Hunt, L. (2001) Heterogeneous grazing causes local extinction of edible perennial shrubs: a matrix analysis. *Journal of Applied Ecology* **38**, 238–252.

Husband, B.C. & Barrett, S.C.H. (1998) Spatial and temporal variation in population size of *Eichhornia paniculata* in ephemeral habitats: implications for metapopulation dynamics. *Journal of Ecology* **86**, 1021–1031.

James, C.D., Landsberg, J. & Morton, S.R. (1999) Provision of watering points in the Australian arid zone: a review of effects on biota. *Journal of Arid Environments* **41**, 87–121.

Keddy, P.A. (1981) Experimental demography of the sand-dune annual, *Cakile edentula*, growing along an environmental gradient in Nova Scotia. *Journal of Ecology* **69**, 615–630.

Keitt, T.H., Lewis, M.A. & Holt, R.D. (2001) Allee effects, invasion pinning, and species' borders. *American Naturalist* **157**, 203–216.

Kluyver, H.N. & Tinbergen, L. (1953) Territory and the regulation of density in titmice. *Archives Neerlandaises de Zoologie* **10**, 265–289.

Krebs, J.R. (1970) Regulation of numbers in the great tit (Aves: Passeriformes). *Journal of Zoology* **162**, 317–333.

Lennon, J.J., Turner, J.R.G. & Connell, D. (1997) A metapopulation model of species boundaries. *Oikos* **78**, 486–502.

Levins, R. (1970) Extinction. In: *Lectures on Mathematics in the Life Sciences, 2* (ed. M. Gerstenhaber), pp. 77–107. American Mathematics Society, Providence, RI.

Levins, R. & Culver, D. (1971) Regional coexistence of species and competition between rare species. *Proceedings of the National Academy of Sciences, USA* **68**, 1246–1248.

Lewis, M.A. & Kareiva, P. (1993) Allee dynamics and the spread of invading organisms. *Theoretical Population Biology* **43**, 141–158.

Lurz, P.W.W., Rushton, S.P., Wauters, L.A., *et al.* (2001) Predicting grey squirrel expansion in North Italy: a spatially explicit modelling approach. *Landscape Ecology* **16**, 407–420.

Memmott, J., Fowler, S.V. & Hill, R.L. (1998) The effect of release size on the probability of establishment of biological control agents: gorse thrips (*Sericothrips staphylinus*) released against gorse (*Ulex europaeus*) in New Zealand. *Biocontrol Science and Technology* **8**, 103–115.

Merkt, J.R. (1981) An experimental study of habitat selection by the deer mouse, *Peromyscus maniculatus*, on Mandarte Island, BC. *Canadian Journal of Zoology* **59**, 589–597.

Moser, M.E. (1988) Limits to the numbers of grey plovers *Pluvialis squatarola* wintering on British estuaries: an analysis of long-term population trends. *Journal of Applied Ecology* **25**, 473–485.

Nash, M.S., Whitford, W.G., de Soyza, A.G., Van Zee, J.W. & Havstad, K.M. (1999) Livestock activity and Chihuahuan Desert annual-plant communities: boundary analysis of disturbance gradients. *Ecological Applications* **9**, 814–823.

Newton, I. (1986) *The Sparrowhawk.* Poyser, Calton, Staffordshire.

O'Connor, R.J. (1982) Habitat occupancy and regulation of clutch size in the European kestrel *Falco tinnunculus. Bird Study* **29**, 17–26.

Okubo, A. (1980) *Diffusion and Ecological Problems: Mathematical Models.* Springer-Verlag, Berlin.

Pollit, M., Cranswick, P., Musgrove, A., *et al.* (2000) *The Wetland Bird Survey 1998–99.* British Trust for Ornithology, Wildfowl and Wetlands Trust, Slimbridge, and Joint Nature Conservation Committee (JNCC), Peterborough.

Pulliam, H.R. (1989) Source, sinks and population regulation. *American Naturalist* **132**, 652–661.

Rushton, S.P., Lurz, P.W.W., Fuller, R. & Garson, P.J. (1997) Modelling the distribution of the red and grey squirrel at the landscape scale: a combined GIS and population dynamics approach. *Journal of Applied Ecology* **34**, 1137–1154.

Shea, K. & Possingham, H.P. (2000) Optimal release strategies for biological control agents: an application of stochastic dynamic programming to population management. *Journal of Applied Ecology* **37**, 77–86.

Sjögren, J. (1991) Extinction and isolation gradients in metapopulations: the case of the pool frog (*Rana lessonae*). *Biological Journal of the Linnean Society* **42**, 135–147.

Skellam, J.G. (1951) Random dispersal in theoretical populations. *Biometrika* **38**, 196–218.

Stephens, P.A. & Sutherland, W.J. (1999) Consequences of the Allee effect for behaviour, ecology and conservation. *Trends in Ecology and Evolution* **14**, 401–405.

Stephens, P.A., Sutherland, W.J. & Freckleton, R.P. (1999) What is the Allee effect? *Oikos* **87**, 185–190.

Thomas, C.D., Thomas, J.A. & Warren, M. S. (1992) Distributions of occupied and vacant butterfly habitats in fragmented landscapes. *Oecologia* **92**, 563–567.

Thompson, K., Hodgson, J.G. & Gaston, K.J. (1998) Abundance–range size relationships in the herbaceous flora of central England. *Journal of Ecology* **86**, 439–448.

Thompson, K., Gaston, K.J. & Band, S.R. (1999) Range size, dispersal and niche breadth in the herbaceous flora of central England. *Journal of Ecology* **87**, 150–155.

Wadsworth, R.A., Collingham, Y.C., Willis, S.G., Huntley, B. & Hulme, P.E. (2000) Simulating the spread and management of alien riparian weeds: are they out of control? *Journal of Applied Ecology* **37**, 28–38.

Watkinson, A.R. (1980) Density-dependence in single-species populations of plants. *Journal of Theoretical Biology* **83**, 345–357.

Watkinson, A.R. (1984) Yield–density relationships: the influence of resource availability on growth and self-thinning in populations of *Vulpia fasciculata*. *Annals of Botany* **53**, 469–482.

Watkinson, A.R. (1985) On the abundance of plants along an environmental gradient. *Journal of Ecology* **73**, 569–578.

Watkinson, A.R. & Sutherland, W.J. (1995) Sources, sinks and pseudo-sinks. *Journal of Animal Ecology* **64**, 126–130.

Watkinson, A.R., Freckleton, R.P. & Forrester, L. (2000) Population dynamics of *Vulpia ciliata*: regional, patch and local dynamics. *Journal of Ecology* **88**, 1012–1029.

Williamson, M. (1996) *Biological Invasions*. Chapman and Hall, London.

Chapter 15
Genetics and the boundaries of species' distributions

Roger K. Butlin, Jon R. Bridle and Masakado Kawata*

Introduction

Species have characteristic geographical ranges. The relationships between range area and other features such as body size or metabolic rate are a central part of macroecology. Patterns of overlap between the geographical ranges of different species underlie variation in diversity, another key topic in macroecological research (see Hubbell & Lake, this volume). Therefore, an understanding of the factors that limit geographical ranges is an important goal. It has become an even more pressing issue in recent years as evidence has accumulated for expansion and contraction of ranges in response to climatic change both in the past (e.g. Hewitt 1999) and in the present (e.g. Thomas *et al.* 2001).

In this chapter, we compare two distinct classes of distribution limits: parapatric boundaries and range margins. At a parapatric boundary, one species, subspecies or race is replaced by another. Range margins typically occur at the latitudinal or altitudinal limits of a species' distribution, where some critical feature of the environment (such as aridity or temperature) passes beyond a threshold level. In such situations, although the presence of other species may have a role in determining range margins, there is not the simple geographical replacement found at parapatric boundaries. It is often difficult to determine the factors that prevent the species from occupying habitats beyond its current margin but, at least in principle, range margins are set by the limits of the species' ecological tolerances (see Woodward and Kelly, this volume). However, from an evolutionary genetics perspective this type of explanation is incomplete. What prevents populations near the range margin from evolving traits that permit them to occupy additional territory? This issue has attracted considerable attention, especially from theorists (Holt & Gomulkiewicz 1997; Kirkpatrick & Barton 1997; Case & Taper 2000; Kawecki 2000; see Lenormand 2002 for a recent review). Here we ask whether studies of parapatric boundaries can

* *Correspondence address: r.k.butlin@leeds.ac.uk*

provide insights into evolutionary explanations for range margins, and introduce an individual-based simulation that allows us to add some of the real-world complexity that is omitted from existing models. First, however, it is helpful to emphasize the variation that exists within these two broad categories of distribution edge.

Types of parapatric boundary

Parapatric boundaries may form as a result of divergence *in situ*, perhaps to the point of complete speciation. These primary contacts are distinct historically from secondary contacts, following range expansion, between populations that have accumulated genetic differences in allopatry. However, it can be extremely difficult to demonstrate whether a given contact is primary or secondary on the basis of its present-day characteristics (Barton & Hewitt 1985). Nevertheless, inferences about the biogeographical history of populations from genetic and palaeoecological data can often provide convincing evidence for secondary contact (e.g. Hewitt 2001). Parapatric boundaries may involve hybridization and introgression of alleles between the two populations or they may not, i.e. they may be boundaries between biological species or between subspecies or differentiated populations. Of the four possible classes of parapatric boundary defined by these two categorizations (primary/secondary, with/without hybridization), secondary boundaries with hybridization are certainly the most widely studied and they appear to be the most common (Barton & Hewitt 1985; Harrison 1993), probably because they remain stable over many generations.

It is difficult to point to good examples of abrupt parapatric boundaries without hybridization. Bridle *et al.* (2001a) describe two possible cases in the grasshopper genus *Chitaura* in Sulawesi, Indonesia. Boundaries of this type may be genuinely rare because they are unstable to either competitive exclusion or character displacement (Schluter 2000). Such character displacement allows range overlap and so replacement no longer directly accounts for the position of the range margin (but see below). There are also few fully convincing examples of primary parapatric boundaries, although it is not clear whether they are actually rare or simply difficult to distinguish from interactions resulting from secondary contact. Perhaps the best examples come from heavy metal tolerance in plants growing on contaminated soil around mines (e.g. *Anthoxanthum*; Antonovics 1976), although differentiation between the populations is limited to a few characteristics in most cases. Examples of more extensive differentiation that may be primary in origin are the morphological forms of the littoral snail *Littorina saxatilis* on various European coasts (England, Wilding *et al.* 2001; Spain, Rolán-Alvarez *et al.* 1997; Sweden, Johannesson & Johannesson 1996). On the northeast coast of England divergence in characters associated with adaptation to a steep environmental gradient is accompanied by assortative mating and genetic incompatibility in hybrids but gene exchange is, nevertheless, only impeded for a small proportion of the genome (Wilding *et al.* 2001). However, although divergence *in situ* seems the most parsimonious explanation for the origin of these morphs (Johannesson 2001), secondary contact cannot be excluded at present.

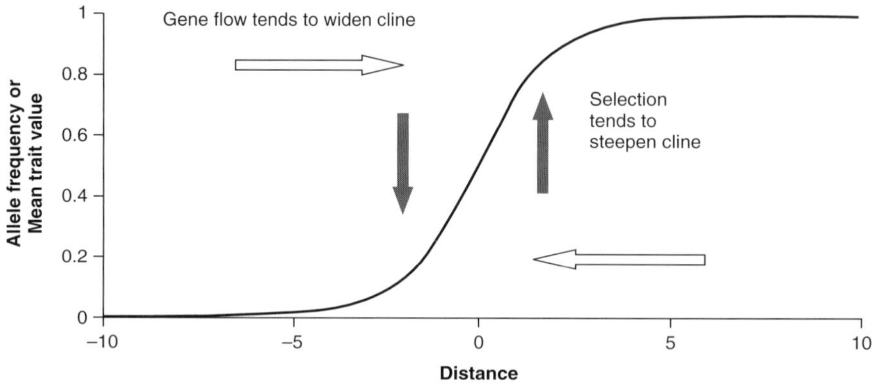

Figure 15.1 The basic tension zone model. This form of cline can result from either an abrupt change in environment at distance = 0 or selection against hybrids.

Hybrid zones

Many parapatric boundaries with hybridization can be understood in terms of the tension zone model (Barton & Hewitt 1985). A tension zone is a hybrid zone maintained by a balance between dispersal and selection against hybrids arising from genetic incompatibilities between the interacting populations. Hybrid zones that form where differently adapted populations meet at a sharp environmental transition share many features with tension zones. In either case, and for a variety of different forms of selection, the resulting pattern is a cline in allele frequency at loci that have diverged between populations that approximates to a sigmoid (tanh) curve (Fig.15.1)(Barton & Gale 1993). Similarly, the mean for a quantitative trait that differs between populations as a result of additive genetic effects is expected to show a sigmoid cline. The width of the cline is proportional to the dispersal distance and inversely proportional to the strength of selection. Even when selection is strong, clines are several dispersal distances wide (e.g. nine times the standard deviation of parent–offspring distances for 10% selection against heterozygotes). This means that parental genotypes do not meet in simple tension zones. Instead, they are separated by an area in which all individuals are of mixed ancestry. Typically, clines in multiple characters are centred in approximately the same position but differ in width.

Tension zones between subspecies of the grasshopper *Chorthippus parallelus* and between the toad species *Bombina bombina* and *B. variegata* are thoroughly studied examples with contrasting characteristics. In *C. parallelus* hybrid zones in the Pyrenees (Butlin 1998, and references therein), clines for several morphological, behavioural and chromosomal characters differ in width and position. Although laboratory F1 hybrid males are completely sterile, it appears that selection in natural hybrid populations has favoured gene combinations that produce fertile males. Selection is weak relative to recombination, and clines are typically wide for many characters. This allows the introgression of alleles for individual traits to vary independently in response to selection. There appears to be no habitat-associated fitness difference between the subspecies.

Bombina bombina and *B. variegata* are much more genetically differentiated than the grasshopper species and they are clearly adapted to different habitat types (MacCallum *et al.* 1998, and references therein). In this situation, strong selection acting on many loci distributed throughout the genome results in steep clines and high levels of linkage disequilibrium (non-random associations between alleles derived from the parental taxa). This causes the clines to be steeper in the zone centre than at the edges and to be similar in width and position because selection is prevented from acting independently on loci for different characters. Selection acts on multilocus genotypes, rather than on the average effect of an allele over many genetic backgrounds. This can allow local populations to deviate from the clinal expectation because parental combinations of traits are favoured in appropriate habitat types, or because differently adapted genotypes can actively choose local habitats where they will have high fitness. In the extreme, the distribution of genotypes in central populations can become bimodal and parental genotypes can meet.

Types of range margin

Range margins also may be classified according to two axes. Firstly, margins may be 'external' or 'internal'. External margins occur at the geographical limits of the species' distribution: they are the edges depicted in field guides. However, within the region delineated by its external margins, a species' distribution is rarely continuous. There are gaps and the edges of these gaps can be described as internal margins. Internal and external margins may be set by the same physiological limits or competitive interactions but the gradients for the critical environmental parameters will typically be steeper at internal margins. The other dimension of classification concerns history: margins may be new or old, with a history of expansion or contraction. Indeed, they may not be at equilibrium if the environment is changing or has recently changed. Populations at the northern edge of the range for Northern Hemisphere temperate species have a history of expansion. As a result, they may have lower genetic variation and greater dispersal capabilities than those at the southern edge and may still be expanding (e.g. Cwynar & MacDonald 1987; Thomas *et al.* 2001).

Chorthippus parallelus populations expanded from refugia in Spain, Italy and the Balkans at the end of the last glaciation (Hewitt 1999). Hybrid zones formed where divergent populations from these refugia met in the Alps and Pyrenees. The external limit of the species' range is set in the south by the Mediterranean coast and in the north probably by the decreasing length of summer. Within this range, *C. parallelus* occurs only in meadows and similar moist grassland habitats. In the south of its range, the grasshoppers are found only in mountain habitats, not on the dry lower slopes. In the north, the species occurs only at low elevations and on south-facing slopes. Thus, the determinants of the internal boundaries vary geographically in response to broader scale environmental variation. These boundaries may not be stable: many orthopteran species have shown marked northward range expansions in recent years (Thomas *et al.* 2001).

Similar descriptions could be provided for most widespread temperate species. The point is that range margins may take different forms, and have different causes, within a single species. They will certainly also differ across taxa.

What determines the positions of distribution boundaries?

Where the environment changes gradually, relative to the dispersal distance of an organism, each local population can adapt to its immediate environment. This generates a 'dispersal independent' cline (Barton & Hewitt 1985) such as the Europe-wide pattern of variation in cyanogenesis in clover (*Trifolium repens*; Daday 1954). In this situation, both the width and the position of the cline depend on the environment. However, where the habitat changes abruptly, dispersal and gene flow prevent local populations from matching the environmental optima. The dispersal-dependent clines that result from such abrupt changes therefore have positions determined by the environment but widths determined by the balance between dispersal and selection. By contrast, in tension zones where selection against hybrids is entirely the result of genetic incompatibilities rather than an interaction with the environment, neither the width nor the position of the resulting cline is determined by the environment: the tension zone is free to move.

What determines the positions of tension zones and other forms of parapatric boundary?

Secondary zones may remain at the point of initial contact between the populations. Where the populations are differentially adapted to the environment, ecotones may directly determine the positions of parapatric boundaries. However, another important factor is the influence of variable population density on gene flow. Given that offspring disperse in random directions from their place of birth, there will be more movement of genes from regions of high population density to regions of low density than vice versa. On a density gradient, there will be a net flow of genes down the gradient. This force will move a cline down a density gradient until it reaches a local minimum, a 'density trough'. Even shallow gradients in density can move clines against selective gradients (Barton & Hewitt 1989). There is good evidence that tension zone positions are determined in this way (e.g. Nichols & Hewitt 1986). Therefore, although the general area in which a zone resides is determined by history, its precise position is the result of an interaction between adaptation to local environments and the effects of asymmetrical gene flow trapping the zone in a density trough.

These considerations for parapatric boundaries have important parallels in range margins. History may be important in many ways, such as in determining which populations colonize an area first or the consequences of temporary barriers to dispersal. In addition, where an environmental gradient is very steep, such as at the coast, it may fix the position of a range margin (although even these boundaries are not insurmountable on evolutionary time-scales, otherwise there would be no life on land or mammals in the sea). Where an environmental gradient is shallow relative to dispersal, the density of a species' population may decline as conditions approach the limits of its tolerance until eventually the margin of the range is reached (Brown 1984). Local population density matches local adaptation in a way that parallels dispersal-independent clines. Where the gradient is steeper, dispersal may maintain 'sink' populations of the species in regions where the environment is outside the set of conditions it requires for a positive population growth rate. This situation is comparable to a dispersal-dependent cline: the position is fixed by the environment but the decline in density is determined by dispersal. Lennon *et al.* (1997) have discussed these alternatives in relation to metapopulation boundaries.

Crucially, the dispersal from high to low density that maintains sink populations also generates directional gene flow. The influx of alleles from populations adapted to conditions in the source population may impede local adaptation in the sink habitat and so prevent expansion of the species' range. This effect was first suggested by Haldane (1956) and recently has been modelled extensively (Holt & Gomulkiewicz 1997; Kawecki 2000). Here, we will concentrate on a model by Kirkpatrick & Barton (1997) that was an important advance over previous theory because it considered evolution of a quantitative trait on an environmental gradient rather than the conditions for spread of alleles in sink populations.

The Kirkpatrick and Barton model

Building on foundations laid previously (Pease *et al.* 1989; Garcia-Ramos & Kirkpatrick 1997), Kirkpatrick & Barton (1997) considered a linear environmental gradient and a single phenotypic trait with constant additive genetic variance. In this model, fitness is maximized when the phenotype of an individual matches its local environmental optimum. Population density is a function of adaptation. A well-adapted central population will have a higher density than a poorly adapted marginal population and the density gradient so created results in differential gene flow. Dispersing individuals tend to encounter habitats in which they are poorly adapted, resulting in a 'migrational load' on the resident population, which is adapting to novel conditions. The load is greatest at the range margin where density is low and many migrants are received from core, high-density populations. Load increases as migration increases or as the environmental gradient becomes steeper, because both cause incoming migrants to have phenotypes further from the local optimum. For a population initiated at the centre of the environmental gradient, three outcomes are possible: it may expand its range to fill the available space, it may have a limited range, or it may become extinct. The parameter space for this model can be reduced to just two dimensions. One describes the 'rate of change of the environment' and depends on the slope of the gradient and the dispersal distance, expressed relative to the strength of selection towards the local optimum. The other describes the 'genetic potential for adaptation': the additive genetic variance, also expressed relative to the strength of selection.

High dispersal distance or steep gradients (increasing rate of change of the environment) both stop the population from expanding to fill the available space, especially where the genetic potential for adaptation is low (Fig. 15.2). The margins are

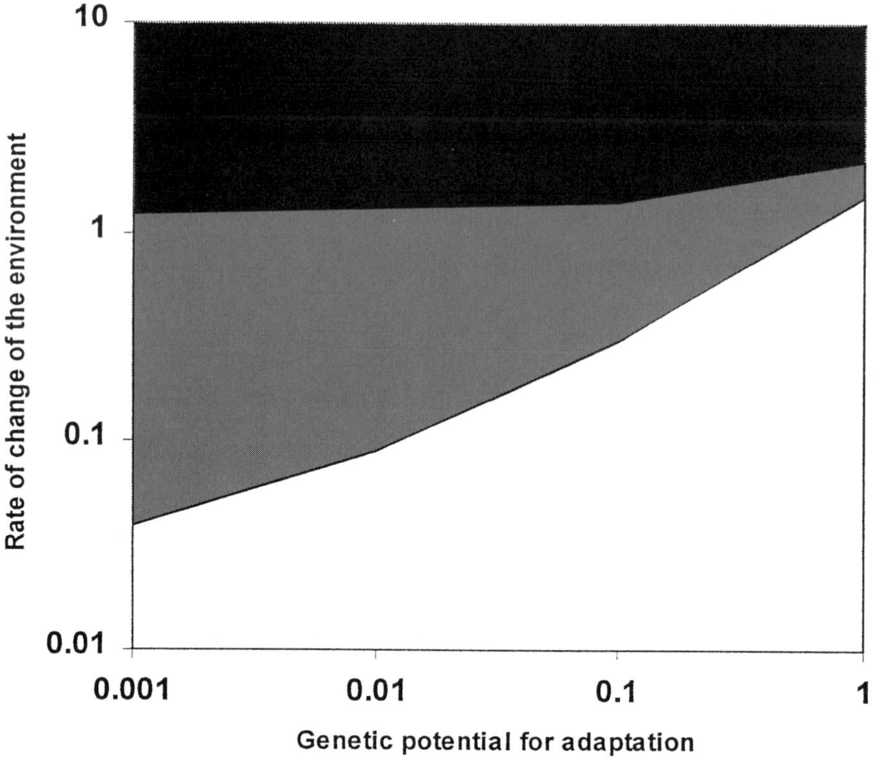

Figure 15.2 Summary of the results of the Kirkpatrick and Barton model. In the white area, the population expands to fill the environment. In the grey area, the population has a limited range. In the black area the population becomes extinct. (Redrawn from Kirkpatrick & Barton 1997, with permission from the University of Chicago Press.)

prevented from expanding by differential gene flow resulting from a density gradient. However, the range of parameter values resulting in stable range margins is small: if the gradient is too steep, or dispersal too great, the population becomes extinct. Unfortunately, Kirkpatrick & Barton (1997) could find very few empirical data suitable to determine whether the parameter values required were realistic (but see Case & Taper (2000) and Barton (2001)).

The Kirkpatrick and Barton model is complex because it involves coupled population dynamics and evolutionary genetics. However, it remains simple in relation to the real world. Although population dynamics are dependent on density and adaptation, relative fitness is density and frequency independent. Phenotypes are strictly genetically determined with no plasticity. The environmental gradient is uniform, genetic variance is constant and stochastic effects on both population dynamics and allele frequencies are ignored. Mating and dispersal are both random. Would adding these further complexities to the model increase or decrease the parameter space in which stable range margins exist?

Density and frequency dependent fitness on an environmental gradient recently has been considered by Doebeli & Dieckmann (2003) but in the context of parapatric speciation rather than range margins. Agrawal (2001) has suggested that phenotypic plasticity can increase the potential of a species to invade new habitat, and has provided some likely examples, but plasticity has not been incorporated into range-margin models. Lenormand (2002) has discussed the evolution of dispersal. Here, we will discuss the evolution of genetic variance and the consequences of non-additive inheritance, non-random mating and non-uniform gradients. The simulation model that we introduce can be modified readily to explore other additions to the basic Kirkpatrick and Barton scenario.

Evolution of genetic variance

Another parallel between tension zones and range margins is instructive here. A population at the centre of a tension zone receives migrants from genetically distinct populations on either side. These migrants bring alleles typical of both parental populations together and so elevate the genetic variance for any quantitative trait that differs between the parental populations (Barton & Gale 1993; Barton 1999). The narrower the cline, relative to dispersal, the more different the migrants from either side will be and so the greater will be the increase in genetic variance. In fact, migrants do not simply bring alleles typical of the parental populations, they bring combinations of alleles at different loci. This causes an increase in linkage disequilibrium in central populations, which is reflected in further elevation of the genetic variance for a single quantitative trait and in associations between traits.

The expected pattern of increased variance in the centre of a hybrid zone and increased covariance between traits has been observed in a contact between the grasshopper species *Chorthippus brunneus* and *C. jacobsi* in northern Spain (Bridle & Butlin 2002). In this example, the variance in number of stridulatory pegs, the variance in song characteristics, and the covariance between these two traits all show greater increases at the zone centre than are expected on the basis of cline widths and estimates of dispersal. This means that additional factors must be involved in elevating linkage disequilibrium such as assortative mating, extinction and recolonization of demes, or habitat preferences.

The basic principle that dispersal between genetically divergent demes increases genetic variance applies equally to populations adapting to an environmental gradient. In the original Kirkpatrick and Barton model, an increase in genetic variance makes range expansion more likely and decreases the parameter space in which stable margins form. Recently, Barton (2001) has shown, as expected from this argument, that allowing evolution of the genetic variance does further restrict the potential for stable range margins. This is true with either of two contrasting genetic models: two alleles per locus at many loci or a continuum of alleles at a few loci. In this revised model, Barton also considered the effect of abrupt steps within the environmental gradient and found that populations were able to spread past them.

It has been argued previously that range margins might be caused by either of two contrasting effects: low genetic variation for relevant traits at the range margin and swamping by gene flow from core populations (Hoffmann & Blows 1994). These effects are contrasting because gene flow helps to maintain genetic variation. Elevated gene flow therefore can aid spread by increasing the response of marginal populations to selection, or prevent it by increasing the migrational load. Barton's (2001) model suggests that the effect on genetic variation tends to outweigh the swamping effect. However, his model does not take into account any loss of variation owing to genetic drift in small marginal populations. This may be a minor factor because the effective sizes of marginal populations are kept large by the influx of migrants from central populations.

Kawecki (2000) has reconciled conflicting conclusions about the impact of dispersal on the initial establishment of beneficial mutations in sink populations by showing that it depends on the magnitude of the effect on fitness. This suggests that details of the genetic architecture of traits required for range expansion are likely to be important.

Species interactions

The Kirkpatrick and Barton model considers adaptation to a constant gradient such as temperature or humidity. Clearly it is possible that the range of a species is limited by interactions with other species rather than with the physical environment alone. It may be reasonable to view some aspects of the biotic environment as a fixed gradient equivalent to an abiotic change. However, when the target species has a significant ecological impact on the species with which it interacts, there is the potential for evolutionary change in the environment as well as in the target species. Symmetrical competitive interactions, for example, cannot be represented by adaptation of one species to a fixed pattern of habitat change.

Case & Taper (2000) extended the Kirkpatrick and Barton model to consider the evolution of two competing species on a habitat gradient. They found a range of parameter values, beyond those for which a single species has a limited range, where both species have stable margins. At equilibrium, the two species show overlapping ranges. In the allopatric ranges, outside the area of overlap, each species follows the habitat gradient closely. However, character displacement in the area of overlap results in neither species' phenotype matching the environmental optimum. In this region, population density is reduced both as a result of departure from the optimum and because of interspecific competition. This enhances the density gradient and thus the migrational load, preventing further expansion. Range overlap is always substantial in relation to dispersal distance, perhaps explaining the rarity of abutting distributions. However, an overlap of 10–20 times the dispersal distance remains narrow in relation to the rest of the species' distributions. The position of the boundary between the two species moves to the centre of a uniform gradient with two exactly equivalent species but is displaced if one species has a higher intrinsic rate of increase, for example. The boundary becomes trapped at the steepest point of a non-uniform gradient.

Under other conditions (less dispersal, shallower gradient or stronger interspecific competition) the two species are both able to fill the environment: there is character displacement but no allopatry or range limitation. Alternatively, with low dispersal and weak interspecific competition, the two species both converge on the environmental optimum and again there is no range limitation.

Overall, this model shows that species interactions can broaden the conditions under which species have limited ranges. Case & Taper (2000) suggested that empirically determined gradients are too shallow to account for range margins in the absence of species interactions but do fall within the range required for boundaries to form in their model. However, Barton (2001) has questioned their interpretation of the data for two reasons:

1 Case & Taper (2000) estimate gradients from clines in morphological traits that may, in fact, be shallower than the clines in environmental optima.

2 If adaptation to a gradient requires evolution of multiple phenotypic traits, the reduction in fitness may be greater than for a single trait and so the gradient required to maintain a range margin may be shallower.

Only further empirical work can resolve this issue.

An individual-based simulation

In order to consider the influence of further complicating factors that might expand the possibilities for range limitation, we have constructed an individual-based simulation. The simulation is based on the model developed by Kawata (2002). It will be described in detail elsewhere.

Briefly, we consider an environment with a maximum extent of 1000×8000 units. There is a habitat gradient along the long axis of this environment that is linear (in the initial runs), with slope G. The organisms occupying this environment have a phenotype z determined by a set of 16 additive biallelic loci that mutate at a rate of 0.0001 per locus per generation. The number of offspring left by a female is determined by her phenotype as follows:

$$W = 2 + r(1 - K/N) - (U_x - z)^2 / 2s$$

where r is the intrinsic rate of increase (set to 1.6 in runs reported here), K is the equilibrium density (set to 7 in runs reported here), N is the number of individuals in a square of side 50 units centred on the focal female, U_x is the phenotypic optimum at the point on the gradient occupied by the female (x), and s measures the rate of decline in fitness for phenotypes that depart from the optimum (set to 2 in runs reported here). This is similar to equation (7) in Kirkpatrick & Barton (1997).

Each female chooses a mate from the males available within her mating area. This is a circle, centred on her position, of size determined by the mating area parameter M. Male fitness is determined by the same function as female fitness. The probability of mating for each male in the area is then directly proportional to his fitness. If no male is available within the mating area, the female leaves no offspring. The standard deviation of mating distances, SM, for given M was determined by saving data on positions of mating partners in representative runs of the simulation.

Offspring of the female disperse to new positions in the habitat with a Gaussian distribution of dispersal distances, mean 0 and standard deviation D, in uniformly distributed random directions. As mating is a form of dispersal by males, the standard deviation of total dispersal is given by $(D^2 + 0.5SM^2)^{0.5}$. This will be called simply 'total dispersal' (Crawford 1984).

Across plants and animals, these two phases of dispersal can differ substantially in their contributions to overall gene flow, and may have different consequences because of their different positions in the life cycle. For example, offspring dispersal by seeds may be much less than mating distance, owing to pollen movement in wind-pollinated plants, but planktonic dispersal of larvae may be much greater than mating distances in sessile barnacles with internal fertilization. If selection for locally adapted genotypes occurs primarily in pre-reproductive stages, then the two phases of dispersal occur before and after selection.

Simulation runs were started with a population of 500 individuals distributed in the central 500 units of the environment. Their phenotypes ranged from $z_{opt} - 2$ to $z_{opt} + 2$, where z_{opt} is the optimum phenotype at the centre of the range. Most simulations were run for 1000 generations, by which time one of three outcomes was generally achieved: extinction, spread to fill the environment, or a stable limited range.

Non-random mating

In this basic form, the simulation closely follows the Kirkpatrick and Barton model. The key difference is the pattern of mating, which is dependent on the size of the mating area and the fitness of males within it. Here, again, a parallel can be drawn with hybrid zones, where mating pattern can potentially have a major influence on the outcome of contact. Jiggins & Mallet (2000) have distinguished between two classes of hybrid zones: bimodal and unimodal zones. In the classic tension zone model, a population at the zone centre is expected to have a unimodal distribution of phenotypes or genotypes, with mean intermediate between the parents, and elevated variance. Parental gene or character combinations are expected to be rare. However, some zones do not conform to this pattern. Central populations may contain parental-like individuals with only a few intermediates. Selection against hybrids is insufficient to explain this pattern, instead it seems to result from strong assortative mating (see examples in Jiggins & Mallet 2000). The persistence of parental genotypes, and their contact at the zone centre, might increase the possibility of continued divergence by reinforcement (Cain *et al.* 1999; Britch *et al.* 2001; but see Turrelli *et al.* 2001).

Deviations from random mating might have similarly profound consequences for range margins. Proulx (1999) found, in a model of the colonization of a distinct island niche from a mainland population, that assortative mating could impede local adaptation because it results in a frequency dependent disadvantage to variant individuals. In the hybrid zone situation, this type of frequency dependence may contribute to the stability of the zone, preventing increased range overlap. On the other hand, a preference for locally adapted males (or greater success of locally adapted males in competition for females) may promote invasion of a marginal island niche (Proulx 1999).

In our simulation, we found an effect of mating area on the ability of the population to spread to fill the environment that is independent of the contribution it makes to total dispersal (Figs 15.3 and 15.4). As expected, the population can spread when total dispersal is low, has a limited range for a small set of parameter values and becomes extinct if total dispersal is too high. On a steeper gradient, this pattern is shifted to lower dispersal distances. However, for a given total dispersal, very low mating areas result in extinction, somewhat larger areas in a stable range boundary and larger areas still in range expansion. Then, when mating area is very large, there is another set of conditions where the range is limited.

Extinction or range limitation at small mating areas is the result of an 'Allee effect': as population density declines at the edge of the range, an increasing proportion of females fails to find a mate (Fig. 15.5) and this reduces population growth rate. Low genetic variation as a result of genetic drift in small marginal populations is another form of Allee effect that is incorporated in our model. Variation is reduced at the margins in our simulations but it is not clear how much this influences the probability of expansion because we have not yet investigated the consequences of varying equilibrium population density. These, and similar, effects of rarity are absent from the Kirkpatrick and Barton model but there is growing empirical evidence for Allee effects (Courchamp *et al.* 1999). Recent work on invasion of new niches (Keitt *et al.* 2001) demonstrates the importance of Allee effects and our results suggest that their impact on range margins should be considered further.

The influence of large mating areas has a more subtle explanation. Females tend to mate with the best-adapted males in their mating area. Male adaptation is a function of the match of his phenotype to his local environment. For a female near the range margin with a large mating area, the best-adapted males available to her are likely to be those in the core of the distribution rather than nearby males at the margin. Thus, the mating pattern tends to increase dispersal of genotypes well adapted to core regions into the margins, increasing the migrational load and preventing spread. There is an interesting contrast here with Proulx's (1999) results: he found that locally adapted male mating advantage increased the chances of invading a new niche but, in his mainland-island model, mating occurred only within habitat types. On the habitat gradient, well-adapted males have high mating success but they are not necessarily adapted to the same habitat as their mating partner.

Where total dispersal is high, a large mating area prevents extinction but does not allow the range to increase. This appears to be because the large mating area results in sexual selection favouring males of intermediate phenotype situated near the centre of the environmental gradient, where they have high fitness. This reduces the migrational load on the central population, allowing it to persist, while simultaneously increasing the load on the marginal populations.

Doebeli & Dieckmann (2003) have incorporated assortative mating into a model of adaptation to a habitat gradient. Their model was not designed to investigate range margins and it differs from models of the Kirkpatrick and Barton type in a.

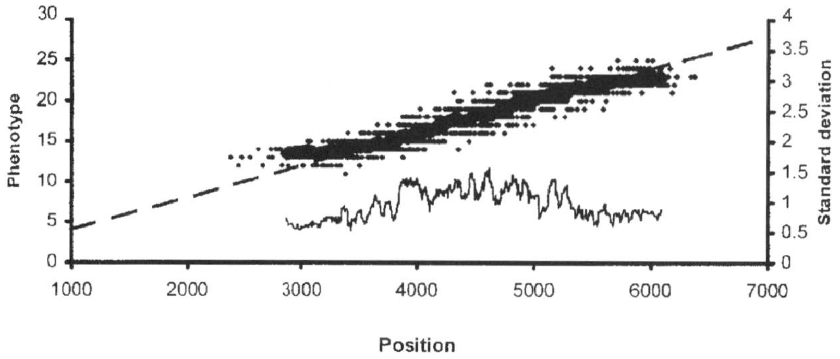

b.

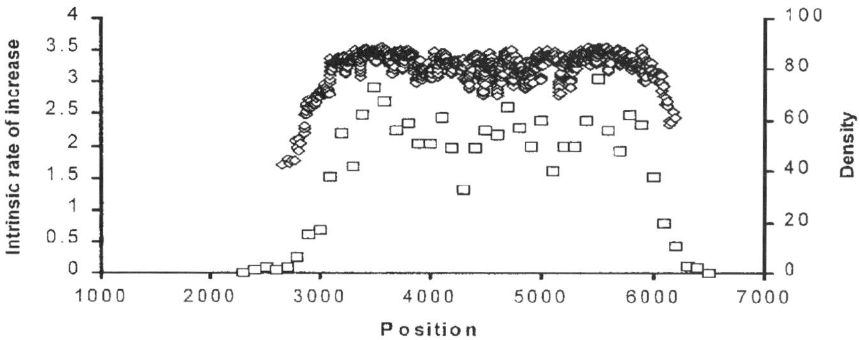

Figure 15.3 (a) An example simulation run with D = 200 and M = 75 showing stable range margins after 1000 generations. Open diamonds, individual phenotypes; filled diamonds, 30-point moving average phenotype; line, 30-point moving standard deviation (as there is no environmental variation, this equals [additive genetic variance]$^{0.5}$). The broken line is the environmental optimum. Note how the marginal populations are poorly adapted and how genetic variance drops off at the margins. (b) Density and fitness in the same run. Density is the number of individuals in a 100 unit range along the environmental gradient (squares). Fitness is expressed as an 11-point moving average of the female fitness in the absence of competitors (diamonds). This is equivalent to the local intrinsic rate of increase. Note how fitness drops off at the ends. It is also interesting that fitness is lower in the centre than near the margins. This is because the centre receives migrants from both sides and so has a higher load.

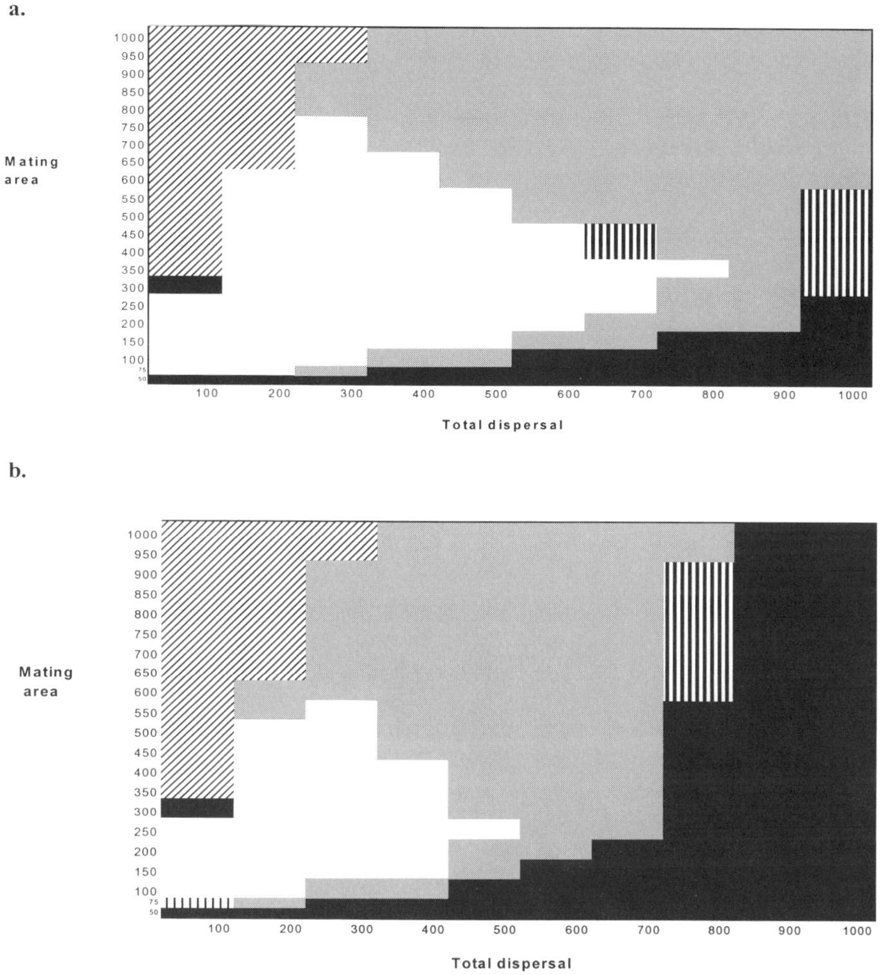

Figure 15.4 A summary of simulation output for two different slopes of a uniform gradient: (a) $G = 0.004$; (b) $G = 0.005$. Black, extinction; grey, stable margin; white, spread to fill environment. Vertical shading indicates areas where two runs of 1000 generations gave different outcomes. Diagonal shaded parameter combinations are not possible.

having frequency and density dependent fitness. However, there are many parallels between the models. Without assortative mating, the population evolves to match the gradient in optimum phenotype but with high variance in the phenotypic trait because intraspecific competition generates disruptive selection. When assortative mating is introduced, as in their non-spatial model of speciation (Dieckmann & Doebeli 1999), the population splits into two overlapping, reproductively isolated species. Neither daughter species is well adapted to the local phenotypic optimum in the area of overlap: in effect there is a pattern of character displacement

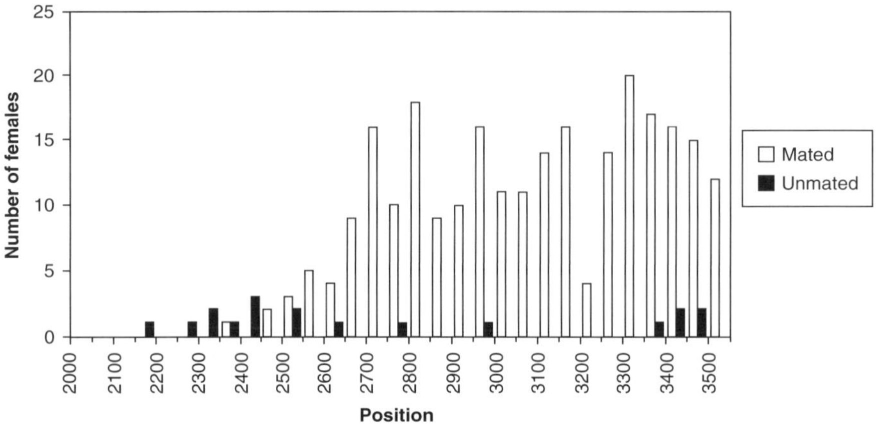

Figure 15.5 The Allee effect at the range margin when mating area is small ($M = 75$). Numbers of females in 50 unit sections of the environmental gradient. Other parameters as in Fig. 15.3.

indistinguishable from the pattern generated by the Case & Taper (2000) model, and the range of each species is limited for the same reasons. So, evolution to fill an environmental gradient can result in limited ranges via a different route: parapatric speciation.

Non-uniform gradients and patchy environments

Patterns of genotypes or phenotypes in hybrid zones often fit very closely to the predictions of the tension zone model. For example, a fitted sigmoid cline explains about 90% of among-population variation in stridulatory peg number in *Chorthippus parallelus* in the Pyrenees (Butlin *et al.* 1991). However, in other hybrid zones the transition from one form to another is much broader and patchier. These zones are called 'mosaic hybrid zones' and the classic examples are two cricket species pairs in North America: *Gryllus firmus* and *G. pennsylvanicus* (Harrison & Rand 1989), and *Allonemobius fasciatus* and *A. firmus* (Britch *et al.* 2001). In both cases, the interacting species have different habitat requirements. More or less pure parental populations occupy their preferred habitat patches, with hybridization limited to patch boundaries (in *Gryllus*) or to intermediate patches occupied by both species (in *Allonemobius*). These are extreme examples. In reality, there is a continuum between simple tension zones and mosaic zones. In some parts of the *Bombina bombina* and *B. variegata* toad hybrid zone, deviations from clinal expectations can be explained by associations with local habitat characteristics (MacCallum *et al.* 1998). However, other processes may also generate patchy patterns of interaction. Bridle *et al.* (2001b) found that, in contrast to the *C. parallelus* example, a sigmoid cline explained only around 70% of among-population variation in stridulatory peg number in the *Chorthippus brunneus* and *C. jacobsi* hybrid zone, but departures from the cline were only very weakly associated with habitat characteristics. They suggested that extinction and recolonization of habitat patches could generate the observed pattern. Hauffe & Searle (1993) provide evidence for the effects of extinction and recolonization on hybrid zones in mice.

Both non-uniform gradients and patchy environments have been considered in the context of range margins. Steep metapopulation margins can be generated on shallow underlying gradients in either the density of available patches or the quality of patches (Lennon *et al.* 1997), and there is evidence that patch network characteristics may prevent populations from expanding to fill suitable habitat (Hill *et al.* 2001). However, both the models and the empirical studies in this area assume that the habitat requirements of the species remain constant. Given the scale on which asymmetrical gene flow influences range margins in the Kirkpatrick and Barton type of model, it is actually more relevant to the expansion of local demes into marginal surrounding habitat than it is to the larger spatial scale of expansion of the metapopulation boundary. As yet, there has been no synthesis of these two views of range margins.

Just as the *Gryllus* and *Allonemobius* mosaic hybrid zones are at the end of a continuum, so metapopulation boundary models are the end of a continuum of non-uniform habitat gradients. Fluctuations in habitat characteristics (such as those in Kawata 2002) have not been considered in range-margin models but they may make range expansion more difficult. Both Case & Taper (2000) and Barton (2001) consider steep sections on an otherwise uniform gradient and find that in some conditions they can trap range margins. Temporal fluctuations in selection pressures also may be important. For example if range margins are determined primarily by rare extreme events such as harsh winters, it may be difficult for selection to be sufficiently sustained to increase overall tolerance.

Our individual-based simulation lends itself to further investigation of these issues. To date, we have considered only the effect of an accelerating rate of change in the habitat. As expected, the initial population expands through the area of relatively constant habitat but reaches a point at which the habitat gradient becomes too steep for further progress (Fig. 15.6). In this simulation run (Fig.15.6), the point on the gradient at which the population stops expanding (where the population growth rate at low density becomes negative, i.e. at the transition from source to sink) has a slope of approximately 0.0056. For the same parameters, a uniform gradient prevents expansion when it exceeds about 0.0068. This discrepancy may be because the shallow slope in the central part of the non-uniform gradient supports a more dense population and so accentuates the asymmetrical gene flow at the margin.

Epistasis

Polygenic models of range margins have concentrated on simple additive genetic effects. However, in hybrid zones, epistatic interactions between alleles at different loci derived from the two parental genotypes are known to be important in reducing hybrid fitness, and in affecting zone structure (Wu & Palopoli 1994; Barton & Shpak 2000). At range margins, it has been argued that genetic interactions among traits

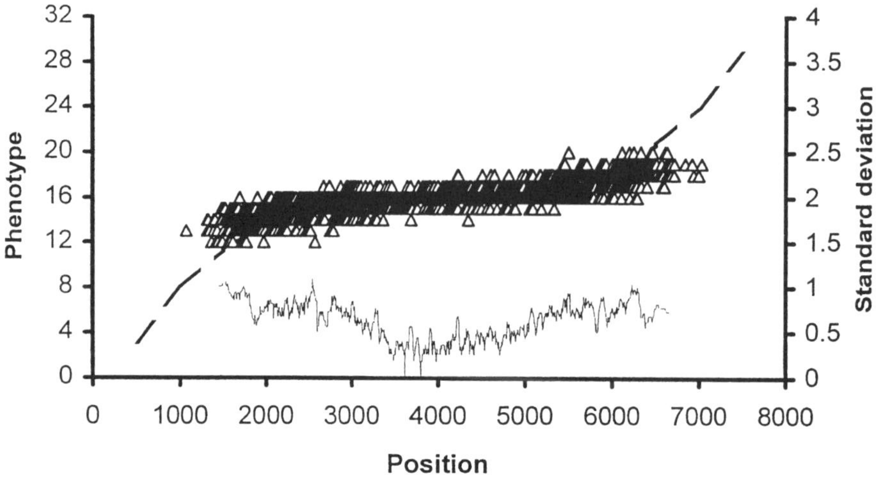

Figure 15.6 An example of a limited range resulting from a steepening gradient. The broken line is the environmental optimum; triangles are individuals at generation 500 with mating area 100 and dispersal 200 (total dispersal 205). The solid line shows the 30-point moving standard deviation as in Fig. 15.3. Note that in this case there is no reduction in variance near the margins.

might be important (Hoffman & Blows 1994). A trade-off between fitness in core and marginal habitats is a key feature of all of the models discussed above. The probability of spread of an allele that increased fitness in the margin without cost in the core populations would be limited only by genetic drift. However, Hoffman & Blows (1994) also suggested that multiple character changes may be needed for range expansion. The implication is that a given allele substitution would not increase fitness unless accompanied by a substitution at another locus: a form of epistasis. It is not clear to us how this differs from their third category of 'genetic trade-offs among fitness traits in marginal conditions'.

In our individual-based simulation, we have incorporated a form of epistasis by making the relationship between genotype and phenotype curvilinear (Fig. 15.7a). This means that close to its starting point, only a small amount of genetic change is required for a given increase in range on the environmental gradient. However, as the range expands, the amount of genetic change needed increases, an effect equivalent to the requirement for simultaneous divergence at multiple traits. As expected, the range expands initially but becomes limited (Fig. 15.7b). The stronger the epistasis, the narrower the range. This effect is equivalent to a steepening gradient, which also requires greater genetic change for unit spatial advance as the population moves away from the starting position. Mutation and recombination can potentially create the necessary genotypes for range expansion so the underlying reason for range limitation is actually unchanged: it is the result of asymmetrical gene flow.

a.

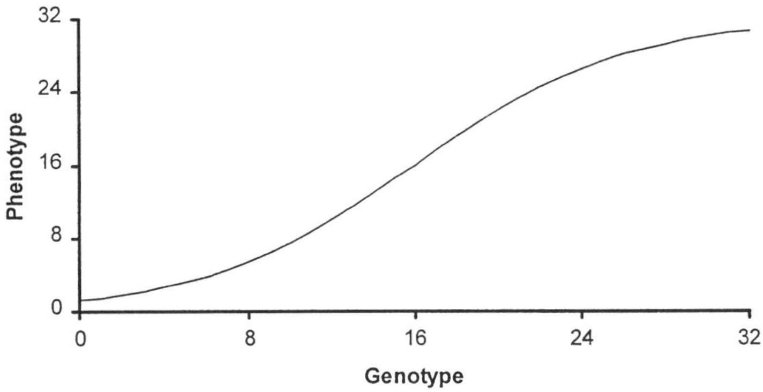

b.

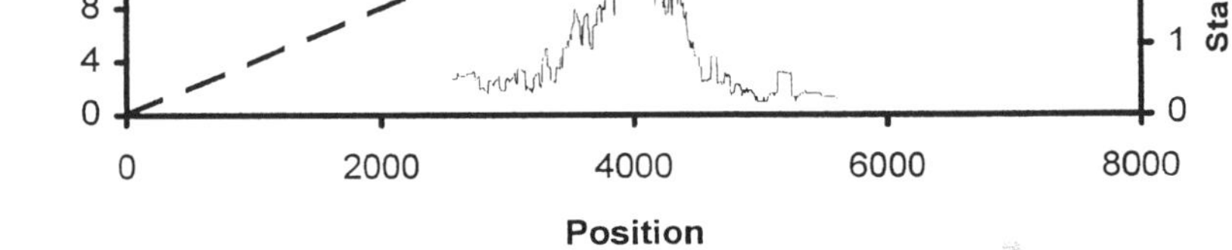

Figure 15.7 An example of a limited range resulting from epistasis. (a) The relationship between phenotype and genotype when strong epistasis is present. (b) Outcome at generation 500 when a stable margin had been reached. Mating area 100, dispersal 200 (total dispersal 205), epistasis as in (a). The solid line is the 30-point moving standard deviation as in Fig. 15.3 except that in this case it cannot be interpreted as a measure of additive genetic variance.

Prospects

We have tried to show that the boundaries to species distributions come in a wide variety of forms. They do not all have the same cause but there are important parallels between them that make comparative studies valuable. In particular, the substantial body of literature on the structure of parapatric boundaries, primarily based on the tension zone model, contains lessons for the development of evolutionary explanations for range margins. Both types of distribution edges are dependent on the effects of asymmetrical gene flow on population density gradients. In both cases, the elevation of genetic variance resulting from gene exchange between divergent populations has important consequences. Departures from simple models such as non-random mating and non-uniform habitat variation may also turn out to be critical for understanding patterns in nature. An important distinction is that spread of genotypes across parapatric boundaries may be impeded by both genotype × genotype interactions and genotype × environment interactions. Furthermore, because genetic differentiation may have accumulated over long periods of time, the negative genetic interactions can cause major reductions in fitness and can involve many loci. At range margins, negative genetic interactions (epistasis) may also influence evolution into new habitats but fewer loci and alleles of smaller effect will typically be involved (because less time has been available for substitutions to accumulate). Just how these kinds of differences in genetic architecture influence the structure of both hybrid zones and range margins is an important topic for the future.

At present, models for range margins are problematic because they suggest that range expansion will occur under many conditions, perhaps too broad a set of parameters to account for the limited distributions we see in nature. A substantial problem is the gap between theory and empirical measurements of key parameters. This gap exists because the parameters are difficult to measure. In particular, environmental gradients have to be measured in the terms experienced by the organisms themselves, in relation to the fitness consequences of imperfect adaptation. It is also critical to know the full set of environmental parameters that change on the gradient and that influence fitness. On top of these demands, it is necessary to have a reliable estimate of dispersal and an understanding of the available genetic variation, and its nature. It is therefore likely to be some time before we can confidently compare theoretical predictions and causal factors with the positions of real margins.

In the meantime, it may be more fruitful to consider whether additional factors can substantially increase the range of conditions that generate stable range margins. There are many candidates, as outlined above. All of them are actually mechanisms that enhance the effects of asymmetrical gene flow, leaving this as the principal underlying cause of limited ranges. This includes the effects of competitors and the Allee Effects, both of which reinforce asymmetrical gene flow by further reducing the sizes of marginal populations.

There is indirect evidence for the importance of gene flow in determining range margins. For example, geographical parthenogenesis can be interpreted as a result of the escape of asexual lineages from range limitation owing to gene flow (Peck *et al.* 1998) and heavy metal tolerant plants evolve selfing to avoid swamping by gene flow from larger surrounding populations on normal soils (Antonovics 1976). In general, it is well known that gene flow between populations in different environments can result in maladaptation (Endler 1977). However, empirical work demonstrating the link between this effect and range margins seems to be lacking. Hoffmann & Blows (1994) reviewed studies comparing fitness in marginal and central environments. These studies typically show that marginal populations are better adapted to marginal conditions and central populations to central conditions. Unfortunately, these studies do not address the prediction from the swamping hypothesis that marginal populations are not as well adapted to their local conditions as are central populations, or test whether this effect can be attributed to gene flow. Although models continue to evolve towards greater sophistication and complexity, empirical work lags behind. The real need is for detailed investigations of range margins using both ecological and genetic approaches at the spatial and temporal scale at which local adaptation and gene flow are likely to interact.

Acknowledgements

We are grateful to NERC and BBSRC for support of our work on hybrid zones and to the Society for Population Ecology and Japanese Society for the Promotion of Science for assisting with our collaboration on this project. We have had very valuable discussions with Chris Thomas, Bill Kunin and Nick Barton. Nick Barton made valuable comments on an earlier draft.

References

Agrawal, A.A. (2001) Phenotypic plasticity in the interactions and evolution of species. *Science* **294**, 321–326.

Antonovics, J. (1976) The nature of limits to natural selection. *Annals of the Missouri Botanical Garden* **63**, 224–247.

Barton, N.H. (1999) Clines in polygenic traits. *Genetical Research* **74**, 223–236.

Barton, N.H. (2001) Adaptation at the edge of a species' range. In: *Integrating Ecology and Evolution in a Spatial Context* (eds J. Silvertown & J. Antonovics), pp. 365–392. Blackwell Science, Oxford.

Barton, N.H. & Gale, K.S. (1993) Genetic analysis of hybrid zones. In: *Hybrid Zones and the Evolutionary Process* (ed. R.G. Harrison), pp. 13–45. Oxford University Press, New York.

Barton, N.H. & Hewitt, G.M. (1985) Analysis of hybrid zones. *Annual Review of Ecology and Systematics* **16**, 113–148.

Barton, N.H. & Hewitt, G.M. (1989) Adaptation, speciation and hybrid zones. *Nature* **341**, 497–503.

Barton, N.H. & Shpak, M. (2000) The effect of epistasis on the structure of hybrid zones. *Genetical Research* **75**, 179–198.

Bridle, J.R. & Butlin, R.K. (2002) Mating signal variation and bimodality in a mosaic hybrid zone between *Chorthippus* grasshopper species. *Evolution* **56**, 1184–1198.

Bridle, J.R., Garn, A.-K., Monk, K.A. & Butlin, R.K. (2001a) Speciation in *Chitaura* grasshoppers (Acrididae: Oxyinae) on the island of Sulawesi: colour patterns, morphology and contact zones. *Biological Journal of the Linnean Society* **72**, 373–390.

Bridle, J.R., Baird, S.J.E. & Butlin, R.K. (2001b) Spatial structure and habitat variation in a grasshopper hybrid zone. *Evolution* **55**, 1832–1843.

Britch, S.C., Cain, M.L. & Howard, D.J. (2001) Spatio-temporal dynamics of the *Allonemobius fasciatus–A. socius* mosaic hybrid zone: a 14-year perspective. *Molecular Ecology* **10**, 627–638.

Brown, J.H. (1984) On the relationship between abundance and distribution of species. *American Naturalist* **124**, 255–279.

Butlin, R.K. (1998) What do hybrid zones in general, and the *Chorthippus parallelus* zone in particular, tell us about speciation? In: *Endless Forms: Species and Speciation* (eds D.J. Howard & S. Berlocher), pp. 367–378. Oxford University Press, New York.

Butlin, R.K., Ritchie, M.G. & Hewitt, G.M. (1991) Comparisons among morphological characters and between localities in the *Chorthippus parallelus* hybrid zone (Orthoptera: Acrididae). *Philosophical Transactions of the Royal Society, London, Series B* **334**, 297–308.

Cain, M.L., Andreasen, V. & Howard, D.J. (1999) Reinforcing selection is effective under a relatively broad set of conditions in a mosaic hybrid zone. *Evolution* **53**, 1343–1353.

Case, T.J. & Taper, M.L. (2000) Interspecific competition, environmental gradients, gene flow, and the coevolution of species' borders. *American Naturalist* **155**, 583–605.

Courchamp, F., Clutton-Brock, T. & Grenfell, B. (1999) Inverse density dependence and the Allee effect. *Trends in Ecology and Evolution* **14**, 405–410.

Crawford, T.J. (1984) The estimation of neighbourhood parameters in plant populations. *Heredity* **52**, 272–283.

Cwynar, L.C. & MacDonald, G.M. (1987) Geographical variation of lodgepole pine in relation to population history. *American Naturalist* **129**, 463–469.

Daday, H. (1954) Gene frequencies in wild populations of *Trifolium repens* (L.). I. Distribution by latitude. *Heredity* **8**, 61–78.

Dieckmann, U. & Doebeli, M. (1999) On the origin of species by sympatric speciation. *Nature* **400**, 354–357.

Doebeli, M. & Dieckmann, U. (2003) Speciation along environmental gradients. *Nature* **421**, 259–264.

Endler, J.A. (1977) *Geographic Variation, Speciation and Clines.* Princeton University Press, Princeton, NJ.

Garcia-Ramos, G. & Kirkpatrick, M. (1997) Genetic models of rapid evolutionary divergence in peripheral populations. *Evolution* **51**, 1–23.

Haldane, J.B.S. (1956) The relation between density regulation and natural selection. *Proceedings of the Royal Society, London, Series B* **145**, 306–308.

Harrison, R.G. (1993) *Hybrid Zones and the Evolutionary Process.* Oxford University Press, New York.

Harrison, R.G. & Rand, D.M. (1989) Mosaic hybrid zones and the nature of species boundaries. In: *Speciation and its Consequences* (eds D. Otte & J.A. Endler), pp. 111–134. Sinauer Associates, Massachusetts.

Hauffe, H.C. & Searle, J.B. (1993) Extreme karyotypic variation in a *Mus musculus–domesticus* hybrid zone—the tobacco mouse story revisited. *Evolution* **47**, 1374–1395.

Hewitt, G.M. (1999) Postglacial recolonisation of European biota. *Biological Journal of the Linnean Society* **68**, 87–112.

Hewitt, G.M. (2001) Speciation, hybrid zones and phylogeography—or seeing genes in space and time. *Molecular Ecology* **10**, 537–549.

Hill, J.K., Collingham, Y.C., Thomas, C.D., *et al.* (2001) Impacts of landscape structure on butterfly range expansion. *Ecology Letters* **4**, 313–321.

Hoffmann, A.A. & Blows, M.W. (1994) Species borders: ecological and evolutionary perspectives. *Trends in Ecology and Evolution* **9**, 223–227.

Holt, R.D. & Gomulkiewicz, R. (1997) How does migration influence local adaptation? A reexamination of a familiar paradigm. *American Naturalist* **149**, 563–572.

Jiggins, C.J. & Mallet, J. (2000) Bimodal hybrid zones and speciation. *Trends in Ecology and Evolution* **15**, 250–255.

Johannesson, B. & Johannesson, K. (1996) Population differences in behaviour and morphology in the snail *Littorina saxatilis*: phenotypic plasticity or genetic differentiation? *Journal of Zoology* **240**, 475–493.

Johannesson, K. (2001) Parallel speciation: a key to sympatric divergence. *Trends in Ecology and Evolution* **16**, 148–153.

Kawata, M. (2002) Invasion of vacant niches and subsequent sympatric speciation. *Proceedings of the Royal Society, London, Series B* **269**, 55–63.

Kawecki, T.J. (2000) Adaptation to marginal habitats: contrasting influence of the dispersal rate on the fate of alleles with small and large effects. *Pro-*

ceedings of the Royal Society, London, Series B **267**, 1315–1320.

Keitt, T.H., Lewis, M.A. & Holt, R.D. (2001) Allee effects, invasion pinning, and species borders. *American Naturalist* **157**, 203–216.

Kirkpatrick, M. & Barton, N.H. (1997) Evolution of a species' range. *American Naturalist* **150**, 1–23.

Lennon, J.J., Turner, J.R.G. & Connell, D. (1997) A metapopulation model of species boundaries. *Oikos* **78**, 486–502.

Lenormand, T. (2002) Gene flow and the limits to natural selection. *Trends in Ecology and Evolution* **17**, 183–189.

MacCallum, C.J., Nurnberger, B., Barton, N.H. & Szymura, J.M. (1998) Habitat preference in the *Bombina* hybrid zone in Croatia. *Evolution* **52**, 227–239.

Nichols, R.A. & Hewitt, G.M. (1986) Population structure and the shape of a chromosomal cline between two races of *Podisma pedestris* (Orthoptera, Acrididae). *Biological Journal of the Linnean Society* **29**, 301–316.

Pease, C.M., Lande, R. & Bull, J.J. (1989) A model of population growth, dispersal and evolution in a changing environment. *Ecology* **70**, 1657–1664.

Peck, J.R., Yearsley, J.M. & Waxman, D. (1998) Explaining the geographic distribution of sexual and asexual populations. *Nature* **391**, 889–892.

Proulx, S.R. (1999) Mating systems and the evolution of niche breadth. *American Naturalist* **154**, 89–98.

Rolán-Alvarez, E., Johannesson, K. & Erlandsson, J. (1997) The maintenance of a cline in the marine snail *Littorina saxatilis*: the role of home site advantage and hybrid fitness. *Evolution* **51**, 1836–1847.

Schluter, D. (2000) Ecological character displacement in adaptive radiation. *American Naturalist* **156**, S4–S16.

Thomas, C.D., Bodsworth, E.J., Wilson, R.J., *et al.* (2001) Ecological and evolutionary processes at expanding range margins. *Nature* **411**, 577–581.

Turelli, M., Barton, N.H. & Coyne, J.A. (2001) Theory and speciation. *Trends in Ecology and Evolution* **16**, 330–343.

Wilding, C.S., Butlin, R.K. & Grahame, J.W. (2001) Differential gene exchange between parapatric morphs of *Littorina saxatilis* detected using AFLP markers. *Journal of Evolutionary Biology* **14**, 611–619.

Wu, C.-I. & Palopoli, M.F. (1994) Genetics of postmating reproductive isolation in animals. *Annual Review of Genetics* **28**, 283–308.

Part 7

Why are there interspecific allometries?

Chapter 16

Intraspecific body size optimization produces interspecific allometries

Jan Kozłowski, Marek Konarzewski and Adam T. Gawelczyk*

Introduction

Peters (1983) argues that interspecific allometric equations are good predictive tools: we cannot study the details of physiology and ecology for each of thousands of species, but we can roughly estimate such data on the basis of the size of individuals alone. This use of interspecific allometries is not controversial, so long as we remember that the 95% confidence limits of the slopes of such equations are many times narrower than the range enclosing 95% of species (Fig. 16.1). This must warn not only those using such estimates, but especially those applying interspecific allometries to evolutionary considerations, and even more those seeking simple and universal explanations of allometries on the basis of structural constraints. Ecological and physiological traits of particular groups of species often deviate substantially from the values predicted from allometries. The metabolic rate of shrews, extremely high for their body size, is a well-known example (Taylor 1998). Such systematic deviations from general allometric relationships indicate that the intraspecific effect of natural selection can overcome the structural constraints postulated for broad taxonomic groups. It is therefore safer to say that the constraints impose allowable ranges for physiological or ecological parameters, not that they determine the slopes and intercepts of interspecific allometries.

Those seeking universal functional explanations for allometries, especially allometries of metabolic rates, usually stick to specific values of their slopes, most often either 0.75 or 0.67. They should be more careful. Recently Lovegrove (2000) analysed the basal metabolic rates of 487 species of mammals. He demonstrated significant differences in allometric slopes of basal metabolic rate (BMR) of mammalian species from six terrestrial zoogeographical zones. He also found significantly different slopes for large and small mammals. Lovegrove distributed his database to other researchers, which allowed us to make new estimates. The general slope for all mammals equals 0.69, significantly different from 0.75 and, thanks to the enormous number of data points, from 0.67 as well. Table 16.1 shows the

* *Correspondence address: kozlo@eko.uj.edu.pl*

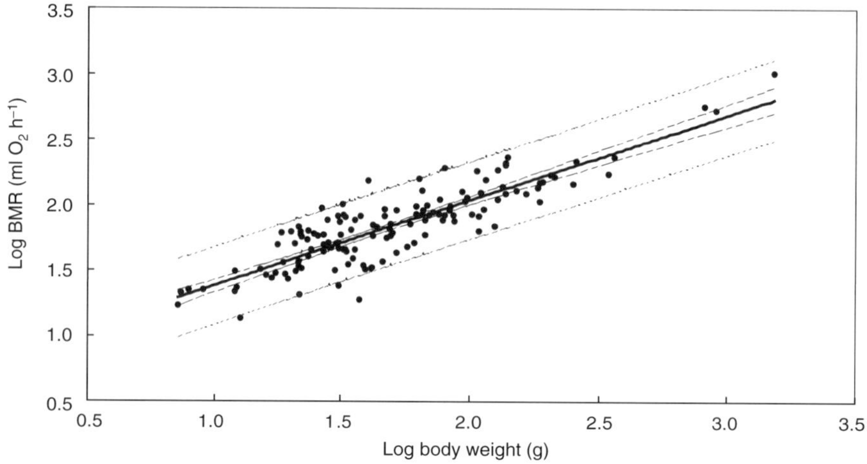

Figure 16.1 Metabolic rate (R) in Muridae as a function of body weight (w). Data according to Lovegrove (2000). Regression line represents the function $\log R = 0.731 + 0.653 \log w$. Dashed curves delimit 95% confidence limits for the slope, and dotted lines delimit the range containing 95% of studied species. Note that the confidence limit for the regression line is much narrower than the range containing 95% of species.

Table 16.1 Slopes for basal metabolic rate (BMR) of mammals classified by systematic position. Data were drawn from Lovegrove's (2000) database. Only groups with 20 or more species are shown.

Order	*n*	Slope	Standard error	Different from 0.75
Rodentia	248	0.668	0.014	Yes
Insectivora	40	0.420	0.050	Yes
Carnivora	24	0.855	0.081	No
Chiroptera	44	0.785	0.039	No
Dasyuromorphia	20	0.741	0.021	No
Diprotodontia	22	0.708	0.018	Yes

slopes calculated from Lovegrove's (2000) data for mammalian orders represented by 20 or more species. The null hypothesis about the uniformity of slopes among orders must be rejected (ANCOVA, $F_{5,386} = 15.16$, $P < 0.0001$). Furthermore, in three out of six orders the slopes significantly differ from 0.75.

Dodds *et al.* (2001) re-examined allometries from classic sources such as Brody (1945), Hemmingsen (1960), Kleiber (1961) and Heusner (1987) in order to reject the null hypotheses of 0.75 or 0.67. For mammals with body weight lower than 10 kg the slope 0.67 must be rejected for Brody's data, and the slope 0.75 for Brody's and Heusner's data. For mammals heavier than 10 kg the slope 0.67 must be rejected for

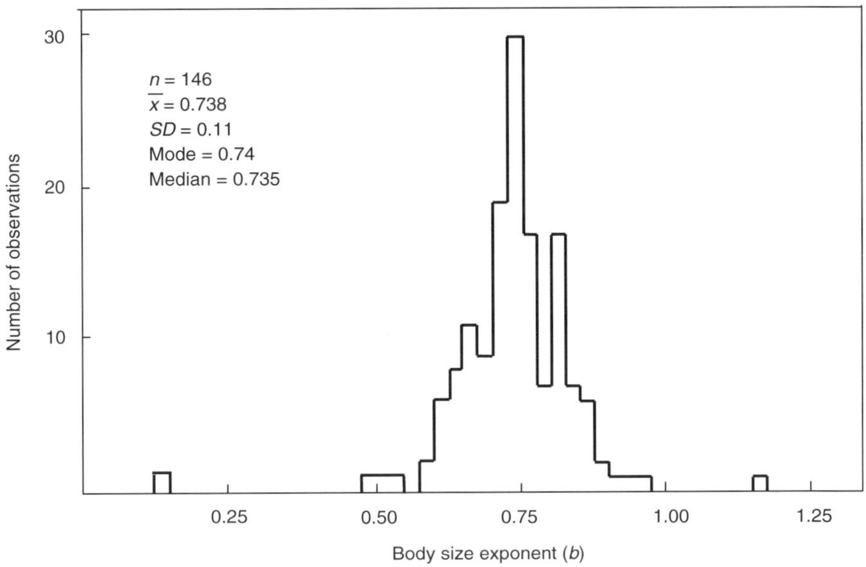

Figure 16.2 Distribution of slopes for metabolic rates of diverse systematic groups. (After Peters 1983, with permission of Cambridge University Press.)

all the sets, and the slope 0.75 for Heusner's data. For the entire mass range the slope 0.67 must be rejected for all the sets, and the slope 0.75 for Brody's and Heusner's data. The slopes that Peters (1983) provided for the metabolic rates of diverse taxa have a broad distribution (Fig. 16.2), which can be explained only partly by measurement errors because many of the exponents he reported differ significantly from 0.75 and/or 0.67.

The conclusion is clear: there is no unique slope for metabolic rate, and any explanation of interspecific allometries must acknowledge this. The explanation that we suggest in this chapter satisfies this condition. It explicitly acknowledges that natural selection works on the intraspecific level, whereas there are good reasons for intraspecific and interspecific allometries to differ substantially. The only fact requiring explanation is that most interspecific slopes have values in a relatively narrow range, which systematically deviates from the range of within-species slopes.

Optimization of body size

In our view, the major obstacle in resolving controversies related to the interpretation of allometric slopes is neglect of the processes affecting optimization of species-specific body size, being the key variable of almost all allometric relationships. The optimal size from the evolutionary point of view is the one that maximizes fitness, that is, lifetime reproductive success, measured here as lifetime reproductive allocation. The two main factors governing optimal size are the production and mortality rates and, more precisely, the dependence of these rates on body size (Kozłowski 1992; Charnov 1993; Perrin & Sibly 1993). Expected future reproduction must be weighted by the probability of surviving to a given age. Adult size is not given to an organism but is developed through growth. Growth is an investment in an increased future reproductive rate. It is obvious, therefore, that the decision whether to reproduce or to keep growing must be mediated by the mortality rate: under heavy mortality the investment in future reproduction through growth is likely to be lost. Larger size requires longer development time if the same growth rate prevails, but organisms that grow faster because of a larger production potential can reach the same size earlier. Optimal investment decisions should take into account both the mortality rate and the production rate. This does not mean that other factors are not important, but that they affect the optimal size by changing these rates. For example, food habits govern the rate of energy acquisition and expenditure and thus the amount of energy available for production (e.g. Arends & McNab 2001), whereas better insulation or a more compact body shape decreases heat loss and thereby increases the amount of resources available for production.

With an optimization model, optimal size is found under a set of simplifying assumptions. Because we are interested not in predicting any precise values but in explaining the relationship between intraspecific and interspecific allometries, an obvious candidate is therefore a model based on allometric relationships. Here we assume that the mortality rate $m(w)$, resource acquisition rate $A(w)$ and metabolic rate $R(w)$ are power functions $y = \alpha w^{\beta}$, whereas the production rate $P(w)$ is the difference $A(w) - R(w)$. Figure 16.3 demonstrates the shape of the size-dependence of the resource acquisition and metabolic rates described by the power functions $A(w) = aw^b$, $R(w) = cw^d$ and $P = A - R$ on linear and logarithmic scales. For the sake of simplicity we also assume that optimization occurs in a constant environment, with continuous reproduction after maturation is reached. Under such assumptions it is optimal to switch from growth to reproduction at such size w, which satisfies the following condition:

$$\frac{\mathrm{d}[P(w)/m(w)]}{\mathrm{d}w} = 1 \qquad (16.1)$$

where $P(w)$ is the size-dependent production rate and $m(w)$ is the size-dependent mortality rate (Perrin & Sibly 1993; Kozłowski 1996a). For size-independent mortality this condition simplifies to

$$\frac{1}{m}\frac{\mathrm{d}P(w)}{\mathrm{d}w} = 1 \qquad (16.2)$$

from which it is clear that the optimal size is lower than the one maximizing the production rate, which must satisfy the condition

$$\frac{\mathrm{d}P(w)}{\mathrm{d}w} = 0 \qquad (16.3)$$

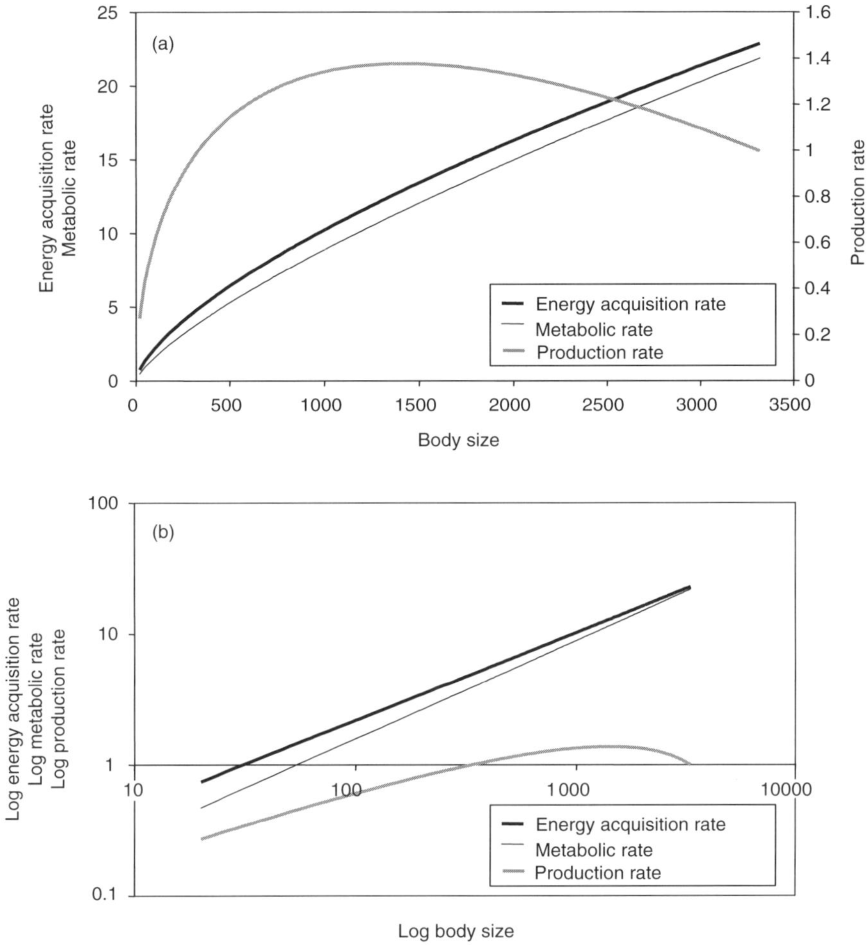

Figure 16.3 The rate of production, being the difference between energy acquisition (assimilation) and metabolic (respiration) rates, on arithmetic (a) and logarithmic (b) scales. Note that the production rate is a humped curve, but on a double logarithmic scale it looks almost linear for a broad range of sizes. Energy acquisition rate equals $0.1w^{0.67}$, and respiration rate $0.05w^{0.75}$, where w is body size in arbitrary units. (After Kozłowski & Gawelczyk 2002.)

Condition (16.2) also shows that optimal size decreases with mortality under a given size-dependence of production $P(w)$ (Fig. 16.4a). If mortality decreases with size, it is possible that optimal size will be located beyond the value that maximizes the production rate (Fig. 16.4b).

Lifetime reproductive output will be maximized at the optimal size w_{opt} satisfying condition (16.1) only if the function $P(w)/m(w)$ is convex (concave downward; increasing with size slower than linearly) in the vicinity of w_{opt} (Kozłowski 1996a).

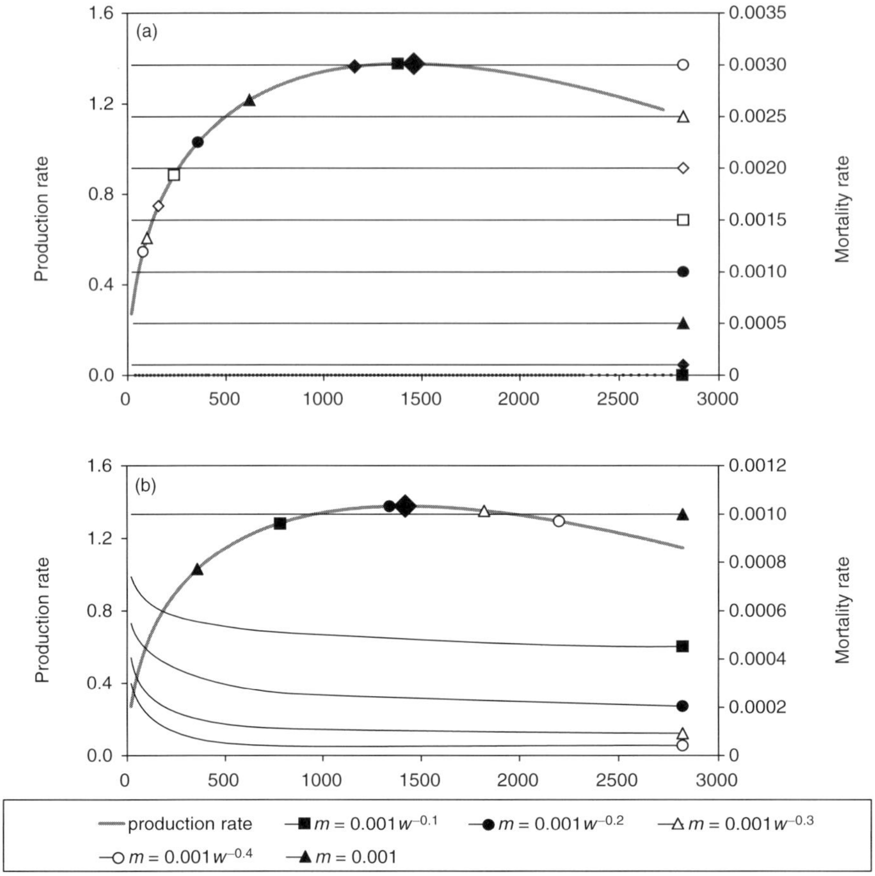

Figure 16.4 Dependence of optimal size (the size maximizing lifetime reproductive output) on size-independent mortality (a) or size-dependent mortality (b). Thick black lines represent production rate, and lines with different marks represent mortality rates. Optimal sizes are shown on the production curves with the same marks as on the line describing the corresponding mortality rate. Observe that for size-independent mortality (a) the optimal sizes are always lower than the size with maximum production rate (indicated by the diamond), even when mortality is almost negligible. For mortality rate decreasing with size (b), optimal sizes also can be located to the right of the hump on the production curve; this is because large size may be a successful way to escape from predation at the cost of lower reproduction. (After Kozłowski & Gawelczyk 2002.)

For size-independent mortality it is enough for the production rate to increase with size slower than linearly, which seems to be the case for all organisms. For a mortality rate decreasing with size the situation is more complex. We can easily imagine the ratio P/m increasing faster than linearly in some ranges of size. In such a case the optimal size cannot be located within the size range, and a species having its adult size there will be under very strong selection pressure to increase in size until the size leaves the range. Punctuated equilibrium with rapid size increase at punctuation may reflect this phenomenon.

Interspecific allometries of ecophysiological parameters

Our logic is generally similar to that of Kozłowski & Weiner's (1997) model, which used the same optimization condition. Kozłowski and Weiner's model assumed normal distributions for all the parameters of the functions, however, meaning they were assumed to be shaped by many unknown additive causes. Here the strategy has been changed. We explicitly assume that the functions embedded in condition (16.1) and representing the size-dependence of the metabolic, resource acquisition, production and mortality rates are shaped by natural selection working within the niche of each species. Because niches differ between species, the parameters of the functions should also vary. However, we assume that the values of the within-species slopes of the allometries describing the dependence of metabolic rate on body size fall between 0.7 and 1, giving a high average, 0.85. Later we provide a biological justification for this assumption. We argue that this restricted variability of the scaling of metabolic rate, together with optimization of body size, produces interspecific allometries with slopes close to the frequently invoked 0.75 or 0.67, but lower than the average slopes for within-species functions.

Figure 16.5 illustrates the functions describing the dependence of the production rate on body size resulting from metabolic rate $R(w)$ with different slopes (Fig. 16.5a) or intercepts (Fig. 16.5b) and a common resource acquisition function $A(w)$. A qualitatively similar picture is obtained if the parameters for $A(w)$ vary and $R(w)$ is common. Each separate production line has, on the log–log axis, characteristics as follows: (i) the dependence is not linear but is very close to linear for a broad range of body sizes; (ii) there is a maximum, its position strongly dependent on the parameters of the $A(w)$ and $R(w)$ functions; (iii) for a given $A(w)$, the quasi-line has a lower slope and peak production is lower and positioned at smaller sizes if $R(w)$ has either a steeper slope (Fig. 16.5a) or a higher intercept (Fig. 16.5b); (iv) for the same $R(w)$ the quasi-line has a higher slope and peak production is higher and positioned at larger sizes if $A(w)$ has either a steeper slope or a higher intercept; (v) the production rate is extremely sensitive to small changes in the $A(w)$ and $R(w)$ parameters in the vicinity of peaks and much less at the quasi-line.

According to condition (16.1), the mortality rate determines optimal size for each production function. The set of optimal sizes for a given mortality level, marked in Fig. 16.5 with squares for high mortality, diamonds for intermediate mortality and triangles for low mortality, forms a quasi-linear function on the log-body-size–log-production-rate plane. The quasi-lines, representing the interspecific allometries of the production rate for taxa having the same mortality rates, are steeper under higher mortality. Note that the lines describing the interspecific dependence of the production rate on body size are almost linear even when they cross the production functions at their non-linear phase (Fig. 16.5). The range of optimal sizes is

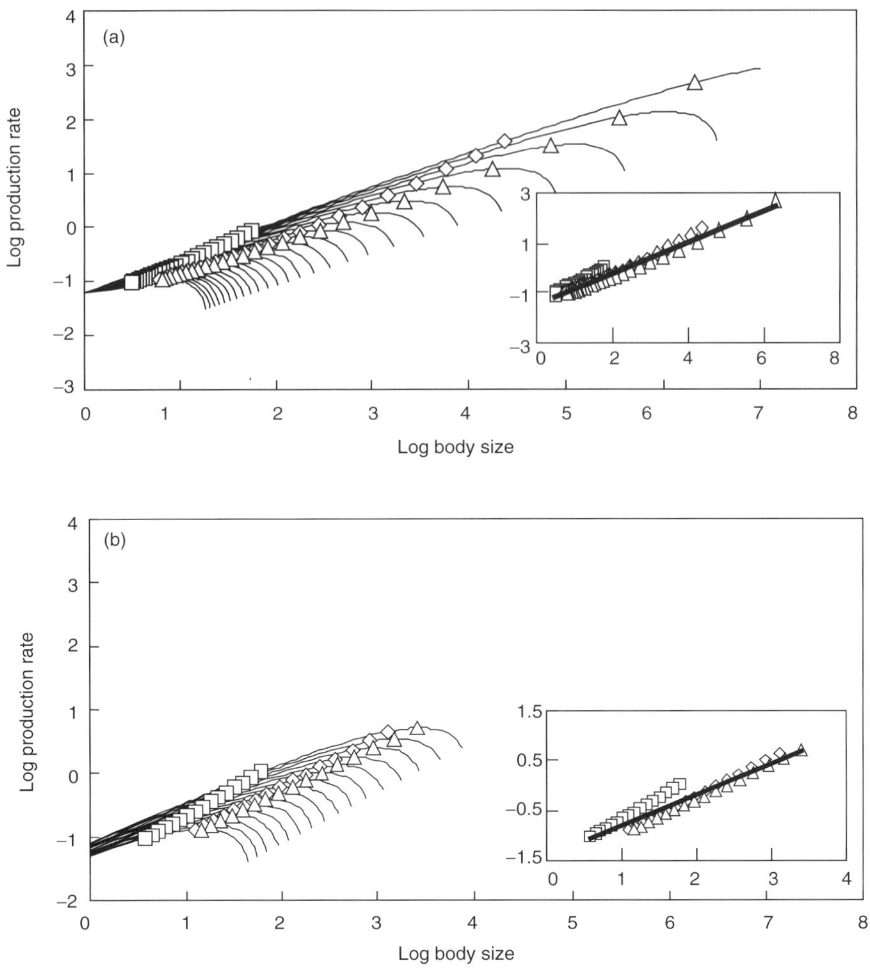

Figure 16.5 Production rate according to the equation $P = A - R = 0.1w^{0.7} - cw^{d}$, where A is resource acquisition rate, R is metabolic rate and w is body size in energy units (lines), and optimal sizes of species under different mortality rates m (squares, $m = 0.01$; diamonds, $m = 0.001$; triangles, $m = 0.0001$). (a) The constant $c = 0.04$ for all lines, and the slope d changes between lines by increments of 0.01, from 0.75 (uppermost line) to 1 (lowest line). (b) The common slope $b = 0.875$ (the average of slopes in (a)), and the constant c changes from 0.02 (uppermost line) to 0.05 (lowest line). Optimal sizes for each production line, calculated according to condition (16.2), form interspecific allometries when all production lines are considered jointly. Interspecific allometries of production rate in (a) have 0.75 slopes for the highest mortality rate, 0.69 for the intermediate mortality rate and 0.63 for the lowest mortality rate, with the common slope equal to 0.61. Interspecific allometries of production rate in (b) have 0.86 slopes for the highest mortality rate, 0.76 for the intermediate mortality rate and 0.71 for the lowest mortality rate, with the common slope equal to 0.61.

much narrower under high than under low mortality, and under lower mortality the system is much more sensitive to changes in the slope of the metabolic rate than to changes in the intercept (Fig. 16.5a versus Fig. 16.5b). Whereas small and moderate sizes can be optimal for different mortality levels, large sizes are optimal only for the less likely combination of high production and low mortality. This can explain right-skewed species size distributions (Kozłowski & Gawelczyk 2002).

The interspecific slope for the production rate equals 0.74 for the highest mortality rate applied in the construction of Fig. 16.5a, 0.69 for intermediate mortality, and 0.63 for the lowest mortality, despite the use of the same set of production functions in the three cases. Now imagine that a higher taxonomic group includes all three groups differing in mortality. The insert in Fig. 16.5a shows the grand regression line for this higher taxon: the slope not only is lower than the average of the separate slopes of particular taxa, but is even lower than the least steep slope. The situation is qualitatively similar in the case illustrated in Fig. 16.5b, although here the slopes for each mortality level are much steeper than in Fig. 16.5a and the common slope is slightly lower.

To study the effect of body size optimization on the interspecific allometry of the metabolic rate, let us assume first that a group of species (taxon) has the same function describing the size-dependence of resource acquisition and mortality. The species differ only in the parameters of the function $R(w)$ describing the size-dependence of the metabolic rate. It is straightforward that conditions (16.1) and (16.2) are satisfied at smaller body sizes for species with steeper slopes because at the same resource acquisition rate the production rate is lower. Figure 16.6a illustrates the connection between the within-species slopes for metabolic costs and the resulting optimal body sizes. An immediate implication of the smaller optimal body sizes of species with steeper slopes is that the line connecting optimal sizes lying at different metabolic rate lines is less steep than the average within-species slopes. We can go a step further and assume that taxa belonging to a higher taxon differ in the function describing the dependence of mortality on body size. Each taxon will be characterized by a separate interspecific allometry for metabolic rates, with a distinct slope (Fig. 16.6a). Interestingly, for exactly the same set of within-species allometries for metabolic rates, taxa differing in mortality will scale interspecifically with different slopes. The grand slope for species belonging to the higher taxon is still lower than the average within-species slope, but higher than the slopes for all particular taxa (insert in Fig. 16.6a).

Figure 16.6b shows how the difference between the intra- and interspecific slopes of the resource acquisition rate originates. The lines represent different slopes, and the marked points represent, similarly to Fig. 16.6a, optimal sizes characteristic for a given mortality level and each slope of the resource acquisition rate. The points form quasi-linear lines representing interspecific allometries, but unlike the metabolic rate their slopes are steeper under low mortality. Also unlike the metabolic rate, the common regression line is steeper than the average within-species lines.

The few examples shown here explain how evolutionary optimization of size translates the within-species dependencies of ecophysiological parameters to inter-

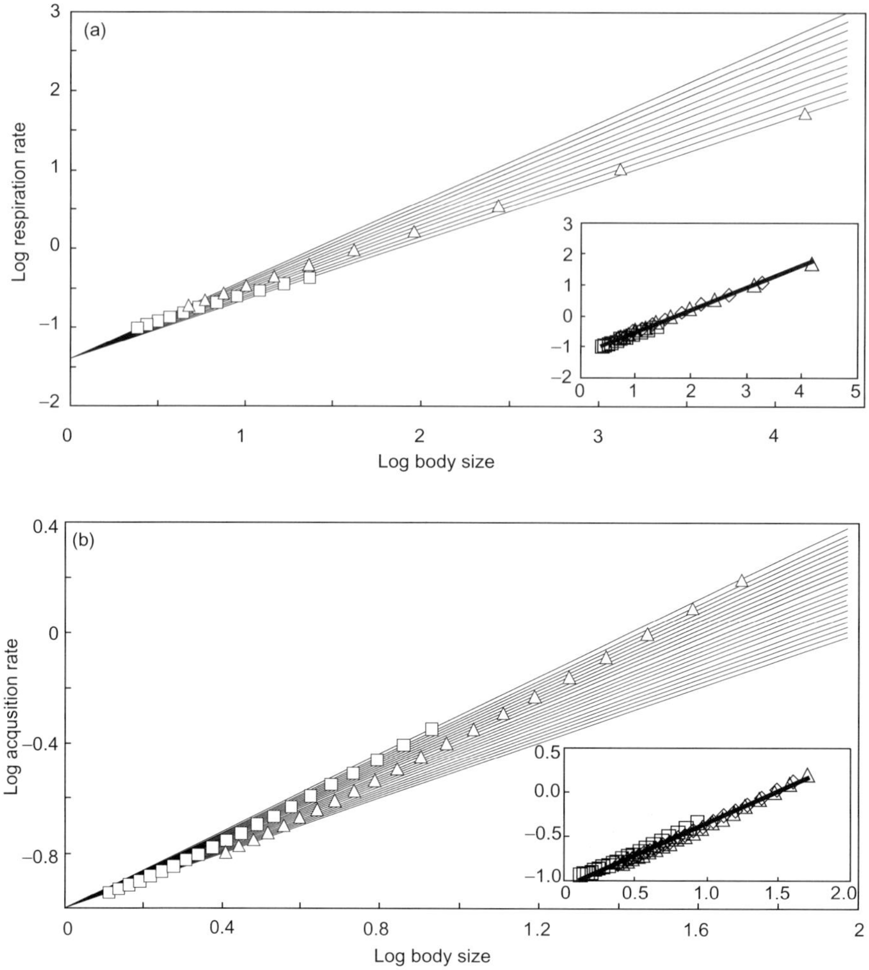

Figure 16.6 Interspecific allometries of ecophysiological parameters resulting from body size optimization in lineages differing in their slopes for the dependence of metabolic rate or resource acquisition rate on body size. (a) Thin black lines depict within-lineage scaling of metabolic rate according to the equation $R(w) = 0.04w^d$, where d changes from 0.75 to 1.0, giving the average 0.875; optimal body sizes were calculated with condition (16.2), assuming that energy acquisition scales as $A(w) = 0.1w^{.67}$ and mortality rate is 0.01 (squares) or 0.0001 (triangles); squares can be well approximated by the regression line $\log R = -1.25 + 0.66 \log w$, and triangles by $\log R = -1.17 + 0.70 \log w$; these slopes, much lower than the average within-lineage slope, represent interspecific allometries for metabolic rate. Insert in (a): common regression line for points marked with squares and triangles in (a), and also optimal sizes for intermediate mortality level 0.001 not shown in (a); grand regression line is $\log R = -1.07 + 0.71 \log w$. (b) Thin black lines depict within-lineage scaling of resource acquisition rate according to the equation $A(w) = 0.01w^b$, where b changes from 0.5 to 0.7, giving the average 0.6; squares and triangles signify same as in (a); optimal body sizes were calculated with condition (16.2) under $R(w) = 0.04w^{0.875}$; squares can be well approximated by the regression line $\log A = -1.04 + 0.73 \log w$, and triangles by $\log A = -1.13 + 0.76 \log w$; these slopes, much higher than the average within-lineage slope, represent interspecific allometries for resource acquisition rate. Insert in (b): common regression line for points marked with squares and triangles in (a), and also optimal sizes for intermediate mortality level 0.001 not shown in (b); grand regression line is $\log R = -1.24 + 0.72 \log w$.

specific allometries. The crux of the matter is that any deviation of the parameters of within-species functions is reflected in optimal size, dependent not only on these parameters but also on the mortality rate. In the case of the production rate, non-linear (on log–log axes) within-species curves are translated to quasi-linear allometric lines. Moreover, the model predicts that within-species allometries of the resource acquisition rate translate to interspecific allometries with steeper slopes. Unfortunately the available data are too few to test this prediction. However, the model also predicts that within-species allometries of metabolic rates should be steeper than interspecific ones, and this is well supported by the values of the slopes Bennett & Harvey (1987) computed for different taxonomic levels in birds. At the level of genus (which is the closest available relationship approximating within-species allometry), avian metabolic rates scale with an average slope of 0.92. The average allometric exponents for families and suborders go down to 0.81 and 0.72, respectively; finally, at the level of class it touches down at 0.68.

Similar comparisons within mammals do not give a consistent picture. To our knowledge, the domestic dog is the only mammal for which enough data exist to resolve the within-species allometry. Its exponent equals 0.88 (Heusner 1991) and does not differ from the allometric slope of 0.85 within Carnivora (Table 16.1). Harvey *et al.* (1991) also failed to demonstrate any consistent relationship between the allometric exponents of metabolic rates and the levels of mammalian taxonomy. Furthermore, the average allometric exponent computed for mammalian families from data of Hayssen & Lacy (1985) equals 0.68, whereas at the level of order and class it is 0.71 and 0.69, respectively. The small sample sizes within mammalian families, and the great heterogeneity of slopes and intercepts (Table 16.1), make these comparisons dubious, though.

When we held some parameters constant and varied the remaining parameters we obtained slopes of resource acquisition rates ranging from 0.65 to 0.88, and for metabolic rates from 0.63 to 0.87. Most likely, however, all the parameters embedded in optimization conditions vary within higher taxa. Interestingly, when all the parameters were free to vary in the parameter space assuring the existence of optimal sizes, the interspecific slopes varied in a much narrower range: from 0.66 to 0.79 for resource acquisition rate, from 0.63 to 0.76 for metabolic rate, and from 0.72 to 0.86 for production rate, with most of the results confined to even more restricted ranges around 0.75. This may explain why slopes close to the oft-quoted value of 0.75 are usually found for higher taxa and the slopes for lower taxa are often much more diversified. Lower taxa are likely to share some parameters.

Allometries of life-history parameters

This chapter is devoted mainly to explaining the nature of allometries of ecophysiological parameters. As Kozłowski & Weiner (1997) and Kozłowski (2000) showed, allometries of life expectancy and of age at maturity, as well as the dependence between these two, are also by-products of body-size optimization. Here we sketch only the reasoning needed to understand the next sections.

Condition (16.1) integrates ecophysiological (production rate) and demographic (mortality rate) properties of species in determining optimal size. Figure 16.4 shows how these two kinds of features co-work in a taxon having the same production curve and mortalities differing between species. Figure 16.5 shows how these two kinds of features co-work in a taxon with production curves differing between species and having the same mortality. Each line joining the optimal sizes resulting from different production curves is calculated for a common mortality rate. It is clear that the ranges of optimal sizes produced by different mortality rates overlap. This means that a high production rate can compensate high mortality. However, larger optimal organisms have, on average, lower mortality. Because mortality is inversely related to life expectancy, a positive relationship between body size and life expectancy is produced on the interspecific level in this way. Harvey & Zammuto (1985) and Purvis & Harvey (1995) showed such a relationship for mammals, Calder (1983) for birds, and Promislow *et al.* (1992) for lizards. Differences in the production rate must cause large residuals around the predicted allometry, as Kozłowski & Weiner (1997) demonstrated with simulations and a semi-analytical method.

Optimal size depends mechanically on the growth rate and the length of the growing period. In other words, a set of species of a given size will contain species maturing early and growing fast, and maturing late and growing slowly, but on average large species mature late. This is the reason for the allometry of age at maturity, again with large residuals (Kozłowski & Weiner 1997; Kozłowski 2000). Sæther (1987) described such a relationship in birds, Purvis & Harvey (1995) in mammals and Promislow *et al.* (1992) in lizards. Very large species require both fast growth and a long growing period. This combination is undoubtedly rare because it requires high production and low mortality.

Positive relationships between body size and life expectancy and between body size and age at maturity, resulting from evolutionary size optimization, imply another positive relationship: between age at maturity and life expectancy (Kozłowski & Weiner 1997; Kozłowski 2000). Such a relationship was described by Harvey & Zammuto (1985) and Promislow & Harvey (1990) for mammals, and by Sæther (1988) for birds. The finding that the same optimal size can result from different combinations of mortality and productivity also implies other relationships after the effect of body size is removed: a negative relationship between age at maturity and production rate, and a negative relationship between life expectancy and production rate (Kozłowski & Weiner 1997; Kozłowski 2000).

Interpreting the functions in the optimization condition

Animals can be classified as determinate or indeterminate growers. The first do not grow or do not grow substantially after reaching maturation. Such a strategy is optimal in an aseasonal environment (Kozłowski 1992; Perrin & Sibly 1993), but also can be adopted in a seasonal environment because of some design constraints (Kozłowski & Wiegert 1987). Indeterminate growers accomplish a substantial part of their growth after maturation. Seasonality is one of the reasons (Kozłowski & Uchmański 1987; Kozłowski & Teriokhin 1999), and not all reasons are known, as some species grow even within one season and others do not. For indeterminate growers the term 'adult size' is meaningless; instead, adult size is usually characterized by size at maturity, asymptotic size and the ratio of the two.

The functions present in optimization condition (16.1) (the dependence of the production and mortality rates on body size) or embedded in the condition (the dependence of the resource acquisition and metabolic rates on body size; the functions shaping the production rate) have been called within-species functions in this chapter. Interpreting them is not so simple, because the functions should apply to adult animals of different sizes. More precisely, they should tell us what metabolic (or resource acquisition, or mortality) rate adult animals of the species are expected to have if selective forces push their size to the right or left. Even more precisely, selection pressure should result from differences in mortality if we are interested in ecophysiological functions. Although this is clear, the problem of how to estimate the functions remains. For indeterminate growers we can assume that adult animals are physiologically similar, which allows within-species allometries for adults to be used as the estimate. Unfortunately, the condition applied in this chapter and the entire approach cannot be used because of the lack of an unequivocal meaning for the term 'adult size'. The problem is tractable, but with a different approach (Kozłowski 1996b). For determinate growers the range of adult sizes is usually very limited. Even worse, we are never sure if this small variability reflects natural variance around the optimum, or if some very small or very large individuals represent a distinct or even defective physiology. It is also risky to apply the functions obtained with selected lines: such lines were not subject to natural selection and their physiology could diverge from natural populations. Even data on animals from different populations must be applied with caution if we are not convinced that the differences in adult sizes were caused by differences in mortality rates. Using ontogenetic functions will also give a biased estimate: physiology is in principle different in growing and non-growing individuals, and changes in body proportions and tissue proportions appear during growth.

Are the within-species functions applied in optimization condition (16.1) only a concept, something that cannot be measured in principle? Even under such a pessimistic view the considerations presented in this chapter have an important merit: they show that the slopes of interspecific allometries cannot be explained by species-level physiological relationships, and they cannot replace difficult-to-measure within-species relationships. This is because interspecific slopes *must* differ from unknown intraspecific slopes. But a sense of something missing remains. We hope to fill this gap, at least for metabolic rate, with considerations on the relationship between cell size and metabolic rate. Instead of within-species relationships, relationships resulting from the within-lineage pattern of adult size increase with respect to cell size and cell number can be used. We arbitrarily restrict the term 'lineage' to a group of related species, alive or extinct, sharing the same pattern of body size change with respect to cell size and cell number, manifested in having a common line describing the dependence of log metabolic rate on log body size. In contrast, 'taxon' here means a group of extant related species that may differ in their slopes for metabolic rate. They may also differ in the intercepts of allometric relationships because of differences in cell membrane permeability, which significantly contributes to the metabolic costs of preserving ionic gradients (Rolfe & Brown 1997).

Cell size and wasteful and frugal strategies

Changes in cell volume indirectly affect body size and general metabolism (Goniakowska 1973; Gregory 2001). In our view this fundamental relationship underlies metabolic rate scaling. Consider the evolutionary changes in the relationships between body size and metabolic rate in hypothetical lineages of species originating from species of similar body size. Body size in a lineage can shrink or expand through changes in cell sizes or cell numbers, as illustrated in Fig. 16.7, and the two mechanisms usually work together. Under body expansion exclusively through cell number, we expect the standard metabolic rate to increase in direct proportion (isometrically, with a slope of 1) to body volume (and body mass as well), because the body is composed of larger numbers of the same units (Fig. 16.7). Under body expansion exclusively through cell size, the cell surface-to-volume ratio decreases. If a large part of metabolic costs are for preserving ionic gradients on cell membranes (Goniakowska-Witalińska 1976; Else & Hulbert 1987; Porter & Brand 1993), the metabolic rate is expected to increase slower than body volume (and body mass as well). For the extreme case when all metabolic costs are proportional to cell surface and when size expansion is exclusively through cell size, the metabolic rate of the whole organism should increase in proportion to body volume raised to the power 0.67. Because only part of the cost is proportional to cell surface and part to cell volume, the power must be substantially higher. If additionally we acknowledge that body size expansion in a lineage is usually realized through simultaneous increases in cell number and cell size, we expect that the standard metabolic rate in a lineage will be proportional to body mass raised to a power closer to 1 than to 0.67, and that the power will differ between lineages because they differ in the participation of cell size and cell number effects in body size changes.

The thin lines in Figs 16.5 and 16.6 should not be interpreted strictly as within-species allometries. Nor do they reflect ontogenetic changes in metabolic rate with body size. Rather they represent the evolutionary trajectories of the relationships between metabolic rate and body size dictated by the changes in cell size and numbers illustrated in Fig. 16.7. Thus, for a particular combination of cell size and numbers they set the slope of the metabolic rate allometry embedded in condition (16.1) or (16.2). As we discussed in the previous section, because the range of body size variation is narrow within most extant species, such allometries for adults cannot be estimated. However, they can be approximated by allometries within narrow taxonomic groups such as genera, roughly conforming to our definition of lineage.

It is not difficult to imagine selective forces toward a high or low metabolic rate. A high metabolic rate is usually correlated with the high rate of food processing,

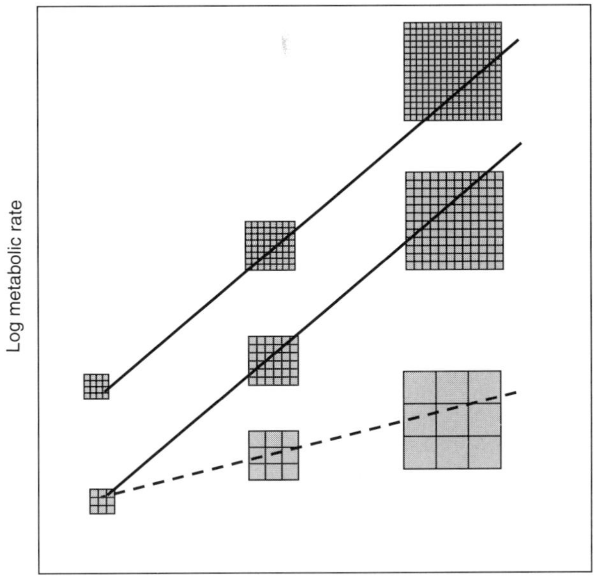

Figure 16.7 The relationship between body size and metabolic rate in hypothetical lineages originating from species of similar body size. Upper solid line represents a lineage in which the increase in body size has been realized entirely through cell number increase. Lower solid line represents a lineage also increasing body size exclusively through cell number increase, but having larger cells and hence a lower metabolic rate. The two solid lines are parallel and have a slope of 1. Dashed line represents a lineage increasing body size entirely through cell size expansion; the slope is much lower because of the lower specific metabolic rate for large cells.

enabling rapid conversion of surplus energy to own tissues or offspring tissues under food excess. A low metabolic rate enables long fasting but precludes assimilation of all available resources at the time they are abundant. This means that a high metabolic rate allows for high offspring production under good food conditions, and a low metabolic rate allows some offspring production even under moderate food availability and ensures better survival under poor food conditions. A low metabolic rate also permits survival in hypoxic conditions. Szarski (1983) calls the first strategy wasteful and the second frugal, with a full continuum in between. He links the wasteful strategy with small cells and frugal strategies with large cells.

The best examples of frugal and wasteful strategies can be found among amphibians and fishes. Anurans have small cells and small genomes, whereas urodels often have very large cells and huge genomes (Roth *et al.* 1994; Jockusch 1997). Some miniature salamanders, especially from the group *Bolitoglossini*, consist of a relatively low number of very large cells, and have some very simplified organs including their brains, a very low metabolic rate and slow development (Roth *et al.* 1994,

1997). Lungfishes, which survive long periods under hypoxic conditions, have very low metabolic rates and the largest genomes of all vertebrates (Gregory 2001), indicating very large cells.

Mammals, and especially birds with their small cells, represent a wasteful strategy, but even within these groups, small genome size, a good measure of cell size (Gregory 2001), correlates with high metabolic rates corrected for body size (Vinogradov 1995, 1997), and in birds also correlates with short development time (Morand & Ricklefs 2001).

The position on the wasteful–frugal strategy axis determines the size-dependence of the production rate $P(w)$, being the difference between the resource acquisition $A(w)$ and metabolic $R(w)$ rates. The standard metabolic rate at a given ambient temperature for an animal of a given size depends mainly on internal properties of the organism such as cell size and cell number. The dependence of the resource acquisition rate on body size may be complex, because it depends not only on the organism's physiology and behaviour but also on food availability. We assume here that this rate can be approximated by a simple function, for example an allometric equation, which is convenient because the slope will measure the rate of change of the acquisition rate with size. We argue that the optimal adult body mass is strongly affected by its position on the wasteful–frugal axis, that is, on the mechanisms determining the metabolic rate and resource acquisition rate, such as cell size. In other words, body mass cannot be treated as an independent variable for the metabolic rate and resource acquisition rate, as these traits co-vary with body size under the pressure of natural selection. The mortality rate is another factor strongly influencing the optimal body size, and its role defines another continuum described in the next section.

Fast and slow life

As seen from Fig. 16.4, the production rate increases with body size to a certain size and then decreases. Even in the increasing phase there are diminishing returns of size increase, which means that the derivative of the production rate with respect to body size is a decreasing function. An animal growing for a longer time to a larger size (but not exceeding the size giving the peak production rate) will be able to produce offspring at a higher speed, or more precisely, to allocate resources to offspring production at a higher speed. This potential increase of the reproductive allocation is not costless: postponing reproduction decreases the chances of survival to reproduction. Thanks to diminishing returns in production, there must exist a size that maximizes the expected lifetime reproductive allocation. This optimal size must be determined not only by the dependence of the production rate on body size, but also by mortality. Higher returns from size increase are necessary to compensate for heavy mortality. This dependence of optimal size on mortality defines the second dimension of life histories: the slow to fast life continuum (Promislow & Harvey 1990).

If we imagine a group of species having exactly the same functions defining the dependence of the metabolic and resource acquisition rates on body size, we still expect variability of optimal body sizes resulting from differences in mortality rates

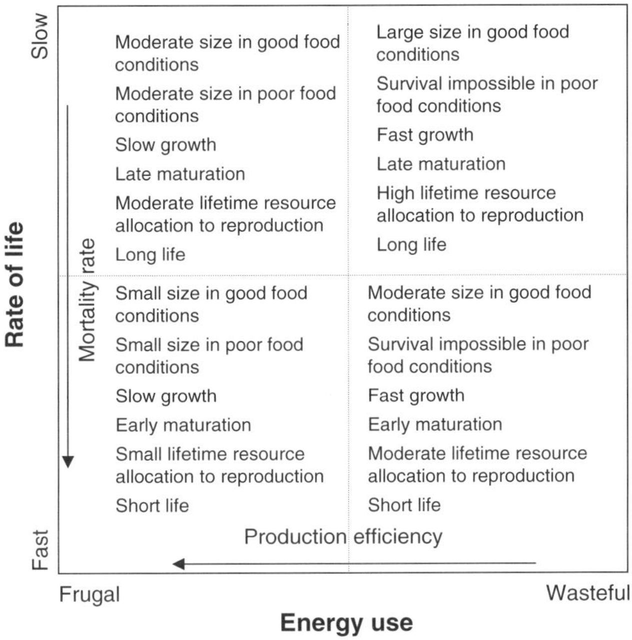

Figure 16.8 Life-history traits expected for frugal–wasteful and fast–slow continua.

(Kozłowski 1992; Charnov 1993). Species under the stress of heavy mortality are expected to be relatively small, to mature young and die early. Their lifetime reproductive allocation expected at maturation is relatively small. Species with a lower mortality rate are expected to grow longer to a relatively greater size, to mature late and die late. Their lifetime reproductive allocation expected at maturation is relatively high. What constitutes relatively small size and relatively low reproductive allocation depends on the position on the wasteful–frugal strategy axis.

Figure 16.8 shows the life-history traits expected at the frugal–fast, frugal–slow, wasteful–fast and wasteful–slow corners of the two-dimensional continuum presented in the previous and present sections. Size-independent mortality was assumed to this point. Under such a mortality pattern, as well as under mortality increasing with size, the optimal size must be located before the size maximizing the production rate because there is no selective pressure pushing a species to a region beyond the size maximizing production. The situation can differ if the mortality rate decreases with size, which is typical for many species in terrestrial ecosystems. Growth beyond the size maximizing the production rate may be optimal as a way of escaping a high mortality rate. There appears a new strategy, not illustrated in Fig. 16.8, which we call relative gigantism. Such a relative giant has a very low mortality rate for its size, late reproduction for its size, and long life. This safety comes at a cost: a relatively low production rate reflected in a relatively low reproductive allocation.

We call such animals 'relative giants' because they do not have to be as huge as whales or mastodons. More often they are small or even very small, but large compared with other taxon members, with sizes located to the left of the size giving the peak production rate.

Conclusions

The parameters for allometric functions are deduced from evolutionary mechanisms common to all cellular organisms, namely, the well-documented dependence of the metabolic rate on cell size. It is predicted, and to some extent experimentally verified, that the within-species relationship between the metabolic rate and body size is steep, and can even reach unity (e.g. Weathers & Siegel 1995; Riisgard 1998; Clarke & Johnston 1999). This is optimization of body size, which makes this dependence flatter on the interspecific level.

The relationship between the resource acquisition rate and body size is undoubtedly more complex, because it depends on both food abundance and physiology. High resource acquisition rates require a large intestinal surface area (Karasov & Diamond 1985), which is expensive to maintain (Konarzewski & Diamond 1995) and therefore finely tuned to prevailing feeding conditions (Secor & Diamond 1998; Starck 1999). It seems possible, therefore, to estimate the within-species allometry of resource acquisition by measuring the absorbing surface of the digestive tract. Because such data are too scarce, we assumed that the slope of the within-species resource acquisition rate varies in an assumed range. We predict that, unlike the metabolic rate slopes, the average within-species slope of the resource acquisition rate should be flatter than its interspecific slope, which is shaped additionally by body size optimization.

In principle, the third determinant of the optimal body size, the relationship between the mortality rate and body size, cannot be studied on the organismal level because it results from the ecological context. Also, this parameter is difficult to measure with sufficient precision in natural populations. In the case of this relationship, unfortunately, studies on the origin of interspecific allometries will have to use parameters drawn at random from an assumed distribution.

Our model does not predict exactly a 0.75 exponent for metabolic rate, but only an exponent in this range, as in the case of real data. We consider this a great advantage of our model. The model excludes too steep a relationship: from considerations on the relationship between cell size and the metabolic rate we do not expect within-species slopes steeper than 1, and optimization makes the interspecific value flatter. On the other hand, a within-species slope lower than 0.75 is not expected even in lineages increasing in size purely through cell-size increase. This means that a very low value is not expected even after flattening through body size optimization.

Size optimization leads to neat interspecific allometries of production rates with relatively high slopes, in sharp contrast to the humped shape of intraspecific production curves. Such interspecific allometries do exist (Peters 1983; Charnov 1993), but it is difficult to imagine the mechanism leading to their origin on the intraspecific level (see Fig. 16.3). However, under size optimization the existence of interspecific allometries for production becomes obvious: species having parameters of resource acquisition and metabolic rate functions that allow for fast conversion of resources to tissues should be larger on average (on average, because mortality also determines optimal size).

The role of the mortality rate in directly shaping optimal sizes and indirectly shaping interspecific allometries and size distributions cannot be overestimated, as seen from Figs 16.4–16.6. The variability of this powerful force together with the diversity of physiological parameters assures niches for animals of an enormous range of sizes. In our view, they also assure that the slopes of between-species allometries, resulting from within-species scaling, are not equal to 0.75 or any other specific, constant value.

Our explanation of the 0.75 power law for metabolic rate is an alternative to the model of West *et al.* (1997, see also Enquist, this volume). The most important difference is in the view of an organism's geometry. West *et al.* assume that the metabolic rate depends on the fractal-like structure of the supplying systems (e.g. the circulatory system in vertebrates). We focus on the cellular structure of organisms, with cell size being one of the determinants of the cell metabolic rate through differences in the area-to-volume ratio. Another important determinant is membrane permeability (Rolfe & Brown 1997), which is likely to differ dramatically between broad systematic groups (e.g. birds versus mammals; endotherms versus ectotherms) but not within narrow systematic groups having similar physiology and biochemistry. Whereas the manner of size change within a lineage with respect to cell size and cell number may be responsible for differences in within-lineage slopes, as shown in this chapter, the differences in membrane permeability may be responsible for the differences in intercepts (in the general level of metabolic rates in a group). If an organism is not extremely small, diffusion processes are not enough to supply resources and remove waste products. Special systems must develop, and it is likely that a fractal-like structure is the most efficient way of filling the space unless the shape of an organism prevents it. In our opinion the size of the finest branch of such a system should depend on a cell metabolism, not vice versa, as West *et al.* (1997) assume.

This brings us to the problem of the proximate physiological factors governing metabolic rate scaling. There is no doubt that a fractal-like supplying system structure operates most of the time at levels of metabolism far exceeding resting metabolic rates. As Darveau *et al.* (2002) rightly pointed out, this suggests that the functioning of a supply system is optimized with respect to rates of metabolism several-fold higher than BMR/SMR (standard metabolic rate), for which the 0.75 power scaling law has been proposed by West *et al.* (1997). It is therefore very unlikely that fractal-like structures can act as a single, rate-limiting step governing the scaling of both resting and active metabolic rates, because in resting conditions they should have unutilized delivery capacity. Furthermore, interspecific scaling of both field metabolic rate (FMR) and peak metabolic rate (PMR) significantly deviates from the 0.75 power (e.g. Ricklefs *et al.* 1996; Bishop 1999). Clearly, this is yet another argument against the major prediction of West *et al.*'s (1997) model – the universality of 0.75 scaling.

In contrast, cellular metabolism contributing to the resting metabolic rate can be interpreted as the obligatory cost of readiness for higher levels of metabolism. Thus, these costs are largely a derivative of active metabolism and should be different for organisms capable of reaching different levels of FMR and/or PMR. Furthermore, cellular metabolism depends on several factors such as cell size and membrane permeability, and therefore it is unlikely that any one of them acts as a single bottleneck governing the resting metabolic rate. For all these physiological proximate reasons, ultimately moulded by evolutionary factors embedded in our model, BMR/SMR scaling varies greatly in nature.

Acknowledgements

We thank M. Jacobs for helping to edit the paper. Figure 16.2 is used with the kind permission of Cambridge University Press, and Figs 16.3 and 16.4 are used with the kind permission of the British Ecological Society.

References

Arends, A. & McNab, B.K. (2001) The comparative energetics of caviomorph rodents. *Comparative Biochemistry and Physiology (A)* **130**, 105–122.

Bennett, P.M. & Harvey, P.H. (1987) Active and resting metabolism in birds: allometry, phylogeny and ecology. *Journal of Zoology, London* **213**, 327–363.

Bishop, C.M. (1999) The maximum oxygen consumption and aerobic scope of birds and mammals: getting to the heart of the matter. *Proceedings of the Royal Society, London, Series B* **266**, 2275–2281.

Brody, S. (1945) *Bioenergetics and Growth.* Reinhold, New York.

Calder, W.A.III. (1983) Body size, mortality, and longevity. *Journal of Theoretical Biology* **102**, 135–144.

Charnov, E.L. (1993) *Life History Invariants. Some Explorations of Symmetry in Evolutionary Ecology.* Oxford University Press, Oxford.

Clarke, A. & Johnston, N.M. (1999) Scaling of metabolic rate with body mass and temperature in teleost fish. *Journal of Animal Ecology* **68**, 893–905.

Darveau, C.-A., Suarez, R.K., Andrews, R.D. & Hochachka, P.W. (2002) Allometric cascade as a unifying principle of body mass effects on metabolism. *Nature* **417**, 166–170.

Dodds, P.S., Rothman, D.H. & Weitz, J.S. (2001) Re-examination of the '3/4-law' of metabolism. *Journal of Theoretical Biology* **209**, 9–27.

Else, P.L. & Hulbert, A.J. (1987) Evolution of mammalian endothermic metabolism: 'leaky' membranes as a source of heat. *American Journal of Physiology* **22**, R1–R7.

Goniakowska, L. (1973) Metabolism, resistance to hypotonic solutions, and ultrastructure of erythrocytes of five amphibian species. *Acta Biologica Cracoviensia, Series Zoologica* **16**, 114–134.

Goniakowska-Witalińska, L. (1976) Effect of ouabin on oxygen consumption and on osmotic swelling of amphibian erythrocytes. *Bulletin de l'Academie Polonaise des Sciences, Classe II* **24**, 221–226.

Gregory, T.R. (2001) Coincidence, coevolution, or causation? DNA content, cell size, and the C-value enigma. *Biological Review* **76**, 65–101.

Harvey, P.H. & Zammuto, R.M. (1985) Patterns of mortality and age at first reproduction in natural populations of mammals. *Nature* **315**, 319–320.

Harvey, P.H., Pagel, M.D. & Rees, J.A. (1991) Mammalian metabolism and life histories. *American Naturalist* **137**, 556–566.

Hayssen, V. & Lacy, R.C. (1985) Basal metabolic rates in mammals: taxonomic differences in the allometry of BMR and body mass. *Comparative Biochemistry and Physiology (A)* **81**, 741–754.

Hemmingsen, A. (1960) Energy metabolism as related to body size and respiratory surfaces, and its evolution. *Reports of the Steno Memorial Hospital and Nordinsk Insulin Laboratorium* **9**, 6–110.

Heusner, A.A. (1987) What does the power function reveal about structure and function in animals of different size? *Annual Review of Physiology* **49**, 121–133.

Heusner, A.A. (1991) Body mass, maintenance and basal metabolism in dogs. *Journal of Nutrition* **121**, S8–S17.

Jockusch, E.L. (1997) An evolutionary correlate of genome size change in plethodontid salamanders. *Proceedings of the Royal Society, London, Series B* **264**, 597–604.

Karasov, W.H. & Diamond, J.M. (1985) Digestive adaptations for fueling the cost of endothermy. *Science* **228**, 202–204.

Kleiber, M. (1961) *The Fire of Life. An Introduction to Animal Energetics.* Wiley, New York.

Konarzewski, M. & Diamond, J. (1995) Evolution of basal metabolic rate and organ masses in laboratory mice. *Evolution* **49**, 1239–1248.

Kozłowski, J. (1992) Optimal allocation of resources to growth and reproduction: implications for age and size at maturity. *Trends in Ecology and Evolution* **7**, 15–19.

Kozłowski, J. (1996a) Optimal initial size and adult size of animals: consequences for macroevolution and community structure. *American Naturalist* **147**, 101–114.

Kozłowski, J. (1996b) Optimal allocation of resources explains interspecific life-history patterns in animals with indeterminate growth. *Proceedings of the Royal Society, London, Series B* **263**, 559–566.

Kozłowski, J. (2000) Does body size optimisation alter the allometries for production and life history traits? In: *Scaling in Biology* (eds J.H. Brown & G.B. West), pp. 237–252. Oxford University Press, Oxford.

Kozłowski, J. & Gawelczyk, A.T. (2002) Why are species body size distributions usually skewed to the right? *Functional Ecology* **16**, 419–432.

Kozłowski, J. & Teriokhin, A.T. (1999) Energy allocation between growth and reproduction: the Pontryagin Maximum Principle Solution for the case of age- and season-dependent mortality. *Evolutionary Ecology Research* **1**, 423–441.

Kozłowski, J. & Uchmański, J. (1987) Optimal individual growth and reproduction in perennial species with indeterminate growth. *Evolutionary Ecology* **1**, 214–230.

Kozłowski, J. & Weiner, J. (1997) Interspecific allometries are byproducts of body size optimisation. *American Naturalist* **149**, 352–380.

Kozłowski, J. & Wiegert, R.G. (1987) Optimal age and size at maturity in annuals and perennials with determinate growth. *Evolutionary Ecology* **1**, 231–244.

Lovegrove, B.G. (2000) The zoogeography of mammalian basal metabolic rate. *American Naturalist* **156**, 201–219.

Morand, S. & Ricklefs, R.E. (2001) Genome size, longevity and development time in birds. *Trends in Genetics* **17**, 567–568.

Perrin, N. & Sibly, R.M. (1993) Dynamic models of energy allocation and investment. *Annual Review of Ecology and Systematics* **24**, 379–410.

Peters, R.H. (1983) *The Ecological Implications of Body Size.* Cambridge University Press, Cambridge.

Porter, R.K. & Brand, M.D. (1993) Body mass dependence of H+ leak in mitochondria and its relevance to metabolic rate. *Nature* **362**, 628–630.

Promislow, D.E.L. & Harvey, P.H. (1990) Living fast and dying young: a comparative analysis of life-history variation among mammals. *Journal of Zoology, London* **220**, 417–437.

Promislow, D., Clobert, J. & Barbault, R. (1992) Life history allometry in mammals and squamate reptiles – taxon-level effects. *Oikos* **65**, 285–294.

Purvis, A. & Harvey, P.H. (1995) Mammal life-history evolution: a comparative test of Charnov's model. *Journal of Zoology, London* **237**, 259–283.

Ricklefs, R.E., Konarzewski, M. & Daan, S. (1996) The relationship between basal metabolic rate and daily energy expenditure in birds and mammals. *American Naturalist* **147**, 1047–1071.

Riisgard, H.U. (1998) No foundation of a '3/4 power scaling law' for respiration in biology. *Ecology Letters* **1**, 71–73.

Rolfe, D.F.S. & Brown, G.C. (1997) Cellular energy utilization and molecular origin of standard

metabolic rate in mammals. *Physiological Reviews* **77**, 731–758.
Roth, G., Blanke, J. & Wake, D.B. (1994) Cell size predicts morphological complexity in the brains of frogs and salamanders. *Proceedings of the National Academy of Sciences, USA* **91**, 4796–4800.
Roth, G., Nishikava, K.C. & Wake, D.B. (1997) Genome size, secondary simplification, and the evolution of the brain in salamanders. *Brain, Behaviour and Evolution* **50**, 50–59.
Sæther, B.-E. (1987) The influence of body weight on the covariation between reproductive traits in European birds. *Oikos* **48**, 79–88.
Sæther, B.-E. (1988) Pattern of covariation between life-history traits of European birds. *Nature* **331**, 616–617.
Secor, S.M. & Diamond, J. (1998) A vertebrate model of extreme physiological regulation. *Nature* **395**, 659–661.
Starck, J.M. (1999) Phenotypic flexibility of the avian gizzard: rapid, reversible and repeated changes of organ size in response to changes in dietary fibre content. *Journal of Experimental Biology* **22**, 3171–3179.
Szarski, H. (1983) Cell size and the concept of wasteful and frugal evolutionary strategies. *Journal of Theoretical Biology* **105**, 201–209.
Taylor, J.R.E. (1998) Evolution of energetic strategies in shrews. In: *The Evolution of Shrews* (eds J.M. Wójcik & M. Wolsan), pp. 309–346. Mammal Research Institute, PAS, Bialowieża.
Vinogradov, A.E. (1995) Nucleotypic effect in homeotherms: body-mass-corrected basal metabolic rate of mammals is related to genome size. *Evolution* **49**, 1249–1259.
Vinogradov, A.E. (1997) Nucleotypic effect in homeotherms: body-mass independent resting metabolic rate of passerine birds is related to genome size. *Evolution* **51**, 220–225.
Weathers, W.W. & Siegel, R.B. (1995) Body size establishes the scaling of avian postnatal metabolic rate: an interspecific analysis using phylogenetically independent contrasts. *Ibis* **137**, 532–542.
West, G.B., Brown, J.H. & Enquist, B.J. (1997) A general model for the origin of allometric scaling laws in biology. *Nature* **276**, 122–126.

Chapter 17
Scaling the macroecological and evolutionary implications of size and metabolism within and across plant taxa

*Brian J. Enquist**

Introduction

A central premise of macroecology is that the study of general statistical patterns within and across ecological systems may provide clues to the operation of 'equally general law-like processes' (Brown 1999, p. 3). Over the past decade there has been a surge in interest in documenting numerous macroecological patterns (see Brown 1995; Rosenzweig 1995; Maurer 1999; Gaston & Blackburn 2000; Hubbell 2001). Although there has been much focus on documenting pattern, macroecology has generally lacked a unified focus on specific mechanisms generating observed pattern and hence lacks a common theoretical framework (Blackburn & Gaston 1997; Brown 1999; Gaston & Blackburn 1999; but see Hubbell 2001; Hubbell & Lake, this volume).

There are two core areas that have long offered the potential for a general mechanistic framework for many macroecological patterns. The first relates to how changes in organismal size (allometry) influences functional, life-history and ecological differences between organisms (Calder 1984; Charnov 1993; Brown 1995). For example, how attributes of organisms are related to their size can be described by the allometric equation

$$Y = Y_0 M^b \qquad (17.1)$$

where Y is the trait of interest, Y_0 is a normalization constant, which may or may not differ between individuals or taxa, M is body mass and b is an exponent that tends to take on unique values (i.e. 3/4, 1/4, 3/8). Across diverse taxa organismal traits such as metabolism and rates of physiological processing, time until reproduction and foraging area scale with characteristic values of Y_0 and b (Peters 1983). A second and related approach has investigated how rates of energy input, metabolic processing and production determine variation in many large-scale patterns such as population abundance, species diversity, trophic level diversity and rates of community turnover and ecosystem production (see Odum 1969; Rosenzweig & Abramsky

* *Correspondence address: benquist@u.arizona.edu*

1993; Post *et al.* 2000, and references therein). For example, variation in species richness, abundance and trophic levels have been thought to be the result of variation in interspecific rates of biological production. Together, these two approaches share a common focus on the role of rates of biomass production—or energetics—in influencing many biological attributes.

Over the past decade several authors have expressed uncertainty over the general importance of energetics in explaining pattern in biology. For example, although there is broad recognition of the multitude of physiological, life history and ecological correlates associated with organismal size some have expressed caution over the ability to infer mechanisms from allometric relationships and how such organismal traits then 'scale-up' to ecological and evolutionary patterns (see Peters 1983; Blackburn *et al.* 1993; Blackburn & Gaston 1997; Kozłowski & Weiner 1997). In addition, others have shown that it is unclear how rates of energy processing and productivity interact to produce large-scale ecological patterns (Rosenzweig & Abramsky 1993; Post *et al.* 2000; Gross *et al.* 2000). Thus, the status of the role of energetics in many aspects of macroecology is uncertain.

This chapter shows how metabolic scaling theory (see also Brown *et al.*, this volume) can explain many emergent patterns and processes in ecology and evolution. I focus on how application of metabolic scaling theory across plant taxa quantitatively predicts how variation in rates of energy processing and allocation has influenced the evolution of plant form, function, ecology and evolution. In particular I contrast prominent physiological and ecological scaling relationships within and across two major plant clades—gymnosperms and angiosperms. The results strongly suggest that the scaling of metabolism has uniformly constrained the evolution and ecology within and across diverse plant groups. The strength of this approach is that it reduces much of the complexity of organisms and ecosystems to simple, but universally applicable, physical and chemical principles. In doing so, this approach reveals common, fundamental axes of diversification towards which organisms have converged by natural selection, and which then govern the tradeoffs around which immense biotic diversity has evolved. Highlighting scaling phenomena has been a rich source of insight into many physical laws (e.g. Bridgman 1963; McMahon & Bonner 1983). The scaling approach advocated here enables ecologists to view the emergent and law-like behaviour of biological systems.

A general model for allometric scaling

Variation in organismal size or mass, M, is probably the most central feature influencing the scaling of biological rates and times (Brown *et al.* 2000). Recently, West, Brown, and Enquist presented a general theoretical framework for the origin of such allometric scaling laws in biology (West *et al.* 1997, 1999a,b, 2001; Brown *et al.* 2000; Enquist *et al.* 2000; see also Gillooly *et al.* 2001; Brown *et al.* this volume), hereafter referred to as the WBE model. The model posits that evolution by natural selection has acted to maximize organismal fitness by maximizing metabolic capacity—the rate at which material resources are taken up from the environment and allocated to some combination of survival and reproduction. This is equivalent to maximizing the scaling of whole-organism metabolic rate, B. The WBE model shows that it then follows that B is limited by the geometry and scaling behaviour of the surface areas, where resources are exchanged with the external or internal environment, and how resources are then distributed within the body (West *et al.* 1999a). Examples of such surface areas include capillary surface area, the leaf surface area, and the total area of mitochondrial inner membranes within cells.

Within the WBE model, body size constrains the scaling of cellular metabolic rate by imposing powerful geometric constraints on biological exchange surfaces and on the distribution networks that transport resources from surface areas to the rest of the body. Thus, the scaling of physiological rates and times (cellular metabolism, rates of physiological processing) simply matches the ability of exchange surfaces to obtain resources from the environment and distribute resources to metabolizing cells. As a result of these general principles organisms tend to obey a common set of 'quarter-power' scaling relationships with body mass or volume (see West *et al.* 1997, 1999a,b, 2000). So, whole-organism metabolic rate approximately scales as body mass as $M^{3/4}$, and mass-specific metabolic rate and most other biological rates and times approximately scale as $M^{-1/4}$.

Applications of the general model: vascular plants

Applying the general principles of the WBE model to the specifics of the plant vascular system predicts numerous attributes of form and function (West *et al.* 1997, 1999a,b; Enquist *et al.* 2000). The application of the WBE model to plants assumes that (i) photosynthetic surface areas are maximized; but yet (ii) the internal transport distances or the total energy required to transport resources through the body is minimized; (iii) the fundamental 'units' involved in resource transport (leaves, terminal xylem size, etc.) are invariant with body size. It is also important to note that any potential deviation from these principles of organismal form and function will lead to calculable deviation in allometric relationships (Enquist *et al.* 2000; Enquist 2002).

Given these assumptions, the WBE model predicts over 20 allometric scaling relationships for anatomical and physiological attributes of vascular plants (either as a function of plant mass, M, or stem diameter, D). These allometric relationships fundamentally govern the scaling of numerous anatomical and physiological attributes of form and function. If these assumptions hold within a given plant during ontogenetic growth or even between plants of differing body size, then intra- and interspecific plant allometric relationships should be characterized by identical exponents. In particular the total number of leaves or leaf mass (M_L) is predicted to scale as $M^{3/4}$. As leaf physiology is approximately independent of plant size (assumption (iii)) then whole-plant rates of resource use should scale as $M^{3/4}$. Further, in order to minimize hydrodynamic resistance, the individual xylem conduits that transport fluid from the roots to the leaves (tracheids and vessels) must taper in cross-sectional area with increasing distance from the leaf. Specifically, xylem conduit area must scale as D raised to an exponent, which is greater than or equal to 1/6.

Such tapering ensures that the total hydrodynamic cost (resistance) of transport is independent of plant size (West *et al.* 1999b).

Broadening the empirical support for the general plant model

Evidence for optimal xylem conduits

If the central assumptions of the WBE model are upheld then many attributes of plant anatomy and physiology should be described by characteristic allometric exponents. In particular, because of its central role in resource transport across divergent plant clades, how attributes of xylem anatomy and physiology scale with plant size provides a critical test of a central prediction of the WBE model. However, angiosperms and gymnosperms primarily rely on two different xylem elements (vessels and tracheids respectively) to transport fluid from the soil to leaves. Enquist (2003) reviews a growing literature on scaling properties of plant xylem and shows that both intra- and interspecific variation in xylem element dimensions, from both gymnosperms and angiosperms, scale in accord with exponents predicted by the WBE model. Such similarities in scaling exponents across clades suggest evolutionary convergence. Further, intraspecific data presented in West *et al.* (1999a) for *Acer saccharum*, and recent analysis by Becker *et al.* (2000), also indicate that vascular element diameters within angiosperms also scale within the predicted bounds of the model. Together, these results support the prediction that total hydrodynamic resistance is invariant with plant size for both angiosperms and gymnosperms.

Scaling whole-plant metabolism: stem sap flux

The WBE model predicts that whole-plant rates of resource use and metabolism should scale as the 3/4 power of plant mass. Interestingly, the allometric scaling of whole-plant rates of resource use has, until recently, not been quantified (but see Hemmingsen (1950) for an early attempt). The movement of fluid through the plant stem per unit time offers one potential means of quantification. Because of the transpiration/photosynthesis compromise, transpirational demands, as indexed by fluid flow through the stem, must match rates of whole-plant photosynthesis ([CO_2] assimilated/per unit time), P. Thus, Q_0, the stem fluid transport rate,

$$Q_0 = c_Q P \propto M^{3/4} \tag{17.2}$$

where c_Q is a mathematical constant, which reflects a multitude of biotic and abiotic influences. The value of c_Q will likely be a function of the local environment and resource availability and may or may not vary across taxa. Here c_Q reflects potential differences in water use efficiency. If c_Q does not scale with plant size then $Q_0 \propto M^{3/4}$ or, converting mass as a function of basal stem diameter, $Q_0 \propto D^2$. Enquist *et al.* (1998) tested this prediction by using maximum values of sap flux as a function of plant size (as basal stem diameter, D) from small herbaceous individuals to large trees and showed that across species sap-flux scales allometrically with an exponent indistinguishable from the predicted value of 2 ($Q_0 \propto D^{1.87}$, 95% CI (confidence interval) for the exponent using Model II reduced major axis (RMA) regression (α_{RMA}), 1.736 – 2.01). Although differences in water use efficiency must be taken into account, rates of transpiration or xylem transport are appropriate, but generally overlooked, indices of plant metabolism (Enquist *et al.* 1998).

Scaling whole-plant metabolism: rates of biomass production

If the annual rate of biomass production, G, is proportional to whole-plant metabolic rate, B, which is also proportional to whole-plant photosynthetic rate P. Values of P will scale isometrically ($b \approx 1.0$) to the total leaf mass, M_L and the total surface area of photosynthetic surfaces, A_P. Therefore, from equation (17.2), G should scale as the 3/4 power of total plant mass, M. Thus, $B \propto G \approx c_P P$, where $P = c_L \beta M^{3/4}$. Therefore,

$$G = c_P c_L \beta M^{3/4} \tag{17.3}$$

where, as will be discussed below, the values of c and β reflect potentially important differences among taxa. Specifically, β indexes the proportion of total plant biomass allocated to photosynthetic mass (see below), c_L reflects the photosynthetic rate per unit mass of leaf area (which will reflect the specific leaf area (SLA); see Lambers *et al.* 1998; Westoby *et al.* 2002) and c_P quantifies the efficiency of conversion of photosynthetic products into new growth.

Niklas & Enquist (2001) compared allometric rates of biomass production across several major plant clades and grades (unicellular algae, a few pteridophytes and numerous gymnosperms and angiosperms). These major plant taxa originated at different times during evolution and include species with significantly differing body plans. Nevertheless, the basic biochemistry of photosynthesis and the structure of chloroplasts across each of these diverse groups has remained essentially unchanged (see Niklas 1997). Further, if as shown in equation (17.3) values of c_L and c_P are approximately constant across these same groups, and potential environmental influence does not systematically vary with size, then allometric rates of biomass production should scale according to a unified allometry. Niklas & Enquist (2001) find robust support for this prediction. Specifically, among all metaphyte species the slope of the reduced major axis exponent α_{RMA} did not differ from the predicted value of 0.75. Further, the allometric intercept did not differ between the major plant groups. Thus, when the data for G and M are pooled and regressed for all plant species, a single allometric scaling formula with a 3/4 exponent was found to span the 20 orders of magnitude of body size represented in the data set (Fig. 17.1). The findings of Niklas & Enquist (2001) are in accord with the qualitative observation that size-specific growth rates are generally highest in annuals and small herbs and lowest in large trees (Grime & Hunt 1975; Tilman 1988) as per mass production (G/M) decreases with increasing size as $G/M = (M^{3/4}/M) \propto M^{-1/4}$.

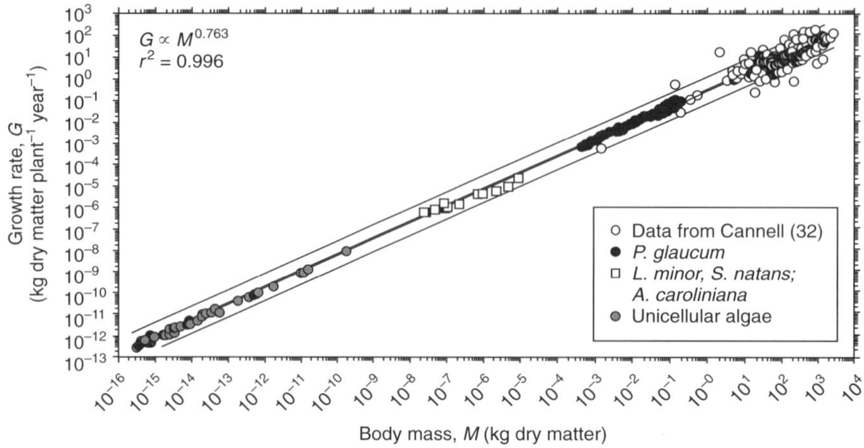

Figure 17.1 Annualized biomass production ('growth') rates G of unicellular and multicellular plants plotted against body mass M (from Niklas & Enquist 2001). Solid lines are reduced major axis (Model II) regression curves (hair lines indicate 95% confidence intervals); scaling exponents (upper left) based on $\log_{10}$-transformed data.

Extending allometric theory: allocation and partitioning, or how do plant taxa differ from one another?

A variety of plant ecological strategies have been recognized by plant ecologists (see reviews in Westoby 1998; Westoby *et al.* 2002). Most aspects of plant life-history variation appear to be described by a few important traits (seed size and number, growth rate, time until reproduction, competitive ability, tolerance for stress, etc.; see Grime 1974; Grime *et al.* 1988; Venable 1996; Westoby *et al.* 2002, and references within). These traits, in turn, influence how species respond to landscape and biotic heterogeneity and influence larger scale variation in species diversity and ecosystem level processes (e.g. Grime 1994; Ritchie & Olff 1999; Niinemets 2001). However, with some notable exceptions (Venable 1996; Reich *et al.* 1997) most attempts to integrate the important axes of differentiation have not been quantitative. Further, attempts to explain observed plant 'strategies' have not been based on synthetic first principles (Westoby *et al.* 2002). As I show below, a general theory of metabolic scaling highlights the axes that form the constraints guiding biological trade-offs. Such a framework is capable of quantifying and even predicting many attributes of variation in plant growth and allocation.

Scaling the partitioning of biomass and metabolic production: comparing angiosperms and gymnosperms and the importance of allocation

The WBE model provides a framework by which to show quantitatively how plants may differ from one another in terms of how they allocate metabolic production and partition biomass between organs and functions. For example, Enquist & Niklas

(2002) show that if plant metabolic rate scales as the 3/4 power of total plant biomass, M, and that metabolic rate scales isometrically with respect to M_L (Niklas & Enquist 2001), the fraction of total biomass that is allocated between the major plant organs (leaf, M_L, stem, M_S, and root, M_R) must then be constrained. Enquist & Niklas (2002) note that $B = \beta_1 M^{3/4} = \beta_1(M_L + M_S + M_R)^{3/4}$, where $B = \beta_2 M_L$. Here values of β may reflect potential taxon-specific variation in tissue level metabolism and biomass allocation, where β_1 and β_2 include units of year^{-1}. Therefore, the total biomass portioned to leaves is $M_L = \beta_3(M_L + M_S + M_R)^{3/4}$, where the value of $\beta_3 = \beta_1 / \beta_2$. Enquist & Niklas (2002) show how each organ (M_L, M_S and M_R) should scale allometrically with each other but may differ with potential variation in proportional allocation (the values of β_1 and β_2). They show how an extension of WBE assumption (ii) (see above) must constrain the biomass allocated between roots and stems so that M_S and M_R must scale proportionally to one another (see also Huxley 1932; Pearsall 1927). Thus, it is possible to derive relationships between all three major plant organs, where (using Enquist & Niklas' (2002) original notation) $M_L = \beta_{12} M_S^{3/4}$, $M_L = \beta_{13} M_R^{3/4}$ and $M_S = (\beta_{12} / \beta_{13})\, M_R$, where β_{12} and β_{13} reflect the amount of biomass partitioned to stems and roots respectively. Thus, according to the WBE model and predictions by Enquist & Niklas (2002) across vascular plant taxa the allometric partitioning of biomass should be described by identical exponents. Comparison of the above-predicted organ partioning scaling relationships across a broad sampling of intraspecific and interspecific relationships, for both gymnosperms and angiosperms, provides robust support for the model (Enquist & Niklas 2002). However, plants may differ in their values of β, which reflect potential differences in proportional biomass allocation.

Allometric similarities and differences in the scaling of leaf partitioning

Although there are several similarities in the allometric scaling of biomass partitioning, there are significant differences between angiosperms and gymnosperms in terms of how they allometrically partition biomass between plant organs (Enquist & Niklas 2002). Here most of the gymnosperms analysed consist of conifers. Utilizing the data from Enquist & Niklas (2002; see also Cannell 1982) I plotted values of stem mass, M_S, root mass, M_R, and leaf mass, M_L, as a function of whole-plant mass. The allometric scaling of leaf mass for both angiosperms and gymnosperms is characterized by the same exponent $M^{3/4}$ ($M_L \propto M^{3/4}$) as predicted by the WBE model (Fig. 17.2a). However, there are important differences in the proportion of biomass that is partitioned to leaves, as reflected in significant differences in the allometric intercept or values of β. For a given plant mass gymnosperms have proportionally *more* leaf mass than angiosperms (Fig. 17.2a; see also eqn 17.3). Thus, a fundamental allometric difference emerges between angiosperms and gymnosperms—gymnosperms allocate about three times more to leaves than do angiosperms, which matches the observation that gymnosperms tend to have on average three cohorts of leaves at any one time (Enquist & Niklas 2002).

Angiosperms and gymnosperms support another important prediction of the WBE model—rates of biomass production scale in direct proportion to leaf area or

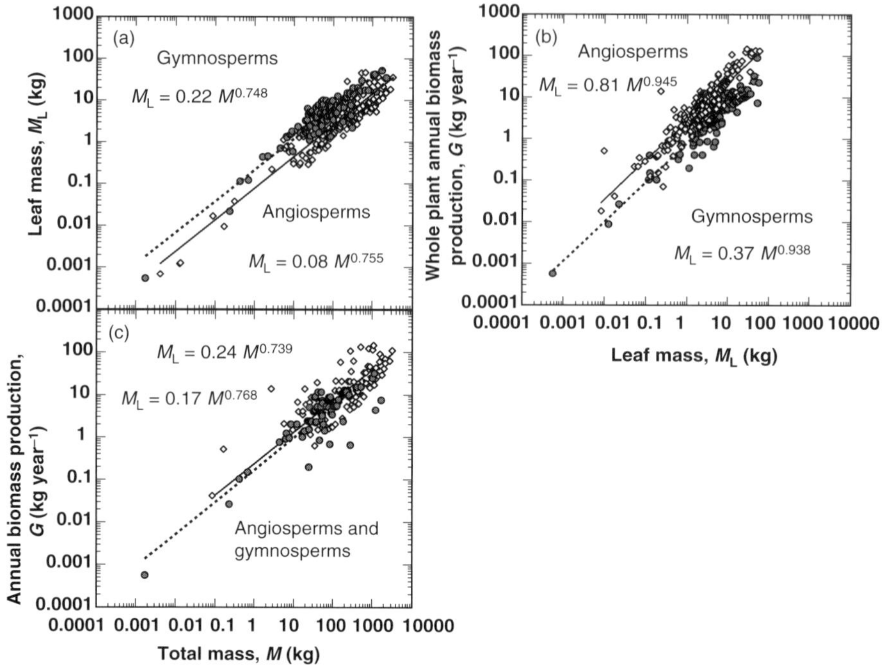

Figure 17.2 (a) Allometric scaling of leaf mass for both angiosperms and gymnosperms (angiosperms $\alpha_{RMA} = 0.755$, 95% CI = 0.803 to 0.707, $\beta_{RMA} = -1.09$, 95% CI = −1.20 to −0.987, r^2 0.839, $F = 891.18$, $n = 173$, $p < 0.0001$; gymnosperms $\alpha_{RMA} = 0.748$, 95% CI = 0.695 to 0.801, $\beta_{RMA} = -0.658$, 95% CI = −0.758 to −0.557, r^2 0.860, $F = 719.18$, $n = 119$, $p < 0.0001$). (b) Rates of biomass production versus leaf mass (gymnosperms: $\alpha_{RMA} = 0.938$, 95% CI = 0.886 to 0.989, $\beta_{RMA} = -0.436$, 95% CI = −0.488 to −0.384, r^2 0.846, $F = 1101$, $n = 202$, $p < 0.0001$; angiosperms: $\alpha_{RMA} = 0.945$, 95% CI = 0.881 to 1.01, β_{RMA} −0.091, 95% CI = −0.150 to −0.033, r^2 0.858, $F = 801.80$, $n = 135$, $p < 0.0001$). (c) Annual rate of biomass production versus total mass (angiosperms: $\alpha_{RMA} = 0.739$, 95% CI = 0.646 to 0.831, $\beta_{RMA} = -0.621$, 95% CI = −0.843 to −0.400, r^2 0.599, $F = 149.64$, $n = 102$, $p < 0.0001$; gymnosperms: $\alpha_{RMA} = 0.756$, 95% CI = 0.664 to 0.846, $\beta_{RMA} = -0.768$, 95% CI = −0.941 to −0.594, r^2 0.764, $F = 239.82$, $n = 76$, $p < 0.0001$).

leaf mass, $G \propto M_L$ (Fig. 17.2b; see also Lambers *et al.* 1998; Niklas & Enquist 2001). However, statistical analysis also reveals another important difference between gymnosperms and angiosperms. Gymnosperms have *lower* rates of biomass production per unit leaf mass, reflecting fundamental differences in c_L and/or c_P (see eqn 17.3; Fig. 17.2b). This observation supports past findings showing that taxa with higher values of c_L or specific leaf area (SLA), such as laminar leaves or angiosperms, tend to have higher rates of photosynthesis per unit leaf mass than species with lower values of c_L or SLA, such as needle-like leaves of conifers (see Field & Mooney 1986; Reich *et al.* 1997; Niinemets 1999; Wright *et al.* 2001). Nevertheless, the decreased production per unit leaf mass, c_L, in conifers is offset by a greater total allocation to leaves, e.g. values of β as shown above (Fig. 17.2c; see also Reich *et al.* 1992; Gower *et al.* 1993; Haggar & Ewel 1995). Therefore, plotting annual rates of biomass production as a function of total mass reveals *no* significant differences between angiosperms and gymnosperms. Thus, rates of production for both gymnosperms and angiosperms scale identically as $M^{3/4}$. Despite fundamental differences in leaf construction and allocation, and in physiology (e.g. differences in c_L and values of β) both clades abide to a common allometric production function (Fig. 17.2c).

These observations are in accord with recent developments in plant ecological physiology, where it is becoming increasingly apparent that allocation to leaf mass and variation in the specific leaf area (photosynthetic area per unit leaf mass or SLA) is a significant determinant of life history and ecological differences among species (see Lambers *et al.* 1998; Lambers & Porter 1992; Westoby *et al.* 2002). Specific leaf area is a key plant trait as variation in SLA enables a plant to expose more or less leaf area to light and CO_2 per given biomass allocated to leaves (Field & Mooney 1986; Lambers & Porter 1992). Conifers and most gymnosperms differ significantly in terms of SLA. Angiosperms have a higher specific leaf area than conifers (see Niklas 1992), which accounts for the higher rate of biomass production per unit angiosperm leaf (Fig. 17.2b). Nevertheless, both groups produce biomass at the same allometric rate (Fig. 17.2c). Together, these results suggest that angiosperms and gymnosperms are similarly constrained in their ability to scale whole-plant metabolic rate as explained by the WBE model, but nevertheless differ in how they proportionately allocate metabolic production.

Scaling the partitioning of biomass and metabolic production: life-history variation and the importance of tissue density

Use of the WBE model thus far has assumed that tissue density within and between species is constant. However, plants often show dramatic differences in tissue specific density (Niklas 1992). Variation in allocation of metabolic production to tissue quality (e.g. tissue density) appears to be an important axis along which plant life-history strategies have diversified. For example, variation in wood density is associated with many anatomical, physiological and ecological differences among species. These in turn limit local ecological distributions and population growth rates (Denslow 1980; Borchert 1986; Sobrado 1986; Loehle 1988; Thomas 1996; Enquist *et al.* 1999). Figure 17.3 shows how variation in wood density influences radial stem growth rates in Amazonian trees (Worbes & Junk 1989). Differences in stem growth rates also reflect important differences in tree life-histories, from fast growing, disturbance specialists with short lifespans, to slower growing emergent trees of mature forest with relatively longer lifespans (Loehle 1988).

Enquist *et al.* (1999; see also Enquist & Niklas 2002; Niklas & Enquist 2002) derive how stem growth rate (dD/dt) and various other allometric relationships are influenced by variation in differences in allocation to tissue density, ρ, within and between plant species. In modelling how plant growth is influenced by differences in allocation to ρ they assume that the rate of biomass production (dM/dt) is directly

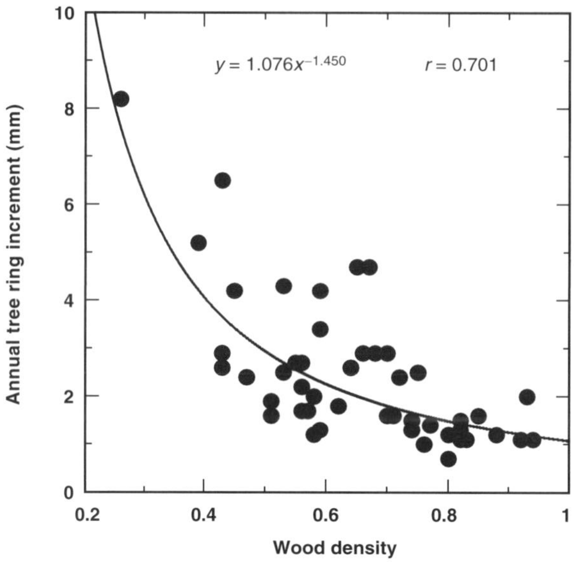

Figure 17.3 Relationship between stem wood tissue density and annual stem growth increment as measured by tropical tree rings (data from Worbes & Junk 1989). The relationship shows a significant negative correlation indicating that light woods have higher rates of radial stem growth than heavy woods.

proportional to whole-plant metabolic rate, B (which again approximates the rate of gross photosynthesis), and scales as plant volume raised to the 3/4 power ($B \propto V^{3/4}$). If ρ is approximately constant, then at any time t, $dM/dt = c_G B$, where $B = c_B(M/\rho)^{3/4}$ and $D = c_D(M/\rho)^{3/8}$, and c_B and c_D are corresponding proportionality constants. Thus the scaling of plant metabolism and the resulting suite of predicted plant allometric relationships (West *et al.* 1999b) are modified accordingly by the inclusion of potential variation in ρ. Enquist & Niklas (2002) and Niklas & Enquist (2002) also show how differences in root, stem and leaf tissue density influence patterns of metabolic allocation and the scaling of biomass partitioning. Incorporating ρ into botanical studies probably will explain a significant degree of residual variation in metabolic and allometric scaling variation.

The above relationships can be combined to give $dD/dt = (3C/2\rho)\,D^{1/3}$, where $C \equiv 1/4 c_G c_B c_D^{2/3}$. Note, here C also reflects many additional biological attributes of a plant that may or may not differ among taxa. Hence, the model provides the basis for the quantitative integration of many functional attributes of plants. This framework allows one to recast the mass-production law as a function of basal stem diameter, $D = dM/dt = (c_G c_B / c_D^2)D^2$. This predicts that across species, dM/dt, for trees of fixed diameter, is *independent* of wood density ρ. That is, if values of c do not vary significantly across species, then variation in stem growth reflects differences in metabolic allocation to stem tissue owing to differences in ρ. Thus, despite differences in allocation *all species should produce biomass at approximately the same allometric rate*, which scales as $M^{3/4}$.

Enquist *et al.* (1999) test these allometric growth predictions in 45 species of tropical trees, which differ in their rates of stem diameter growth and wood tissue density. Functional differences in stem growth rate dD / dt followed the predicted influence of wood tissue density. Further, despite significant variation in dD / dt among species, species-specific rates of biomass production (dM / dt), as predicted, did not appear to differ as all species fell on the same biomass production line described by $dM / dt \propto M^{3/4}$. Enquist *et al.* (1999) further utilize this general form of the growth equation and prior life-history theory (Charnov 1993) to predict how the scaling of life-history events (time until reproduction, average lifespan, etc.) is influenced by variation in allocation as indexed by wood tissue density. Thus, differences in allocation of metabolic production to tissue density quantitatively can be shown to influence variation in important plant life-history characters, which in turn influence local ecological distributions and dynamics.

How the constraints of metabolic scaling ramify to influence large-scale patterns in ecology and evolution

Perhaps the most powerful attribute of a general model for metabolic scaling is that it provides the basis by which to predict quantitatively how variation in plant size, rates of resource use, and patterns of allocation and biomass partitioning influence larger scale ecological patterns and processes (Enquist *et al.* 1998,1999; Enquist & Niklas 2001; Niklas & Enquist 2001). Biological and physical principles imposed upon vascular networks and differences in allocation of metabolic production constrain the physical space over which plants utilize resources and the rates of whole-plant resource use and allocation. If population densities and rates of production are ultimately limited by rates of resource use then there are reasons to suspect that the allometric constraints at the level of the individual in turn constrain allometric and energetic attributes of populations, ecological communities and ecosystems (see also Damuth 1981).

Allometric scaling of plant population density

Enquist *et al.* (1998) proposed a simple model for how metabolic scaling at the level of the individual influences the scaling of population density. They assume that: (i) sessile plants compete for spatially limiting resources; (ii) their rate of resource use will scale proportionally to metabolic rate, where $B \propto M^{3/4}$; and (iii) plants grow until they are limited by resource supply rate, R, so that the rate of resource use by plants approximates resource supply. Assumption (iii) is generally supported by the observation that the rate of biomass production in terrestrial plant communities is directly proportional to the annual rainfall and degree of light intercepted in water and light-limited environments respectively (Niklas 1994; see also Rosenzweig 1968). The maximum number of individuals, N_{max}, that can be supported per unit area is related to the average whole-plant metabolic rate per individual, B, by noting that

$$R \approx B_{tot} \approx (N_{max})(B) \approx (N_{max})(c_P c_L \beta M^{3/4}) \quad (17.4)$$

where, from above, $B \propto G = c_P c_L \beta M^{3/4}$ and B_{tot} is the total amount of resources used by all individuals within a given area. Here the values of c are allometric constants reflecting tissue-specific metabolic demand (see eqn 17.3). So that at equilibrium, when rates of resource use approximate rates of resource supply, R is constant and it follows that

$$N_{max} \approx [R/(c_P c_L \beta)]M^{-3/4} \quad (17.5)$$

Thus, when plants have grown such that rates of use of a limiting resource approximate its rate of supply, there should be an inverse relationship between population density and plant size. The specific exponent ($N_{max}^{-4/3}$) is different from that predicted by the simple geometric model of the 'thinning law' ($N_{max}^{-3/2}$) (Yoda *et al.* 1963). Further, residual variation in the scaling of plant density is predicted to be primarily the result of variations in c and R (and potentially temperature). This shows how variation in mass-specific rates of metabolism and rates of resource supply from the environment can influence variation in population density. As presented in Enquist *et al.* (1999) and Niklas & Enquist (2001) there appears to be little variation in the allometric scaling of growth, indicating that the value of $c_P c_L \beta$ is an approximate constant across major plant taxa. If correct, most residual variation in the scaling of plant density will mainly be attributable to R (and potentially temperature). Thus, environments where rates of limiting resource supply are higher will be characterized by more dense populations for a given plant size and there is strong empirical support for these predictions (Enquist *et al.* 1998; see also Weller 1987; Lonsdale 1990).

Comparing the allometric scaling of population densities across angiosperms and gymnosperms and phytoplankton

The invariance of rates of biomass production across plants that differ by over 20 orders of magnitude of body mass (Fig. 17.1) implies that each also should be described by similar scaling of ecological densities. If equation (17.5) is correct and if allometric rates of resource use (as indexed by $c_P c_L \beta$) are similar then the population densities of gymnosperms, angiosperms and even unicellular plants should fall on the same allometric line. However, if the scaling of population density is more influenced by the geometric constraints associated with scaling of leaf mass, as highlighted in Fig. 17.2, then one might expect that the scaling of population densities also should be significantly offset between plant taxa.

Plotting the relationship between total plant mass and the total number of angiosperm and gymnosperm individuals per unit area shows that population density scales as $M^{-3/4}$ (Fig. 17.4). However, angiosperms and gymnosperms fall on the same scaling function as there is no statistically significant difference between each fitted allometric function. Similarly, unicells and metaphytes fall on the same allometric

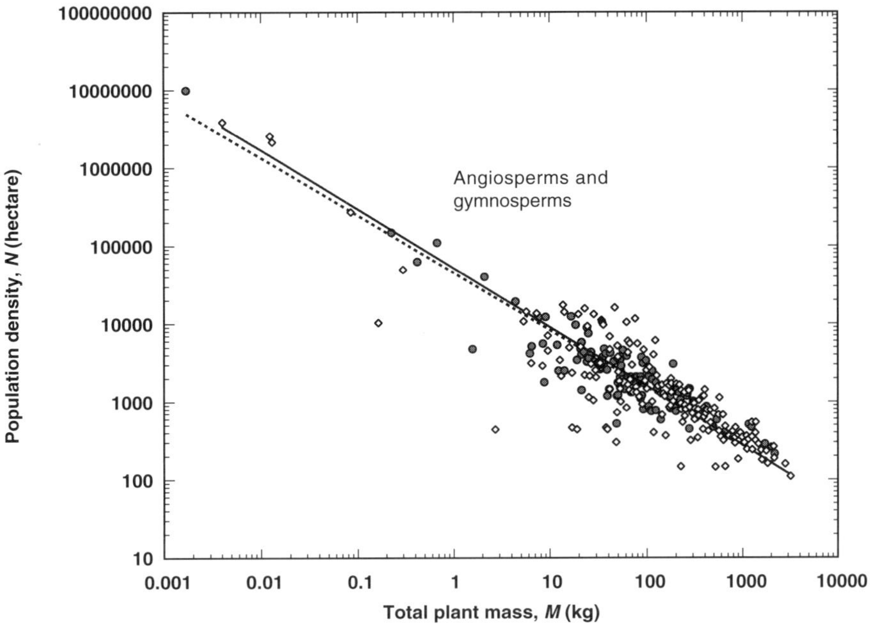

Figure 17.4 Relationship between whole-plant size (roots, stems and leaves) and number of individuals per unit area for both angiosperm (open symbol) and gymnosperm trees (filled symbol; see also Enquist & Niklas 2002). Gymnosperms: $\alpha_{RMA} = -0.737$, 95% CI = -0.790 to -0.684, $\beta_{RMA} = 4.65$, 95% CI = 4.55 to 4.76, r^2 0.845, $F = 637.70$, $n = 119$, $p < 0.0001$. Angiosperms: $\alpha_{RMA} = -0.758$, 95% CI = -0.813 to -0.702, β_{RMA} 4.715, 95% CI = 4.589 to 4.84, r^2 0.753, $F = 539.65$, $n = 179$, $p < 0.0001$.

production function, indicating that (according to eqn 17.5) the scaling of their respective population densities also should fall on the same allometric line. If true this would provide strong evidence for the importance of metabolic scaling in constraining ecological systems both on land and in water. Analysis of population density and size within marine phytoplankton (Belgrano *et al.* 1999) shows that scaling of population densities of terrestrial plants provides reasonable predictions of the maximum population densities of many phytoplankton species, despite the latter being many orders of magnitude smaller (Fig. 17.5).

A general model for how metabolism structures plant communities

One of the most prominent allometric patterns observed in plant and animal communities is the inverse relationship between body mass and abundance (e.g. De Liocourt 1898; Morse *et al.* 1985; Marquet *et al.* 1990). Because an inverse relationship between size and abundance reflects how biomass and productivity are partitioned among individuals, it offers considerable insight into the mechanisms structuring ecological communities in general.

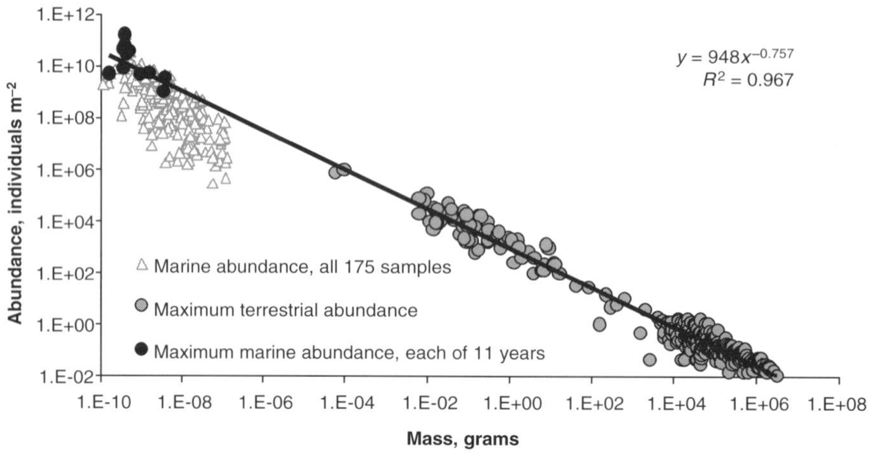

Figure 17.5 Effects of plant size (log mass in grams) on individual density (log individuals m^{-2}) for marine phytoplankton (symbols) and terrestrial gymnosperms and angiosperms including herbaceous, shrubby and arboreal species. (Data from Enquist *et al.* (1998) and Enquist and Niklas (2002).) (Shaded circles; data from Belgrano *et al.* 2002). The line is a Type II regression model fitted to the terrestrial plant data.

As outlined above, extensions of metabolic theory predict that the total number of individuals, N, in any size class m equals $c_m M_m^{-3/4}$, where c_m is the number of individuals per unit area normalized to a given size class m, and M_m is the body mass of size class m. The general allometric model for plant vascular systems also predicts that M is proportional to the 8/3 power of stem diameter, D, for any size class (i.e. $M \propto D^{8/3}$), such that N will scale as $N \propto M^{-3/4} \propto D^{-2}$. If these scaling laws hold for communities, organismal traits can be used to link larger scale properties of communities across different ecosystems. Enquist & Niklas (2001) show how the total standing community biomass, M_{tot}, is given by the formula

$$M_{tot} = \int_a^b MN(M)dm = \int_a^b M(c_m M^{-3/4})dm = \frac{4}{5} c_m \left(M_a^{5/4} - M_b^{5/4}\right) \qquad (17.6)$$

where the subscripts a and b denote, respectively, the minimum and maximum body masses within a given community. As a and b are largely insensitive to species composition or latitude (see Enquist & Niklas 2001; and results below), any variation in M_{tot} will be determined by variation in c_m. However, for closed canopy forests, theory and observation suggest that c_m varies little, such that M_{tot} is expected to vary little across communities. Specifically, the rate of resource use per size class $R_m \approx Q_m \approx c_m B_m$, where the average metabolic rate of a given size class $B_m = c_B A_m$. Here, A_m is leaf or root exchange surface area, and c_B is the rate of resource use per unit area, which can vary across species. As allometric theory and empirical data show that $A_m = c_A (M/\rho)^{3/4}$, where ρ is the bulk tissue density and c_A is a constant of proportionality reflecting the species-specific amount of leaves or roots per individual per unit area, the following formula is derived

$$c_m \approx \frac{R_m}{c_A c_B (M_m/\rho)^{3/4}} \tag{17.7}$$

which shows how numerous biological and abiotic factors can influence plant population density per body size class. Nonetheless, biometric and physiological data indicate no significant differences in the mean values of c_B, c_A and ρ across tropical and temperate tree species or with variation in species richness (Whittaker & Woodwell 1968; Brown 1997; Enquist *et al.* 1998, 1999). This invariance indicates that total community biomass is likely to be relatively insensitive to species diversity (in contrast to recent niche model predictions of Tilman *et al.* 1997) even though c_m can vary with various environmental factors (such as temperature, precipitation, etc.) that influence R_m. Furthermore, if c_m, M_a and M_b do not vary across communities then it follows that variation in M_{tot} and rates of total biomass production by a plant community, G_{tot}, is more influenced by ecological factors that reduce the capacity of metabolic production (e.g. abiotic and biotic features of ecosystems that influence the extent to which plants can maximally transpire water and assimilate CO_2) than by species-specific physiological capacities or variation in species diversity.

The invariance of metabolic constants and exponents (reflecting fundamental similarities among taxa) should ultimately determine body size distributions. Enquist & Niklas (2001) tested this prediction using macroecological data that span taxonomically and physiognomically diverse plant communities (Gentry 1988, 1993), from near-monospecific stands to some of the most biodiverse forested communities on Earth. Species diversity has *no* effect on total standing biomass, although tree density rapidly asymptotes with respect to diversity. Finally, as predicted, the number of individuals per sample area scales as the −2 power of stem diameter or as the −3/4 power of body mass within and across communities (Fig. 17.6). Interestingly, the −2 scaling rule also holds with increasing geographical sampling areas, including continental and global samples (see Enquist & Niklas 2001). Latitude and species number do not contribute greatly to the variance observed in local size distribution exponents. Furthermore, the size frequency distribution exponent does not appear to be correlated with annual precipitation (Enquist & Niklas 2001).

Conclusions

In this chapter, I have presented three central findings that have important implications for the study and integration of plant macroecology and macroevolution.

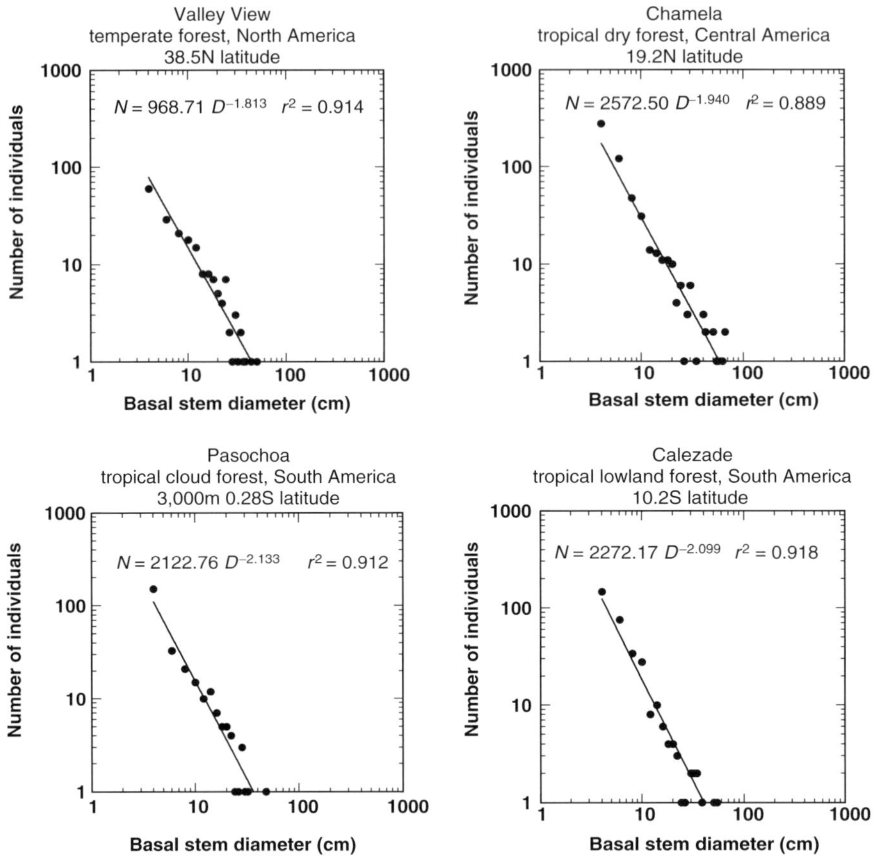

Figure 17.6 Relationship between plant size and number of individuals within a sampling of closed canopy forest communities (see also Enquist & Niklas 2001). Although there are impressive differences in community composition and diversity between communities at the observed scaling, the exponents for each of the fitted allometric relationships are indistinguishable from the predicted value of D^{-2} or $M^{-3/4}$.

1 *Within and across major plant taxa (unicells, gymnosperms and angiosperms) rates of energy processing and biomass production scale as* $M^{3/4}$.

Such invariance probably reflects fundamental similarity in biochemical energetics across plants (see West *et al.* 1997; Niklas & Enquist 2001). A major implication of this finding is that allometric rates of resource use across diverse clades ramify in similar ways to influence larger scale ecological patterns.

2 *Within the context of a general theory for metabolic scaling, taxa differ in terms of how they allocate metabolic production.*

Taxon-specific patterns of biomass partitioning and allocation between roots, stems, leaves and reproductive biomass, in addition to differences in allocation to tissue density, in turn ramify to influence the scaling of many life-history rates and times (Charnov 1993; Enquist *et al.* 1999). It is presumed that natural selection will act to favour a suite of certain integrated allocation strategies depending on the local biotic and abiotic environment (e.g. Grime *et al.* 1988; Venable 1996; Westoby *et al.* 2002, and references therein). By allowing for different allocation strategies a general model for metabolic scaling provides a base line by which to assess patterns of residual allometric variation.

3 *Similarity in rates of energy processing at the level of the individual ramifies similarly to constrain macroecological patterns.*

Despite broad differences in plant form and function (unicells, angiosperms, gymnosperms) evidence is provided for similarity in the size distribution exponent for both populations and communities. Invariance of the exponent describing size, abundance and even total community biomass is in accordance with allometric theory but contrasts in many important ways with speculations and recent niche-based theoretical predictions. Metabolic theory suggests that variation in plant species composition is instead associated with concomitant changes in the degree of partitioning of a limited amount of resources rather than increases (or decreases) in community biomass and, potentially, depending on the local environment, productivity (Tilman *et al.* 1997). Such partitioning most probably is reflected in life-history trade-offs in the allocation of metabolic production (Enquist *et al.* 1999; Enquist & Niklas 2002).

The central hypothesis of the work presented here is that rates of energy processing provide a fundamental constraint on variation in scaling of biological rates and times (see also Brown *et al.*, this volume). Physiological rates and times then influence many organismal attributes. Allometric 'rules' dictate how metabolic production and biomass are partitioned among different body parts at the level of the individual plant. As such they provide a quantitative basis for drawing mechanistic connections between numerous features of organismal biology, ecology, ecosystem studies and evolutionary biology. The constraints on metabolic scaling appear to provide the basis for mechanistically connecting emergent statistical patterns in ecology and evolution (see also Brown *et al.*, this volume). Although aspects of the theory still need to be tested critically and assumptions assessed, empirical data appear to support many of the theoretical predictions. Thus, a common body of allometric and metabolic theory promises to provide a general framework for explaining many attributes of organismal form, function, diversity and evolution.

Acknowledgements

This chapter is dedicated to Dr James Enderson (The Colorado College), an inspirational teacher and biologist, who ultimately laid the seed for this work by teaching my first college biology class from an allometric perspective and by showing me that not all big questions in biology had been solved. I thank K. Gaston, D. Kerkhoff and C. Price for comments and especially an anonymous outside reviewer for providing numerous helpful comments and suggestions. In particular, I wish to thank T. Blackburn and K. Gaston for organizing a stimulating and rewarding meeting. This work was funded by a National Science Foundation CAREER award, NSF DEB 0133974 to B.J.E. and by a fellowship from the Center for Applied Biodiversity Science, Conservation International.

References

Becker, P., Gribben, R.J. & Lim, C.M. (2000) Tapered conduits can buffer hydraulic conductance from path-length effects. *Tree Physiology* **20**, 965–967.

Belgrano, A., Lindahl, O. & Hernroth, B. (1999) North Atlantic Oscillation (NAO) primary productivity and toxic phytoplankton in the Gullmar Fjord, Sweden (1985–1996). *Proceedings of the Royal Society, London, Series B* **266**, 425–430.

Belgrano, A., Allen, Enquist, B.J. & Gillooly (2002) Allometric scaling of maximum population density: a common rule for marine phytoplankton and terrestrial plants. *Ecology Letters* **5**, 611–613.

Blackburn, T.M. & Gaston, K.J. (1997) A critical assessment of the form of the interspecific relationship between abundance and body size in animals. *Journal of Animal Ecology* **66**, 233–249.

Blackburn, T.M., Lawton, J.H. & Pimm, S.L. (1993) Nonmetabolic explanations for the relationship between body-size and animal abundance. *Journal of Animal Ecology* **62**, 694–702.

Borchert, R. (1986) Soil and stem water storage determine phenology and distribution of Dry Tropical forest trees. *Ecology* **75**, 1437–1449.

Bridgman, P.W. (1963) *Dimensional Analysis.* Yale University Press, New Haven.

Brown, J.H. (1995) *Macroecology.* University of Chicago Press, Chicago.

Brown, J.H. (1999) Macroecology: progress and prospect. *Oikos* **87**, 3–14.

Brown, J.H., West, G.B. & Enquist, B.J. (2000) Scaling in biology: patterns and processes, causes and consequences. In: *Scaling in Biology* (eds J.H. Brown & G.B. West), pp. 1–24, Oxford University Press, Oxford.

Brown, S. (1997) *Estimating Biomass and Biomass Change of Tropical Forests.* Forestry Paper 134, Food and Agriculture Organization of the United Nations, Rome.

Calder, W.A. III (1984) *Size, Function, and Life History.* Dover Publications, Mineola, MN.

Cannell, M.G.R. (1982) *World Forest Biomass and Primary Production Data.* Academic Press, New York.

Charnov, E.L. (1993) *Life History Invariants: some Explorations of Symmetry in Evolutionary Ecology.* Oxford University Press, Oxford.

De Liocourt, F. (1898) *De l'amenagement des sapinieres.* Bulletin de la Societe Forestiere de Franche-Comteet Belfort, Besancon.

Damuth, J. (1981) Population density and body size in mammals. *Nature* **290**, 699–700.

Denslow, J.S. (1980) Gap partitioning among tropical rain forest trees. *Biotropica* (Supplement) **12**, 47–55.

Enquist, B.J. (2002) Universal scaling in tree and vascular plant allometry: toward a general quantitative theory linking plant form and function from cells to ecosystems. *Tree Physiology* **22**, 1045–1064.

Enquist, B.J. (2003) Copes Rule and the evolution of long distance transport in vascular plants: allometric scaling, biomass partitioning, and optimization. *Plant Cell and Environment* **26**, 151–161.

Enquist, B.J. & Niklas, K.J. (2001) Invariant scaling relations across tree-dominated communities. *Nature* **410**, 655–660.

Enquist, B.J. & Niklas, K.J. (2002) Global allocation rules for patterns of biomass partitioning in seed plants. *Science* **295**, 1517–1520.

Enquist, B.J., Brown, J.H. & West, G.B. (1998) Allometric scaling of plant energetics and population density. *Nature* **395**, 163–165.

Enquist, B.J., West, G.B., Charnov, E.L. & Brown, J.H. (1999) Allometric scaling of production and life-history variation in vascular plants. *Nature* **401**, 907–911.

Enquist, B.J., West, G.B. & Brown, J.H. (2000) Quarter-power scaling in vascular plants: functional basis and ecological consequences. In: *Scaling in Biology* (eds J.H. Brown & G.B.

West), pp. 167–199. Oxford University Press, Oxford.

Field, C.B. & Mooney, H.A. (1986) The photosynthesis–nitrogen relationship in wild plants. In: *On the Economy of Plant Form and Function* (ed. T.J. Givnish), pp. 25–55. Cambridge University Press, Cambridge.

Gaston, K.J. & Blackburn, T.M. (1999) A critique for macroecology. *Oikos* **84**, 353–368.

Gaston, K.J. & Blackburn, T.M. (2000) *Pattern and Process in Macroecology*. Blackwell Science, Oxford.

Gentry, A.H. (1988) Changes in plant community diversity and floristic composition on environmental and geographic gradients. *Annals of the Missouri Botanical Garden* **75**, 1–34.

Gentry, A.H. (1993) Diversity and floristic composition of lowland tropical forest in Africa and South America. In: *Biological Relationships Between Africa and South America* (ed. P. Goldblatt), pp. 500–547. Yale University Press, New Haven.

Gillooly, J.F., Brown, J.H., West, G.B., Savage, V.M., & Charnov, E.L. (2001) Effects of size and temperature on metabolic rate. *Science* **293**, 2248–2251.

Gower, S.T., Reich, P.B. & Son, Y. (1993) Canopy dynamics and aboveground production of five tree species with different leaf longevities. *Tree Physiology* **12**, 327–345.

Grime, J.P. (1974) Vegetation classification by reference to strategies. *Nature* **250**, 26–31.

Grime, J.P. (1994) The role of plasticity in exploiting environmental heterogeneity. In: *Exploitation of Environmental Heterogeneity in Plants: Ecophysiological Processes Above—and Below—Ground* (eds M.M. Caldwell & R.W. Pearcy), pp. 1–19. Academic Press, New York.

Grime, J.P. & Hunt, R. (1975) Relative growth-rate: its range and adaptive significance in a local flora. *Journal of Ecology* **63**, 393–422.

Grime, J.P. Hodgson, J.G. & Hunt, R. (1988) *Comparative Plant Ecology*. Unwin-Hyman, London.

Gross, K.L., Willig, M.R., Grough, L. Inouye, R. & Cox, S.B. (2000) Patterns of species diversity and productivity at different scales in herbaceous plant communities. *Oikos* **89**, 417–427.

Haggar, J.P. & Ewel, J.J. (1995) Establishment, resource acquisition, and early productivity as determined by biomass allocation patterns of three tropical tree species. *Forestry Science* **41**, 689–708.

Hemmingsen, A.M. (1950) The relation of standard (basal) energy metabolism to total fresh weight of living organisms. *Reports of the Steno Memorial Hospital and the Nordisk Insulinlaboratorium* **4**, 7–51.

Hubbell, S.P. (2001) *The Unified Neutral Theory of Biodiversity and Biogeography*. Princeton University Press, Princeton.

Huxley, J.S. (1932) *Problems of Relative Growth*. Methuea, Methuea.

Kozłowski, J. & Weiner, J. (1997) Interspecific allometries are by products of body size optimization. *American Naturalist* **149**, 352–380.

Lambers, H. & Porter, H. (1992) Inherent variation in growth rate between higher plants: A search for physiological causes and ecological consequences. *Advances in Ecological Research* **22**, 187–261.

Lambers, H., Chapin, F.S. & Pons, T.L. (1998) *Plant Physiological Ecology*. Springer-Verlag, New York.

Loehle, C. (1988) Tree life history strategies: the role of defences. *Canadian Journal of Forestry Research* **18**, 209–222.

Lonsdale, W.M. (1990) The self-thinning rule: dead or alive? *Ecology* **71**, 1373–1388.

Maurer, B.A. (1999) *Untangling Ecological Complexity: the Macroscopic Perspective*. University of Chicago Press, Chicago.

Marquet, P.A., Navarrete, S.A. & Castilla, J.C. (1990) Scaling population density to body mass in rocky tidal communities. *Science* **250**, 1125–1127.

McMahon, T.A. & Bonner, J.T. (1983) *On Size and Life*. Scientific American Library, New York.

Morse, D.R., Lawton, J.H., Dodson, J.H. & Williamson, M.M. (1985) Fractal dimension of vegetation and the distribution of arthropod body lengths. *Nature* **314**, 731–733.

Niinemets, U. (1999) Components of leaf dry mass per area—thickness and density—alter photosynthetic capacity in reverse directions in woody plants. *New Phytologist* **144**, 35–47.

Niinemets, U. (2001) Global-scale climatic controls on leaf dry mass per area, density, and thickness in trees and shrubs. *Ecology* **82**, 453–469.

Niklas, K.J. (1992) *Plant Biomechanics: an Engineering Approach to Plant Form and Function*. University of Chicago Press, Chicago.

Niklas, K.J. (1994) *Plant Allometry*. University of Chicago Press, Chicago.

Niklas, K.J. (1997) *The Evolutionary Biology of Plants.* University of Chicago Press, Chicago.
Niklas, K.J. & Enquist, B.J. (2001) Invariant scaling relationships for interspecific plant biomass production rates and body size. *Proceedings of the National Academy of Sciences, USA* **98**, 2922–2927.
Niklas, K.J. & Enquist, B.J. (2002) On the vegetative biomass partitioning of seed plant leaves, stems, and roots. *American Naturalist* **159**, 482–497.
Odum, E.P. (1969) The strategy of ecosystem development. *Science* **164**, 262–270.
Pearsall, W.H. (1927) Growth studies. VI. On the relative sizes of growing plant organs. *Annals of Botany* **41**, 549–556.
Peters, R.H. (1983) *The Ecological Implications of Body Size.* Cambridge University Press, Cambridge.
Post, D.M., Pace, M.L. & Hairston Jr., N.G. (2000) Ecosystem size determines food-chain length in lakes. *Nature* **405**, 1047–1049.
Reich, P.B., Walters, M.B. & Ellsworth, D.S. (1992) Leaf life-span in relation to leaf, plant, and stand characteristics among diverse ecosystems. *Ecological Monographs* **62**, 365–392.
Reich, P.B., Walters, M.B. & Ellsworth, D.S. (1997) From tropics to tundra: global convergence in plant functioning. *Proceedings of the National Academy of Sciences, USA* **94**, 13730–13734.
Ritchie, M.E. & Olff, H. (1999) Spatial scaling laws yield a synthetic theory of biodiversity. *Nature* **400**, 557–560.
Rosenzweig, M.L. (1968) Net primary productivity of terrestrial communities: prediction from climatological data. *American Naturalist* **102**, 67–74.
Rosenzweig, M.L. (1995) *Species Diversity in Space and Time.* Cambridge University Press, Cambridge.
Rosenzweig, M.L. & Abramsky, Z. (1993) How are diversity and productivity related? In: *Species Diversity in Ecological Communities: Historical and Geographical Perspectives* (eds R.E. Ricklefs & D. Schluter), pp. 52–65. University of Chicago Press, Chicago.
Sobrado, M.A. (1986) Aspects of tissue water relations of evergreen and seasonal changes in leaf water potential components of evergreen and deciduous species coexisting in tropical forests. *Oecologia* **68**, 413–416.
Tilman, D. (1988) *Plant Strategies and the Dynamics and Structure of Plant Communities.* Princeton University Press, Princeton.
Tilman, D., Lehman, C.L. & Thomson, K.T. (1997) Plant diversity and ecosystem productivity. *Proceedings of the National Academy of Sciences, USA* **94**, 1857–1861.
Thomas, S.C. (1996) Reproductive allometry in Malaysian rain forest trees: biomechanics versus optimal allocation. *Evolutionary Ecology* **10**, 517–530.
Venable, D.L. (1996) Packaging and provisioning in plant reproduction. *Philosophical Transactions of the Royal Society, London, Series B* **351**, 1319–1329.
Weller, D.E. (1987) A re-evaluation of the –3/2 power rule of plant self-thinning. *Ecological Monographs* **57**, 23–43.
West, G.B., Brown, J.H. & Enquist, B.J. (1997) A general model for the origin of allometric scaling laws in biology. *Science* **276**, 122–126.
West, G.B., Brown, J.H. & Enquist, B.J. (1999a) The fourth dimension of life: fractal geometry and allometric scaling of organisms. *Science* **284**, 1677–1679.
West, G.B., Brown, J.H. & Enquist, B.J. (1999b) A general model for the structure and allometry of plant vascular systems. *Nature* **400**, 664–667.
West, G.B., Brown, J.H. & Enquist, B.J. (2001) A general model for ontogenetic growth. *Nature* **413**, 628–631.
Westoby, M. (1998) A leaf-height seed (LHS) plant ecology strategy scheme. *Plant Soil* **199**, 213–227.
Westoby, M., Falster, D., Moles, A., Vesk, P. & Wright, I. (2002) Plant ecological strategies: some leading dimensions of variation between species. *Annual Review of Ecology and Systematics* **33**, 125–159.
Whittaker, R.H., & Woodwell, G.M. (1968) Dimension and production relations of trees and shrubs in the Brookhaven forest, New York. *Journal of Ecology* **57**, 1–25.
Worbes, M. & Junk, W.J. (1989) Dating tropical trees by means of C-14 from bomb tests. *Ecology* **70**, 503–507.
Wright, I.J., Reich, P.B. & Westoby, M. (2001) Strate-

gy shifts in leaf physiology, structure and nutrient content between species of high- and low-rainfall and high- and low-nutrient habitats. *Functional Ecology* **15**, 423–434.

Yoda, K., Kira, T., Ogawa, H. & Hozumi, K. (1963) Self-thinning in overcrowded pure stands under cultivated and natural conditions. *Journal of Biology Osaka City University* **14**, 107–129.

Part 8
Why is macroecology important?

Chapter 18
Macroecology and conservation biology

Kevin J. Gaston and Tim M. Blackburn*

Introduction

The geographical ranges of species are not scattered uniformly across Earth. Rather the distribution is such that they give rise to a richly textured surface of variation in biodiversity, with peaks of high species richness (often at low latitudes), valleys of low richness (often at high latitudes), and sometimes wide plains in between. Likewise, the individuals of a species are not distributed uniformly within the bounds to its occurrence, but can again be visualized as forming a landscape with its own unique patterning of peaks and valleys, sculpted in part by variation in the abundances of other species, serving as resources, competitors, predators or parasites. The superimposing of numerous abundance surfaces constitutes the structure of the local assemblages with which the terrestrial-bound observer is more familiar, with this structure altering as these surfaces ebb and flow through time with changes in environmental regimes and drivers of the evolutionary process.

Macroecology is concerned perhaps foremost with understanding the patterns of abundance and distribution of species at these large spatial scales, and over timespans that reflect their geographical dynamics (Brown & Maurer 1989; Brown 1995; Blackburn & Gaston 1998, this volume; Gaston & Blackburn 1999, 2000; Maurer 1999). Conservation biology is concerned with maintaining into the future, preferably *in situ*, the biodiversity of which these patterns are comprised (e.g. Hunter 1996). Given that maintenance is best founded on mechanistic understanding (as anyone will know who runs a car or other complex piece of machinery), the two fields are intimately linked. In this chapter, we examine selected facets of that linkage. In particular, we focus on and explore the way in which the largely pure science and heuristic concerns of macroecology provide much of the context in which the applied problems confronting conservation biology have to be resolved, and the perspective this should give on the latter.

We are aware that at times what follows may stray beyond the bounds of what some would strictly see as the preserves of either macroecology or of conservation

* *Correspondence address: k.j.gaston@sheffield.ac.uk*

biology. We make no apologies for this. Ultimately, the bounds of fields are matters of convenience, to be broken where this is useful.

Taking the long view

The principal achievement of macroecology has been to encourage ecologists to consider the influence of processes operating on long timescales and/or over large spatial scales on the structure of the local assemblages that can more readily be studied and conserved (Gaston & Blackburn 2000). This is simply because that influence potentially can be highly significant. Taking the long view inevitably forces one to think about the influence that human activities have had on the environment. This is a core issue for conservation biology, but has received substantially less attention than it should from macroecologists, who have tended to have a preoccupation with the pristine (or what they presume to be so).

For much of their history, humans have been shaping the world around them in ways that have markedly changed the abundance and distribution of other species, altering forever the landscapes of species richness and assemblage structure. Thus, as prehistoric human populations established and spread across the land masses (probably following a positive density–range-size relationship, as documented for other species; Gaston & Blackburn 2000), they wielded each of what have come to be recognized as the four main drivers of species extinction (Diamond 1989; Gaston 2002).

1 Overexploitation. Although there is continued debate as to the extent to which exploitation by early humans was responsible for the late Quaternary extinction of many large-bodied bird and mammal species that seems to have been temporally associated with the human colonization of major continental land masses, the evidence seems to us highly suggestive (Martin 1984, 2001; MacPhee 1999; Miller *et al.* 1999; Flannery 2001; but see Grayson 2001). Certainly, there is no doubt as to the potential for small-scale human societies markedly to reduce the abundances of prey species and cause local extinctions (Terborgh 1999; Holdaway & Jacomb 2000; Alroy 2001; Grayson 2001), and there is widespread agreement that such exploitation contributed to the loss of species from smaller land masses (Olson & James 1982; Cassels 1984; Holdaway 1989; Milberg & Tyrberg 1993; Pimm *et al.* 1995; Steadman 1995, 1997; Pimm 1996; Worthy 1997; Burney *et al.* 2001; Grayson 2001; Reis & Garong 2001; Duncan *et al.* 2002). The best examples doubtless arise from the spread of the Polynesians across the islands of the Pacific, an expansion that began perhaps 30 000 years before the present [BP] and was almost complete by 1000 year BP. The combined effects of resource exploitation, deforestation and the introduction of alien species are estimated to have resulted in the loss of perhaps 8000 species or populations of sea and landbirds (Steadman 1995). More generally, agriculture may have developed at least in part as a response to overexploitation of food sources by hunter–gatherers (Diamond 1998).

2 Habitat destruction and fragmentation. Dramatic reshaping of the distribution of habitats or vegetation types has been a feature of much of the history of humankind, with habitat change as a consequence of the activities of prehistoric populations having been reported on numerous occasions (McGlone 1983; Kershaw 1986; McGlone & Basher 1995; Diamond 1998; Krech 1999; Sadler 1999; Pudjoarinto & Cushing 2001). Indeed, it repeatedly has been discovered that what had been held to be 'natural' landscapes had actually been much transformed by earlier human activities (for discussion see, e.g. Isenberg 2000; Wilcove 2000). Such changes doubtless altered the abundances and distributions of species, and contributed to species losses. Thus, habitat use has been shown to be a significant predictor of prehistoric extinction risk for species in the avifauna of Marfells' Beach, on the South Island of New Zealand, over and above the effect of hunting by Maori, with species occupying terrestrial rather than aquatic habitats being more vulnerable (Duncan *et al.* 2002). Extensive forest clearance by fire occurred in this area shortly after Maori arrival (McGlone 1983; McGlone & Basher 1995) and this habitat loss may have selectively reduced populations of terrestrial birds.

3 Introductions. Non-domesticated species have, since prehistoric times, been introduced intentionally or accidentally by humans into areas in which they did not previously occur, breaching many natural barriers to their dispersal. The earliest known instance involves the introduction of a marsupial, the gray cuscus *Phalanger orientalis*, to New Ireland about 19 000 years ago (Grayson 2001). However, the diversity of species that were introduced by human colonists included a wide range of plants, invertebrates and vertebrates (see Grayson (2001) for many examples). Particularly when made to islands, these introductions resulted, through competition, predation, disturbance and/or disease transmission, in the extinction of native species. For example, the kiore *Rattus exulans* was introduced to the islands of the New Zealand archipelago within the past 2000 years by Polynesians (Anderson 1991; Holdaway 1996). The subsequent disappearance of many small-bodied native birds, especially species laying small eggs and nesting on the ground, strongly implicates predation by kiore as a primary driver of extinction in this avifauna (Holdaway 1999; Duncan & Blackburn, unpubl.).

4 Extinction cascades. The extinction of one species may lead to the extinction of others. Indeed, this is inevitable where this species provides critical resources for others, such as specialist herbivores, parasites, or predators, or perhaps itself acts as a specialist pollinator or dispersal agent. Thus, for example, in New Zealand, the giant eagle *Harpagornis moorei* almost certainly preyed on moas, and its extinction probably resulted when moas declined in numbers as a result of the hunting by the Maori that led to their demise (Cassels 1984; Worthy 1997; Holdaway 1999; Holdaway & Jacomb 2000). Likewise, the extinction of most bird and mammal species was accompanied by the extinction of associated specialist lice and ticks (Stork & Lyal 1993). More complex sets of interactions also may result in cascades of extinctions, as evidenced by the dramatic, and often extensive, changes in floral and faunal composition that can result from changes in the abundance and occurrence of key species (e.g. large-bodied predators and herbivores; e.g. Terborgh 1988; Owen-Smith 1989; Crooks & Soulé 1999; Jackson 2001; Terborgh *et al.* 2001).

The continued action of all four of these processes through to the present, albeit with varying intensities and increasingly sophisticated technologies, and the associated extinctions, has been well documented (e.g. Hannah *et al.* 1995; Smith *et al.* 1993; Klein Goldewijk 2001; Pimentel 2001). Without doubt they destroy any notion that conservation can have as a goal the maintenance or the restoration of 'natural' ecosystems, and equally emphasize how much the maintenance of present patterns of biodiversity may rest on the continuation of human management regimes. There are scarcely any ecological systems with structures that do not bear the imprint of human activities.

One outcome of human activities has been to alter the form of many of the basic patterns on which macroecology has been centred. Thus, for example:

1 Species–body-size distributions. The shapes of these distributions for many taxonomic groups and for assemblages in many regions have been restructured through extinctions and introductions. Commonly, extinctions have fallen foremost on larger bodied species, to the point where the largest species in a number of taxa and areas have been lost (e.g. the largest bodied recent earwig, lizard, bird and terrestrial mammal species have all become extinct as a direct or indirect consequence of human activities). For example, the discovery and colonization of New Zealand by Polynesians was followed by the extinction (for reasons mentioned above) of 62 of the 132 bird species listed by Holdaway (1999) as known to be breeding on the three main islands of this archipelago. Species of a range of body masses have been driven extinct, but the large-bodied have clearly suffered disproportionately (Fig. 18.1a). The geometric mean body mass of species in the avifauna nearly halves when extinct species are excluded, and the skewness of the distribution drops. The drop in mean mass is largely attributable to the extinction of 11 species of moa (Dinornithiformes): excluding these, species that have become extinct are not significantly larger than those that are still extant (Gaston & Blackburn 2000; Cassey 2001). Thus, although large-bodied species tend to be lost from faunas, they do not do so alone. Of course, human colonization has also resulted in 40 additional species being successfully introduced to the New Zealand fauna. Introduced species span a similar range of body masses to the extant native fauna, with a geometric mean of 222 g, and there is no significant difference between the body masses of native and introduced species (Fig. 18.1b; unpaired t-test, $t = 0.36$, d.f. = 108, $p = 0.74$). The introduced species increased the skewness value for the overall body mass distribution to −0.12, from −0.39 for extant natives alone, normalizing it somewhat, but had much less effect overall on the distribution than did extinctions. A similar restructuring of a species–body-size distribution is observed for mammals of Britain (Blackburn & Gaston, unpubl.).

2 Species–abundance distributions. The abundances of different species have obviously been greatly altered by human activities, with some declining and others increasing. Empirical estimates of such changes over long timescales are scarce. Subfossil remains from New Zealand suggest that hunting by early human settlers must have had substantial effects on the abundances of species in the prehistoric avifauna, as shown by Duncan *et al.* (2002). Their analyses demonstrated that prehistoric bird extinctions in New Zealand derived from selective hunting of favoured

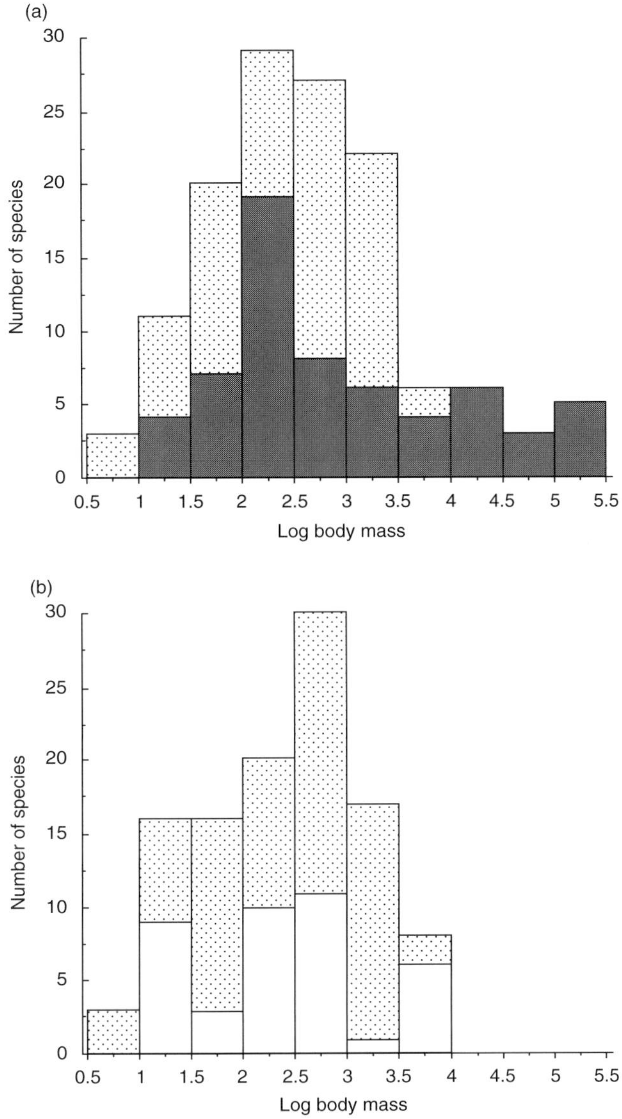

Figure 18.1 (a) The frequency distribution of body masses (g) of the New Zealand pre-human avifauna (n = 132, geometric mean = 444 g, skew = 0.61). The dark-stippled areas represent those species that went extinct from the three main islands (North, South and Stewart) after human arrival (n = 62), although some of these persist on small inshore islets (e.g. stitchbird *Notiomystis cincta*) and offshore islands (e.g. common diving-petrel *Pelecanoides urinatrix*); the original fauna minus these extinctions has geometric mean = 250 g, skew = –0.39. Light-stippled bars should be read from the top of the dark-stippled bars, not from the abscissa. (b) The frequency distribution of body masses (g) of the current New Zealand avifauna (n = 110). The non-stippled areas represent those species introduced to the three main islands after human arrival, and currently persisting in a naturalized state (n = 40). Stippled bars should be read from the top of the non-stippled bars, not from the abscissa. (From data in Holdaway (1999) and Heather & Robertson (1996).)

species, and that they were not, therefore, simply a consequence of some species being intrinsically more susceptible to extinction (e.g. through slower population growth rates) under a scenario of random harvesting. The implication is that, by differentially targeting certain species, human activities can alter species abundances (to the extent that targeted species can be driven to extinction), and perhaps also affect the form of species–abundance distributions more generally.

Humans may also affect species–abundance distributions by introducing exotic species. Of 217 species in the British breeding avifauna, the populations of 17 derive largely or entirely from individuals of captive origin (see appendix III of Gaston & Blackburn 2000). The populations of these introductions are markedly lower than those of native species (geometric mean sizes of 3634 and 14 419 individuals, respectively). The frequency distribution of logarithmically transformed population sizes for native species is left-skewed (skew = –0.516), whereas that for introductions is right-skewed (skew = 0.487; Fig. 18.2), indicating that few introduced species in Britain are common even in comparison with other introductions. Introduced species lower the mean and reduce the left skew in the overall species–abundance distribution of British breeding birds (geometric mean = 12 943, skew = –0.446), albeit not to any great degree. However, effects may be more marked in other regions. For example, it seems clear that many of the most common bird species in New Zealand are introduced.

3 Species–range-size distributions. Given the general positive interspecific relationship between population size and range size (Fig. 18.3), and the likelihood that this is a causal link (Warren & Gaston 1997; Gaston & Blackburn 2000), changes in the structure of species–abundance distributions have unsurprisingly been accompanied by changes in the structure of species–range-size distributions. Lomolino & Channell's (1995) collation of historic and present range sizes for species of non-volant terrestrial mammals reveals how great the reductions in global ranges of some species have been (Fig. 18.4). In island systems, range reductions caused by colonists and/or later settlers have often markedly disrupted the distributions of extant species (e.g. Steadman 1993), in the extreme creating, for example, what Grayson (2001) termed 'pseudo-endemics' (species apparently restricted to single islands but only because of human-driven extinctions elsewhere). For New Zealand, data suggest that serious range declines have been experienced by at least 69% of the species present in the avifauna at the time of first human colonization (Duncan & Blackburn, unpubl.), resulting in the extinction of 47% (Holdaway 1999). Although many species have undoubtedly spread as a consequence of human activities, the general impression is of a net decline in mean range size, with this decline having commonly been disproportionately for larger bodied species (e.g. Smith & Quin 1996).

4 Species richness gradients. The species richness of much of Earth has been reduced or inflated by human activities, changing observed patterns. Thus, for example, lower numbers of seabird species and individuals nesting on tropical/subtropical compared with temperate/subantarctic Pacific islands is at least in part a consequence of the losses caused by Prehistoric colonists (Steadman 1995).

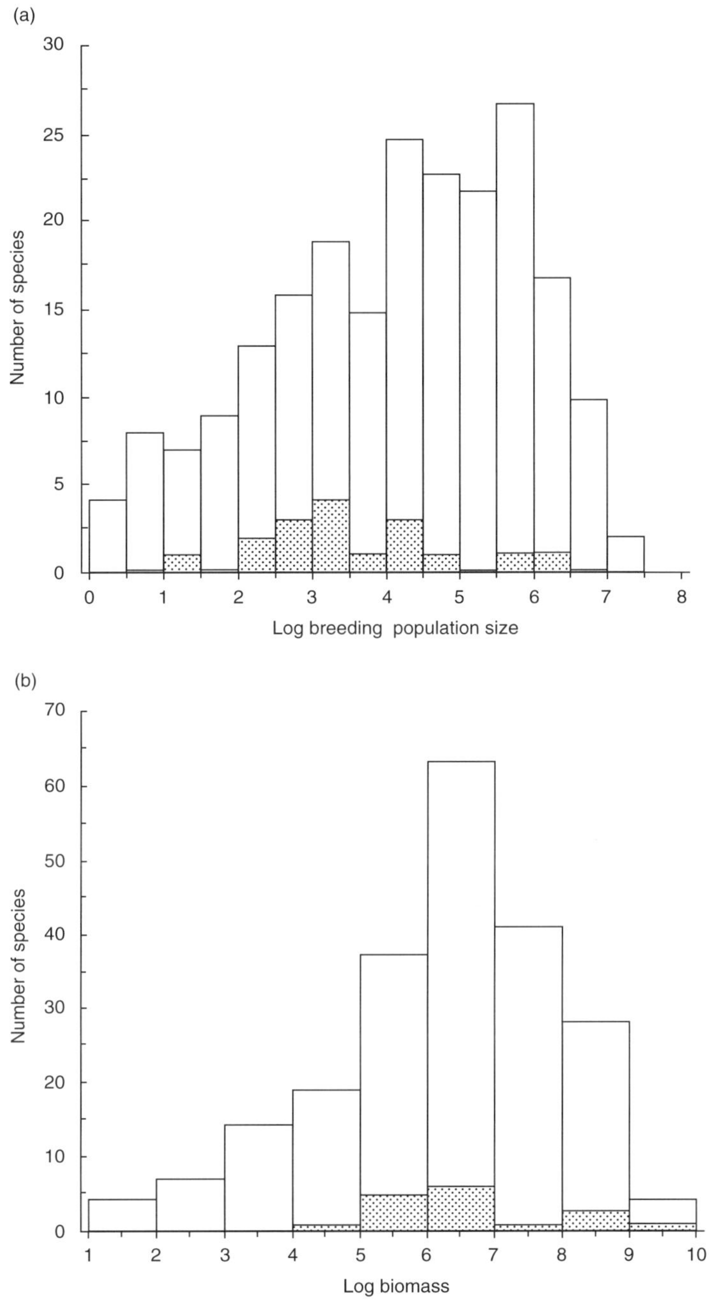

Figure 18.2 The frequency distributions of (a) breeding population sizes and (b) biomasses (mean mass (g) × breeding population size) for British breeding birds, distinguishing those species populations derived from captive origin (stippled areas). Non-stippled bars should be read from the top of the stippled bars, not from the abscissa. (From data in Gaston & Blackburn 2000.)

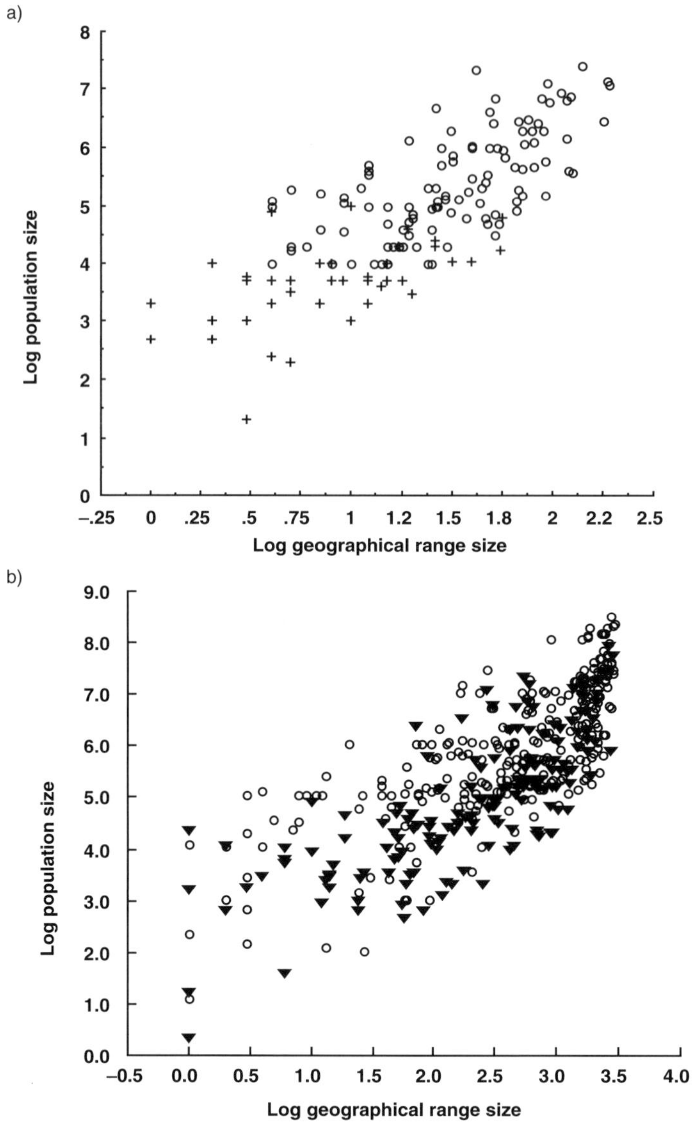

Figure 18.3 Interspecific population-size–range-size relationships for (a) wildfowl of the world, (b) European bird species (triangles, species of conservation concern; circles, species not considered to be of conservation concern). Conservation concern was defined on the basis of a species' SPEC (Species of European Conservation Concern) categorization. Species in categories 1–3 are considered to be of concern (1, species of global conservation concern; 2, species with global populations concentrated in Europe, and which have unfavourable conservation status there; 3, species with global populations not concentrated in Europe, but which have unfavourable conservation status in Europe), whereas species in category 4 (populations concentrated in Europe but with favourable conservation status) or not on the SPEC list are considered not to be of conservation concern. (From data sources listed in Gaston & Blackburn (1996), and in BirdLife International/European Bird Census Council (2000).)

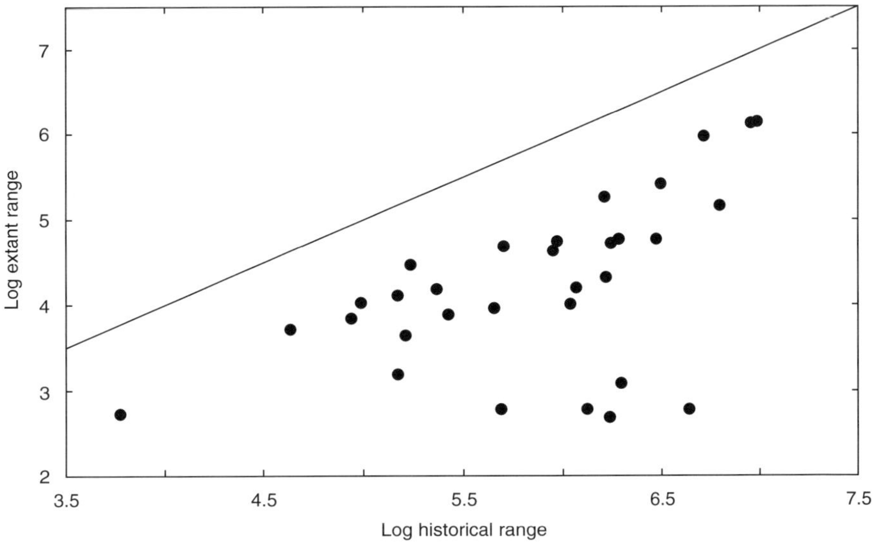

Figure 18.4 The relationship between the historical and extant geographical ranges (km^2) of 31 species of terrestrial non-volant mammals. The solid line is that of equality. (From data in Lomolino & Channell 1995.)

Similarly, Pimm *et al.* (1995) showed that the avian species richness of certain Pacific islands was inversely related to island area, and positively related to the island's isolation from the source pool of avian colonists. They argued that these patterns of richness variation oppose those normally found because the larger and less isolated islands were colonized earlier by humans, and have as a consequence lost greater proportions of their native avifaunas. Such striking examples of richness changes driven by human activities are probably unusual, as it seems more likely that these activities will serve to modify rather than reverse existing gradients. For example, an approximate comparison of the species–area relationships for land and freshwater birds on five New Zealand islands at the time of first human colonization and at the time of European arrival, suggests that extinction has steepened the species–area relationship (Fig. 18.5). The most wide-ranging anthropogenic effects on natural ecosystems have probably been wrought at temperate rather than tropical latitudes, and on small (e.g. island) rather than large (e.g. continental) land masses. Broad-scale richness patterns would not be disrupted completely by such effects. Nevertheless, there is evidence that some continental areas which have had long histories of human occupation have experienced surprisingly low numbers of recent extinctions, suggesting that substantial species losses may have occurred much earlier (e.g. Greuter 1995). The ultimate outcome of the widespread conversion of natural ecosystems to agricultural monocultures that is currently underway will be that much of Earth's biodiversity will be eradicated. Natural patterns in the distribution of biodiversity inevitably will disappear too.

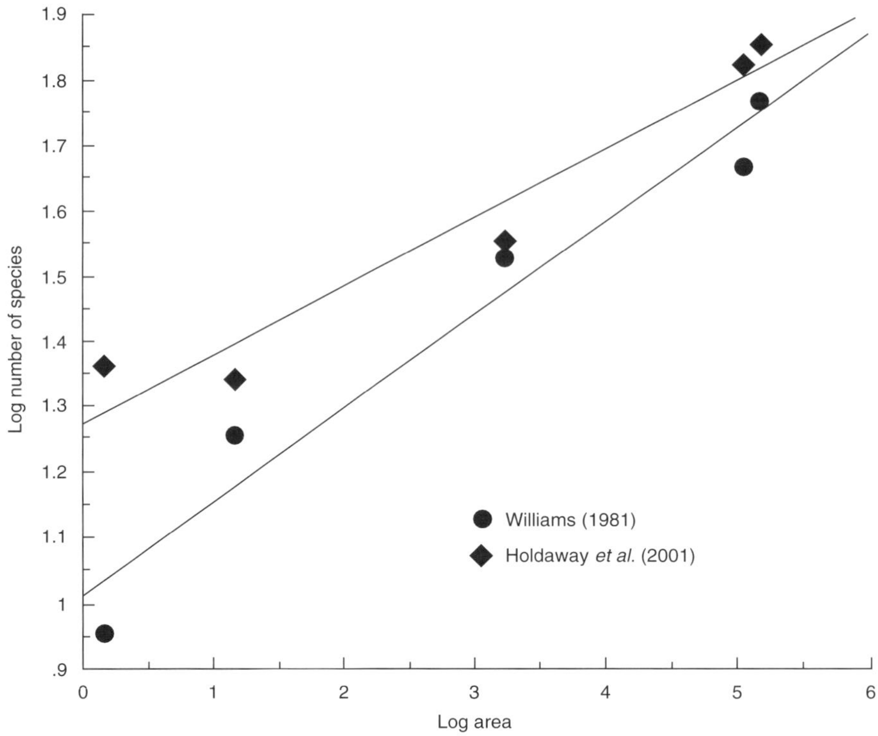

Figure 18.5 Estimates of the species–area relationships for land and freshwater birds on five islands in the New Zealand archipelago (in ascending order of area, Stephens, Codfish, Stewart, North and South) at the time of first human colonization ($S = 18.79\,A^{0.106}$; data from Holdaway *et al.* 2001), and the time of European discovery ($S = 10.23\,A^{0.144}$; data from Williams 1981). This comparison is not precise, as the lists at the two times differ not only because of prehistorically extinct species, but also because of taxonomic revision amongst extant and recently extinct species.

Even where the basic shape of macroecological patterns has apparently remained approximately constant over long periods (e.g. Murray & Dickman 2000), the internal structure of these patterns has typically altered, with species no longer making the same relative contributions (e.g. common species becoming rare, rare becoming common). Such restructuring following the impact of humans begs the question of precisely what it is macroecologists, or ecologists in general for that matter, are trying to explain (a point to which we will return). Certainly they suggest that attempting to explain current macroecological patterns may require a millennial scale view of processes and cannot ignore human influences, although this has been the usual approach. Equally, they underline the notion that conservation arguably is not a crisis discipline in the sense that many seem to have used the phrase (Wilson 1985; Eldredge 1999). One definition of a crisis is 'a turning point in the progress of anything' or 'a state of affairs in which a decisive change for better or worse is imminent' (Shorter Oxford English Dictionary). In terms of the loss of species and changes in their abundances and ranges, such changes have been a continual feature of biodiversity throughout much of human history. Conservation needs to be motivated by what humans have done to biodiversity in the long-term, and not just by recent impacts. Indeed, conservation may thus perhaps best be likened to a struggle to retain the increasingly dim reflection of the biodiversity that pertained before human activities, in the knowledge that much of this biodiversity is essential for the provision of the ecosystem services on which we all depend, and that the role of much of the rest remains unknown.

Taking the wider view

If the broad temporal view of macroecology provides something of a long-term perspective on the current status of conservation, the broad spatial scale should help provide perspective on the magnitude of the task that conservation presently faces. That is, it should provide the background against which levels of threat are judged or, conversely, the context within which species are assumed to be relatively immune to imminent extinction. Here, we focus on one particular example, namely the abundance of birds.

Total numbers

Starting at the largest spatial scale, there have been several published estimates of the total number of individual wild birds in the world: 100 billion (Fisher & Paterson 1964), 100 billion (Wood 1982), 300 billion (de Juana 1992) and 200–400 billion (Gaston & Blackburn 1997). The origin of some of these figures is unclear, and the basis of most is not made explicit. Most recently, Gaston *et al.* (unpubl.) estimated the global breeding bird population by calculating the total number of individual breeding birds found in 16 different land-use classes derived from the History Database of the Global Environment (HYDE) (Klein Goldewijk 2001), combined with data on the local densities of birds. They derived three numbers for each land-use class, representing likely typical, low and high average totals. This approach gave an estimate of the global breeding bird population of 86.7 billion individuals, with low and high estimates of 39.34 and 134.04 billion. Certainly the global population cannot be substantially larger. Densities of *c.*2500 birds km^{-2} might be expected for a relatively pristine moist tropical forest, although frequently they may be considerably less (see Terborgh *et al.* 1990; Thiollay 1994), which, if the whole world were to be covered in this vegetation type, would give a total of *c.*340 billion birds. Temperate forests have total bird densities of perhaps half this level, and again often much less, and those of no other widespread vegetation types regularly approach densities in tropical forests (e.g. Udvardy 1957; Wiens & Dyer 1975; Terborgh *et al.* 1990; Wiens 1991). In such broad brush calculations, seabirds can largely be ignored. Shuntov (1974) estimates the number of seabirds at 1–3 billion individuals. The apparent high abundance of seabirds is partly an illusion deriving from their tendency to breed in large and often conspicuous colonies. Most individual birds live thinly spread over the land masses of the world, rather than densely packed around their edges.

A total of 86.7 billion birds is actually a rather small number. With a global human population of around 6 billion individuals, this amounts to only about 15 breeding birds per person, which perhaps helps put the problems of avian conservation into some kind of perspective. Put like this, many higher taxa of birds comprise surprisingly small numbers of individuals even at a global scale. For example, there are estimated to be approximately 7 million albatrosses, 65 million penguins (del Hoyo *et al.* 1992), 115 million auks (del Hoyo *et al.* 1996), and 100–200 million individual wildfowl (Webb *et al.* 2001). Thus, there is less than 1 wild duck for every 30 people.

Of course, all these estimates ignore the complications of seasonal fluctuations in numbers (which doubtless differ between latitudes and biogeographical regions), and any temporal trends that have resulted from human activities. Remsen (1995) calculates, on the basis of minimum estimates of the rate of destruction of tropical forest of 76 000 km^2 $year^{-1}$ and a bird density of 1900 km^{-2}, that 144.4 million individual birds may be lost annually as a consequence of tropical deforestation. Gaston *et al.* (unpubl.) calculate that 18.5% (range 15.6–19.3) of the individuals in the global bird population inhabit the human-modified land-use classes of cropland and pasture. Estimates from HYDE of the original areas of land-use classes before conversion allowed them to estimate the original bird populations of these areas, and hence the population changes that land-use conversion has precipitated over time. This suggested a loss in individual bird numbers of 22.1% (range 21.2–24.7).

Threatened species

The most recent listing of globally threatened bird species (BirdLife International 2000) includes population estimates for 1165 species. As many of the population data for threatened species consist of estimates of the likely upper and lower limits to population size, it is impossible to derive a definitive number for the total number of individuals represented by them. However, the methods described in Table 18.1 give an estimate of approximately 52.7 million individuals summed over all threatened bird species. Thus, although threatened species comprise 12% (1165 out of 9700) of the extant species of birds, they thus comprise only around 0.06% (52.7 million out of 86.7 billion) of the overall total number of individuals. Clearly, the very rarest comprise a rather small proportion of the global total number of individual birds.

Species–abundance distributions

Estimates of total numbers of birds have not explicitly addressed how these are distributed amongst species. Between the extremes of global population size, it is well established that there are many more species that are rare than are common. The detailed form the distribution takes is more debatable. At a regional scale, distributions of population sizes for bird species tend to be approximately log-normal, albeit that most distributions show a longer than expected tail towards rare species, and hence are left-skewed when abundances are logarithmically transformed (Nee *et al.* 1991;

Table 18.1 The number of species (percentage in parentheses) that would be lost if the populations of all species were to undergo declines at the same rate. We derived starting population sizes for the real data from BirdLife International (2000): these are the geometric mean of the minimum and maximum estimates if given, otherwise the single number recorded (ignoring greater than or less than signs). Model starting population sizes derive from the power fraction model with $k = 0.15$ applied to a global avian population of 86.7 billion individuals.

	Number of species lost		
Rate at which individuals lost	25 years	50 years	100 years
Real data			
25 $year^{-1}$	294 (3.0)	327 (3.4)	467 (4.8)
50 $year^{-1}$	327 (3.4)	467 (4.8)	889 (9.2)
100 $year^{-1}$	467 (4.8)	889 (9.2)	1066 (11.0)
Model estimates			
100 $year^{-1}$	375 (3.9)	578 (6.0)	871 (9.0)
500 $year^{-1}$	990 (10.2)	1444 (14.9)	2039 (21.0)
1000 $year^{-1}$	1444 (14.9)	2039 (21.0)	2777 (28.6)

Gregory 1994, 2000; Gaston & Blackburn 2000; Hubbell 2001). The largest area for which a species–abundance distribution can be calculated is for Europe, which distribution is significantly left-skewed under logarithmic transformation (Gregory 2000). However, this significance disappears if 10 species of predominantly Asian or African distribution that have small populations in Europe are ignored.

The global species–abundance distribution for birds could potentially depart substantially from this shape because it combines the avian assemblages in different biogeographical regions, some of which share species but most of which do not to any marked degree. Although we do not know the shape of the distribution, we do know that threatened species basically comprise the bulk of the left-hand tail. We can thus model the overall distribution and use the data for threatened species to validate the form of the left-hand tail. This allows us at least to assess what the global distribution might look like given that our model reasonably estimates the pattern of abundances amongst rare species. We modelled the global avian species–abundance distribution using Tokeshi's 'power fraction' model (see Marquet *et al.*, this volume) to divide estimates of the global total number of individual birds amongst the global total number of bird species, thereby deriving a global species abundance distribution. The abundances of the rarest species could be reasonably well modelled assuming a global avian population size of 86.7 billion individuals and with the power fraction $k = 0.15$ (Fig. 18.6).

Figure 18.7 shows how the abundances of all bird species would be distributed relative to the global avian population size if the power fraction model for globally threatened species applied more generally. The frequency distribution of log-

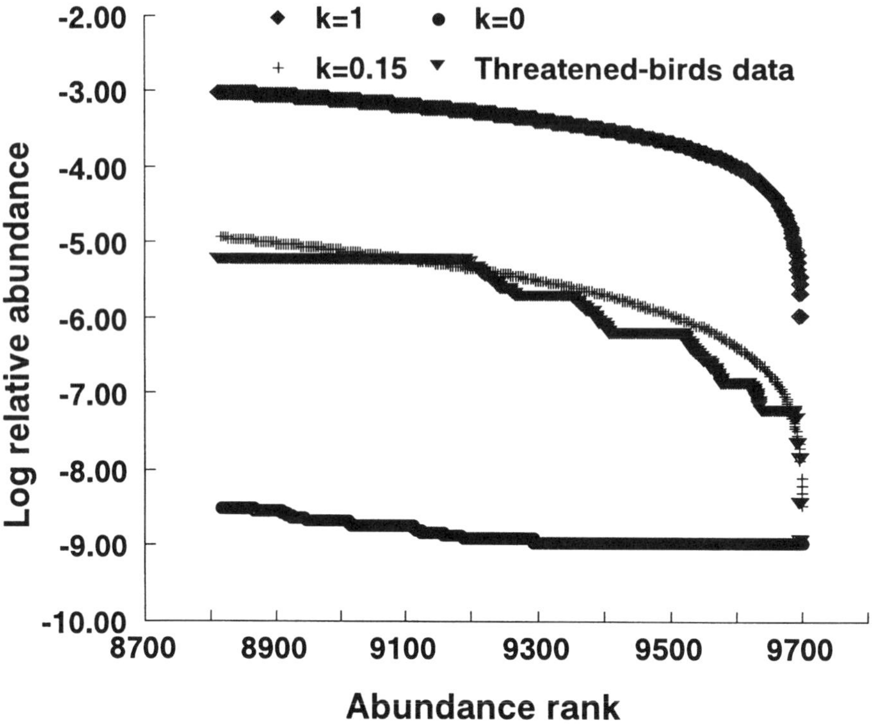

Figure 18.6 Real and simulated relative abundance plots for the rarest birds of the world. The real data derive from the BirdLife International (2000) list of threatened birds of the world, and are the estimated abundances for the rarest 889 bird species (all those with an estimated global population size of 5000 individuals or less). The data are step-like, because many of the population size estimates are given within broad limits (e.g. 500–5000 individuals) that are the same for many species. The simulated data derive from a power fraction model of species' abundances (Tokeshi 1996) with different values of k, assuming a global total bird population of 86.7 billion individuals distributed amongst 9702 species. Tokeshi (1996, 1999) showed that a range of empirical species–abundance distributions could be approximated by k in the range 0 to 0.2. The curves for $k = 0, 0.15$ and 1 are based on 10 repetitions each of the model. If $k = 0$, the probability that a species' abundance is chosen for division is independent of population size, and if $k = 1$ the probability is proportional to population size. These models have the additional constraint that population sizes of 1 individual cannot be chosen for division, which accounts for the unusual shape of the curve for $k = 0$. The y axis is the log of the percentage of the total global bird population size represented by each species.

transformed population sizes is approximately normal, but with significant left skew (Fig. 18.7). The median avian population size under this model is 450 000, and the geometric mean is 376 000. More than one-fifth (21%) of species have global population sizes below 50 000. This suggests that a high proportion of the world's bird species may have population sizes not far from the level below which species are considered vulnerable to extinction (BirdLife International 2000).

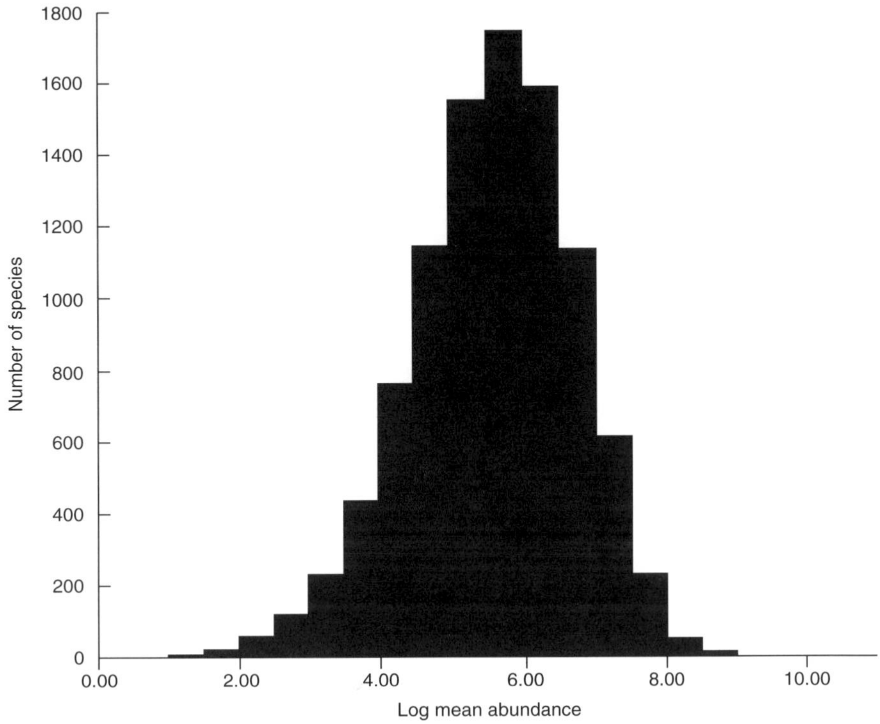

Figure 18.7 The simulated frequency distribution of log population sizes for all the world's bird species, derived using a power fraction model of species' abundances (Tokeshi 1996) with $k = 0.15$, assuming a global total bird population of 86.7 billion individuals distributed amongst 9702 species. Mean ± SD = 5.58 ± 1.14, skewness = −0.396 ($t = 15.9, p < 0.001$).

The power fraction model with $k = 0.15$ applied to a global avian population of 86.7 billion individuals is inaccurate in that it predicts a population size for the most abundant bird species of around 30 billion individuals. This greatly exceeds estimates for the most abundant bird species in recent historical times, which generally is regarded to have been the passenger pigeon *Ectopistes migratorius*, which when Europeans reached North America is estimated to have had a population of perhaps 3–5 billion individuals (Schorger 1955). This difference probably arises because we model the global avifauna as one population, whereas a more realistic treatment would consider the populations on different continents or other evolutionarily distinct landmasses separately. It seems likely that this alternative treatment would result in a model in which more species had relatively low population sizes. Simply, if the global population sizes of birds are distributed as studies would lead us to expect, most bird species will be surprisingly rare.

Just how rare most bird species are if this power fraction model is correct can be seen if we use it to predict the consequences for the global avifauna of sustained annual reductions in the population sizes of extant species (Table 18.1). The actual

Table 18.2 The suggested relative importance of entry and exit rules and transformations in shaping some continental scale multi-species macroecological patterns, scaled from white (low importance) to black (high importance).

Pattern	Entry rules	Exit rules	Transformations
Species–body-size distribution			
Species–abundance distribution			
Species–range-size distribution			
Latitudinal species richness gradient			

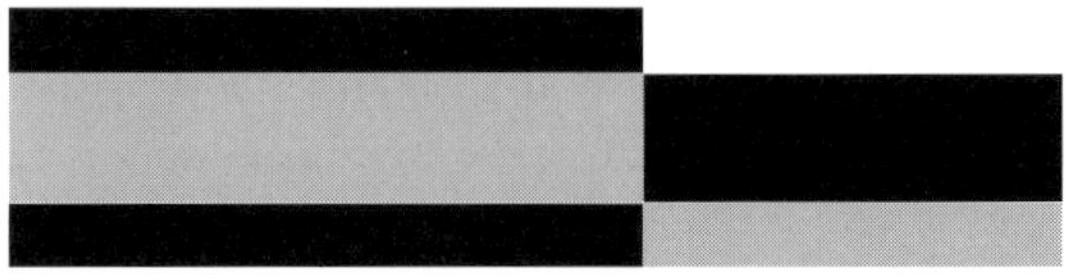

population size data for threatened birds show that even a sustained annual loss of only 25 individuals per species would result in extinction of around 5% of the global avifauna in the next 100 years. A loss of 100 individuals per species per year would see a loss of more than 10% over the next century. If that number were raised to 1000, the loss would be almost one-third of all bird species. One thousand individuals per species per year may seem a large loss, but combine global rates of tropical deforestation with the observation that 1 km^2 of pristine tropical forest may house 2500 individual birds.

Playing by the rules

The 'rules' that govern the form of large-scale patterns in ecology are shared concerns of both macroecology and conservation biology, as they are not only important to theoretical understanding of those patterns but also may be of significance to management issues. Many multispecies/interspecific macroecological patterns can be thought of as resulting from (i) 'entry' rules—biases in the processes that determine who joins a set of species through speciation or immigration, (ii) 'exit' rules—biases in the processes that determine who leaves a set of species through extinction or emigration, and/or (iii) 'transformations'—changes caused by processes, such as behaviour and evolution, that act on species when they are members of a set (Kunin 1997).

The relative importance of these three sets of processes is likely to vary markedly from one macroecological pattern to another (Table 18.2). Thus, species–body-size distributions are likely to be structured principally by entry and exit rules, because of the rather limited temporal change (transformation) that occurs in body size during the lifetime of a species compared with the differences in body size between species (Maurer *et al.* 1992; Owens & Bennett 1995). In contrast, the shape of species–range-size distributions is likely to be strongly influenced by the changes (transformation) in the size of a range that occur between speciation (its birth) and extinction (its death), which may result in the area of occupancy becoming several orders of magnitude larger than at speciation (Gaston 1998; Gaston & Chown 1999; Webb & Gaston 2000).

Plainly human activities have influenced entry and exit rules and transformations. They have affected entry rules through the high levels of introduction of species to areas in which previously they did not occur, and the biases in the kinds of species that are introduced (Crawley *et al.* 1997; Daehler 1998; Pyšek 1998; Lockwood 1999; McKinney & Lockwood 1999; Lockwood *et al.* 2000; Blackburn & Duncan 2001). They may also have had some impact on speciation by changing patterns of allopatry and sympatry, and increasing peripheral isolate formation (Myers & Knoll 2001). Humans have obviously affected exit rules inasmuch as they have increased rates of extinction far above normal background levels (Smith *et al.* 1993; Lawton & May 1995), with these extinctions again being biased in favour of species with particular kinds of traits (Diamond 1984; Gaston & Blackburn 1995, 1997; Bennett *et al.* 2001). Finally, transformations have been altered by human activities through the changes that have been wrought on the distributions and abundances of individual species, and on the body-size distributions of their populations (e.g. through size-selective harvesting).

Given the magnitude of the changes humans have wrought on Earth, and the way in which agriculture has freed them from constraints pertaining to all other species, it is easy to imagine that a fundamental reshaping of the processes giving rise to macroecological patterns may well have occurred. There are, however, two counter-arguments (Gaston & Blackburn 2000; but see Jablonski *et al.*, this volume). First, these activities have not served greatly to affect the broad climatic and environmental patterns that must contribute to explanations of many macroecological patterns (e.g. temperature gradients, biome areas). More generally, the basic laws of mathematics and physics that mediate many of the interactions between organisms and their environments (see Brown *et al.*, this volume) remain unchanged. Second, human activities have not served to alter the fundamental life-history traits exhibited by most species. Thus, for example, patterns of investment by species in reproduction, trading off current and future investment against a background of adult and juvenile mortality, remain largely unchanged.

The extent to which humans have reshaped the processes that determine macroecological patterns may thus rest on the relative importance of the broad-scale environmental gradients and the rules that govern life histories in influencing entry and exit rules and transformations, and on whether other rules are changed by human activities, or just have less material on which to act. The importance of such effects may therefore depend on what features of patterns are of relevance. For example, human activities seem unlikely on the whole to have generated a general trend for more species to be found in larger areas, but such activities may have affected the statistics of such relationships (e.g. Fig. 18.5). Human activities are unlikely to be responsible for the positive relationship between population size and geographical range size, and may also not have affected the statistics of such relationships (increasing and decreasing species seem to move along such relationships, rather than deviate far from them; Gaston & Blackburn 2000), but may perhaps have truncated such relationships at their upper end, or raised the density of species clustered towards their lower end (Fig. 18.3). Thus, humans seem more likely to have had quantitative rather than qualitative effects on such relationships. If so, they may not have altered the broad patterns of the distribution of life, but may have significantly hampered the search for quantitative theories of the mechanisms underlying such patterns. For example, the attempt to explain the species–body-mass distribution in terms of the patterns of speciation and extinction that lead to a log right-skewed distribution will need to know whether this skew is a natural feature of such distributions, or a consequence of size bias in human-induced extinctions. Similarly, given that much attention recently has been focused on the left skew of species–abundance distributions (see Hubbell 2001; Hubbell & Lake, this volume), it would be useful to know whether this is a consequence of human effects on species abundances.

Given the rapidity with which humans have transformed Earth, their influence on the rules that structure macroecological patterns is of significance to conservation biology. This is because it is likely to influence how resilient species are to these changes. For example, if ordinarily rare species attain their rarity relatively slowly they will have a chance to evolve responses to that rarity—but if most currently rare species have attained that circumstance more quickly they may not have done so, and may thus prove far less resilient to this state (Webb *et al.* 2001). In that regard, it is interesting that Duncan & Blackburn (unpubl.) found that one crude measure of rarity in the pre-human avifauna of New Zealand (whether a species was endemic to one or both of the two main islands of New Zealand) does not predict extinction in this fauna, once life-history differences among species are accounted for. In other words, it was not rarity that determined likelihood of extinction, but how susceptible a species was to the predation and habitat change effects that accompanied human colonization of the archipelago.

In conclusion

In a letter to one of his many correspondents, Charles Darwin (1856–58) observed 'I have lately been especially attending to Geograph. Distrib., & most splendid sport it is,—a grand game of chess with the world for a Board.' Useful as this metaphor may have been, unfortunately from long before Darwin, and ever since, people have been playing this game for real, moving some pieces to places they would not naturally occupy, removing others from places that they would, and removing many pieces from the board entirely. As larger and rarer pieces are likely to go first, we lose the kings and queens, leaving a board housing little more than pawns. We move a grand game of chess towards a commonplace game of draughts, with the concomitant loss of complexity and possibility that that entails. Understanding the present distribution of pieces demands a long-term view of the consequences of human activities, which also serves to highlight the extent of simplification of the game that has taken place. Just how much the rules of the game have changed accordingly is unclear, but the rules undoubtedly have been changed.

Large-scale patterns in the distribution and diversity of life existed long before humans appeared on the scene to interfere with the processes that determine them (e.g. Crane & Lidgard 1989; Crame 2000). Such large-scale patterns still exist today, despite the extent of that interference. Are these the same patterns, driven by the same processes, or have humans had such a significant effect that the present distribution of biodiversity is principally driven by our activities? Either way, we have something worthy of study. Large-scale patterns are worth understanding, whatever the processes that are generating them. For effective conservation, that understanding may be essential. If it turns out that the key processes relate to human influences on the environment, that would be just about as interesting (or frightening) as any result we could imagine.

Acknowledgements

We are grateful to S. Gaston for discussion and comments, and R. Gregory for supplying data on the geographical range and population sizes of European birds. This work was supported in part by the Darwin Initiative of the U.K. Department for Environment, Food and Rural Affairs.

References

Alroy, J. (2001) A multispecies overkill simulation of the end-Pleistocene megafaunal mass extinction. *Science* **292**, 1893–1896.

Anderson, A.J. (1991) The chronology of colonization in New Zealand. *Antiquity* **65**, 767–795.

Bennett, P.M., Owens, I.P.F. & Baillie, J.E.M. (2001) The history and ecological basis of extinction and speciation in birds. In: *Biotic Homogenization* (eds J.L. Lockwood & M.L. McKinney), pp. 201–222. Kluwer Academic/Plenum Publishers, New York.

BirdLife International (2000) *Threatened Birds of the World.* Lynx Edicions and BirdLife International, Barcelona and Cambridge.

BirdLife International/European Bird Census Council (2000) *European Bird Populations: Estimates and Trends.* BirdLife International, Cambridge.

Blackburn, T.M. & Duncan, R.P. (2001) Establishment patterns of exotic birds are constrained by non-random patterns in introduction. *Journal of Biogeography* **28**, 927–939.

Blackburn, T.M. & Gaston, K.J. (1998) Some methodological issues in macroecology. *American Naturalist* **151**, 68–83.

Brown, J.H. (1995) *Macroecology.* University of Chicago Press, Chicago, IL.

Brown, J.H. & Maurer, B.A. (1989) Macroecology: the division of food and space among species on continents. *Science* **243**, 1145–1150.

Burney, D.A., James, H.F., Pigott Burney, L., *et al.* (2001) Fossil evidence for a diverse biota from Kaua'i and its transformation since human arrival. *Ecological Monographs* **71**, 615–641.

Cassels, R. (1984) The role of prehistoric man in the faunal extinctions of New Zealand and other Pacific islands. In: *Quaternary Extinctions: a Prehistoric Revolution* (eds P.S. Martin & R.G. Klein), pp. 741–767. Arizona University Press, Tucson.

Cassey, P. (2001) Determining variation in the success of New Zealand land birds. *Global Ecology and Biogeography* **10**, 161–172.

Crame, J.A. (2000) The nature and origin of taxonomic diversity gradients in marine bivalves. In: *The Evolutionary Biology of the Bivalvia* (eds E.M. Harper, J.D. Taylor & J.A. Crame). *Geological Society, London, Special Publications* **177**, 347–360.

Crane, P.R. & Lidgard, S. (1989) Angiosperm diversification and paleolatitudinal gradients in Cretaceous floristic diversity. *Science* **246**, 675–678.

Crawley, M.J., Harvey, P.H. & Purvis, A. (1997) Comparative ecology of the native and alien floras of the British Isles. In: *Plant Life Histories: Ecology, Phylogeny and Evolution* (eds J. Silvertown, M. Franco & J.L. Harper), pp. 36–52. Cambridge University Press, Cambridge.

Crooks, K.R. & Soulé, M.E. (1999) Mesopredator release and avifaunal extinctions in a fragmented system. *Nature* **400**, 563–566.

Daehler, C.C. (1998) The taxonomic distribution of invasive angiosperm plants: ecological insights and comparisons to agricultural weeds. *Biological Conservation* **84**, 167–180.

Darwin, C. (1856–58) [publ. 1975] *Charles Darwin's Natural Selection: being the second part of his big species book written from 1856–1858* (ed. R.C. Stauffer). Cambridge University Press, Cambridge.

De Juana, E. (1992) Class Aves (birds). In: *Handbook of the Birds of the World*, Vol. 1 (eds J. del Hoyo, A. Elliott & J. Sargatal), pp. 36–73. Lynx Edicions, Barcelona.

Del Hoyo, J., Elliott, A. & Sargatal, J. (eds) (1992) *Handbook of the Birds of the World*, Vol. 1. *Ostrich to Ducks.* Lynx Edicions, Barcelona.

Del Hoyo, J., Elliott, A. & Sargatal, J. (eds) (1996) *Handbook of the Birds of the World*, Vol. 3. *Hoatzin to Auks.* Lynx Edicions, Barcelona.

Diamond, J.M. (1984) 'Normal' extinctions of isolated populations. In: *Extinctions* (ed. M.H. Nitecki), pp. 191–246. University of Chicago Press, Chicago, IL.

Diamond, J. (1989) Overview of recent extinctions. In: *Conservation for the Twenty-first Century* (eds D. Western & M. Pearl), pp. 37–41. Oxford University Press, New York.

Diamond, J. (1998) *Guns, Germs and Steel: a Short History of Everybody for the Last 13 000 Years.* Vintage, London.

Duncan, R.P., Blackburn, T.M. & Worthy, T.H. (2002) Prehistoric bird extinctions and human hunting. *Proceedings of the Royal Society, London, Series B* **269**, 517–521.

Eldredge, N. (1999) Cretaceous meteor showers, the human ecological 'niche', and the sixth extinction. In: *Extinctions in Near Time: Causes, Contexts and Consequence* (ed. R.D.E. MacPhee), pp. 1–15. Kluwer Academic/Plenum, New York.

Fisher, J. & Paterson, R.T. (1964) *The World of Birds: a Comprehensive Guide to General Ornithology.* MacDonald, London.

Flannery, T. (2001) *The Eternal Frontier: an Ecological History of North America and its Peoples.* Heinemann, London.

Gaston, K.J. (1998) Species-range size distributions: products of speciation, extinction and transformation. *Philosophical Transactions of the Royal Society, London, Series B* **353**, 219–230.

Gaston, K.J. (2002) Extinction. In: *Encyclopedia of Evolution*, vol. 1 (ed. M. Pagel), pp. 344–349. Oxford University Press, New York.

Gaston, K.J. & Blackburn, T.M. (1995) Birds, body size, and the threat of extinction. *Philosophical Transactions of the Royal Society, London, Series B* **347**, 205–212.

Gaston, K.J. & Blackburn, T.M. (1996) Global scale macroecology: interactions between population size, geographic range size and body size in the Anseriformes. *Journal of Animal Ecology* **65**, 701–714.

Gaston, K.J. & Blackburn, T.M. (1997) How many birds are there? *Biodiversity and Conservation* **6**, 615–625.

Gaston, K.J. & Blackburn, T.M. (1999) A critique for macroecology. *Oikos* **84**, 353–368.

Gaston, K.J. & Blackburn, T.M. (2000) *Pattern and Process in Macroecology.* Blackwell Science, Oxford.

Gaston, K.J. & Chown, S.L. (1999) Geographic range size and speciation. In: *Evolution of Biological Diversity* (eds A.E. Magurran & R.M. May), pp. 236–259. Oxford University Press, Oxford.

Grayson, D.K. (2001) The archaeological record of human impacts on animal populations. *Journal of World Prehistory* **15**, 1–68.

Gregory, R.D. (1994) Species abundance patterns of British birds. *Proceedings of the Royal Society, London, Series B* **257**, 299–301.

Gregory, R.D. (2000) Abundance patterns of European breeding birds. *Ecography* **23**, 201–208.

Greuter, W. (1995) Extinctions in Mediterranean areas. In: *Extinction Rates* (eds J.H. Lawton & R.M. May), pp. 88–97. Oxford University Press, Oxford.

Hannah, L., Carr, J.L. & Lankerani, A. (1995) Human disturbance and natural habitat: a biome level analysis of a global data set. *Biodiversity and Conservation* **4**, 128–155.

Heather, B. & Robertson, H. (1996) *The Field Guide to the Birds of New Zealand.* Oxford University Press, Oxford.

Holdaway, R.N. (1989) New Zealand's pre-human avifauna and its vulnerability. *New Zealand Journal of Ecology* **12** (Supplement), 11–25.

Holdaway, R.N. (1996) Arrival of rats in New Zealand. *Nature* **384**, 225–226.

Holdaway, R.N. (1999) Introduced predators and avifaunal extinction in New Zealand. In:

Extinctions in Near Time: Causes, Contexts and Consequences (ed. R.D.E. MacPhee), pp. 189–238. Kluwer Academic/Plenum, New York.

Holdaway, R.N. & Jacomb, C. (2000) Rapid extinction of the moas (Aves: Dinornithiformes): model, test, and implications. *Science* **287**, 2250–2254.

Holdaway, R.N., Worthy, T.H. & Tennyson, A.J.D. (2001) A working list of breeding bird species of the New Zealand region at first human contact. *New Zealand Journal of Zoology* **28**, 119–187.

Hubbell, S.P. (2001) *The Unified Neutral Theory of Biodiversity and Biogeography*. Princeton University Press, Princeton, NJ.

Hunter, M.L. Jr. (1996) *Fundamentals of Conservation Biology*. Blackwell Science, Cambridge, MA.

Isenberg, A.C. (2000) *The Destruction of the Bison: an Environmental History, 1750–1920*. Cambridge University Press, Cambridge.

Jackson, J.B.C. (2001) What was natural in the coastal oceans? *Proceedings of the National Academy of Sciences, USA* **98**, 5411–5418.

Kershaw, A.P. (1986) Climatic change and Aboriginal burning in north-east Australia during the last two glacial/interglacial cycles. *Nature* **322**, 47–49.

Klein Goldewijk, K. (2001) Estimating global land use change over the past 300 years: the HYDE database. *Global Biogeochemical Cycles* **15**, 417–433.

Krech, S. III. (1999) *The Ecological Indian: Myth and History*. Norton, New York.

Kunin, W.E. (1997) Introduction: on the causes and consequences of rare–common differences. In: *The Biology of Rarity: Causes and Consequences of Rare–Common Differences* (eds W.E. Kunin & K.J. Gaston), pp. 3–11. Chapman and Hall, London.

Lawton, J.H. & May, R.M. (eds) (1995) *Extinction Rates*. Oxford University Press, Oxford.

Lockwood, J.L. (1999) Using taxonomy to predict success among introduced avifauna: Relative importance of transport and establishment. *Conservation Biology* **13**, 560–567.

Lockwood, J.L., Brooks, T.M. & McKinney, M.L. (2000) Taxonomic homogenization of the global avifauna. *Animal Conservation* **3**, 27–35.

Lomolino, M.V. & Channell, R. (1995) Splendid isolation: patterns of geographic range collapse in endangered mammals. *Journal of Mammalogy* **76**, 335–347.

MacPhee, R.D.E. (ed.) (1999) *Extinctions in Near Time: Causes, Contexts, and Consequences*. Kluwer Academic/Plenum, New York.

Martin, P.S. (1984) Prehistoric overkill: the global model. In: *Quaternary Extinctions: a Prehistoric Revolution* (eds P.S. Martin & R.G. Klein), pp. 354–403. Arizona University Press, Tucson.

Martin, P.S. (2001) Mammals (Late Quaternary), extinctions of. In: *Encyclopedia of Biodiversity*, Vol. *3* (ed. S.A. Levin), pp. 825–839. Academic Press, San Diego, CA.

Maurer, B.A. (1999) *Untangling Ecological Complexity*. University of Chicago Press, Chicago.

Maurer, B.A., Brown, J.H. & Rusler, R.D. (1992) The micro and macro in body size evolution. *Evolution* **46**, 939–953.

McGlone, M.S. (1983) Polynesian deforestation of New Zealand: a preliminary synthesis. *Archaeology in Oceania* **18**, 11–25.

McGlone, M.S. & Basher, L.R. (1995) The deforestation of the upper Awatere catchment, Inland Kaikoura Range, Marlborough, South Island, New Zealand. *New Zealand Journal of Ecology* **19**, 63–66.

McKinney, M.L. & Lockwood, J.L. (1999) Biotic homogenization: a few winners replacing many losers in the next mass extinction. *Trends in Ecology and Evolution* **14**, 450–453.

Milberg, P. & Tyrberg, T. (1993) Naïve birds and noble savages—a review of man-caused prehistoric extinctions of island birds. *Ecography* **16**, 229–250.

Miller, G.H., Magee, J.W. Johnson, B.J., *et al.* (1999) Pleistocene extinction of *Genyornis newtoni*: human impact on Australian megafauna. *Science* **283**, 205–208.

Murray, B.R. & Dickman, C.R. (2000) Relationships between body size and geographical range size among Australian mammals: has human impact distorted macroecological patterns? *Ecography* **23**, 92–100.

Myers, N. & Knoll, A.H. (2001) The biotic crisis and the future of evolution. *Proceedings of the National Academy of Sciences, USA* **98**, 5389–5392.

Nee, S., Harvey, P.H. & May, R.M. (1991) Lifting the veil on abundance patterns. *Proceedings of the Royal Society, London, Series B* **243**, 161–163.

Olson, S.L. & James, H.F. (1982) Fossil birds from the Hawaiian Islands: evidence for wholesale extinction by man before western contact. *Science* **217**, 633–635.

Owens, I.P.F. & Bennett, P.M. (1995) Ancient ecological diversification explains life-history variation among living birds. *Proceedings of the Royal Society, London, Series B* **261**, 227–232.

Owen-Smith, N. (1989) Megafaunal extinctions: the conservation message from 11 000 years B.P. *Conservation Biology* **3**, 405–412.

Pimentel, D. (2001) Agricultural invasions. In: *Encyclopedia of Biodiversity*, Vol. 1 (ed. S.A. Levin), pp. 71–83. Academic Press, San Diego, CA.

Pimm, S.L. (1996) Lessons from a kill. *Biodiversity and Conservation* **5**, 1059–1067.

Pimm, S.L., Moulton, M.P. & Justice, L.J. (1995) Bird extinctions in the central Pacific. In: *Extinction Rates* (eds J.H. Lawton & R.M. May), pp. 75–87. Oxford University Press, Oxford.

Pudjoarinto, A. & Cushing, E.J. (2001) Pollen-stratigraphic evidence of human activity in Dieng, Central Java. *Palaeogeography, Palaeoclimatology, Palaeoecology* **171**, 329–340.

Pyšek, P. (1998) Is there a taxonomic pattern to plant invasions? *Oikos* **82**, 282–294.

Reis, K.R. & Garong, A.M. (2001) Late Quaternary terrestrial vertebrates from Palawan Island, Philippines. *Palaeogeography, Palaeoclimatology, Palaeoecology* **171**, 409–421.

Remsen, J.V. Jr. (1995) The importance of continued collecting of bird specimens to ornithology and bird conservation. *Bird Conservation International* **5**, 145–180.

Sadler, J.P. (1999) Biodiversity on oceanic islands: a palaeoecological assessment. *Journal of Biogeography* **26**, 75–87.

Schorger, A.W. (1955) *The Passenger Pigeon: its Natural History and Extinction.* University of Wisconsin Press, Madison.

Shuntov, V.P. (1974) *Sea Birds and the Biological Structure of the Ocean.* Translated for Bureau of Sport Fisheries and Wildlife, U.S. Department of the Interior and the National Science Foundation, Washington, DC.

Smith, A.P. & Quin, D.G. (1996) Patterns and causes of extinction and decline in Australian conilurine rodents. *Biological Conservation* **77**, 243–267.

Smith, F.D.M., May, R.M., Pellew, R., Johnson, T.H. & Walter, K.S. (1993) Estimating extinction rates. *Nature* **364**, 494–496.

Steadman, D.W. (1993) Biogeography of Tongan birds before and after human impact. *Proceedings of the National Academy of Sciences, USA* **90**, 818–822.

Steadman, D.W. (1995) Prehistoric extinctions of Pacific Island birds: biodiversity meets zooarchaeology. *Science* **267**, 1123–1131.

Steadman, D.W. (1997) The historic biogeography and community ecology of Polynesian pigeons and doves. *Journal of Biogeography* **24**, 737–753.

Stork, N.E. & Lyal, C.H.C. (1993) Extinction or 'co-extinction' rates? *Nature* **366**, 307.

Terborgh, J. (1988) The big things that run the world—a sequel to E.O. Wilson. *Conservation Biology* **2**, 402–403.

Terborgh, J. (1999) *Requiem for Nature.* Island Press, Washington.

Terborgh, J., Robinson, S.K., Parker, T.A. III, Munn, C.A. & Pierpont, N. (1990) Structure and organization of an Amazonian forest bird community. *Ecological Monographs* **60**, 213–238.

Terborgh, J., Lopez, L., Nuñez V.P., *et al.* (2001) Ecological meltdown in predator-free forest fragments. *Science* **294**, 1923–1926.

Thiollay, J.-M. (1994) Structure, density and rarity in an Amazonian rainforest bird community. *Journal of Tropical Ecology* **10**, 449–481.

Tokeshi, M. (1996) Power fraction: a new explanation of relative abundance patterns in species-rich assemblages. *Oikos* **75**, 543–550.

Tokeshi, M. (1999) *Species Coexistence.* Blackwell Science, Oxford.

Udvardy, M.D.F. (1957) An evaluation of quantitative studies in birds. *Cold Spring Harbor Symposium* **22**, 301–311.

Warren, P.H. & Gaston, K.J. (1997) Interspecific abundance–occupancy relationships: a test of mechanisms using microcosms. *Journal of Animal Ecology* **66**, 730–742.

Webb, T.J. & Gaston, K.J. (2000) Geographic range size and evolutionary age in birds. *Proceedings of the Royal Society, London, Series B* **267**, 1843–1850.

Webb, T.J., Kershaw, M. & Gaston, K.J. (2001) Rarity and phylogeny in birds. In: *Biotic Homogenization* (eds J.L. Lockwood & M.L. McKinney), pp. 57–80. Kluwer Academic/Plenum Publishers, New York.

Wiens, J.A. (1991) The ecology of desert birds. In:

The Ecology of Desert Communities (ed. G.A. Polis), pp. 278–310. University of Arizona Press, Tucson.
Wiens, J.A. & Dyer, M.I. (1975) *Rangeland avifaunas: their composition, energetics, and role in the ecosystem.* USDA Forest Service General Technical Report **WO-1**, 146–181.
Wilcove, D.S. (2000) *The Condor's Shadow: the Loss and Recovery of Wildlife in America.* Anchor Books, New York.
Williams, G.R. (1981) Aspects of avian island biogeography in New Zealand. *Journal of Biogeography* **8**, 439–456.
Wilson, E.O. (1985) The biological diversity crisis: a challenge to science. *BioScience* **35**, 700–706.
Wood, G.L. (1982) *The Guinness Book of Animal Facts and Feats,* 3rd edn. Guinness Superlatives, Enfield.
Worthy, T.H. (1997) What was on the menu? Avian extinction in New Zealand. *New Zealand Journal of Archaeology* **19**, 125–160.

Chapter 19

Evolutionary macroecology and the fossil record

David Jablonski, Kaustuv Roy and James W. Valentine*

Introduction

From the very outset of the recent burst of work in macroecology it has been clear that major macroecological patterns have a strong historical underpinning. Many spatial relationships bear the imprint of Pleistocene climatic changes, and patterns underlain by differential speciation and extinction have their roots even deeper in geological time. Although efforts to infer long-term dynamics from the topology of phylogenetic trees are increasing, many macroecological analyses inevitably yield only static snapshots of biotic patterns. The fossil record offers a rich archive of natural experiments in the *dynamic* relationships among many of the variables of interest to macroecologists, including how those variables respond to perturbations of all magnitudes. Thanks to the temporal scope and thus range of phenomena encompassed by the fossil record, these perturbations range from modest temperature changes to extreme glacial–interglacial cycles, and from subtle shifts in productivity that track long-term changes in the carbon cycle to sudden upheavals triggered by the impact of objects 10 km in diameter. Thus, although as with any large and heterogeneous database care must be taken to avoid sampling and other artefacts, palaeontology can provide direct empirical data on the dynamics underlying macroecological patterns: spatial shifts, origination, extinction and trends over spans of 10^3 to 10^7 years (see also Clarke & Crame, this volume). At the same time, palaeontologists need to incorporate macroecological insights into their research, particularly regarding the linkage and covariation found among the important variables (Lawton 1999; Gaston & Blackburn 2000; Blackburn & Gaston 2001).

Spatially explicit palaeobiological research has been somewhat neglected in favour of synoptic, global-scale analyses, but a growing body of palaeobiological work with a strong spatial component is a welcome development that will further promote the integration of disciplines into what might be termed evolutionary macroecology.

* *Correspondence address: djablons@midway.uchicago.edu*

Targets for palaeobiological analysis have included, among others: the relationship between intrinsic biological traits and geographical distributions as well as durations of taxa (reviewed e.g. by Stanley 1979, 1990; Jablonski 1995; McKinney 1997a,b; Kammer *et al.* 1998); the shape and dynamics of body-size distributions (Jablonski 1996, 1997; Alroy 1998; Roy *et al.* 2001a); the packing of species and higher taxa into regions and provinces, through time and along gradients of latitude, longitude and depth (Valentine 1973; Valentine *et al.* 1978; Stanley 1979; Sepkoski 1991; Jablonski 1993); and ecological responses to changes in the physical environment, in resource availability, and in biotic interactions (reviews by Vermeij 1994; Jablonski 1995, 2001, 2002; Jablonski & Sepkoski 1996; Miller 1998; Erwin 2001).

Marine macroecology, an essential bridge to the fossil record

Palaeontologists have studied macroecological aspects of the fossil record for decades, just as some ecologists did macroecology long before the term was introduced (see Lawton 1999). The richness and density of the marine fossil record has promoted a predominantly marine invertebrate approach to evolutionary macroecology. Certainly, there are a number of important studies in the evolutionary macroecology of terrestrial organisms. However, owing to sampling and other obstacles, these studies are restricted mainly to North American and European mammals, plants and insects. Forging stronger ties between palaeobiology and ecology on one hand, and establishing the generality of macroecological 'rules' on the other, will clearly require a serious effort in the macroecology of the modern marine biota.

Gaston & Blackburn (1999, p. 355) list a set of variables important to macroecology, including 'species richness, abundance, range size, body size, trophic or functional group, life history, and reproductive traits'. Interrelationships among some of these variables, such as abundance, geographical range and body size, have been a central focus of terrestrial macroecology. These variables are still not well known for marine organisms (e.g. Chapman 1999), nor have their interactions been much addressed in palaeobiological analyses. However, the compilation of distributional data and a number of other key variables is proceeding for major groups.

Marine and terrestrial environments differ profoundly in terms of key biotic and abiotic processes, ranging from patterns of primary productivity to structures of food webs to modes of dispersal and aspects of life histories (Clarke 1993; Steele *et al.* 1993; Cohen 1994; Roughgarden *et al.* 1994). One of the most intriguing research agendas is simply to test whether these differences in processes lead to different macroecological patterns in the sea. For example, the reproductive strategies of most marine invertebrates select for life histories that differ in many respects from those of the land vertebrates. Dispersal tends to be more extensive, and dispersal stages much more abundant, in marine invertebrates (e.g. Strathmann 1990). Further, fecundity usually scales positively with body size rather than negatively as in birds and mammals, a contrast that may be important in understanding patterns in size–frequency distributions and the role of body size in invasion success (e.g. Roy *et al.* 2000a, 2001b).

Body size

It has become a cliché that body size affects almost every aspect of the biology of a species, from physiology to life history, and that it plays an important role in the organization of ecological communities. Here we address only two aspects of body size in the marine fossil record: its relation to the volatility of geographical ranges in response to changing climates and other perturbations, and the behaviour of clades relative to modal body sizes over geological timescales.

Range expansions

The west coast of North America contains one of the world's best-studied Pleistocene (Valentine 1961, 1989) and Recent molluscan faunas (see Roy *et al.* 2001a, 2002). Comparisons of Pleistocene and modern distributions of these species have shown that the range limits of many of these taxa shifted significantly in response to past climatic changes (both warming and cooling). Body size appears to have played an important role in mediating the responses of these molluscan species to climate change. The extralimital species, as a group, have significantly larger body sizes than the rest of the Pleistocene species pool (Roy *et al.* 2001b; Fig. 19.1). Interestingly, the same size bias is also seen in marine bivalve species with geographical ranges that have expanded in historical times thorough human-mediated introductions (Roy *et al.* 2001b). In the latter case, the selectivity is evident for regional assemblages and in a global analysis of the mussels (Family Mytilidae), chosen because it is the clade with the greatest number of (non-commercial) invasive species (Roy *et al.* 2002) (Fig. 19.2). Among the invasive bivalves of the temperate northeastern Pacific coast, large invasive species are also significantly more widespread in their new ranges compared with small ones (Roy *et al.* 2002). Finally, on a geological timescale, large-bodied bivalve genera predominated in biotic interchanges following the end-Cretaceous (K–T) mass extinction, 65 million years ago (Ma). The K–T event was one of the large five mass extinctions of the Phanerozoic, and although extinction intensities were surprisingly homogeneous worldwide, the recovery, which lasted several million years at least, differed among continents (Jablonski 1998). The faunal recovery in the Gulf Coast of the USA was more strongly driven by invading species than in the other regions. As shown in Fig. 19.3, the post-Cretaceous invaders in the Gulf Coast are significantly larger than the taxa indigenous to the region. Although the data in this last example are at the generic level, the patterns match those in the Pleistocene and Recent examples.

All of these observations together suggest that range limits of large-bodied marine bivalves tend to be more volatile compared with those of smaller species. These patterns are different from terrestrial case studies, involving birds and mammals, where no consistent relationship between body size and invasion success has emerged (Veltman *et al.* 1996; Forys & Allen 1999; Duncan *et al.* 2001). One hypothesis for this contrast between bivalves and higher vertebrates is that the difference derives from the positive relationship between body size and fecundity in marine bivalves, as opposed to the negative relationship seen for interspecific comparisons in mammals and birds (Roy *et al.* 2002). The difference is seen even when larval

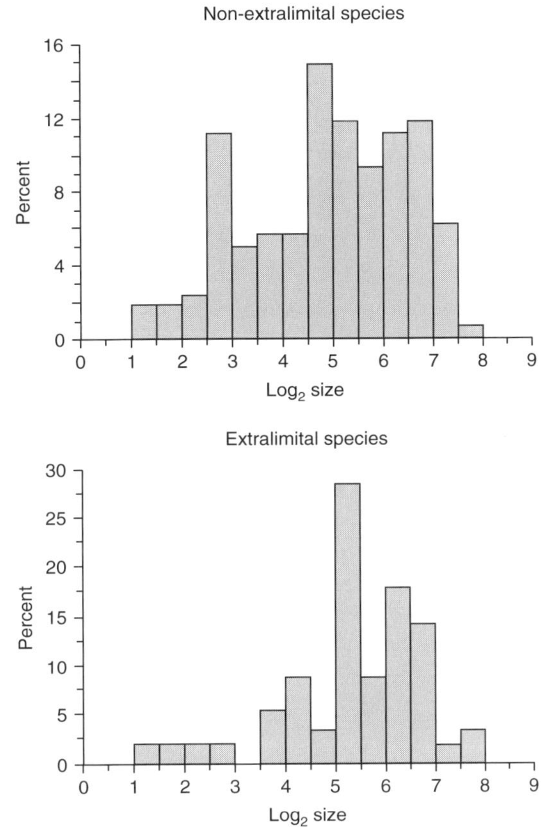

Figure 19.1 Size–frequency distributions of species with Pleistocene occurrences that do not fall outside their present-day range (top; $n = 160$), compared with species known to have occurred at least 1° latitude outside their present-day range limits during the Pleistocene (bottom; $n = 56$). The extralimital species are significantly larger in size (body size is taken as the geometric mean of length and height (mm); size data $\log_2$-transformed before analysis; $p < 0.01$, Mann-Whitney *U* test). (After Roy *et al.* 2001b.)

mode, and thus dispersal ability, is held constant (Roy *et al.* 2002). In any event, the concordance of modern, Pleistocene and Cretaceous results, despite the very different sampling issues, different modes of transport and different kinds of perturbations in the recipient provinces, suggest a general rule relating invasiveness to body size for marine bivalves, but one that may not be applicable to terrestrial vertebrates.

Body size evolution

The fossil record permits us to track the evolutionary dynamics underlying macroecological patterns in body size, which have been quite controversial (see reviews by Jablonski 1996; Gaston & Blackburn 2000), and holds some surprises.

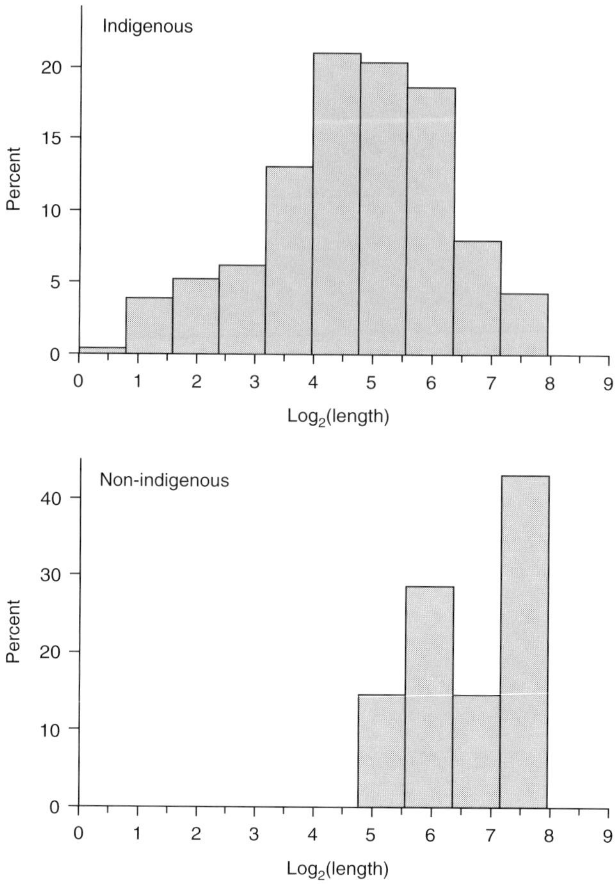

Figure 19.2 Size–frequency distributions for the intertidal and shelf-depth mytilid bivalve species of the world. Non-mariculture species that have been introduced outside their native ranges through human activity (below, $n = 7$) are significantly larger than the species not known to occur outside their native ranges (indigenous species, above, $n = 292$); $p = 0.0001$, Mann–Whitney *U* test. (After Roy *et al.* 2002.)

Roy *et al.* (2000a) found remarkable constancy of body size distributions over four biotic provinces arrayed along the northeastern Pacific shelf from the Equator to the Arctic Ocean, despite a fourfold decrease in species richness, an almost complete turnover in species, and a considerable shift in family-level composition. A more detailed analysis revealed latitudinal trends in body size within individual provinces but no net trend from the Equator to the North Pole (Roy & Martien 2001). In addition, marine bivalve size–frequency distributions tend to be log-normal or slightly left-skewed (Roy *et al.* 2000a) as opposed to the right skew seen in most vertebrate clades (e.g. Gaston & Blackburn 2000).

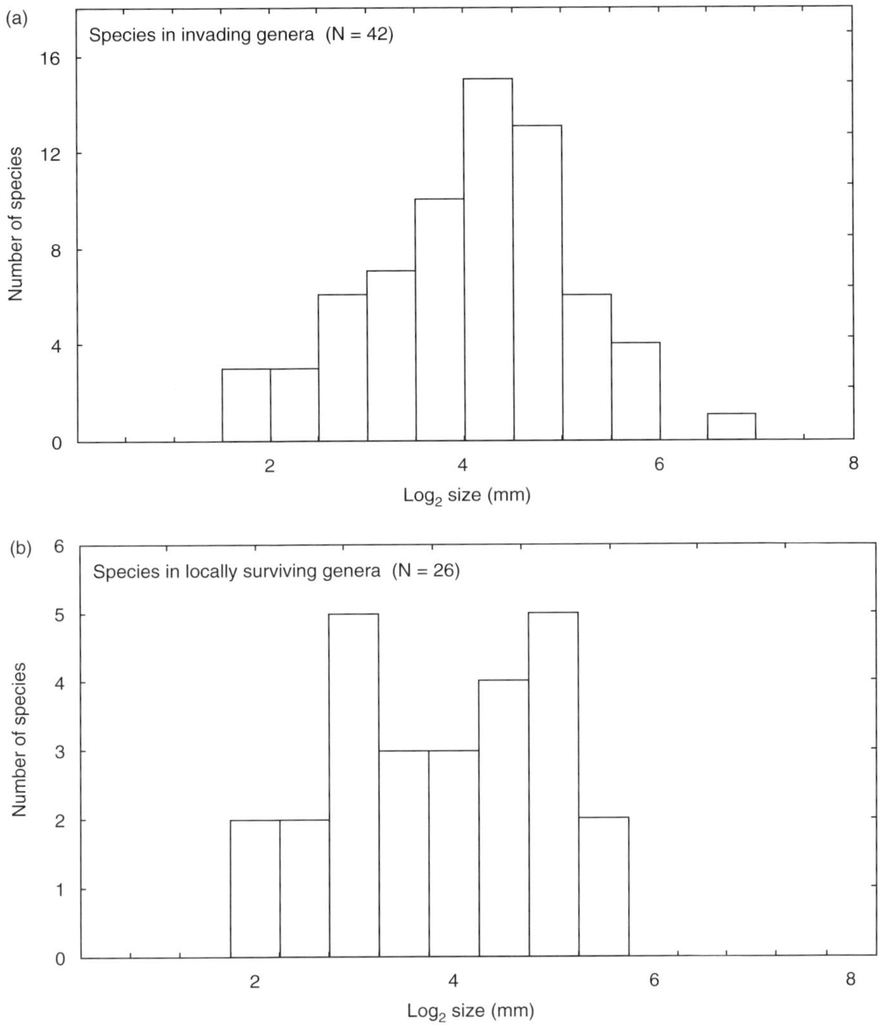

Figure 19.3 Bivalve invasions of the North American Coastal Plain province after the end-Cretaceous mass extinction were size-selective. Species belonging to genera that invaded the province (a) were significantly larger than species belonging to genera that survived locally (b) ($p = 0.02$, Mann–Whitney *U* test).

The energetic model of Brown *et al.* (1993), developed for mammalian faunas, successfully predicts the modal size of the northeastern Pacific bivalve distributions, thereby suggesting a role of energetics in the evolution of marine bivalve size distributions (Roy *et al.* 2000a). However, despite the stability of the body-size mode over four provinces spanning more than 75 degrees of latitude, the fossil record provides no evidence that the mode operates as an evolutionary attractor (Roy *et al.* 2000a). A more realistic alternative, that the mode is indeed a clade-wide optimum but entry is generally blocked by established species (e.g. Maurer *et al.* 1992; Brown 1995), is also undermined by the palaeontological evidence for the highly dynamic behaviour of bivalve lineages relative to the modal size (Fig. 19.4). The failure of these molluscan lineages to exhibit any kind of organized dynamics relative to the mode suggests that the temporal and spatial constancy of that mode involves species sorting, that is differential origination and extinction within and among size classes, rather than directional trends in individual lineages (Roy *et al.* 2000a). Again, speciation and extinction rates operating over millions of years are implicated in the shaping of macroecological patterns.

Geographical range and evolutionary dynamics

The relationship between geographical range size and evolutionary dynamics—speciation rates, extinction rates and species durations—figures into many macroecological models and analyses (for reviews see Chown 1997: Gaston & Chown 1999; Gaston & Blackburn 2000; Hubbell 2001). Palaeontological data can be used to test these relationships directly, in systems where sampling and preservation are unlikely to overwhelm the biological signal.

Extinction

An inverse relationship between geographical range and extinction rate, or a positive relationship between geographical range and species duration, has been documented in a number of fossil molluscan assemblages (Jackson 1974; Hansen 1978, 1982; Stanley 1979; Jablonski 1986a, 1987, 1995; see also McKinney 1997a; Gaston & Blackburn 2000, p. 120), and is also generally supported by ecological data and theory (e.g. Maurer & Nott 1998; Gaston & Blackburn 2000, p. 174, and references therein). Analysis of a revised and updated data set for the Late Cretaceous gastropods of the Gulf and Atlantic Coastal Plain of North America shows a significant positive relationship between maximum geographical range of species at any one time and their durations in millions of years (Fig. 19.5). The geographical and stratigraphical ranges of these fossil species should not be taken as absolute values, because they are subject to incomplete sampling, but as a rank-order array of ranges and durations useful for interclade comparisons (e.g. Jablonski 1987; Jablonski & Valentine 1990). This relationship between geographical range and geological duration appears to be a general one, and in at least some situations can be shown to be robust to the kinds of sampling effects discussed by Russell & Lindberg (1988) (see Jablonski 1988; Marshall 1991; Smith 1994).

The relationship between geographical range and extinction-resistance at the species level should be viewed separately from related patterns at the clade level. However, during times of background extinction the duration of a genus is directly related to the geographical ranges of its constituent species (and this interacts positively with species richness of the genus; Jablonski 1986b). During the end-Cretaceous mass extinction, this positive effect across hierarchical levels from species to clade survival is disrupted, and survivorship is promoted instead by

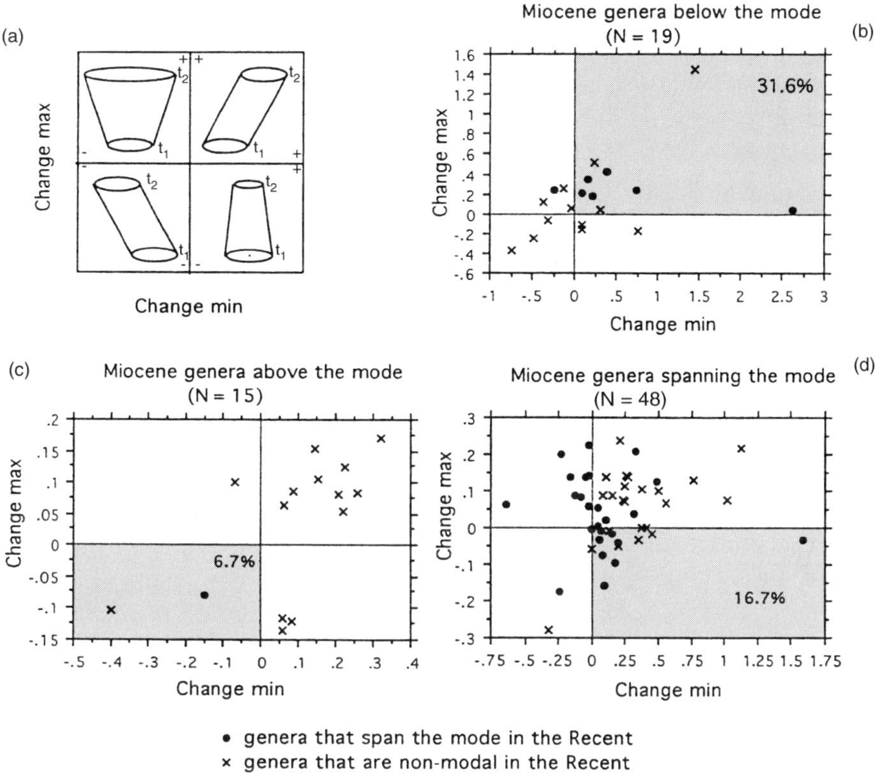

Figure 19.4 Body size evolution in Miocene–Recent bivalve genera. (a) Graphical model for body-size evolution, with each quadrant representing a different evolutionary pattern; as in Jablonski (1987), the vertical axis is the change in the upper bound of the size distribution of the species in a lineage and the horizontal axis is the change in the lower size bound of the distribution. (b–d) Evolutionary size change in 82 Miocene–Recent genera and subgenera of eastern Pacific bivalves, partitioned according to their starting position relative to the modal size; for each plot, the shaded quadrant is the one that would be the most heavily occupied if the modal size was an evolutionary attractor, with the proportion of genera actually falling into that quadrant. Of the genera that started *above* the mode, only 7% showed a directional trend towards it. Of the genera starting *below* the mode, 32% showed a directional shift towards the mode, but that is not significantly different from the 21% of clades that started below the mode but increased their size range by expansion in both directions. Of the larger number of clades that bracketed the mode in the Miocene, only 17% narrowed their size range. An equal percentage of modal clades expanded both upper and lower bounds, and almost 50% show a directional shift *away* from the mode, with most of those actually leaving the modal size class completely. (From Roy *et al.* 2000a.)

broad geographical range at the level of the clade, regardless of the within-province geographical ranges of its species (Jablonski 1986b; Jablonski & Raup 1995). This positive relationship between survivorship and geographical range at the clade level is one of few general rules that have emerged from extensive work on the 'Big Five' ex-

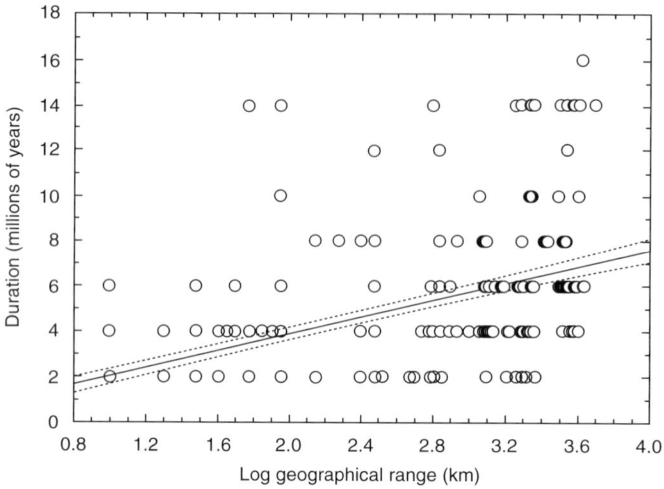

Figure 19.5 Positive relationship between geographical range and stratigraphical duration in Cretaceous gastropod species ($N = 397$). Simple linear regression, with 95% confidence interval, shown for comparative purposes (Pearson's $r = 0.62, p < 0.001$); a more appropriate Spearman's rank-order test is also highly significant ($R = 0.78, p < 0.001$).

tinction events of the geological past (see Jablonski 1995, 2001). The relationship is probably also effective during background times, but is harder to detect then owing to other factors that also influence durations (Jablonski 1995).

Speciation

The relationship between geographical range and speciation rate is much more contentious. The notion of a positive relationship between geographical range and speciation rate or speciation probability is generally derived from the reasonable argument that, all other factors being equal, broad geographical ranges are most likely to be broken by barriers, leading to higher speciation rates (see Rosenzweig 1995; Chown 1997; Gaston 1998; Maurer & Nott 1998; Maurer 1999, pp. 186–189; Hubbell 2001 adopts a similar argument but emphasizes that this model probably applies most strongly to vicariant, rather than peripatric (peripheral-isolate) speciation, and Endler 1977, p. 175, applies this logic to parapatric speciation).

All other factors are rarely equal, however. As many authors have also argued, the factors that promote broad geographical ranges, such as relatively high dispersal abilities, also tend to make them relatively insensitive to barriers and thereby damp speciation rates (e.g. Mayr 1963; Jablonski 1986a; and other papers cited by Gaston & Chown 1999; see also Maurer & Nott 1998; Hubbell 2001, p. 194; Vogler & Ribera, this volume). Conversely, species with limited dispersal ability tend to have more fragmented populations that will make them more vulnerable to both vicariant and peripheral-isolate speciation (as also argued by Maurer & Nott 1998). The direct linkages among broad geographical range, larval dispersal ability and low speciation rates have been documented in several palaeontological analyses of marine gastropods (reviewed by Jablonski & Lutz 1983; see also Hansen 1978, 1982; Jablonski 1986a, 1995; Scheltema 1989, 1992; Gili & Martinell 1994; and see Budd & Johnson (2001) on late Cenozoic corals). Although exceptions exist, the positive relationships among dispersal, gene flow and geographical range are generally well-supported for marine and terrestrial taxa (Bohonak 1999; Kittiwattanawong 1999; Pechenik 1999; Jablonski 2000; Collin 2001). There is also increasing evidence from marine taxa that species with limited dispersal capabilities tend to show more spatial structuring in phylogeographical data compared with those with high-dispersal larvae (e.g. Hellberg *et al.* 2001).

A new analysis of Late Cretaceous gastropods shows a strong inverse relation between the geographical ranges of species and speciation rate per species per million years (Jablonski & Roy 2003) (Fig. 19.6a). If this relationship is underlain by mechanisms for species cohesion such as gene flow, then, as with the extinction patterns, it is important to distinguish processes at the species level from those at the clade level (which are mingled indiscriminately by Gaston 1998; Gaston & Chown 1999; but see several species-level botanical studies with results consistent with those shown here, cited by Chown 1997, p. 97). Whatever the heritability (Jablonski 1987) or phylogenetic inertia (Harvey & Pagel 1991) between species, subdivision or budding of widespread genera may tend to give rise to more descendant genera relative to endemics. Such a relationship was found by Budd & Coates (1992) in a group of Cretaceous corals and Roy (1994) in a group of Mesozoic–Cenozoic gastropods.

We can also test whether the total number of species produced by a clade is a positive or negative function of the geographical ranges of its constituent species. Even if per-species speciation rates are lower in widespread species, it may be that the low per-species speciation rates are offset by the longer durations of their species (Gaston & Chown 1999; Gaston & Blackburn 2000, p. 119). The Cretaceous data, however, do show a weak but significant inverse relationship between geographical ranges of species and the total number of species produced by a clade (Fig. 19.6b). It is not surprising the relationship is weaker: restricted species have shorter durations, so many will die young and leave few descendants. This would yield much scatter at the left end of the plot. Wagner & Erwin (1995) analysed two Neogene clades of planktonic foraminifera and an Ordovician family of marine gastropods, with mixed results. Longer lived taxa showed some tendency to leave more descendants, but Gaston's (1998) reanalysis of the gastropods found no significant partial correlation between number of descendants and geographical range size, controlling for differences in longevity.

The palaeontological data are certainly not free of biases. Perhaps most importantly, they must lack many of the rarest species, and, depending on the spatial scale of sampling relative to the smallest viable geographical ranges, are also likely to lack some of the most spatially restricted species (Raup 1979; McKinney 1997b; Jablonski 1995; Kidwell 2001), an effect that will be accentuated if rarity and geographical range are correlated, as occurs in terrestrial organisms (e.g. Gaston &

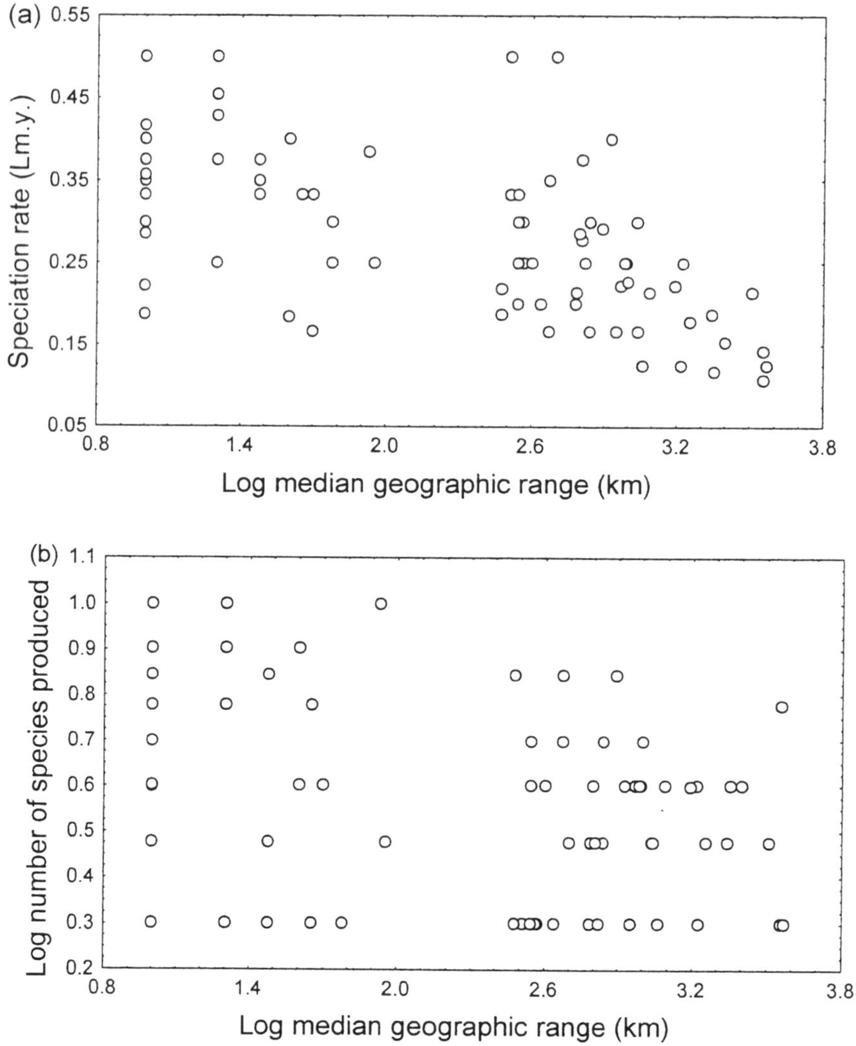

Figure 19.6 Geographical range and speciation in Cretaceous gastropods. (a) Inverse relationship between geographical range and per-species speciation rate in Cretaceous gastropods ($n = 90$ genera, Pearson's $r = -0.66$, Spearman's R for a rank-order test is -0.68, $p < 0.00001$ for both tests); Lm.y. = per-species speciation rate in Lineage-million years (Raup 1985). (b) Weak inverse relationship between geographical range and total numbers of species produced in gastropod lineages over an 18-million-year interval of the latest Cretaceous ($n = 90$ genera, Pearson's $r = -0.24$, $p = 0.025$; Spearman's $R = -0.17$, $p = 0.10$). (After Jablonski & Roy 2003.)

Blackburn 2000) but has yet to be tested in the oceans. If these unrecorded rare species are even more ephemeral than the shortest ranging species captured by palaeontological sampling, as seems likely, this would increase their probability of disappearing before speciating. Adding them back into Fig. 19.6a would produce a downturn of the relationship between number of descendants and geographical range at the lowest ranges, giving the peaked distribution suggested by Gaston & Chown (1999).

Morphology and taxonomy through time

The fossil record provides time-series in more than just taxonomic richness and relative abundances. It permits us to analyse the *morphological* deployment of clades or communities through time. An increasingly sophisticated set of analytical methods and models have become available for the analysis of taxa within a multivariate morphospace (reviews in Foote 1996, 1997; Roy & Foote 1997; McGhee 1999; see Ciampaglio *et al.* 2001 for comparative application of different methods).

Palaeontologists applying these methods to fossil and living organisms have shown that temporal or spatial patterns of morphological diversity need not correlate with species richness, and the times and places where those two metrics of biodiversity are strongly decoupled are especially interesting. This discordance alone is sufficient to demonstrate that such analyses of morphologically defined species are not circular. More importantly, these analyses permit us directly to quantify spatial and temporal trends in specific sets of ecologically or functionally important traits. Such characters are often convergent or plesiomorphic in nature and hence not used to define individual species. For example, morphological disparity—the dispersion of taxa within a morphospace—significantly outpaced taxonomic diversity in the early Palaeozoic diversification of blastozoan echinoderms (see Foote 1996, 1997, 1999). Wills *et al.* (1994) found a similar pattern for Cambrian arthropods: the morphospace occupied by what must be a very incompletely known arthropod fauna (mainly from the Burgess Shale) was roughly equal in volume to that occupied by all living arthropod classes and subclasses, despite the vastly greater species richness and more extensive sampling today. The pattern of morphospace occupation during the intervening 500 million years is a fascinating and largely unexamined issue. Such an early burst of morphological disparity is a common pattern (see e.g. Niklas 1997; Lupia 1999; Thomas *et al.* 2000), although exceptions are known (see Foote 1997). The pattern is attributed most often to the rapid filling of a newly accessible adaptive zone, although ideas involving developmental constraints have staunch advocates (see Valentine 1995; Knoll & Carroll 1999; Valentine *et al.* 1999).

Macroecological analyses of Recent taxa have focused almost exclusively on the origin and maintenance of species richness. As a result, the relationship between macroecological processes and patterns of morphological and functional diversity is largely unknown. The application of palaeontological methods of morphospace analysis to macroecological problems would be a valuable step. Such an interaction would be even more interesting with the addition of a stronger functional compo-

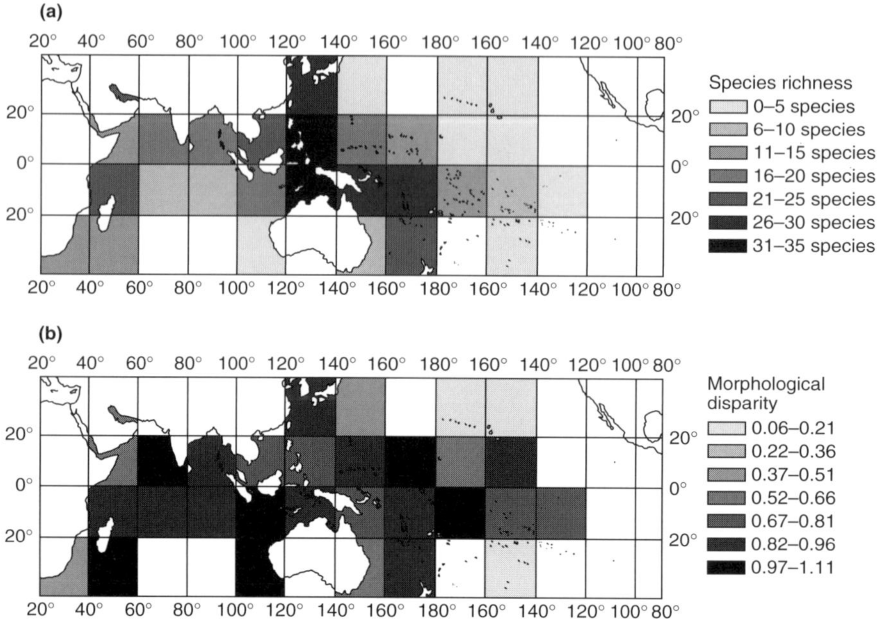

Figure 19.7 Taxonomic (a) and morphological (b) diversity in strombid gastropods of the Indo-West Pacific. Morphological diversity is measured here as disparity, i.e. the geometric mean of the variance of scores on the first six axes of a principal component analysis. (From Roy *et al.* 2001a.)

nent to the palaeo-morphospace work, yielding a synthesis of ecomorphology (e.g. Ricklefs & Miles 1994; Hulsey & Wainwright 2002) and macroecology on evolutionary timescales.

Just as morphological and taxonomic diversity can be decoupled over time, they can be decoupled spatially in the living biota. For example, Roy *et al.* (2001a) showed that spatial patterns of morphological diversity cannot be predicted from data on species richness alone in a large clade of Indo-Pacific gastropods, the Strombidae; regions with relatively few species can still harbour an impressive array of morphologies (Fig. 19.7). The total volume of morphospace occupation correlates fairly well, although non-linearly, with species richness (Fig. 19.8a), but is much more poorly correlated with the dispersion of species within that morphospace (Fig. 19.8b). These plots show that at low species richness the species can be close together or far apart in morphospace, but at high species richness both the total morphospace volume occupied by the clade *and* the spacing among species increases.

Spatial patterns of functional diversity in living molluscs along the northeastern Pacific shelf are also decoupled from taxonomic trends. Species richness of marine bivalves declines by a factor of four from the tropics to the Arctic but the ratio of infaunal species (burrowers such as cockles and razor shells) to epifaunal species

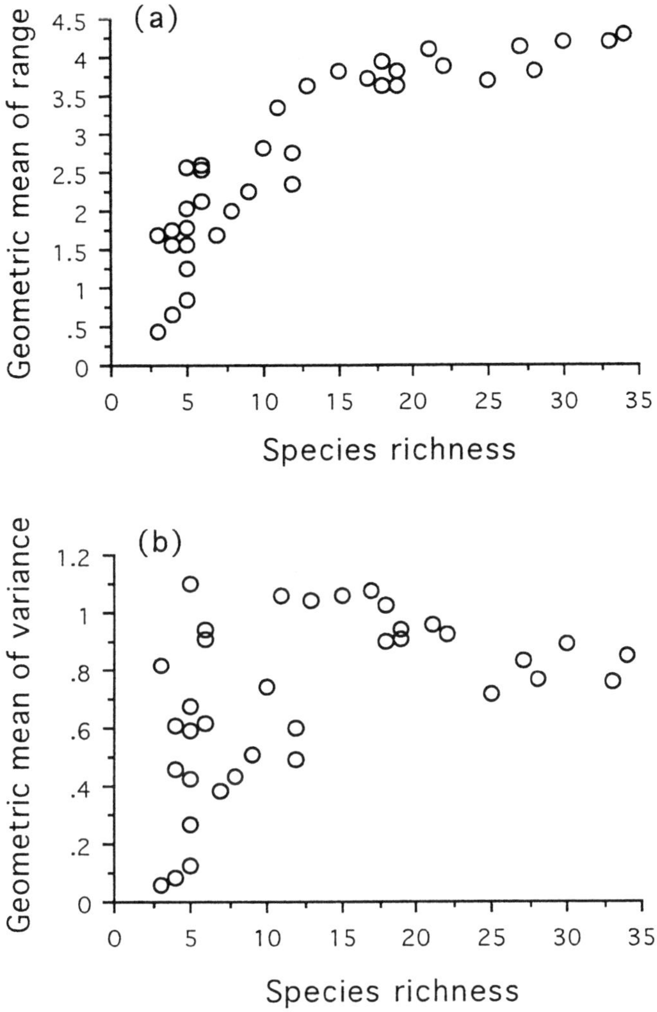

Figure 19.8 Two plots of morphological versus taxonomic diversity in strombids. (a) Morphological diversity as the geometric mean of the *range* of scores on the first six principal components, i.e. the volume of morphospace occupation. (b) Morphological diversity as the geometric mean of the *variance* of scores on the first six axes of a principal component analysis, i.e. the dispersion of morphospace occupation, or disparity. (From Roy *et al.* 2001a.)

(surface-dwellers such as scallops and mussels) increases with latitude (Roy *et al.* 2000b). This contrasts strikingly with the negative slope of the infaunal/epifaunal ratio seen in latest Jurassic bivalves (Tithonian Stage, about 145–150 Ma) (Crame 1996, 2002). Spatial trends in functional diversity can clearly change and even reverse over time, although the underlying processes remain poorly known.

In marine gastropods, trends in functional diversity also tend to be decoupled from trends in species richness (Valentine *et al.* 2002). The steep latitudinal gradient in species richness (Roy *et al.* 1998) contrasts with the non-linear trend in the ratio of carnivorous to non-carnivorous species (Fig. 19.9a). As in bivalves, the processes underlying these trends in gastropod functional diversity remain poorly understood but macroecological processes may play an important role. For example, the fine-scale change in the ratio of gastropod feeding groups is not seen in the ratio at the level of provinces (Fig. 19.9b), indicating more rapid spatial turnover of non-carnivorous species in the temperate zone relative to carnivores there, or relative to either group in the tropics (a given degree of temperate latitude includes fewer overlapping geographical ranges for non-carnivores than carnivores, driving the ratio down on a per-degree basis but keeping it high at the province scale). This, of course, raises the question of whether these differences in distributional patterns largely reflect historical contingencies such as differential extinctions and/or range shifts during Neogene time (Todd *et al.* 2002) or whether they are maintained by ecological processes. Nor do we know much about how the major macroecological variables mentioned above vary among functional groups or how they relate to morphological traits. Palaeontological data have the potential to provide the most direct understanding of the dynamics underlying such large-scale spatial patterns.

A Pleistocene baseline?

Human impacts are increasingly altering every ecosystem on the planet. These changes have produced the shifting baseline syndrome (Pauly 1995; Jackson 1997; Dayton *et al.* 1998), where the ecosystems studied by each generation of ecologists are more degraded than those seen by the previous generation. This syndrome is widespread both on land and in the oceans, from deep-water fisheries to intertidal communities (Jackson 2001; Jackson *et al.* 2001).

Given the magnitude of this environmental disturbance, the Pleistocene and Holocene fossil record may often be our best source for quantifying the relationships among the variables mentioned at the beginning of this chapter. Clearly, the taxonomic richness and abundance structure of many present-day communities are very different from those of just a millennium, a few centuries, or even a few decades ago. This is perhaps most clear for oceanic islands, where the fossil record shows that Polynesians, Melanesians and other 'first arrivals' had a major impact on vertebrate species richness and body-size–frequency distributions (e.g. Steadman 1995; Burney *et al.* 2001; Gaston & Blackburn, this volume). Gaston & Blackburn (2000; p. 297) outline some aspects of macroecology that may be robust to human disruption, such as body-size–life-history trade-offs and other broad features of the biology of species and their interactions with the physical environment. But accurately detecting even these may be problematic in marine ecosystems. For example, intraspecific size distributions of species are being drastically altered by size-selective human predation on marine fish (Jackson *et al.* 2001) and intertidal invertebrates (Pombo & Escofet 1996; Griffiths & Branch 1997; Lindberg *et al.* 1998). These changes are oc-

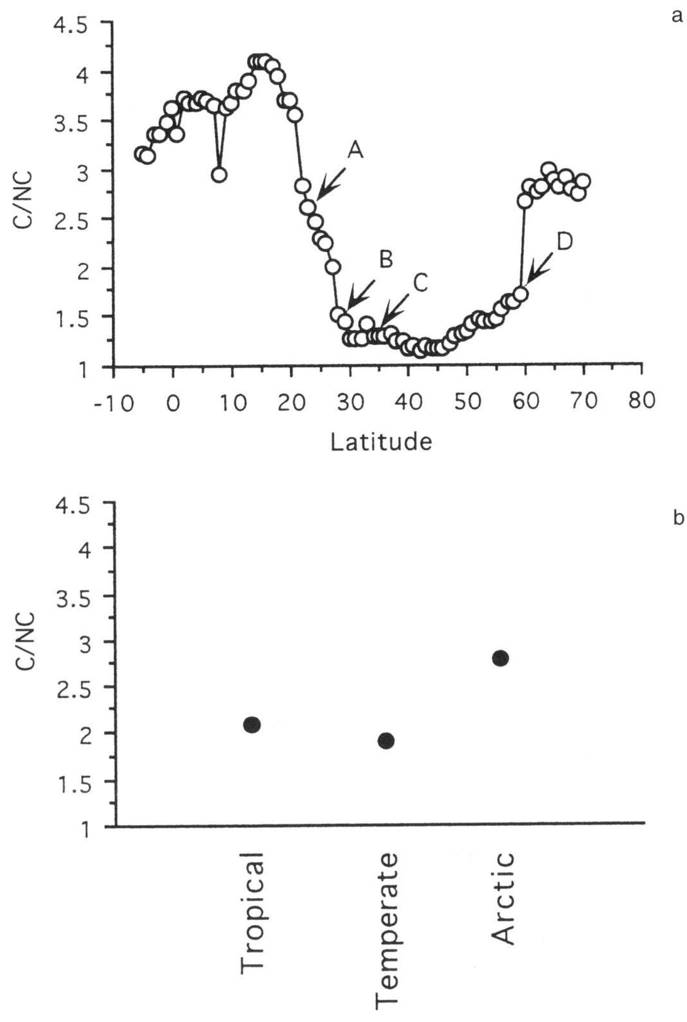

Figure 19.9 Ratio of carnivorous (C) to non-carnivorous (NC) shelled marine gastropods along the eastern Pacific shelf. (a) Plotted for 1° bins; arrows indicate boundaries between molluscan provinces as follows: A, Panamic–Surian; B, Surian–Californian; C, Californian–Oregonian; D, Oregonian–Arctic (after Valentine *et al.* 2002). (b) Same data plotted for major climate zones.

curring in concert with changes in species abundances (Jackson *et al.* 2001). The most effective tests of macroecological hypotheses have been quantitative and not qualitative (e.g. Gaston & Blackburn 1999, 2000; Lawton 1999), hence changes in abundance, body size and maturation age over the past century owing to pressure by fisheries (e.g. Law 2000; Reynolds, this volume) represent distortions in the values of macroecologically important variables that may obscure fundamental relationships, both among biological variables and between biological and physical parameters. These effects are almost certainly present among terrestrial vertebrates as well, but historical trends are yet to be quantified. Finally, if as the data suggest the most severe perturbations are non-randomly distributed along key environmental gradients (e.g. latitude, altitude and bathymetry), then the stability of latitudinal and other gradients in environmental factors such as temperature mean and variance does not guarantee undistorted correlations to biotic patterns. For many taxa the late Pleistocene and Holocene fossil record provides a pre-human baseline and a fuller picture of the envelope of natural variation in the focal variables than can be gleaned from short-term observations on the present-day biota. Incorporation of such historical data would go a long way towards avoiding the pitfalls of doing macroecology in a world dominated by 'unnatural' ecosystems (*sensu* Jackson 2001).

The fidelity with which fossil assemblages reflect the living associations from which they are recruited is being studied, and the results indicate that it is greater than has been generally realized. Macroecologically important parameters are captured by fossils right across the ecological hierarchy. For example, in the Pleistocene of the northeastern Pacific, the species composition of the entire living molluscan fauna is well-represented (e.g. Valentine 1989), and the bioprovincial framework and community-level associations are preserved (Valentine 1961). Although fossil faunas are time-averaged (i.e. represent the accumulation of individuals over many generations), specific morphological parameters, including measures of allometric ratios and of the variability found in living populations, are commonly preserved (Bush *et al.* 2002). Even more encouraging, Kidwell's (2001) meta-analysis of species abundance data shows that dead shells capture the abundance structure of the corresponding live community. Indeed, the fossil assemblages probably constitute a better sample of regional to local biotas, with their large complement of rare species, than do short-term censuses of live individuals. Thus there are aspects of time-averaging that actually enhance macroecological interpretations.

Conclusion

Not only does a stronger partnership between palaeontology and macroecology have great potential, it is likely to be essential for answering many important questions. Macroecology has largely been dominated by the detection and interpretation of correlations among variables. A deeper understanding of causality requires the analysis of dynamics (as advocated, for example, by Gaston 1998). For many groups, palaeobiological evidence provides the most direct way of quantifying dynamics, by providing data on origination, extinction and other features such as shifts in geographical ranges and morphological diversity, to produce an evolutionary macroecology. The fossil record will help to choose between plausible alternatives, as in the relationship between geographical range and speciation rate (Fig. 19.6). The record is sure to offer surprises as well, such as uncovering the greater volatility of geographical range in large-bodied bivalves (Figs 19.1–19.3), the indifference of clade evolutionary trajectories to modal body sizes (Fig. 19.4), and the discordance between taxonomic and morphological diversity in time and space (Figs 19.7 and 19.8). On the other hand, modern neontological research draws upon the full range of organisms within the biosphere, and can provide the indispensable knowledge of the physiological and genetic underpinnings of the macroecological parameters, which is the only hope for achieving a fundamental understanding of fossil patterns. The fields of palaeobiology and macroecology are clearly destined to be conjoined.

Summary

For decades, palaeontologists have been studying macroecological aspects of the fossil record, from explicitly dynamic and hierarchical perspectives. The integration of macroecological and macroevolutionary fields into what might be called evolutionary macroecology will be valuable, because dynamics underlying present-day macroecological patterns tend to operate on palaeontological timescales through such processes as speciation, extinction and range shifts. We present examples where the fossil record yields insight into these processes, including the relationship between range shifts and body size, the dynamics of body-size evolution relative to modal values, extinction and geographical range, speciation and geographical range, and the non-linear relationships between functional or morphological diversity and species richness.

Building stronger ties between palaeobiology and neontology will require a serious effort to explore the macroecology of the modern marine biota. It is not clear that marine patterns correspond to those observed on land for the main macroecological variables such as body size, geographical range and abundance; terrestrial and marine processes may be sufficiently distinct that qualitatively different patterns emerge from similar analyses.

The fossil record also demonstrates that species richness and species associations in many regions — islands and mainlands alike — have changed significantly over the past 10 000 years, and even more dramatically over long timescales, owing to extinctions and migrations. This raises serious doubts about the stability of patterns observed on neontological timescales. Palaeontological data thus offer an important baseline for relationships among body size, geographical range, species richness, abundance and other macroecologically important variables subject to alteration by human activities.

Acknowledgements

We thank T. M. Blackburn, S. M. Kidwell and an anonymous reviewer for helpful reviews. The research reported here was supported by the U.S. National Science Foundation. D.J. is an Honorary Research Fellow of the Natural History Museum, London.

References

Alroy, J. (1998) Cope's rule and the dynamics of body mass evolution in North American fossil mammals. *Science* **280**, 731–734.

Blackburn, T.M. & Gaston, K.J. (2001) Linking patterns in macroecology. *Journal of Animal Ecology* **70**, 338–352.

Bohonak, A.J. (1999) Dispersal, gene flow, and population structure. *Quarterly Review of Biology* **74**, 21–45.

Brown, J.H. (1995) *Macroecology*. University of Chicago Press, Chicago, IL.

Brown, J.H., Marquet, P.A. & Taper, M.L. (1993) Evolution of body size: consequences of an energetic definition of fitness. *American Naturalist* **142**, 573–584.

Budd, A.F. & Coates, A.G. (1992) Nonprogressive evolution in a clade of Cretaceous *Montastraea*-like corals. *Paleobiology* **18**, 425–446.

Budd, A.F. & Johnson, K.G. (2001) Contrasting patterns in rare and abundant species during evolutionary turnover. In: *Evolutionary Patterns* (eds J.B.C. Jackson, S. Lidgard & F.K. McKinney), pp. 295–325. University of Chicago Press, Chicago, IL.

Burney, D.A., James, H.F., Burney, L.P., *et al.* (2001) Fossil evidence for a diverse biota from Kaua'i and its transformation since human arrival. *Ecological Monographs* **71**, 615–641.

Bush, A.M., Powell, M.G., Arnold, W.S., Bert, T.M. & Daley, G.M. (2002) Time-averaging, evolution, and morphologic variation. *Paleobiology* **28**, 9–25.

Chapman, M.G. (1999) Are there adequate data to assess how well theories of rarity apply to marine invertebrates? *Biodiversity and Conservation* **8**, 1295–1318.

Chown, S.L. (1997) Speciation and rarity: separating cause from consequence. In: *The Biology of Rarity* (eds W.E. Kunin & K.J. Gaston), pp. 91–109. Chapman and Hall, London.

Ciampaglio, C.N., Kemp, M. & McShea, D.W. (2001) Detecting changes in morphospace occupation patterns in the fossil record: characterization and analysis of measures of disparity. *Paleobiology* **27**, 695–715.

Clarke, A. (1993) Temperature and extinction in the sea: a physiologist's view. *Paleobiology* **19**, 499–518.

Cohen, J.E. (1994) Marine and continental food webs: three paradoxes. *Philosophical Transactions of the Royal Society, London, Series B* **343**, 57–69.

Collin, R. (2001) The effects of mode of development on phylogeography and population structure of North American *Crepidula* (Gastropoda: Calyptraeidae). *Molecular Ecology* **10**, 2249–2262.

Crame, J.A. (1996) Antarctica and the evolution of taxonomic diversity gradients in the marine realm. *Terra Antarctica* **3**, 121–134.

Crame, J.A. (2002) Evolution of taxonomic diversity gradients in the marine realm: a comparison of Late Jurassic and Recent bivalve faunas. *Paleobiology* **28**, 184–207.

Dayton, P.K., Tegner, M.J., Edwards, P.B. & Riser, K.L. (1998) Sliding baselines, ghosts, and reduced expectations in kelp forest communities. *Ecological Applications* **8**, 309–322.

Duncan, R.P., Bomford, M., Forsyth, D.M. & Conibear, L. (2001) High predictability in introduction outcomes and the geographical range size of introduced Australian birds: a role for climate. *Journal of Animal Ecology* **70**, 621–632.

Endler, J.A. (1977) *Geographic Variation, Speciation, and Clines*. Princeton University Press, Princeton, NJ.

Erwin, D.H. (2001) Lessons from the past: biotic recoveries from mass extinctions. *Proceedings of the National Academy of Sciences, USA* **98**, 5399–5403.

Foote, M. (1996) Models of morphological diversification. In: *Evolutionary Paleobiology* (eds D. Jablonski, D.H. Erwin & J.H. Lipps), pp. 62–86. University of Chicago Press, Chicago, IL.

Foote, M. (1997) The evolution of morphological diversity. *Annual Review of Ecology and Systematics* **28**, 129–152.

Foote, M. (1999) Morphological diversity in the evolutionary radiation of Paleozoic and post-Paleozoic crinoids. *Paleobiology Memoir* 1 (Supplement to *Paleobiology* **25**(2)).

Forys, E.A. & Allen, C.R. (1999) Biological invasions and deletions: community change in south Florida. *Biological Conservation* **87**, 341–347.

Gaston, K.J. (1998) Species–range size distributions: products of speciation, extinction and

transformation. *Philosophical Transactions of the Royal Society, London, Series B* **353**, 219–230.
Gaston, K.J. & Blackburn, T.M. (1999) A critique for macroecology. *Oikos* **84**, 353–368.
Gaston, K.J. & Blackburn, T.M. (2000) *Pattern and Process in Macroecology.* Blackwell Science, Oxford.
Gaston, K.J. & Chown, S.L. (1999) Geographic range size and speciation. In: *Evolution of Biological Diversity* (eds A.E. Magurran & R.M. May), pp. 237–259. Oxford University Press, Oxford.
Gili, C. & Martinell, J. (1994) Relationship between species longevity and larval ecology in nassariid gastropods. *Lethaia* **27**, 291–299.
Griffiths, C.L. & Branch, G.M. (1997) The exploitation of coastal invertebrates and seaweeds in South Africa: historical trends, ecological impacts and implications for management. *Transactions of the Royal Society of South Africa* **52**, 121–148.
Hansen, T.A. (1978) Larval dispersal and species longevity in Lower Tertiary gastropods. *Science* **199**, 885–887.
Hansen, T.A. (1982) Modes of larval development in early Tertiary neogastropods. *Paleobiology* **8**, 367–372.
Harvey, P.H. & Pagel, M.D. (1991) *The Comparative Method in Evolutionary Biology.* Oxford University Press, Oxford.
Hellberg, M.E., Balch, D.P. & Roy, K. (2001) Climate-driven range expansion and morphological evolution in a marine gastropod. *Science* **292**, 1707–1710.
Hubbell, S.P. (2001) *The Unified Neutral Theory of Biodiversity and Biogeography.* Princeton University Press, Princeton, NJ.
Hulsey, C.D. & Wainwright, P.C. (2002) Projecting mechanics into morphospace: disparity in the feeding system of labrid fishes. *Proceedings of the Royal Society, London, Series B* **269**, 317–326.
Jablonski, D. (1986a) Larval ecology and macroevolution of marine invertebrates. *Bulletin of Marine Science* **39**, 565–587.
Jablonski, D. (1986b) Background and mass extinctions: the alternation of macroevolutionary regimes. *Science* **231**, 129–133.
Jablonski, D. (1987) Heritability at the species level: analysis of geographic ranges of Cretaceous mollusks. *Science* **238**, 360–363.
Jablonski, D. (1988) Estimates of species durations. *Science* **240**, 969.
Jablonski, D. (1993) The tropics as a source of evolutionary novelty: the post-Palaeozoic fossil record of marine invertebrates. *Nature* **364**, 142–144.
Jablonski, D. (1995) Extinction in the fossil record. In: *Extinction Rates* (eds R.M. May & J.H. Lawton), pp. 25–44. Oxford University Press, Oxford.
Jablonski, D. (1996) Body size and macroevolution. In: *Evolutionary Paleobiology* (eds D. Jablonski, D.H. Erwin & J.H. Lipps), pp. 256–289. University of Chicago Press, Chicago.
Jablonski, D. (1997) Body-size evolution in Cretaceous molluscs and the status of Cope's rule. *Nature* **385**, 250–252.
Jablonski, D. (1998) Geographic variation in the molluscan recovery from the end-Cretaceous extinction. *Science* **279**, 1327–1330.
Jablonski, D. (2000) Micro- and macroevolution: scale and hierarchy in evolutionary biology and paleobiology. *Paleobiology* **26** (Supplement to Number 4), 15–52.
Jablonski, D. (2001) Lessons from the past: evolutionary impacts of mass extinctions. *Proceedings of the National Academy of Sciences, USA* **98**, 5393–5398.
Jablonski, D. (2002) Survival without recovery after mass extinctions. *Proceedings of the National Academy of Sciences, USA* **99**, 8139–8144.
Jablonski, D. & Lutz, R.A. (1983) Larval ecology of marine benthic invertebrates: paleobiological implications. *Biological Reviews* **58**, 21–89.
Jablonski, D. & Raup, D.M. (1995) Selectivity of end-Cretaceous marine bivalve extinctions. *Science* **268**, 389–391.
Jablonski, D. & Roy, K. (2003) Geographic range and speciation in fossil and living mollusks. *Proceedings of the Royal Society, London, Series B* **270**, 401–406.
Jablonski, D. & Sepkoski, J.J., Jr. (1996) Paleobiology, community ecology, and scales of ecological pattern. *Ecology* **77**, 1367–1378.
Jablonski, D. & Valentine, J.W. (1990) From regional to total geographic ranges: testing relationship in Recent bivalves. *Paleobiology* **16**, 126–142.
Jackson, J.B.C. (1974) Biogeographic consequences of eurytopy and stenotopy among marine bivalves and their evolutionary consequences. *American Naturalist* **108**, 541–560.
Jackson, J.B.C. (1997) Reefs since Columbus. *Coral Reefs* **16**, S23–S32.
Jackson, J.B.C. (2001) What was natural in the

coastal oceans? *Proceedings of the National Academy of Sciences, USA* **98**, 5411–5418.

Jackson, J.B.C., Kirby, M.X., Berger, W.H., *et al.* (2001) Historical overfishing and the recent collapse of coastal ecosystems. *Science* **293**, 629–638.

Kammer, T.W., Baumiller, T.K. & Ausich, W.I. (1998) Evolutionary significance of differential species longevity in Osagean–Meramecian (Mississippian) crinoid clades. *Paleobiology* **24**, 155–176.

Kidwell, S.M. (2001) Preservation of species abundance in marine death assemblages. *Science* **294**, 1091–1094.

Kittiwattanawong, K. (1999) The relation of reproductive modes to population differentiation in marine bivalves and gastropods. *Phuket Marine Biological Center Special Publication* **19**, 129–138.

Knoll, A.H. & Carroll, S.B. (1999) Early animal evolution: emerging views from comparative biology and evolution. *Science* **284**, 2129–2137.

Law, R. (2000) Fishing, selection, and phenotypic evolution. *ICES Journal of Marine Science* **57**, 659–668.

Lawton, J.H. (1999) Are there general laws in ecology? *Oikos* **84**, 177–192.

Lindberg, D.R., Estes, J.A. & Warheit, K.I. (1998) Human influences on trophic cascades along rocky shores. *Ecological Applications* **8**, 880–890.

Lupia, R. (1999) Discordant morphological disparity and taxonomic diversity during the Cretaceous angiosperm radiation: North American pollen record. *Paleobiology* **25**, 1–28.

Marshall, C.R. (1991) Estimation of taxonomic ranges from the fossil record. In: *Analytical Paleobiology. Short Courses in Paleontology 4* (eds N.L. Gilinsky & P.W. Signor), pp. 19–38. Paleontological Society, Knoxville, TN.

Maurer, B.A. (1999) *Untangling Ecological Complexity.* University of Chicago Press, Chicago, IL.

Maurer, B.A. & Nott, M.P. (1998) Geographic range fragmentation and the evolution of biological diversity. In: *Biodiversity Dynamics* (eds M.L. McKinney & J.A. Drake), pp. 31–50. Columbia University Press, New York.

Maurer, B.A., Brown, J.H. & Rusler, R.D. (1992) The micro and macro in body size evolution. *Evolution* **46**, 939–953.

Mayr, E. (1963) *Animal Species and Evolution.* Harvard University Press, Cambridge, MA.

McGhee, G.R. (1999) *Theoretical Morphology.* Columbia University Press, New York.

McKinney, M.L. (1997a) Extinction vulnerability and selectivity: combining ecological and paleontological views. *Annual Review of Ecology and Systematics* **28**, 495–516.

McKinney, M.L. (1997b) How do rare species avoid extinction? A paleontological view. In: *The Biology of Rarity* (eds W.E. Kunin & K.J. Gaston), pp. 110–129. Chapman and Hall, London.

Miller, A.I. (1998) Biotic transitions in global marine diversity. *Science* **281**, 1157–1160.

Niklas, K.J. (1997) *The Evolutionary Biology of Plants.* University of Chicago Press, Chicago, IL.

Pauly, D. (1995) Anecdotes and the shifting baseline syndrome of fisheries. *Trends in Ecology and Evolution* **10**, 430.

Pechenik, J.A. (1999) On the advantages and disadvantages of larval stages in benthic marine invertebrate life cycles. *Marine Ecology Progress Series* **177**, 269–297.

Pombo, O.A. & Escofet, A. (1996) Effect of exploitation on the limpet *Lottia gigantea*: a field study in Baja California (Mexico) and California (U.S.A.). *Pacific Science* **50**, 393–403.

Raup, D.M. (1979) Biases in the fossil record of species and genera. *Bulletin of the Carnegie Museum of Natural History* **13**, 85–91.

Raup, D.M. (1985) Mathematical models of cladogenesis. *Paleobiology* **11**, 42–52.

Ricklefs, R.E. & Miles, D.B. (1994) Ecological and evolutionary inferences from morphology: an ecological perspective. In: *Ecological Morphology* (eds P.C. Wainwright & S.M. Reilly), pp. 13–41. University of Chicago Press, Chicago, IL.

Rosenzweig, M.L. (1995) *Species Diversity in Space and Time.* Cambridge University Press, Cambridge.

Roughgarden, J., Pennington, T. & Alexander, S. (1994) Dynamics of the rocky intertidal zone with remarks on generalization in ecology. *Philosophical Transactions of the Royal Society, London, Series B* **343**, 79–85.

Roy, K. (1994) Effects of the Mesozoic Marine Revolution on the taxonomic, morphologic and biogeographic evolution of a group: Aporrhaid gastropods during the Mesozoic. *Paleobiology* **20**, 274–296.

Roy, K. & Foote, M. (1997) Morphological approaches to measuring biodiversity. *Trends in Ecology and Evolution* **12**, 277–281.

Roy, K. & Martien, K.K. (2001) Latitudinal distribution of body size in north-eastern Pacific marine bivalves. *Journal of Biogeography* **28**, 485–493.

Roy, K., Jablonski, D., Valentine, J.W., & Rosenberg, G. (1998) Marine latitudinal diversity gradients: tests of causal hypotheses. *Proceedings of the National Academy of Sciences, USA* **95**, 3699–3702.

Roy, K., Jablonski, D. & Martien, K.K. (2000a) Invariant size–frequency distributions along a latitudinal gradient in marine bivalves. *Proceedings of the National Academy of Sciences, USA* **97**, 13150–13155.

Roy, K., Jablonski, D. & Valentine, J.W. (2000b) Dissecting latitudinal diversity gradients: functional groups and clades of marine bivalves. *Proceedings of the Royal Society, London, Series B* **267**, 293–299.

Roy, K., Balch, D.P. & Hellberg, M.E. (2001a) Spatial patterns of morphological diversity across the Indo-Pacific: analyses using strombid gastropods. *Proceedings of the Royal Society, London, Series B* **268**, 2503–2508.

Roy, K., Jablonski, D. & Valentine, J.W. (2001b) Climate change, species range limits and body size in marine bivalves. *Ecology Letters* **4**, 366–370.

Roy, K., Jablonski, D. & Valentine, J.W. (2002) Body size and invasion success in marine bivalves. *Ecology Letters* **5**, 163–167.

Russell, M.P. & Lindberg, D.R. (1988) Real and random patterns associated with molluscan spatial and temporal distributions. *Paleobiology* **14**, 322–330.

Scheltema, R.S. (1989) Planktonic and non-planktonic development among prosobranch gastropods and its relationship to the geographic range of species. In: *Reproduction, Genetics and Distribution of Marine Organisms* (eds J.S. Ryland & P.A. Tyler), pp. 183–188. Olsen & Olsen, Fredensborg, Denmark.

Scheltema, R.S. (1992) Passive dispersal of planktonic larvae and the biogeography of tropical sublittoral invertebrate species. In: *Marine Eutrophication and Population Dynamics* (eds G. Colombo *et al.*), pp. 195–202. Olsen and Olsen, Fredensborg, Denmark.

Sepkoski, J.J., Jr. (1991) A model of onshore–offshore change in faunal diversity. *Paleobiology* **17**, 58–77.

Smith, A.B. (1994) *Systematics and the Fossil Record.* Blackwell Science, Oxford.

Stanley, S.M. (1979) *Macroevolution.* W.H. Freeman, San Francisco.

Stanley, S.M. (1990) The general correlation between rate of speciation and rate of extinction: fortuitous causal linkages. In: *Causes of Evolution* (eds R.M. Ross & W.D. Allmon), pp. 103–127. University of Chicago Press, Chicago, IL.

Steadman, D.W. (1995) Prehistoric extinctions of Pacific island birds: biodiversity meets zooarchaeology. *Science* **267**, 1123–1131.

Steele, J.H., Carpenter, S.R., Cohen, J.E., Dayton, P.K. & Ricklefs, R.E. (1993) Comparing terrestrial and marine ecological systems. In: *Patch Dynamics* (eds S.A. Levin, T.M. Powell & J.H. Steele), pp. 1–12. Springer-Verlag, Berlin, New York.

Strathmann, R.R. (1990) Why life histories evolve differently in the sea. *American Zoologist* **30**, 197–207.

Thomas, R.D.K., Shearman, R.M. & Stewart, G.W. (2000) Evolutionary exploitation of design options by the first animals with hard skeletons. *Science* **288**, 1239–1242.

Todd, J.A., Jackson, J.B.C., Johnson, K.G., *et al.* (2002) The ecology of extinction: molluscan feeding and faunal turnover in the Caribbean Neogene. *Proceedings of the Royal Society, London, Series B* **269**, 571–577.

Valentine, J.W. (1961) Paleoecologic molluscan geography of the Californian Pleistocene. *University of California Publications in Geolological Sciences* **34**, 309–442.

Valentine, J.W. (1973) *Evolutionary Paleoecology of the Marine Biosphere.* Prentice-Hall, Englewood Cliffs, NJ.

Valentine, J.W. (1989) How good was the fossil record? Clues from the Californian Pleistocene. *Paleobiology* **15**, 83–94.

Valentine, J.W. (1995) Why no new phyla after the Cambrian? Genome and ecospace hypotheses revisited. *Palaios* **10**, 190–194.

Valentine, J.W., Foin, T.C. & Peart, D. (1978) Provincial model of Phanerozoic diversity. *Paleobiology* **4**, 55–66.

Valentine, J.W., Jablonski, D. & Erwin, D.H. (1999) Fossils, molecules and embryos: new perspectives

on the Cambrian explosion. *Development* **126**, 851–859.
Valentine, J.W., Roy, K. & Jablonski, D. (2002) Carnivore/noncarnivore ratios in northeastern Pacific marine gastropods. *Marine Ecology Progress Series* **228**, 153–163.
Veltman, C.J., Nee, S. & Crawley, M.J. (1996) Correlates of introduction success in exotic New Zealand birds. *American Naturalist* **147**, 542–557.
Vermeij, G.J. (1994) The evolutionary interaction among species: selection, escalation, and coevolution. *Annual Review of Ecology and Systematics* **25**, 219–236.
Wagner, P.J. & Erwin, D.H. (1995) Phylogenetic patterns as tests of speciation models. In: *New Approaches to Speciation in the Fossil Record* (eds D.H. Erwin & R.L. Anstey), pp. 87–122. Columbia University Press, New York.
Wills, M.A., Briggs, D.E.G. & Fortey, R.A. (1994) Disparity as an evolutionary index: a comparison of Cambrian and Recent arthropods. *Paleobiology* **20**, 93–130.

Chapter 20
Comparative methods for adaptive radiations

Robert P. Freckleton, Mark Pagel and Paul H. Harvey*

Introduction

The rationale for macroecology is that generalization about the factors that determine the ecology of species at a large scale can be achieved by demonstrating relationships or common patterns among key variables across large numbers of species (Brown 1995; Gaston & Blackburn 2000). Using this approach it has proved possible to reveal the existence of relationships between traits such as body size and range size (Brown & Maurer 1989) or range size and abundance (Gaston *et al.* 2000), and to begin to explore the detailed factors that determine the composition of ecological groups such as British birds at both national and local scales (Gaston & Blackburn 2000). The basis for macroecology is thus the comparative approach (Harvey & Pagel 1991).

An important development of the comparative approach has been the realization that incorporation of phylogenetic information is key to the accurate statistical testing of comparative data, because phylogenetically closely related species share characteristics as a consequence of their shared ancestry (e.g. Felsenstein 1985; Grafen 1989; Harvey & Pagel 1991; Harvey *et al.* 1995). This similarity generates non-independence of data that needs to be accounted for in the statistical analysis of comparative data; individual species cannot be regarded as representing independent data points. Unless this non-independence is accounted for, the results of comparative analysis may be statistically incorrect (Grafen 1989; Harvey & Pagel 1991; Martins & Garland 1991; Harvey & Rambaut 1998). To cope with such non-independence, a variety of phylogenetic comparative tests have been developed (e.g. Felsenstein 1985; Grafen 1989; Gittleman & Kot 1990; Harvey & Pagel 1991; Lynch 1991) and the use of phylogenetic comparative approaches have become widespread in analysing macroecological data (Blackburn & Gaston 1998; Gaston & Blackburn 2000; Bennett & Owens 2002). More recently, tests have been developed in order to determine which data sets require phylogenetic analysis and to adjust data according to phylogenetic dependence (Freckleton *et al.* 2002; Blomberg *et al.* in press).

* *Correspondence address: robert.freckleton@zoology.oxford.ac.uk*

Despite the fact that the importance of phylogenetic non-independence is well recognized, the nature of such non-independence is rarely looked at in detail. The problem of non-independence is typically regarded as a statistical nuisance and comparative methods are often viewed as a statistical correction. Ultimately, however, the reason why closely related species show similar characteristics is that they share a common evolutionary history. Evolution will determine the degree to which traits diverge following splitting of species from common ancestors. Although it previously has been emphasized that the mode of evolution is important in the application of comparative techniques (Pagel & Harvey 1989; Harvey & Purvis 1991), this has been largely ignored and most applications of comparative techniques have relied on Brownian motion as a model for character evolution (see below for a more detailed outline). This very simple model has the advantage that it is readily applied through the method of independent contrasts (Felsenstein 1985; Harvey & Pagel 1991).

Although the Brownian model is very simple, it is uncertain how relevant this model will be in many ecological data sets. To illustrate this point, consider two recent theories for the evolution of species' traits during adaptive radiations. An adaptive radiation occurs when the invasion of a new habitat by a species is subsequently followed by speciation and ecological diversification, thereby generating an array of ecologically distinct species (see Schluter 2000). Such radiations include Darwin's finches on the Galapagos (Lack 1947), the Hawaiian silversword alliance (Baldwin 1997), and *Anolis* lizards of the Caribbean (Williams 1972). Firstly, the ecological theory of adaptive radiation states that niche differentiation and speciation are closely linked as a consequence of intense competition within species for resources (Schluter 2000). Secondly, the 'unified neutral theory' states that all species are ecologically equivalent and that the distribution of traits across species is unconstrained by the limiting similarity imposed by niche differentiation (Hubbell 2001). It is clear that these two models for trait evolution are very different and will generate different distributions of traits across species. For this reason the Brownian model is likely to be inadequate for describing many ecological data sets because it will not allow such ecologically contrasting models to be distinguished. Indeed when applied to models for adaptive radiations comparative approaches based on the Brownian model may yield incorrect results (Harvey & Rambaut 2000).

The aim of this article is to explore how comparative methods can be applied to the analysis of data from adaptive radiations in order to characterize different modes of trait evolution. After outlining the main features of the Brownian model, extensions of this basic model to incorporate features such as time-varying and punctuated forms of evolution are described. Although these approaches characterize some forms of non-Brownian evolution that may occur in adaptive radiation, we then describe how they are inadequate to describe the evolution of traits in a range of newly developed models that incorporate ecological processes. We finish by describing initial attempts to characterize such non-Brownian forms of character evolution.

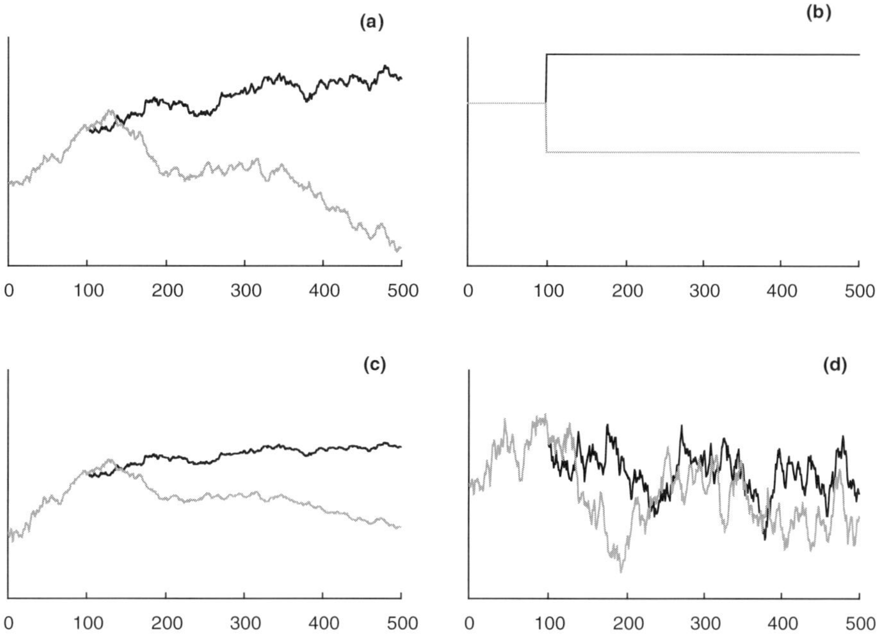

Figure 20.1 Examples of a continuous trait evolving according to different models (see text for details). (a) Brownian motion; (b) punctuated (speciational) evolution; (c) decelerating evolution; (d) stabilizing selection (Ornstein–Uhlenbeck process).

Brownian and Brownian-like models for comparative data

The model for character evolution specifies how traits change through time, and how a given trait changes as a function of one or more other traits. In the following we restrict ourselves to the discussion of bivariate trait evolution. This is because in ecological studies we are generally interested in how species trade off exploitation across small numbers of key resource–exploitation axes. Even when faced with more than two key traits, there are commonly such close correlations between variables that the dimensionality of trait space is effectively low, such that the bivariate case will often be a good approximation.

There is a developing literature on the inference of ancestral character states from data on extant species (e.g. Mooers *et al.* 1999; Webster & Purvis 2002). We are concerned primarily with how the mode of trait evolution affects the distribution of extant values and how this affects comparative tests. However, the issues of how to infer ancestral character states and determining models for character evolution in comparative analyses are closely related (see below).

The basic models we deal with in this section are summarized in Fig. 20.1. We can fit any of the models in Fig. 20.1 to a given data set by maximum likelihood techniques (Hansen 1997; Pagel 1997, 1999). The model with the fewest explicit parameters is the Brownian model and more complex models can be compared with the Brownian model using log-likelihood ratio tests in order to determine whether additional parameters significantly improve the fit of the model to the data.

Brownian model

This model assumes that variance in trait values increases as a linear function of time. As shown in Fig. 20.1a, following a speciation event the traits of a pair of species diverge, and the amount of variance across species increases through time. At any point in time there is no overall direction to the evolutionary process, and changes in traits are equally as likely to be increases or decreases. In the long term the process is like an expanding cloud, and there is no restriction on the values of traits. Although Fig. 20.1a shows the evolution of a single trait, the same principles apply to the correlated evolution of more than one trait. For a pair of traits covariance increases linearly through time, such that the correlation between traits remains constant.

The Brownian model is most commonly fitted to comparative data using the method of independent contrasts (Felsenstein 1985; Harvey & Pagel 1991). This method uses ancestral character state reconstruction to estimate amounts of trait divergence between pairs of sister species throughout the given phylogeny. The correlation between the amounts of divergence for different characters estimates the correlation between traits. This is a key feature of the Brownian model in terms of comparative analyses: the distribution of current traits is built up from a series of small changes during the process of evolution, and the correlation between these changes estimates the correlation between the traits.

In terms of modelling data from adaptive radiations, the Brownian model has a number of limitations. The rate of evolution is constant (on average) and does not change through time. The increase in variance across species is inexorable and there is neither any bound to the values that traits may adopt, nor any halt to the evolutionary process. In reality species' traits may be tied to evolutionary optima required to exploit particular niches. For this reason several alternative models have been developed, and applied to comparative data.

Punctuated evolution

The simplest alternative to the Brownian model is a version in which evolution is punctuated. In this model variance does not increase as a simple linear function of the time since the separation of sister species, but is a constant per speciation event. An example of this process is shown in Fig. 20.1b. In this model the variance increases as a linear function of the number of speciation events. The model may be fitted by applying a power transformation to the branch lengths of the phylogeny (Garland *et al.* 1992; Pagel 1994, 1997, 1999).

This model may be ecologically interesting because the implication of a zero value of the branch length transformation is that change in character states is associated only with speciation events and that traits do not evolve between speciation events. Thus this model may reveal links between speciation and trait divergence. However, this model of evolution can be criticized on two of the same grounds as the

Brownian model described above, namely that trait values are unconstrained and the variance in trait values increases linearly with the number of speciation events.

Slowing down and speeding up of evolution

In the two models described above, the rate of accumulation of variance is constant either through time or per speciation event and does not change. An alternative model is one in which the rate of evolution of traits increases or decreases through time. Thus, Pagel (1997, 1999) suggested a model in which the rate of accumulation of variance in traits is a power function of tree height. This power (δ) determines whether the rate of accumulation of variance increases ($\delta > 1$) or decreases ($\delta < 1$) through evolutionary time. Figure 20.1c shows an example in which the rate of accumulation of variance decreases through time.

This model is potentially extremely useful in studies of adaptive radiations because changes in the rate of evolution can occur, for example, when environmental changes lead to the creation of ecological niches either early or late during the course of an adaptive radiation. However, in practice the model may be difficult to fit. As noted by Grafen (1989) transformations of this sort are likely to be biased, and extensive simulations show that the shape of the phylogenetic tree is extremely important in determining whether this parameter can be estimated accurately (R. Freckleton, unpubl.). These simulations indicate that low ($\delta < 1$) values of δ are extremely suggestive of non-Brownian modes of evolution. However, large ($\delta > 1$) values occur reasonably commonly even when evolution of characters is Brownian and should be interpreted with caution.

Constraints on trait space

The models outlined above can be criticized in that they all assume an unrestricted trait space. There is therefore no bound to the possible values that traits may adopt. This assumption is ecologically unrealistic, and we may expect the distribution of trait values to be restricted. For this reason a model for the analysis of comparative data has been suggested in which the rate of evolution depends on the value of the trait and in which extreme trait values are selected against (Hansen & Martins 1996; Hansen 1997; Martins 2000). This model is called the Ornstein–Uhlenbeck (OU) model and Fig. 20.1d shows an example of traits evolving according to this process. The OU model incorporates a parameter (α) that determines the rate at which selection forces trait values back to the long-term average (μ). The larger the value of α, the stronger the selective pressure forcing traits back to the average. A value of α equal to 0 is Brownian motion. The value of μ is presumed to represent an overall average or optimum around which species vary.

The OU model produces a distribution of trait values that is constrained in trait space, in contrast to the models described above in which traits may adopt any value. This is arguably more realistic because most ecological traits (e.g. body size, food size, range size) are likely to be constrained in some way. In practical terms such modes of evolution are difficult for comparative analyses, because the forcing of trait values away from extremes during the course of evolution tends to eradicate at least some of the history of trait evolution (e.g. Fig. 20.1d). Thus under the OU model it may be impossible to reconstruct ancestral character states with reasonable accuracy, or precisely to infer the nature of the underlying evolutionary model, unless some fossil material is available on non-extant trait values in order to provide information on some of the variance in trait values later eradicated by selection.

One consequence of the loss of history under the OU model is that data generated under this process tend to exhibit less phylogenetic dependence than traits generated under Brownian motion. This is not typical of ecological characters where close phylogenetic dependence would be expected to result from phylogenetic niche conservatism (Harvey & Pagel 1991) and this results in strong phylogenetic dependence in ecological data (Freckleton *et al.* 2002). Although not all ecological characters show such strong phylogenetic dependence as predicted by the Brownian process (Freckleton *et al.* 2002), most characters are strongly phylogenetically dependent, and thus it is questionable whether the kind of stabilizing selection assumed by the OU model is of general importance. However, adaptive constraints on the range of trait values may still be important, and in the next section we outline a model incorporating adaptive constraints and phylogenetic niche conservatism that yields strong phylogenetic dependence.

Implications for comparative methods

Modern comparative analyses are robust under the range of evolutionary models described above (Felsenstein 1985; Martins & Garland 1991; Diaz-Uriarte & Garland 1996, 1998; Harvey & Rambaut 1998; Martins *et al.* 2002). Moreover, if traits are evolving under these processes, the *potential* exists to distinguish between these different forms. As noted above, however, some caution has to be exercised. The performance of some techniques depends on the amount of data available and more complex transformations such as the δ transformation and the OU model may be seriously biased or very difficult to estimate for small (<50 tips) phylogenies (R. Freckleton, unpubl.). Both the OU model and values of δ greater than unity imply some loss of precision to infer ancestral values during the evolutionary process. However, in all of the above cases comparative analyses based on a Brownian model (or a transformed Brownian model in the case of the OU model; Martins *et al.* 2002) perform adequately and generally perform better than across-species analysis, given appropriate diagnostics checks (Garland *et al.* 1992). In the next section we describe a model in which this is not the case and in which the Brownian model fails.

Non-Brownian models for adaptive radiations

Conventional parametric statistical approaches can be used to distinguish the different models of trait evolution outlined above, and to construct comparative analyses that allow the nature of evolution to be inferred, as well as ensuring phylogenetically controlled statistical tests. However, there is no guarantee that the actual pattern of evolution followed one of these models. In an adaptive radiation, for example, we might expect trait evolution and speciation to be a function of the number of species present. In this section we describe models for trait evolution that include this form of evolution and how this affects comparative methods.

Ecological model for trait evolution

The models we outline in this section assume that the values of traits determine the niche that each species exploits. Species with similar trait values thus exploit similar niches. The first model we outline in this section is an analytical extension to the Brownian model that includes three components to trait change. The first component is Brownian drift, such that in the absence of other factors species traits are unconstrained and evolve continuously. The second component is an adaptive constraint on trait values, as in the OU model. The rationale for this component is that ecological niches are not infinite, with the result that the distribution of trait values should be constrained. The third component is competition between species. In this component of the model it is assumed that the change in trait values of a given species is restricted by the presence of other species. The effect of species on each other declines with distance, and the effect of species on each other is one of repulsion, i.e. if species become similar then they tend to evolve in opposite directions.

The behaviour of this model is determined by the offset of these three processes. Brownian drift promotes change in traits. However, if species are too close together in trait space, selection acts to move them apart, and effectively the two species repel each other. This tendency of the species to move apart from each other is opposed by the effects of trait constraints. Thus the distribution of traits across species balances the effects of adaptive constraints and interactions between species. Figure 20.2 shows an example of a pair of traits evolving according to this process. Unlike the Brownian and OU models, change in this model is minimal and the distribution of traits across species tends to approach an equilibrium state. In this state the opposing forces of selection for species to move apart and the constraints on traits balance each other and absorb the Brownian component.

There are basically four important characteristics of traits evolving under this model.

1 As outlined above, traits do not change over long periods of evolutionary time.
2 Changes in trait values are concentrated at speciation, because at speciation the parent and daughter species rapidly evolve ecological differences as a consequence of repulsion between ecologically similar species.
3 At speciation change is concentrated in the daughter species. This is because change in the parent species is opposed by the existing other species, and the daughter species is repelled by the parent species and other species and rapidly changes until its state equilibrates with the existing species.
4 The rate of evolution decreases through time.

Niche-filling models for adaptive radiation

The model of trait change is only one component of the process of adaptive radiation. The process of speciation is also important particularly as this may be affected

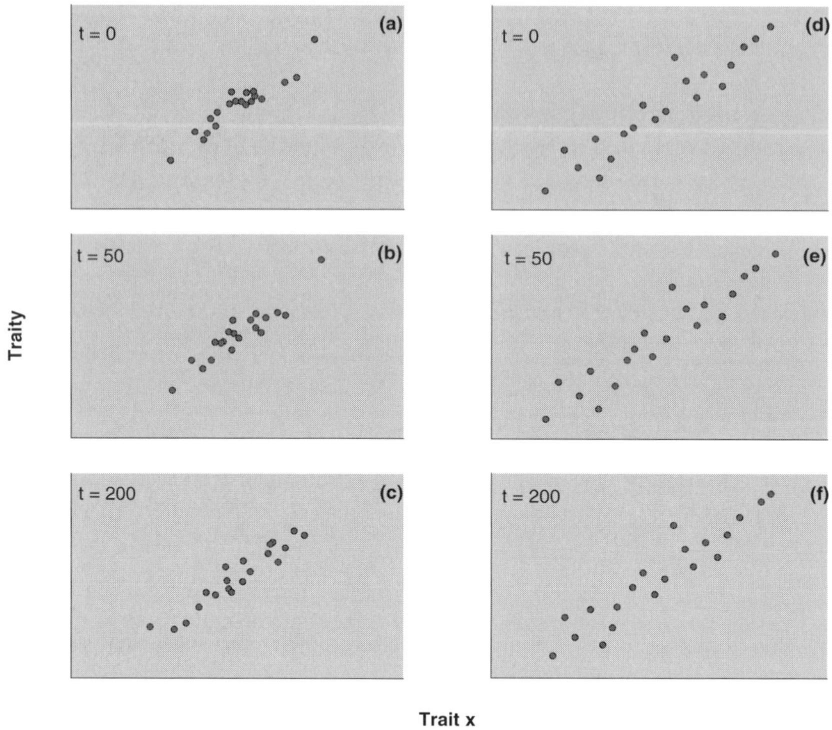

Figure 20.2 Example of a pair of traits evolving through time according to either constrained Brownian motion (a–c) or according to a Brownian process including constraints on trait values and negative interactions between species (d–f). The Brownian drift parameter is the same in both models and the graphs show initial trait values ($t = 0$) and trait values after 50 and 200 time units.

by trait values. To illustrate the importance of this interaction, in this section we outline two models (inspired by Price's 1997 formalization) for the evolution of adaptive radiations that include the mode of trait change predicted by the model described above.

In both models the distribution of traits is limited and is constrained, for example, as given by the distribution shown in Fig. 20.3. For convenience it is assumed that niches are distributed such that the distribution of traits is given by a bivariate normal distribution. Only one species can exploit each niche and each species thus has a unique combination of trait values. Evolution proceeds when new niches become available. The two models differ in terms of how new niches are invaded by the evolution of existing species.

The first model is termed the 'sequential niches model' and models the invasion of a completely empty habitat by a group, for example a newly created volcanic island (this may be thought of as the 'Krakatoa' model). The model is summarized in Fig.

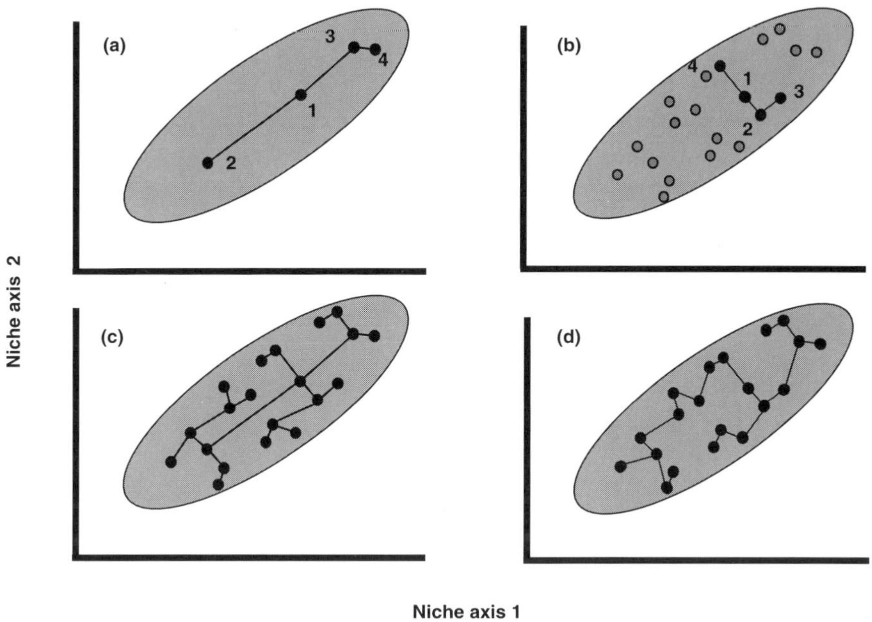

Figure 20.3 Evolution of traits according to the sequential niches (a, c) and empty niches models (b, d). In the sequential niches model, the trait space is initially empty. Niches arise sequentially and are invaded by the closest existing species in niche space. Thus in (a) the first species to invade is species 1, followed by species 2. When niche 3 arises, this is closest to species 1, and hence invaded by species 1. Similarly when niche 4 arises, this is closest to species 3 and is invaded by species 3. In the empty niches model (b) all niches are present originally, and one of these is invaded at random. Subsequently speciation occurs with the closest species to an empty niche invading that niche. The pattern of invasion thus occurs with species 1 invading niche 2; then species 2 invades niche 3 followed by the invasion of niche 4 by species 1. (c, d) The two models iterated for a number of speciation events.

20.3 (a and c). The niche space is initially empty. A new niche arises, for example, through invasion of the empty habitat by a species of plant, with the niche being represented by adaptations required by a herbivore to feed on that plant. The niche is invaded by the herbivore, which adapts to feed on the plant. A second niche arises, for example, by the invasion of the habitat by a second species of plant, there is selection for the herbivore to feed on the new niche, and the initial species speciates and adapts to feed on the new resource. Subsequently new niches arise at random within the trait space and each time a new niche arises it is invaded by the closest existing species in trait space.

In the second model, termed the 'empty niches' model (or which might be thought of as the 'Galapagos' model), all niches are present initially, for instance representing an established island flora which becomes invaded by a herbivore (see Fig.

20.3, b and d). The initial species exploits one of these niches and subsequently, during the radiation, speciation takes place through the invasion of an empty niche by the existing species that is closest to an empty niche.

These two models share some important features. They both incorporate constraints on the distribution of possible trait values, but phylogenetic niche conservatism is explicitly incorporated and thus the phylogenetic distribution of traits is more realistic than that predicted by the OU model. In both, changes in character states from one speciation event to the next are non-Brownian and dominated by the relative orientation of niches within the niche space.

There is also an important difference between the models. In the sequential niches model new niches arise at random within the niche space with respect to existing species. Thus at any point in time any existing species within niche space may be the next to speciate. By contrast, in the empty niches model speciation occurs when the species nearest to an empty niche invades that niche, thus existing species within areas of niche space that are full are less likely to speciate than those at the edge of the already occupied space. Therefore, in the empty niches model the probability of speciation is dependent on trait values, and this probability will change as the radiation proceeds and the trait space becomes full.

Niche-filling models and comparative methods

Current comparative methods perform poorly under this kind of niche-filling model, as they do in the models described by Price (1997) and Harvey & Rambaut (2000). If comparative methods based on the Brownian model (e.g. the method of contrasts) are used to estimate the correlation between traits, then estimates of the correlation are biased and significance tests have high Type I and II error rates. For example, when there is no correlation between traits, comparative methods have a high Type I error rate and the null hypothesis is rejected more frequently than is statistically acceptable. This is despite the fact that data clearly show phylogenetic structure and it would be expected that comparative methods should be necessary.

The reason for the breakdown of the comparative method in the case of the niche-filling model is that the niche-filling model violates the fundamental characteristic of Brownian motion. As noted above, the Brownian model assumes that characters evolve through the accumulation of small changes, and that for pairs of traits the correlation of these changes is the same as the correlation of the traits themselves. It is clear that this is not the case for the niche-filling model. The distribution of changes from one speciation event to the next is clearly not the same as the overall distribution of traits, but instead is a function of the relative orientation of niches in niche space. It is this difference that compromises the performance of the method of independent contrasts.

Although simulations have not been performed, the performance of other comparative methods, such as phylogenetic-subtractive methods, which attempt to dissociate phylogenetic and non-phylogenetic components of trait distributions (e.g. Cheverud *et al.* 1985; Gittleman & Kot 1990; Lynch 1991), also will be compromised because the non-Brownian component of trait change is also highly phylogenetically dependent. The OU model predicts low phylogenetic dependence resulting from stabilizing selection. Thus the OU model also is not appropriate for analysing data on traits that have evolved in this way.

Analysis

Data from niche-filling models cannot be transformed to yield data that are consistent with the assumptions of Brownian motion, as can be done with the OU model, for example. Thus the likelihood approach for comparative analysis will not work and new comparative analyses are required. The two forms of niche-filling model described above make it clear that two processes may be of particular significance: non-Brownian distributions of traits consistent with niche filling, and phylogenetic structures consistent with differential speciation of species at the edge and centre of the occupied trait space.

Non-Brownian trait evolution

In an initial attempt to characterize non-Brownian modes of trait evolution we have explored the use of randomization techniques. The rationale for our analysis is that Brownian motion is the sum of many small changes. In the case of Brownian trait evolution, trait values in extant species result from the accumulation of variance along the branches of a phylogenetic tree. If the mode of evolution is Brownian, then if we knew the distribution of the changes that generated the observed data we should be able to generate random data with the same statistical properties as the actual data. If we have a phylogeny on to which data have been mapped, this distribution is estimated empirically by the distribution of phylogenetic contrasts. Our approach is to randomize contrasts on the tree and compare statistics measured on the randomized data with those of the original data. Figure 20.4 shows this test applied to data on feeding traits in a radiation of Old World leaf warblers (Richman & Price 1992). The frequency distributions are the distributions of variances of the two traits and the covariance between them generated by randomizing the empirically estimated changes on the tree. If the data were generated by Brownian motion we would expect the values measured on the data to be in the centre of the randomized values (R. Freckleton, unpubl.). In fact the values estimated on the data are very much higher than those generated by randomization, indicating major deviations from the Brownian model, and more consistent with a niche-filling model. It is very important to note that standard diagnostics (e.g. Garland *et al.* 1992) failed to reveal any deviation from the Brownian model.

Analysis of tree structure

The two niche-filling models differ in the way that speciation is modelled and, as noted above, the empty niches model assumes that speciation is less likely in densely occupied areas of niche space. Comparing Figs 20.3c and 20.3d it is clear that this affects tree structure. The arrangement of niches is identical in Figs 20.3c and 20.3d, but the order in which they were invaded is very different between the models.

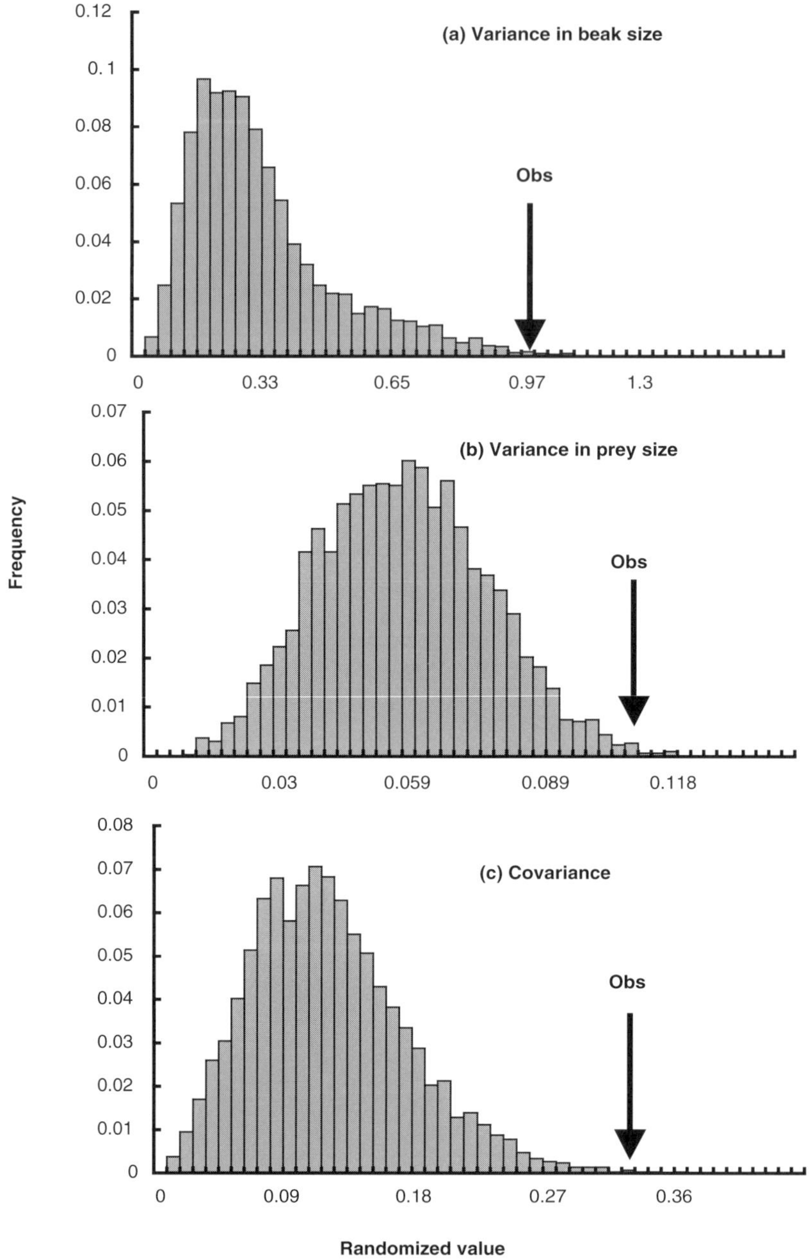

Figure 20.4 Randomization test for deviations from Brownian motion applied to data on feeding in Old World leaf warblers (see text for details). (a) Randomized variance in beak size; (b) randomized variance in prey size; (c) randomized covariance between beak and prey size.

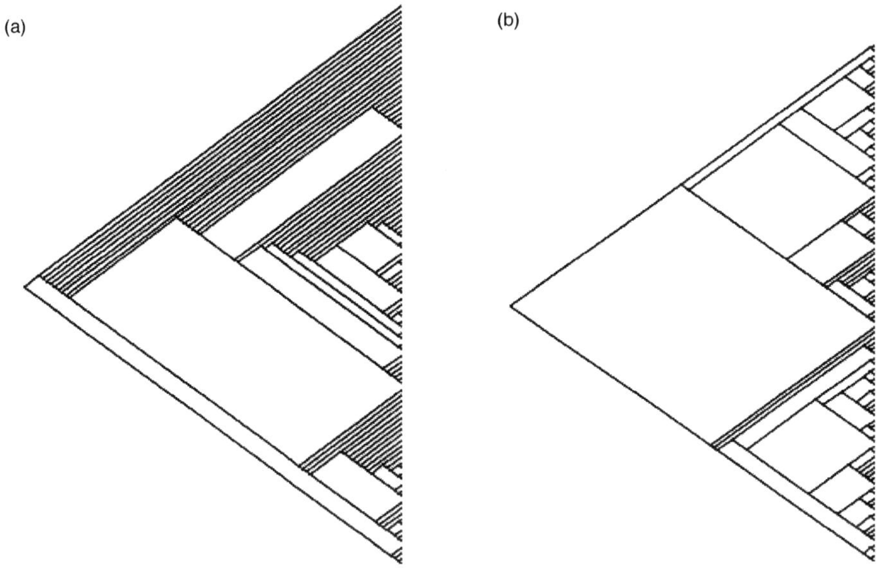

Figure 20.5 Examples of tree structures produced by the niche-filling models. (a) Empty niches model; (b) sequential niches (or constant birth–death) model.

Specifically, in the empty niches model, species at the edge of trait space are most likely to speciate, but having speciated once are less likely to do so again because the neighbouring trait space has become more full. Thus in Fig. 20.3d most species invade only one new niche. A consequence of this for tree shape is shown in Fig. 20.5. Trees produced by the empty-niches model are considerably more pectinate (comb-like or imbalanced) than those produced by either the sequential niches model or a birth–death process.

One metric of tree structure is to measure how imbalance changes through a phylogeny (A. Purvis, unpubl.). This is done by examining how an imbalance index varies with the number of tips subtended by each node on a phylogeny, termed node density. As shown in Fig. 20.6a and b there is an increase in the degree of imbalance with increasing node density in the case of the empty niches model, but no change in imbalance with node density in the sequential niches model, or indeed a constant birth–death process model.

Figure 20.6c–f shows corresponding plots for phylogenies representing adaptive radiations of cichlids (Fig. 20.6c; Goodwin *et al.* 1998), carnivores (Fig. 20.6d; Bininda-Emonds *et al.* 1999), wading birds (Fig. 20.6e; Charadrii; Szekely *et al.* 2000) and neotropical primates (Fig. 20.6f; von Dornum & Ruvolo 1999). Only in the case of the primates is there no significant positive relationship between imbalance and node density. Thus in three out of the four cases the phylogeny is consistent with the structure of tree generated by the empty niches model.

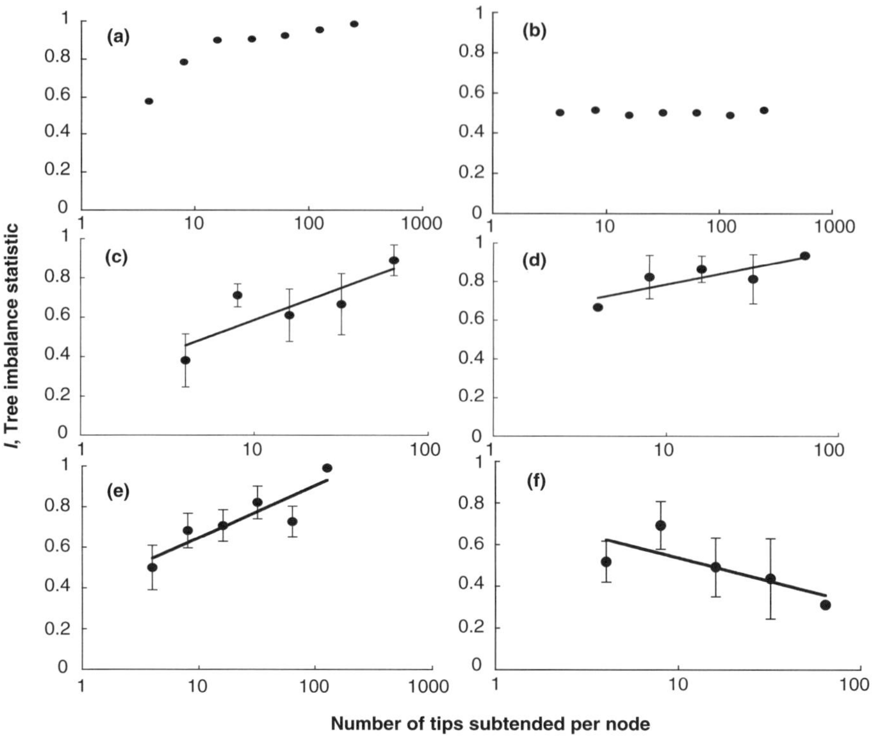

Figure 20.6 Patterns of imbalance in phylogenies for adaptive radiations. The graphs show the relationship between a measure of imbalance (0 = perfectly symmetric; 1 = pectinate) and node density (i.e. the number of species subtended per node) for internal nodes in each phylogeny. (a) Phylogeny generated by the empty niches model in Fig. 20.5a. (b) Phylogeny generated by the sequential niches model in Fig. 20.5b. (c) Cichlids (Goodwin *et al.* 1998); (d) carnivores (Bininda-Emonds *et al.* 1999); (d) Charadrii (Szekely *et al.* 2000); (e) New World primates (von Dornum & Ruvolo 1999).

Concluding remarks

The analyses shown in Figs 20.4 and 20.6 represent diagnostic tests rather than parametric inferential models. They have the drawback, therefore, that although we are able to disprove that the data were generated by conventional models, we do not exactly identify the mode of evolutionary change. On the other hand it is clear that the evolution of traits in adaptive radiations may be more complex than the conventional Brownian model. The fact that it has been possible to identify cases where the assumptions of the conventional model fail should provide justification for analysing more complex models for adaptive radiations.

As outlined above, most phylogenetic comparative techniques have been developed within a likelihood framework (e.g. Felsenstein 1973, 1985; Gittleman & Kot 1990; Lynch 1991; Pagel 1994, 1997, 1999). The advantage of this framework is that the statistical theory of likelihood is well characterized. The drawbacks are that parameters may be biased and that for many types of model (e.g. the OU model and δ transformation) large amounts of data may be required in order to obtain consistent unbiased estimators. Of course some data sets used in comparative analyses are large (e.g. Purvis 1995; Kelly & Woodward 1996; Bininda-Emonds *et al.* 1999). However, in many adaptive radiations the maximum sample sizes are low (e.g. Darwin's finches, 14 species; *Dendroica* warblers, 12 species; *Phylloscopus* warblers, 15 species; Hawaiian silverswords, 25 species). For this reason likelihood methods may be too data-intensive for investigating many such systems, and other methods will be required.

There is a clear relationship between the form of model used in comparative analysis and ancestral state reconstruction, which plays an important role in some analyses of adaptive radiations (e.g. Schluter *et al.* 1997). Ancestral states may be reconstructed using one of the models of evolution outlined above. However, there are important differences between the two problems. Felsenstein (1973) derived phylogenetic contrasts as a method to avoid the inversion of large matrices in likelihood problems. In the likelihood approach a model of evolution specifies the expected distribution of traits across a group of species, given the phylogeny. The 'ancestral states' used in the calculation of phylogenetic contrasts (Felsenstein 1985) also may be regarded as weighted trait means at internal nodes (Felsenstein 1973). In the case of Brownian motion this is a semantic difference. However, in the case of models such as the OU model, the two are very different. The reason is that in the OU model extreme values are selected against, with the consequence that the range of values observed in extant species frequently will be less than existed in the past. Because of this it is impossible to infer ancestral values accurately solely from traits measured on extant species. However, the likelihood model specifies the *expected* distribution of traits at the tips, given the model of evolution. In the case of the OU model this is the same as a Brownian model where node heights have been exponentially transformed. Using this transformed tree (and a large enough data set) one can fit an estimate of the model using the method of contrasts. However, this model will yield incorrect estimates of ancestral character states. Similarly Grafen (1989) gives an example where long-term drift may yield ancestral character state reconstructions that are incorrect, but where a comparative analysis would yield correct results. Thus the failure of ancestral state reconstruction need not imply that comparative analyses are invalid, or that evolutionary models have been incorrectly estimated; in contrast, we frequently expect this to be the case.

The models and results presented above suggest that it should be possible to extract information on the processes that drive community formation during the process of adaptive radiation, given both trait and phylogenetic information. On the one hand the failure of conventional comparative methods based on the Brownian model, when applied to data from the niche-filling models, would appear to present a statistical problem (e.g. statistical tests are biased and Type I and II error rates are unacceptable). On the other hand gross deviations from the Brownian model will be indicative of the occurrence of other ecological phenomena, such as niche filling.

Rather than arguing against the use of phylogenetic comparative methods, these models make clear that more complex models are required, and reinforces the need for a phylogenetic perspective in the analysis of data from many ecological communities.

Acknowledgement

This work was funded by the Natural Environment Research Council, UK (grant no. GR 3 / 12939 to P.H.H. and M.P.).

References

Baldwin, B.C. (1997) Adaptive radiation of the Hawaiian silversword alliance: congruence and conflict of phylogenetic evidence from molecular and non-molecular investigations. In: *Molecular Evolution and Adaptive Radiation* (eds T.J. Givnish & K.J. Sytsma), pp. 103–128. Cambridge University Press, Cambridge.

Bennett, P.M. & Owens, I.P.F. (2002) *The Evolutionary Ecology of Birds.* Oxford University Press, Oxford.

Bininda-Emonds, O.R.P., Gittleman, J.L. & Purvis, A. (1999) Building large trees by combining phylogenetic information: a complete phylogeny of the extant Carnivora (Mammalia). *Biological Reviews* **74**, 143–175.

Blackburn, T.M. & Gaston, K.J. (1998) Some methodological issues in macroecology. *American Naturalist* **151**, 68–83.

Blomberg, S.P., Ives, A.R. & Garland, T. (in press) Detecting phylogenetic signals in comparative data. *Evolution.*

Brown, J.H. (1995) *Macroecology.* University of Chicago Press, Chicago, IL.

Brown, J.H. & Maurer, B.A. (1989) Macroecology: the division of food and space among species on continents. *Science* **243**, 1145–1150.

Cheverud, J.M., Dow, M.M. & Leutenegger, W. (1985) The quantitative assessment of phylogenetic constraints in comparative analyses: sexual dimorphism in body-weight among primates. *Evolution* **39**, 1335–1351.

Diaz-Uriarte, R. & Garland, T. (1996) Testing hypotheses of correlated evolution using phylogenetically independent contrasts: sensitivity to deviations from Brownian motion. *Systematic Biology* **45**, 27–47.

Diaz-Uriarte, R. & Garland, T. (1998) Effects of branch length errors on the performance of phylogenetically independent contrasts. *Systematic Biology* **47**, 654–672.

Felsenstein, J. (1973) Maximum-likelihood estimation of evolutionary trees from continuous characters. *American Journal of Human Genetics* **25**, 471–492.

Felsenstein, J. (1985) Phylogenies and the comparative method. *American Naturalist* **126**, 1–25.

Freckleton, R.P., Harvey, P.H. & Pagel, M. (2002) Phylogenetic dependence and ecological data: a test and review of evidence. *American Naturalist* **160**, 712–726.

Garland, T.J., Harvey, P.H. & Ives, A.R. (1992) Procedures for the analysis of comparative data using phylogenetically independent contrasts. *Systematic Biology* **41**, 18–32.

Gaston, K.J. & Blackburn, T.M. (2000) *Pattern and Process in Macroecology.* Blackwell Science, Oxford.

Gaston, K.J., Blackburn, T.M., Greenwood, J.J.D., *et al.* (2000) Abundance–occupancy relationships. *Journal of Applied Ecology* **37**, 39–59.

Gittleman, J.L. & Kot, M. (1990) Adaptation, statistics and a null model for estimating phylogenetic effects. *Systematic Zoology* **39**, 227–241.

Goodwin, N.B., Balshine-Earn, S. & Reynolds, J.D. (1998) Evolutionary transitions in parental care in cichlid fish. *Proceedings of the Royal Society, London, Series B* **265**, 2265–2272.

Grafen, A. (1989) The phylogenetic regression. *Philosophical Transactions of the Royal Society, London, Series B* **326**, 119–157.

Hansen, T.F. (1997) Stabilizing selection and the

comparative analysis of adaptation. *Evolution* **51**, 1341–1351.

Hansen, T.F. & Martins, E.P. (1996) Translating between microevolutionary process and macroevolutionary patterns: the correlation structure of interspecific data. *Evolution* **50**, 1404–1417.

Harvey, P.H. & Pagel, M.D. (1991) *The Comparative Method in Evolutionary Biology*. Oxford University Press, Oxford.

Harvey, P.H. & Purvis, A. (1991) Comparative methods for explaining adaptations. *Nature* **351**, 619–624.

Harvey, P.H. & Rambaut, A. (1998) Phylogenetic extinction rates and comparative methodology. *Proceedings of the Royal Society, London, Series B* **265**, 1691–1696.

Harvey, P.H. & Rambaut, A. (2000) Comparative analyses for adaptive radiations. *Philosophical Transactions of the Royal Society, London, Series B* **355**, 1599–1606.

Harvey, P.H., Read, A.F. & Nee, S. (1995) Why ecologists need to be phylogenetically challenged. *Journal of Ecology* **83**, 535–536.

Hubbell, S. (2001) *The Unified Neutral Theory of Biodiversity and Biogeography*. Princeton University Press, Princeton, NJ.

Kelly, C.K. & Woodward, F.I. (1996) Ecological correlates of plant range size: taxonomies and phylogenies in the study of plant commonness and rarity in Great Britain. *Philosophical Transactions of the Royal Society, London, Series B* **351**, 1261–1269.

Lack, D. (1947) *Darwin's Finches*. Cambridge University Press, Cambridge.

Lynch, M. (1991) Methods for the analysis of comparative data in evolutionary biology. *Evolution* **45**, 1065–1080.

Martins, E.P. (2000) Adaptation and the comparative method. *Trends in Ecology and Evolution* **15**, 296–299.

Martins, E.P. & Garland, T. (1991) Phylogenetic analyses of the correlated evolution of continuous characters: a simulation study. *Evolution* **45**, 534–557.

Martins, E.P., Diniz, J.A.F. & Housworth, E.A. (2002) Adaptive constraints and the comparative method: a computer simulation test. *Evolution* **56**, 1–13.

Mooers, A.Ø., Vamosi, S.M. & Schluter, D. (1999) Using phylogenies to test macro-evolutionary hypotheses of trait evolution in Cranes (Gruinae). *American Naturalist* **154**, 249–259.

Pagel, M. (1994) Detecting correlated evolution of phylogenies—a general method for the comparative analysis of discrete characters. *Proceedings of the Royal Society, London, Series B* **255**, 37–45.

Pagel, M. (1997) Inferring evolutionary processes from phylogenies. *Zoologica Scripta* **26**, 331–348.

Pagel, M. (1999) Inferring the historical patterns of biological evolution. *Nature* **401**, 877–884.

Pagel, M.D. & Harvey, P.H. (1989) Comparative methods for examining adaptation depend on evolutionary models. *Folia Primatologica* **53**, 203–220.

Price, T. (1997) Correlated evolution and independent contrasts. *Philosophical Transactions of the Royal Society, London, Series B* **352**, 519–529.

Purvis, A. (1995) A composite estimate of primate phylogeny. *Philosophical Transactions of the Royal Society, London, Series B* **348**, 405–421.

Richman, A.D. & Price, T. (1992) Evolution of ecological differences in the old-world leaf warblers. *Nature* **355**, 817–821.

Schluter, D. (2000) *The Ecology of Adaptive Radiations*. Oxford University Press, Oxford.

Schluter, D., Price, T., Mooers, A.Ø. & Ludwig, D. (1997) Likelihood of ancestor states in adaptive radiation. *Evolution* **51**, 1699–1711.

Szekely, T., Reynolds, J.D. & Figuerola, J. (2000) Sexual size dimorphism in shorebirds, gulls and alcids: the influence of sexual and natural selection. *Evolution* **54**, 1404–1413.

Von Dornum, M. & Ruvolo, M. (1999) Phylogenetic relationships of the New World monkeys (primates, Platyrrhini) based on nuclear G6PD DNA sequences. *Molecular Phylogenetics and Evolution* **11**, 459–476.

Webster, A.J. & Purvis, A. (2002) Testing the accuracy of methods for reconstructing ancestral states of continuous characters. *Proceedings of the Royal Society, London, Series B* **269**, 143–149.

Williams, E.E. (1972) The origin of faunas. Evolution of lizard congeners in a complex island fauna: a trial analysis. *Evolutionary Biology* **6**, 47–89.

Chapter 21

The next step in macroecology: from general empirical patterns to universal ecological laws

James H. Brown, James F. Gillooly, Geoffrey B. West and Van M. Savage*

"*A big problem in ecology is that we humans are about the same body size and we operate on about the same temporal and spatial scales as the things that we study. This makes it difficult to make observations that enable us to distinguish the general or universal from the particular or idiosyncratic.*"

P. A. Marquet (1992) graduate student at the University of New Mexico in conversation with J. H. Brown

"*An advantage of ecology is that you can see your particles. You can observe the structure, function, and behaviour of organisms to understand how they generate the emergent properties of complex dynamic systems.*"

L. M. Simmons, Jr. (1992) Vice President for Research at the Santa Fe Institute in conversation with J. H. Brown

"*Body size and energetics provide a promising vehicle for making connections between macroecological patterns and underlying mechanisms at the level of individual organisms. Body size places severe constraints on morphology, physiology, and behavior. Allometric equations describe these general empirical relationships at the level of the individual. They provide a basis for making mathematical models of the ecological effects of body size on use of food resources and space by individuals, on the density, dispersion, and demography of populations, and on the transfer of energy and nutrients within ecosystems. These size-specific attributes of species, in turn, have important biogeographic, macroevolutionary, and practical, conservation-related implications.*"

J. H. Brown (1995) *Macroecology* pp. 100–101

What is macroecology?

Over the past decade or so the progress of macroecological research has been rapid. Many ecologists, especially younger ones, have adopted the conceptual framework and empirical approach of macroecology. There have been hundreds of scientific

* *Correspondence address: jhbrown@unm.edu*

papers, several notable books (e.g. Brown 1995; Rosenzweig 1995; Maurer 1999; Gaston & Blackburn 2000; Hubbell 2001), and now this important symposium.

So, what has all the excitement been about? Just what is macroecology? The first point to be made is that macroecology as a research programme or subdiscipline of ecology is hardly new. The roots are to be found nearly a century ago in the work of Arrhenius (1921) on species–area relations, Grinnell (1922) on accumulation of accidental bird occurrences, and Willis (1922) on the distributions of sizes of geographical ranges. Macroecology, although it was not called that, had a high point in the 1950s and 1960s with the influential work of Hutchinson & MacArthur (1959), Fischer (1960), Taylor (1961), Preston (1960, 1962), MacArthur & Wilson (1963, 1967), Williams (1964), Simpson (1964), MacArthur (1957, 1965), Whittaker (1967), and many others. Then, eclipsed by the advocacy of experimental approaches to ecology, it languished (but see MacArthur 1972). Activity was revived in the 1980s and 1990s, when the limitations of the small spatial and temporal scales and the simplified ANOVA-based designs of manipulative experiments began to become apparent. It is not surprising that the small-scale experiments by themselves are inadequate to answer important questions about regional and global environmental change. It also has become apparent, however, that small-scale manipulation of variables singly and in factorial combinations may not be the best way to understand the causes and consequences of the complex correlation structures and emergent dynamics that are pervasive properties of real ecological systems (see also Blackburn & Gaston, this volume).

Brown and Maurer can be credited with having coined the term 'macroecology' (Brown & Maurer 1989) and playing some role in the revival (e.g. Brown 1981, 1984, 1987; Brown & Maurer 1987, 1989; Maurer & Brown 1988; Maurer 1994). There were many others, however, who made complementary prior or simultaneous research contributions (e.g. May 1978, 1986; Hubbell 1979; Taylor *et al.* 1980; Hengeveld & Haeck 1981, 1982; Bock & Ricklefs 1983; Rapoport 1983; Bock 1984, 1987; Taylor 1986; Dial & Marzluff 1988, 1989; Gaston & Lawton 1988a,b, 1990; Morse *et al.* 1988; Stevens 1989, 1992; Lawton 1990; to cite just a few up to the early 1990s). Looking back at this body of research, it is hard to decide just how many and which ones of the contributions should be classified as macroecology.

This is because, even after the term was coined and began to be widely used, it has taken on a variety of meanings. We can identify two especially important themes. On the one hand, some have taken the 'macro' to refer primarily to large spatial and long temporal scales. Their research has focused on the patterns and processes where ecology meets and overlaps with biogeography and with palaeoecology and palaeontology. This has been the theme of several recent books (Rosenzweig 1995; McKinney & Drake 1998; Maurer 1999; Gaston & Blackburn 2000). This part of macroecology has both contributed to and benefited from the recent scientific emphasis and public concern about 'global change'.

On the other hand, some investigators have taken the 'macro' to refer primarily to large numbers of 'ecological particles'. Their research has focused on the patterns and processes revealed by the statistical distributions of variables among large collections of equivalent, but not identical, units such as individual organisms within species or species within communities and biogeographical regions. The absolute scale need not be large. For example, a model of thinning was used to characterize a constraint envelope for survival as a function of density of plants on 0.25-m^2 plots (Guo *et al.* 1998). Much of the work that has taken this statistical approach has re-examined the very general macroecological patterns that Robert MacArthur, Frank Preston, C. B. Williams, L. R. Taylor, and others described and tried unsuccessfully to explain nearly a half-century ago. This has been the theme of most of Brown's (e.g. 1995, 2001; Brown & West 2000; Brown *et al.* 2001) recent research and Hubbell's new book (2001). It is the kind of macroecology that will be the subject of the remainder of this chapter.

Do general patterns reflect universal processes?

By the 1990s the statistical patterns of abundance, distribution and diversity that had so intrigued Preston (1948, 1960, 1962), Williams (1964), MacArthur (1972) and others had been shown to be very general if not universal. The patterns included frequency distributions of abundances, body sizes and geographical range areas among species, species–area and species–time relationships, correlations between abundance and distribution, food-web statistics, and geographical and temporal patterns of species diversity. They were documented in diverse taxonomic groups of plants, animals, and sometimes even microbes, in many different terrestrial, freshwater and marine environments, and over large expanses of geographical space and geological time. These empirical advances largely resulted from the accumulation of much new and better data as well as the application of computer-based analytical methods. We will not try to cite the many relevant studies here. They are referenced in recent books (e.g. Brown & Lomolino 1998; Gaston & Blackburn 2000; Hubbell 2001), and in many other chapters in this volume.

Satisfactory mechanistic explanations proved elusive, however. There has been a tendency to refer to the patterns themselves as 'ecological laws' (see Lawton 1999). So, for example, we have Bergmann's Rule (Bergmann 1847), Preston's canonical distribution of abundance (Preston 1962), Taylor's power law of variability in abundance (Taylor *et al.* 1980), Rapoport's rule (Rapoport 1983; Stevens 1989), and the thinning law of plant ecology (Yoda *et al.* 1963; Miyanishi *et al.* 1979). But no matter how frequently they may be documented and how precisely they may be described statistically, these are still just empirical patterns. They may, however, reflect the operation of more fundamental laws—of underlying universal mechanistic processes and principles*. To see this, note that the circular rotation of hurricanes,

* There is considerable discussion among both scientists and philosophers of science about just what constitutes a 'scientific theory' and a 'scientific law'. Here, we refer to our conceptual framework, which can be expressed in simple mathematical equations, as metabolic scaling theory, and we suggest that it is a universal ecological law. We use these terms deliberately to imply that both the mechanistic processes and the mathematics that describe them are universal or very nearly so. What often have been called 'laws' or 'rules' in ecology are simply empirical patterns that are observed frequently. We do not regard them as

tornados and dust devils and the near-planar alignment of the planets of the solar system and moons of Jupiter are not physical laws, even though they do indeed reflect the operation of the laws of motion and gravity.

Our own recent research is based on the premise that the general statistical patterns of macroecology, like the physical patterns mentioned above, are emergent phenomena of complex ecological systems that do indeed reflect the operation of universal law-like mechanisms. In one sense our search for mechanism is a reductionist endeavour. It does not, however, attempt to reduce all of ecology to molecular biology or subatomic particle physics. Instead, it seeks to explain the statistical patterns of macroecology in terms of some well-understood first principles of physics, chemistry and biology, because these are fundamental processes that powerfully constrain the structure and dynamics of even the most complex ecological systems. We believe that we have made considerable progress toward elucidating one set of such laws: those that govern the metabolism of individual organisms and thereby determine the distributions and fluxes of energy and materials in ecological systems. These mechanisms are necessary but not sufficient to explain many of the emergent statistical patterns of macroecology. There must be additional ecological laws. We are confident that some will be discovered within the next decade or so. Much progress is being made by participants in this symposium, and exciting developments in macroecological theory are highlighted in their contributions to this volume.

We devote the remainder of this chapter to describing progress of one macroecological research programme. It is one that has dominated our own research activities in recent years. It represents the contributions of a large and increasing number of collaborators, including individuals who by their training and past research are both biologists and physicists, and theoreticians and empiricists. It is one that we believe to be revealing law-like processes that cause many—but by no means all—of the very general statistical patterns of macroecology.

Metabolic scaling theory

Metabolism and ecology

Our current macroecological research programme and this chapter focus on the metabolic basis of ecology. Metabolism is the process of transforming energy and materials in order to sustain life. Our premise is that the metabolism of individual organisms is a fundamental process that underlies many macroecological patterns, in much the same way that the physical laws of motion and gravity underlie the macroscopic statistical patterns of particles in a thunderstorm or stars in a galaxy.

laws, because they are not universal—most have numerous, well-documented exceptions—and they do not necessarily imply causation by some single underlying process or mechanism. Ecologists will need to determine whether metabolic scaling theory is able to provide a universal mechanistic explanation for a large class of ecological phenomena. If it does, then philosophers will need to decide whether it should properly be called a law, a theory, both, or neither.

Understanding how individual organisms acquire metabolic resources from their environments and allocate them to maintenance, growth and reproduction is key to explaining the emergent statistical behaviours of organisms in populations, communities and ecosystems.

The most fundamental biological rate process is the metabolic rate. The metabolic rate can be measured in terms of heat production or any other currency (e.g. oxygen consumption or carbon dioxide production in aerobic heterotrophs, carbon fixation in photoautotrophs, or phosphorus or nitrogen requirements of higher plants and animals) that reflects the stoichiometric constraints on whole-organism metabolism. Metabolic rate can be expressed on either a whole organism (per individual) or mass-specific (per unit body mass) basis.

The metabolic rate is central to nearly all of ecology because it is the rate of resource uptake and allocation. Metabolism sets the demand that individual organisms place on their environment for energy and material resources. So it limits the rate of population growth, determines resource competition, and influences the flux of energy and the distribution of many compounds and elements in ecosystems. Metabolism simultaneously sets the rate of expenditure of resources on maintenance, growth and survival. So it powerfully constrains reproductive effort, ontogenetic growth, timing of life history events, and trade-offs between survival and reproduction.

Effects of body size and temperature on metabolic rate

Many factors influence the rate of energy use and the requirements of different kinds of organisms for particular material resources. However, because all aerobic eukaryotes utilize the same basic biochemical pathways—glycolysis and the tricarboxylic acid cycle—their metabolism is fundamentally very similar. Consequently, variation in metabolic rate among individuals and species of eukaryotes is attributed primarily to two factors: body size, which affects the rate of uptake and distribution of resources, and temperature, which affects the kinetics of biochemical reactions.

Body size affects metabolic rate by imposing powerful geometric constraints on biological exchange surfaces and distribution networks. Resources are taken up from the environment and wastes are excreted into the environment across the membranes that separate the inside of the organism from the outside. Energy and materials are distributed within the body of the organism through vascular systems or other supply networks. Organisms have used fractal geometry to alter the relationships between length (or time), surface area and volume (or mass) that would be expected on the basis of Euclidean geometry. The central role of fractal-like designs is seen in the highly elaborated surfaces and hierarchically branched networks of nearly all organisms.

The result is that whole-organism metabolic rate, B, and most biological rates, R, and times, D, scale with quarter powers of body mass, M (or volume), rather than with the third powers that would be expected from simple Euclidean geometry (West *et al.* 1997, 1999a,b). So, whole-organism metabolic rate:

$$B \propto M^{3/4} \tag{21.1}$$

mass-specific metabolic rate and most other biological rates:

$$R \propto M^{-1/4} \tag{21.2}$$

and biological times:

$$D \propto M^{1/4} \tag{21.3}$$

With a few conspicuous exceptions, such as biomechanical studies, this fractal-based allometric theory appears to explain most of the size-related variation in biological structure and function from molecules (West *et al.* 2002) to ecosystems (Enquist *et al.* 1998; Enquist & Niklas 2001, 2002).

The rate of metabolism and other biological processes is also strongly influenced by temperature because of its effect on the kinetics of biochemical reactions. This effect is characterized by the exponential Boltzmann factor, well-known in physical chemistry but not widely applied to macroecology:

$$R \propto e^{-E/kT} \tag{21.4}$$

where E is the average rate-limiting activation energy for the biochemical reactions, k is Boltzmann's constant and T is temperature in K.

The effects of size and temperature are multiplicative. Body size affects the rate of supply of substrates to the biochemical reactions through allometric constraints on exchange surfaces and distribution networks. Temperature determines the proportion of molecules with kinetic energies that exceed the activation energy for the biochemical reactions. The combined effects of body size and temperature on biological processes can be described by two simple equations: for mass-specific metabolic rate and most other biological rates

$$R \propto M^{-1/4} e^{-E/kT} \tag{21.5}$$

and for biological times

$$D \propto M^{-1/4} e^{E/kT} \tag{21.6}$$

These two fundamental equations explain most of the variation in biological scaling, from rates of mass-specific or cellular metabolism (Gillooly *et al.* 2001) and molecular evolution (Gillooly, Allen, West, Savage and Brown, unpubl.) to rates of ontogenetic development (Gillooly *et al.* 2002) and population growth (Savage, Gillooly, West, Charnov and Brown, unpubl.). These relationships appear to hold for a very wide range of organisms, including both plants and animals and unicellular and multicellular organisms, from a wide range of environments, including terrestrial, freshwater and marine, and temperate and tropical ecosystems.

Applications to macroecology

So far, applications of metabolic scaling theory to macroecology have been limited. Nevertheless, we are excited about the potential of the theory to explain many macroecological patterns. By mechanistically relating ecological processes to the physical, chemical and biological mechanisms of metabolism, metabolic scaling affects all levels of ecological organization. It links energy and resource use of organisms to the environmental interactions and performances of these organisms in populations, communities and ecosystems.

Well before we began our work on metabolic scaling theory, mechanistic hypotheses based on resource limitation and energetic constraints had been proposed to account for many macroecological phenomena (e.g. Brown 1995; Gaston & Blackburn 2000; Hubbell 2001). These hypotheses were not usually cast in the form of quantitative mathematical models and not based explicitly on the new metabolic scaling theory. Nevertheless, to the extent that these mechanistic hypotheses have been evaluated empirically, they have been supported—or at the very least not rejected. Simple common sense dictates that metabolic mechanisms must be necessary, at least in part, to explain many macroecological patterns. For example, it is hard to believe that the increase of home range size (McNab 1963; Kelt & VanVuren 1999; Haskell *et al.* 2002) and the decrease of population density with increasing body size (Damuth 1981, 1987) do not reflect allometric scaling of resource requirements.

Considering how recently metabolic scaling theory has been developed, it has already led to impressive advances. The fractal-like basis for quarter-power scaling of metabolism was first published in 1997. There have been several applications to macroecology. Many of these applications have been to plants; resource-based models have been used to predict growth rates of forest trees (Enquist *et al.* 2000), resource allocation to plant parts (Enquist & Niklas 2002), and the structure, dynamics and productivity of forests (Enquist *et al.* 1999; Enquist & Niklas 2001, 2002; Niklas & Enquist 2001). This work, pioneered by Brian Enquist, is described in detail in the works cited, and is synthesized in Enquist's chapter in this volume. We will mention only one—perhaps the most compelling—example here. The fact that whole-plant metabolic rate scales with size (mass, M) as $M^{3/4}$ (eqn 21.1: West *et al.* 1999b) and with stem diameter (S), as S^2, can be used to develop a resource-based thinning model (Enquist *et al.* 1998, unpubl.). This model predicts that plant density should scale as $M^{-3/4}$ and S^{-2}. Interestingly, this prediction seems to hold not only for traditional thinning of single-species plantations, but also for single and multiple species in steady-state forests throughout the world (Fig. 21.1). The same body of theory predicts that the productivity of terrestrial ecosystems, although varying predictably with temperature, light, water and nutrients, should be independent of the size of the dominant plants. This prediction appears to hold over an amazing 12 orders of magnitude in plant size from duckweed to Douglas firs (Enquist *et al.* 1998).

The theory characterizing the combined effects of body size and temperature on metabolic rate and other biological rates and times has begun to appear only very recently, in 2001. Nevertheless, we have made considerable progress in applying this model (i.e. eqns 21.5 and 21.6) to predict ecological rates and times. Initial applications of this model to life history and demography have successfully pre-

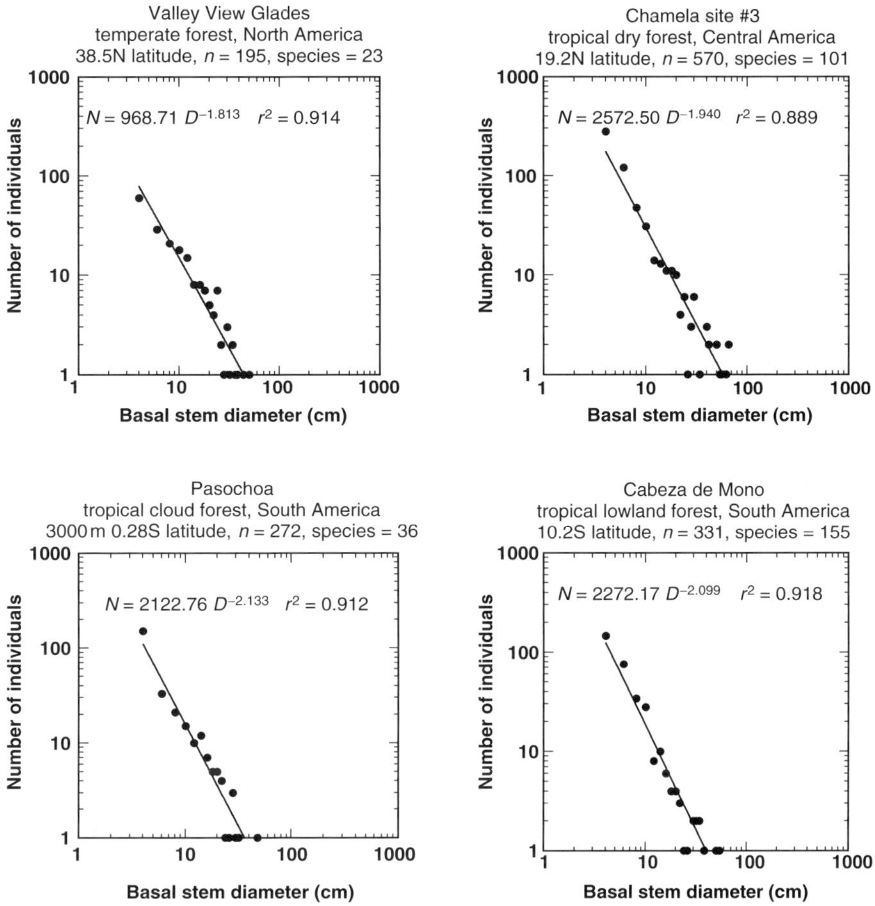

Figure 21.1 Relationship on logarithmic scales between number of stems and stem diameter for all of the trees in four 0.1-ha plots of mature forest: temperate deciduous forest in Missouri, tropical dry forest in Mexico, tropical cloud forest in Ecuador, and tropical rain forest in Peru. Note that although the number of individuals (n) varies from 195 to 570 and the number of species from 23 to 155, the data for each forest fit a power function (straight line) with an exponent (slope) very close to the theoretically predicted value of –2.0. (Figure from Brown (2001); data from the Gentry database (*http://www.mbot.org*); analysis by B.J. Enquist, K. Niklas and J.P. Haskell.)

dicted rates of ontogenetic growth and development (West *et al.* 2001; Gillooly *et al.* 2002), patterns of survivorship (Fig. 21.2), and maximal population growth rates (Savage, Gillooly, West, Charnov and Brown, unpubl.). Logarithmic plots of temperature-corrected rates as a function of mass usually have slopes (allometric exponents) very close to the predicted value of $-1/4$. Another recent application has focused on kinetic theory and the effects of environmental temperature on biogeographical patterns of species diversity (Allen *et al.* 2002; see also Rohde 1992). The

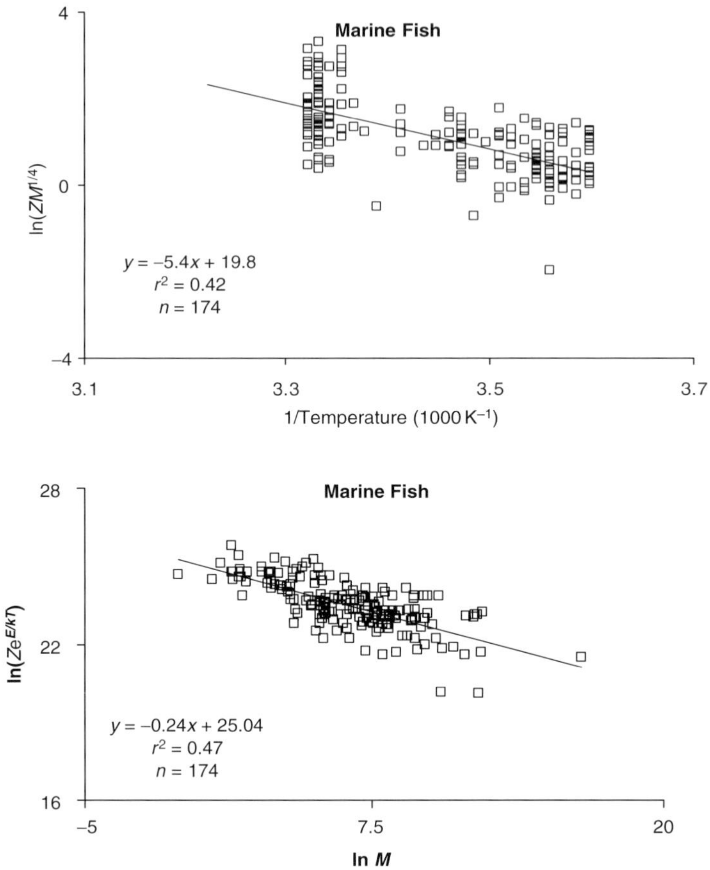

Figure 21.2 Effects of body size and temperature on instantaneous mortality rate for 174 stocks of marine fishes.) Upper: relationship of mass-corrected instantaneous mortality rate ($ZM^{1/4}$) to temperature T, plotted here as (1000 K^{-1}), so that temperature decreases to the right. When plotted in this way, the slope reflects the activation energy (E in eqn 21.4). The value here (−5.4) is somewhat less than the typical value of −7, corresponding to 0.6 eV, the average activation energy for the reactions of aerobic metabolism (see Gillooly *et al.* 2001). Lower: relationship of temperature-corrected instantaneous mortality rate ($Ze^{E/kT}$) to body mass M (μg). When plotted in this way, slope is equal to the temperature-corrected allometric exponent. The value here (−0.24) is very close to the predicted exponent of −1/4. Both lines were fit using Type I linear regression. (Data from Pauly 1980; analysis and figure from V.M. Savage, J.F. Gillooly, G.B. West and J.H. Brown, unpubl.)

Boltzmann relationship can account for most of the variation in species richness of several terrestrial and marine organisms in gradients of both latitude and elevation (e.g. Fig. 21.3). One important result, supporting the central role of metabolism, is that graphs of mass-corrected rates as a function of temperature (1000/T, as in Figs 21.2 and 21.3) appear to reflect the average activation energy for the biochemical re-

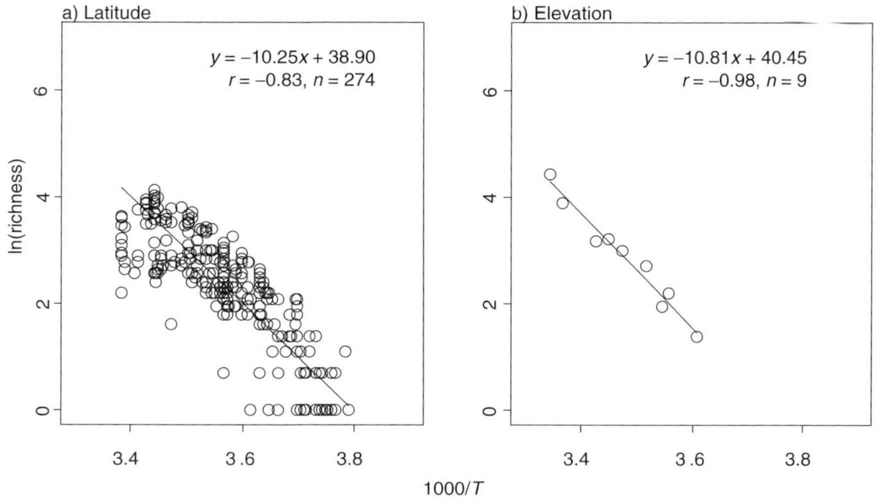

Figure 21.3 Species diversity as a function of environmental temperature for amphibians in two environmental gradients with abundant water: (a) a latitudinal gradient in North America (data from Currie & Fritz 1993), and (b) an elevational gradient in the Andes of Ecuador (data from Duellman 1988). In these graphs, ln(richness) is plotted as a function of mean ambient temperature, expressed as ($1000\,K^{-1}$). The slopes here (−10.25 and 10.81) are somewhat higher than the value of approximately −7 that would be predicted based on the activation energy of the reactions of aerobic metabolism. Note, however, that both the slopes and intercepts are nearly identical, implying that amphibian species diversity shows the same temperature dependence in both latitudinal and elevational gradients. (Reprinted with permission from Allen *et al.* (2002) Global biodiversity, biochemical kinetics and the energetic-equivalence rule. *Science* **297**, 1545–1548. Copyright 2002 American Association for the Advancement of Science.)

actions of metabolism. The slopes typically range from −5 to −10 and cluster around −7, which corresponds to an activation energy of approximately 0.6 eV (Gillooly *et al.* 2001, 2002; Gillooly, Allen, West, Savage and Brown, unpubl.).

These successes suggest that metabolic scaling theory captures the essence of some universal law of ecology. This law is expressed mathematically in equations (21.5) and (21.6), which not only describe the effects of body size and temperature on metabolic rate and many other biological rates and times, but also provide a theoretical basis for understanding the pervasive effects of metabolism on ecology.

Future prospects

The development of metabolic scaling theory and its application to ecology may offer some lessons that will be useful in future efforts to apply this theory and to search for additional laws of ecology. With some trepidation, we share some thoughts about prospects for further advances in macroecology.

First, metabolic scaling theory is still incomplete. Body size and temperature may indeed account for most of the variation in ecological rates and times, but these two factors certainly do not explain all of the variation. Conspicuous by its absence is any explicit incorporation of the influence of chemical stoichiometry. Metabolism creates and maintains differences in chemical composition between organisms and their environments. Recent research shows how stoichiometry influences many aspects of ecological structure and dynamics (e.g. Sterner *et al.* 1992; Sterner 1995; Elser *et al.* 2000), including residual variation in ontogenetic growth rates not explained by body size and temperature (Gillooly *et al.* 2002). The power and generality of metabolic scaling theory would be increased by explicitly incorporating the mechanisms that determine the chemical resource requirements of organisms and the stoichiometric relationships between organisms and their environment.

Second, our research has focused on trying to explain ecological phenomena in terms of first principles of chemistry, physics and biology. The principles from these disciplines that appear to be most useful for making progress in ecology are already understood. They include such things as energetics and thermodynamics, kinetics, mechanics and fluid dynamics, and the basic biochemistry of aerobic metabolism. In basing theory on such well-established and universal principles, we have tried to characterize law-like mechanisms that must influence the organization of ecological systems at all levels of organization. Much of what passes for ecological theory is 'phenomenological' in the sense that it is based on patterns and processes that have no explicit relationship to the laws of physics and chemistry or the fundamental features of molecular, cellular and organismal biology. Examples include theory based on concepts such as environmental stress, density dependence, competition coefficient, food-web connectance, keystone species, river continuum and so on. Such concepts may be useful in capturing important features of ecological organization and guiding the search for underlying mechanisms. But they are likely to remain imprecise and controversial unless they can be firmly rooted in universal scientific principles.

Third, our development of metabolic scaling theory suggests the need for new, innovative statistical approaches and methods. Much of recent ecology has involved adopting statistical methods, such as regression and analysis of variance (ANOVA), developed in the experimental sciences and applying them to describe the structure and dynamics of complex ecological systems. Peters (1991) advocated an extreme version of this approach, suggesting that ecologists should be content with quantitative description and give up the search for mechanism. Some colleagues have complained that our theoretical models and empirical analyses of the effects of body size and temperature on biological processes were nothing new; they already had been documented using ANOVA. Indeed, ANOVA frequently had been used to show statistically significant effects of these two variables. Such analyses, however, only show that rates decrease with size and increase with temperature—basic facts that indeed have been known for nearly a century. By contrast, our model (eqn 21.5), based on mechanistic first principles of allometry and kinetics, predicts the precise functional form for the separate and joint effects of size and temperature. Furthermore, the body-size term is a power function and the temperature term is an exponential. This means that one would be unlikely to come up with equation (21.5) by standard statistical curve-fitting procedures. Still other statistical issues are raised by questions of how to 'test' or evaluate the predictions of metabolic scaling theory. At a minimum, we believe that the emphasis should shift from what fitted function best describes the data to whether analysis of the data gives results that are consistent or inconsistent with the model.

Fourth, our research suggests that it may be possible to develop mechanistic theories of ecology in a sequential fashion, with subsequent advances building on previous ones. Thus, having a theoretical basis for the effects of body size and being able to predict quarter-power scaling of rates and times made it easier, first to elucidate the effects of temperature, and second to characterize the joint effects of size and temperature. Note that in Fig. 21.2, we show these joint effects by 'correcting' for one variable on the *y* axis (i.e. using the theoretically predicted function from eqn 21.2 or 21.4), and plotting this adjusted value as a function of the second variable. Plotting the data in this way not only reduces the scatter (residual variation) in both plots, but also gives quantitative values for the slopes that are very close to values of $-1/4$ predicted by quarter-power allometric theory (Fig. 21.2 lower) and -7.0 predicted from kinetic theory based on the empirically measured activation energy of metabolism (Fig. 21.2 upper).

Fifth, although applications of metabolic scaling theory to macroecology have produced promising results (see above), we believe that these initial studies have only scratched the surface. Because body size and temperature so powerfully constrain metabolism and metabolism is so central to so much of ecology, there are many and diverse potential applications. In particular, metabolic scaling theory can seemingly explain many of the linkages between the performance of individual organisms and the roles of those organisms in communities and ecosystems. For example, equations (21.5) and (21.6) suggest the prospects for developing a more predictive, quantitative understanding of trophic and successional dynamics, potentially clarifying explanations for 'inverted pyramids of biomass' in aquatic as compared with terrestrial ecosystems, and for variation in times of succession across terrestrial ecosystems that differ in environmental temperature or size of dominant plants. As metabolic scaling theory is based on first principles of physics, chemistry and biology, it should facilitate a more rigorous and quantitative approach to ecological theory and practice. For example, applications of the theory show the need to distinguish carefully between standing stocks and turnover rates to understand the effects of resource limitation on structure and dynamics of populations, communities and ecosystems. Additionally, findings that metabolic scaling theory seemingly can be extended to predict rates of molecular evolution (Gillooly, Allen, West, Savage and Brown, unpubl.) and rates of female fertility in modern human societies (Moses and Brown, unpubl.) suggest that it reflects some universal law of nature that has extremely broad applications.

Finally, we readily admit that neither body size and temperature nor a more general metabolic theory of ecology that incorporates stoichiometry will be sufficient to explain all macroecological patterns. If metabolic scaling theory is a law of ecology—and we believe that it is—then there presumably must be other equally universal laws that account for other very general emergent properties of ecological systems. Non-metabolic mechanisms will be required to account for Taylor's power law relating mean and variance of abundance of a species over space or time, for variation in abundances of organisms of similar size and trophic position among species within communities, for species–area and species–time relationships, and for some network properties of food webs. We believe that the time is ripe for their discovery.

Acknowledgements

We are grateful to the members of the 'biological scaling group' at the Santa Fe Institute and the 'ecological complexity' group at the University of New Mexico for help and encouragement throughout the development of metabolic scaling theory. Special thanks go to Brian Enquist who has led the work on plants and contributed in many other ways, and to Ric Charnov, Drew Allen and Melanie Moses, whose collaborative contributions are mentioned in this chapter. We gratefully acknowledge a Packard Interdisciplinary Science Grant, NSF Biocomplexity Grant DEB-0083422, NSF Grant PHY-9873638, the Thaw Charitable Trust, the Santa Fe Institute, the ESPRC, and the Wellcome Trust for their support, and the Department of Mathematics, Imperial College, London for the hospitality extended to G.B.W.

References

Allen, A.P., Brown, J.H. & Gillooly, J.F. (2002) Global biodiversity, biochemical kinetics and the energetic-equivalence rule. *Science* **297**, 1545–1548.

Arrhenius, O. (1921) Species and area. *Journal of Ecology* **9**, 95–99.

Bergmann, C. (1847) Über die Verhältnisse der Wärmeökonomie der Tiere zu ihren Grösse. *Göttinger Studien* **1**, 595–708.

Bock, C.E. (1984) Geographical correlates of abundance vs rarity in some North American winter landbirds. *Auk* **101**, 266–273

Bock, C.E. (1987) Distribution–abundance relationships of some Arizona landbirds: a matter of scale. *Ecology* **68**, 124–129.

Bock, C.E. & Ricklefs, R.E. (1983) Range size and local abundance of some North-American songbirds: a positive correlation. *American Naturalist* **122**, 295–299.

Brown, J.H. (1981) Two decades of homage to Santa Rosalia: toward a general theory of diversity. *American Zoologist* **21**, 877–888.

Brown, J.H. (1984) On the relationship between abundance and distribution of species. *American Naturalist* **124**, 255–279.

Brown, J.H. (1995) *Macroecology*. University of Chicago Press, Chicago.

Brown, J.H. (1999) Macroecology: progress and prospect. *Oikos* **87**, 3–14.

Brown, J.H. (2001) Concluding remarks. In: *Ecology: Achievement and Challenge* (eds M.C. Press, N.J. Huntly & S. Levin), pp. 387–396. Blackwell Science, Oxford.

Brown, J.H. & Lomolino, M.V. (1998) *Biogeography*, 2nd edn. Sinauer, Sunderland, MA.

Brown, J.H. & Maurer, B.A. (1987) Evolution of species assemblages: effects of energetic constraints and species dynamics on the diversification of the North American avifauna. *American Naturalist* **130**, 1–17.

Brown, J.H. & Maurer, B.A. (1989) Macroecology: the division of food and space among species on continents. *Science* **243**, 1145–1150.

Brown, J.H. & West, G.B. (eds) (2000) *Scaling in Biology*. Oxford University Press, Oxford.

Brown, J.H., Ernest, S.K.M., Parody, J.M. & Haskell, J.P. (2001) Regulation of diversity: maintenance of species richness in changing environments. *Oecologia* **126**, 321–332.

Currie, D.J. & Fritz, J.T. (1993) Global patterns of animal abundance and species energy use. *Oikos* **67**, 56–68.

Damuth, J. (1981) Population-density and body-size in mammals. *Nature* **290**, 699–700.

Damuth, J. (1987) Interspecific allometry of population-density in mammals and other animals: the independence of body-mass and population energy-use. *Biological Journal of the Linnean Society* **31**, 193–246.

Dial, K.P. & Marzluff, J.M. (1988) Are the smallest organisms the most diverse? *Ecology* **69**, 1620–1624.

Dial, K.P. & Marzluff, J.M. (1989) Nonrandom diversification within taxonomic assemblages. *Systematic Zoology* **38**, 26–37.

Duellman, W.E. (1988) Patterns of species-diversity in anuran amphibians in the American tropics. *Annals of the Missouri Botanical Garden* **75**, 79–104.

Elser, J.J., Sterner, R.W., Gorokhova, E., *et al.* (2000) Biological stoichiometry from genes to ecosystems. *Ecology Letters* **3**, 540–550.

Enquist, B.J., Brown, J.H. & West, G.B. (1998) Allometric scaling of plant energetics and population density. *Nature* **395**, 163–165.

Enquist, B.J. & Niklas, K.J. (2001) Invariant scaling relations across tree-dominated communities. *Nature* **410**, 655–660.

Enquist, B.J. & Niklas, K.J. (2002) Global allocation rules for patterns of biomass partitioning in seed plants. *Science* **295**, 1517–1520.

Enquist, B.J., West, G.B. & Brown, J.H. (1999) Quarter-power allometric scaling in vascular plants: functional basis and ecological consequences. In: *Scaling in Biology* (eds J.H. Brown & G.B. West), pp. 167–198. Oxford University Press, Oxford.

Enquist, B.J., West, G.B., Charnov, E.L. & Brown, J.H. (2000) Allometric scaling of production and life-history variation in vascular plants. *Nature* **408**, 750.

Fischer, A.G. (1960) Latitudinal variation in organic diversity. *Evolution* **14**, 64–81.

Gaston, K.J. & Blackburn, T.M. (2000) *Pattern and Process in Macroecology*. Blackwell Science, Oxford.

Gaston, K.J. & Lawton, J.H. (1988a) Patterns in body size, population dynamics, and regional distribution of bracken herbivores. *American Naturalist* **132**, 662–680.

Gaston, K.J. & Lawton, J.H. (1988b) Patterns in the abundance and distribution of insect populations. *Nature* **331**, 709–712.

Gaston, K.J. & Lawton, J.H. (1990) Effects of scale and habitat on the relationship between regional distribution and local abundance. *Oikos* **58**, 329–335.

Gillooly, J.F., Brown, J.H., West, G.B., Savage, V.M. & Charnov, E.L. (2001) Effects of size and temperature on metabolic rate. *Science* **293**, 2248–2251.

Gillooly, J.F., Charnov, E.L., West, G.B., Savage, V.M. & Brown, J.H. (2002) Effects of body size and temperature on developmental time. *Nature* **417**, 70–73.

Grinnell, J. (1922) The role of the 'accidental'. *Auk* **39**, 373–380.

Guo, Q., Brown, J.H. & Enquist, B.J. (1998) Using constraint lines to characterize plant performance. *Oikos* **83**, 237–245.

Haskell, J.P., Ritchie, M.E. & Olff, H. (2002) Fractal geometry predicts varying body size scaling relationships for mammal and bird home ranges. *Nature* **418**, 527–530.

Hengeveld, R. & Haeck, J. (1981) The distribution of abundance. II. Models and applications. *Proceedings of the Koninkluke Nederlandse Akademie Van Wetenschappen Series C, Biological and Medical Sciences* **84**, 257–284.

Hengeveld, R. & Haeck J. (1982) The distribution of abundance. 1. Measurements. *Journal of Biogeography* **4**, 303–316.

Hubbell, S.P. (1979) Tree dispersion, abundance, and diversity in a dry tropical forest. *Science* **203**, 1299–1309.

Hubbell, S.P. (2001) *The Unified Neutral Theory of Biodiversity and Biogeography*. Princeton University Press, Princeton, NJ.

Hutchinson, G.E. & MacArthur, R.H. (1959) A theoretical ecological model of size distributions among species of animals. *American Naturalist* **93**, 117–125.

Kelt, D.A. & Van Vuren, D. (1999) Energetic constraints and the relationship between body size and home range area in mammals. *Ecology* **80**, 337–340.

Lawton, J.H. (1990) Species richness and population dynamics of animal assemblages. Patterns in body size: abundance and space. *Philosophical Transactions of the Royal Society, London, Series B* **330**, 283–291.

Lawton, J.H. (1999) Are there general laws in ecology? *Oikos* **84**, 177–192.

MacArthur, R.H. (1957) On the relative abundance of bird species. *Proceedings of the National Academy of Sciences, USA* **43**, 293–295.

MacArthur, R.H. (1965) Patterns of species diversity. *Biological Reviews* **40**, 510–533.

MacArthur, R.H. (1972) *Geographical Ecology*. Harper & Row, New York.

MacArthur, R.H. & Wilson, E.O. (1963) An equilibrium theory of insular zoogeography. *Evolution* **17**, 373–387.

MacArthur, R.H. & Wilson, E.O. (1967) *The Theory of Island Biogeography*. Princeton University Press, Princeton, NJ.

Maurer, B.A. (1994) *Geographic Population Analysis: Tools for Analysis of Biodiversity*. Blackwell Scientific, Oxford.

Maurer, B.A. (1999) *Untangling Ecological Complexity*. University of Chicago Press, Chicago, IL.

Maurer, B.A. & Brown, J.H. (1988) Distribution of energy use and body mass among species of North American terrestrial birds. *Ecology* **69**, 1923–1932.

May, R.M. (1978) The dynamics and diversity of insect faunas. In: *Diversity of Insect Faunas* (eds L.A. Mound & N. Waloff), pp. 188–204. Blackwell, New York.

May, R.M. (1986) The search for patterns in the balance of nature: advances and retreats. *Ecology* **67**, 1115–1126.

McKinney, M.L. & Drake, J.A. (1998) *Biodiversity Dynamics*. Columbia University Press, New York.

McNab, B.K. (1963) Bioenergetics and the determination of home range size. *American Naturalist* **97**, 133–140.

Miyanishi, K., Hoy, A.R. & Cavers, P.B. (1979) Generalized law of self-thinning in plant-populations. *Journal of Theoretical Biology* **78**, 439–442.

Morse, D.R., Stork, N.E. & Lawton, J.H. (1988) Species number, species abundance and body length relationships for arboreal beetles in Bornean lowland rain forest trees. *Ecological Entomology* **13**, 25–37.

Niklas, K.J. & Enquist, B.J. (2001) Invariant scaling relationships for interspecific plant biomass production and body size. *Proceedings of the National Academy of Sciences, USA* **98**, 2922–2927.

Pauly, D. (1980) On the interrelationships between natural mortality, growth-parameters, and mean environmental temperature in 175 fish stocks. *Journal du Conseil* **39**, 175–192.

Peters, R.H. (1991) *A Critique for Ecology*. Cambridge University Press, Cambridge.

Preston, F.W. (1948) The commonness, and rarity, of species. *Ecology* **29**, 254–283.

Preston, F.W. (1960) Time and space and the variation of species. *Ecology* **41**, 612–627.

Preston, F.W. (1962) The canonical distribution of commonness and rarity: parts I and II. *Ecology* **43**, 185–215, 410–432.

Rapoport, E.H. (1983) *Areography*. Pergamon Press, New York.

Rohde, K. (1992) Latitudinal gradients in species diversity: the search for the primary cause. *Oikos* **65**, 514–527.

Rosenzweig, M.L. (1995) *Species Diversity in Space and Time*. Cambridge University Press, New York.

Simpson, G.G. (1964) Species density of North American recent mammals. *Systematic Zoology* **13**, 57–73.

Sterner, R.W. (1995) Elemental stoichiometry of species in ecosystems. In *Linking Species and Ecosystems* (eds C.G. Jones & J.H. Lawton), pp. 240–252. Chapman and Hall, London.

Sterner, R.W., Elser, J.J. & Hessen, D.O. (1992) Stoichiometric relationships among producers and consumers in food webs. *Biogeochemistry* **17**, 49–67.

Stevens, G.C. (1989) The latitudinal gradients in geographic range: how so many species coexist in the tropics. *American Naturalist* **132**, 240–256.

Stevens, G.C. (1992) The elevational gradient in altitudinal range: an extension of Rapoport's latitudinal rule to altitude. *American Naturalist* **133**, 240–256.

Taylor, L.R. (1961) Aggregation, variance and the mean. *Nature* **198**, 732–735.

Taylor, L.R. (1986) Synoptic dynamics, migration and the Rothamsted insect survey. *Journal of Animal Ecology* **55**, 1–38.

Taylor, L.R., Woiwod, I.P. & Perry, J.N. (1980) Variance and the large scale spatial stability of aphids, moths and birds. *Journal of Animal Ecology* **49**, 831–854.

West, G.B., Brown, J.H. & Enquist, B.J. (1997) A general model for the origin of allometric scaling laws in biology. *Science* **276**, 122–126.

West, G.B., Brown, J.H. & Enquist, B.J. (1999a) The fourth dimension of life: fractal geometry and allometric scaling of organisms. *Science* **284**, 1677–1679.

West, G.B., Brown, J.H. & Enquist, B.J. (1999b) A general model for the structure and allometry of plant vascular systems. *Nature* **400**, 664–667.

West, G.B., Brown, J.H. & Enquist, B.J. (2001) A general model for ontogenetic growth. *Nature* **413**, 628–631.

West, G.B., Woodruff, W.H. & Brown, J.H. (2002) Allometric scaling of metabolic rate from molecules and mitochondria to cells and mammals. *Proceedings of the National Academy of Sciences, USA* **99**, 2473–2478.

Williams, C.B. (1964) *Patterns in the Balance of Nature and Related Problems in Quantitative Ecology*. Academic Press, New York.

Willis, J.C. (1922) *Age and Area*. Cambridge University Press, Cambridge.

Whittaker, R.H. (1967) Gradient analysis of vegetation. *Biological Reviews* **42**, 207–264.

Yoda, K., Kira, T., Ogawa, H. & Hozumi, K. (1963) Self-thinning in overcrowded pure stands under cultivated and natural conditions. *Journal of Biology, Osaka City University* **14**, 107–129.

Index

Page numbers in **bold** refer to tables and in *italics* to figures, those with suffix 'n' refer to notes